ENERGY AND FUEL SYSTEMS INTEGRATION

GREEN CHEMISTRY AND CHEMICAL ENGINEERING

Series Editor: Sunggyu Lee

Ohio University, Athens, Ohio, USA

GREEN CHEMISTRY AND CHEMICAL ENGINEERING

ENERGY AND FUEL SYSTEMS INTEGRATION

Yatish T. Shah

CRC Press
Taylor & Francis Group
Boca Raton London New York

CRC Press is an imprint of the
Taylor & Francis Group, an **informa** business

CRC Press
Taylor & Francis Group
6000 Broken Sound Parkway NW, Suite 300
Boca Raton, FL 33487-2742

First issued in paperback 2019

© 2016 by Taylor & Francis Group, LLC
CRC Press is an imprint of Taylor & Francis Group, an Informa business

No claim to original U.S. Government works

ISBN-13: 978-1-4822-5306-1 (hbk)
ISBN-13: 978-0-367-37734-2 (pbk)

Library of Congress Cataloging-in-Publication Data

Shah, Yatish T., author.
  Energy and fuel systems integration / author, Yatish T. Shah.
    pages cm. -- (Green chemistry and chemical engineering)
  Includes bibliographical references and index.
  ISBN 978-1-4822-5306-1 (hardcover : acid-free paper) 1. Power resources. I. Title.

TJ163.13.S475 2015
621.31--dc23   2015026065

Visit the Taylor & Francis Web site at
http://www.taylorandfrancis.com

and the CRC Press Web site at
http://www.crcpress.com

I would like to dedicate this book to my eight grandchildren, Samuel, Savita, Isaac, Meghan, Olivia, Eamon, Sydney, and Ethan.

Contents

Series Preface: Green Chemistry and Chemical Engineering

The subjects and disciplines of chemistry and chemical engineering have encountered a new landmark in the way of thinking about, developing, and designing chemical products and processes. This revolutionary philosophy, termed "green chemistry" and "chemical engineering," focuses on the designs of products and processes that are conducive to reducing or eliminating the use and/or generation of hazardous substances. In dealing with hazardous or potentially hazardous substances, there may be some overlaps and interrelationships between environmental chemistry and green chemistry. While environmental chemistry is the chemistry of the natural environment and the pollutant chemicals in nature, green chemistry proactively aims to reduce and prevent pollution at its very source. In essence, the philosophies of green chemistry and chemical engineering tend to focus more on industrial application and practice rather than academic principles and phenomenological science. However, following the basic philosophy of chemistry and chemical engineering, green chemistry and associated chemical engineering are derived from and build upon organic chemistry, inorganic chemistry, polymer chemistry, fuel chemistry, biochemistry, analytical chemistry, physical chemistry, environmental chemistry, thermodynamics, chemical reaction engineering, transport phenomena, chemical process design, separation technology, automatic process control, and more. In short, green chemistry and chemical engineering is the rigorous use of chemistry and chemical engineering for pollution prevention and environmental protection.

The Pollution Prevention Act of 1990 in the United States established a national policy to prevent or reduce pollution at its source whenever feasible. And adhering to the spirit of this policy, the Environmental Protection Agency (EPA) launched its Green Chemistry Program in order to promote innovative chemical technologies that reduce or eliminate the use or generation of hazardous substances in the design, manufacture, and use of chemical products. The global efforts in green chemistry and chemical engineering have recently gained a substantial amount of support from the international communities of science, engineering, academia, industry, and government in all phases and aspects.

Some of the successful examples and key technological developments include the use of supercritical carbon dioxide as a green solvent in separation technologies; application of supercritical water oxidation for the destruction of harmful substances; process integration with carbon dioxide sequestration steps; solvent-free synthesis of chemicals and polymeric materials; exploitation of biologically degradable materials; use of aqueous hydrogen peroxide for efficient oxidation; development of hydrogen proton exchange membrane (PEM) fuel cells for a variety of power generation needs; advanced biofuel productions; devulcanization of spent tire rubber; avoidance of the use of chemicals and processes causing the generation of volatile organic compounds (VOCs); replacement of traditional petrochemical processes by microorganism-based bioengineering processes; replacement of chlorofluorocarbons (CFCs) with nonhazardous alternatives; advances in the design of energy efficient processes; use of clean, alternative, and renewable energy sources in manufacturing; and much more. Even though this list is only a partial compilation, it is undoubtedly growing exponentially.

This book series on green chemistry and chemical engineering by CRC Press/ Taylor & Francis Group is designed to meet the new challenges of the twenty-first century in the chemistry and chemical engineering disciplines by publishing books and monographs based on cutting-edge research and development to the effect of reducing adverse impacts upon the environment by chemical enterprise. And in achieving this, the series will detail the development of alternative sustainable technologies that will minimize the hazard and maximize the efficiency of any chemical choice. The series aims at delivering the readers in academia and industry with an authoritative information source in the field of green chemistry and chemical engineering. The publisher and its series editor are fully aware of the rapidly evolving nature of the subject and its long-lasting impact upon the quality of human life in both the present and future. As such, the team is committed to making this series the most comprehensive and accurate literary source in the field of green chemistry and chemical engineering.

Sunggyu Lee

Preface

There are ten sources of energy and fuel in this world: coal, oil, gas, biomass, waste, nuclear, solar, wind, geothermal, and water. These can also be classified into three categories as fossil, nuclear, and renewables. Each of these sources has many subcontents, such as: (a) There are numerous ranks of coal such as anthracite, bituminous, subbituminous, and lignite, each with different reactivities and levels of ash and mineral contents; (b) there are numerous types of crude oils with different compositions as well as unconventional oils such as heavy oil, tar sands, bitumen, and oil shale; (c) there are different types of gas such as natural gas, biogas, shale and other unconventional gas, and propane; (d) there are different types of biomass and wastes; (e) there are different types of nuclear reactors; (f) there are numerous methods for capturing solar, wind, and geothermal energy; and, finally, (g) water can act as a source of fuel or energy.

While each source of fuel and energy has been and continues to be developed and harnessed, during the past several decades, the energy and fuel portfolio has been heavily dominated by fossil fuels; to this date, more than 80% of the energy and fuel demand is satisfied by the fossil energy. In recent years, numerous external factors—such as (a) the effects of carbon dioxide and other pollutant emissions coming from fossil fuels, (b) depletions of conventional oil and gas production, and (c) desire for more use of renewable sources of energy—have forced us to examine our strategy for the development of future energy landscape. While the demand for energy is rapidly growing globally, we need to develop a different strategy to increase the energy and fuel supply: the one that does not harm the environment and the one that is not so heavily dependent upon fossil energy. We need to realign our energy portfolio such that fossil, nuclear, and renewable energy contributions are more balanced.

Everybody agrees that future growth in energy and fuel supply should have a significantly larger contribution of renewable energy and even carbon-free nuclear energy, both of which help reduce carbon emissions in the environment. The development of renewable energy sources, however, faces a few challenges: (a) They will require a significant capital investment for their commercialization and they may require new infrastructure, (b) solar and wind energy are time and location dependent and their intermittency will prohibit them as a steady and sustainable source for energy supply, (c) the infrastructure for the development of advanced geothermal systems is not yet developed and it will also take significant investment, (d) conversions of biomass to fuel and energy will require easily accessible large quantities of raw materials (this will be difficult due to their transportation and storage problems), (e) the technologies to use water as source for fuel and energy are not yet fully commercialized and they will also require large capital investments, (f) there are numerous types of wastes and they have to be treated in a distributed manner that will not allow the best use of economy of scale, and (g) political and social acceptance of nuclear energy will require its use for nonelectrical applications along with power generation. This needs to be further developed both at large and small scales.

Against all of these challenges, fossil energy benefits from (a) available infrastructure, (b) easy processing at a large scale where the unit price of energy is more attractive, (c) recent increase in gas and oil productions due to the "fracking process," allowing power plants to be transformed from more harmful coal to more economical and less harmful natural gas, and (d) high mass and energy densities that facilitate their transport and storage. Considering all these factors, a larger and faster penetration of renewable and nuclear energies in the fossil-dominated energy world will require a strategy that (a) allows renewable energy to be operated at large enough scale such that the unit price of energy production is competitive with fossil energy, (b) allows the use of renewable source of fuel and energy to be utilized in a sustainable manner, (c) allows more use of nuclear and renewable energies for nonelectrical applications, (d) takes advantage of smart grid and energy storage to replace the need for fossil energy in the growing market of power production, and (e) allows larger penetrations of renewable and nuclear energy in the distributed and mobile energy market, particularly for electrical energy.

The best method for the implementation of the previously described strategy is the integration of different sources of fuel and energy. Such integration can be carried at three different levels.

LEVEL 1: USE OF MULTIPLE RAW FUELS FOR POWER AND SYNTHETIC FUEL GENERATIONS

This approach entails the integration of raw fuels to generate synthetic fuel or energy through classic thermochemical, biochemical, and catalytic processes. The use of multifuels like coal and biomass/waste, biomass–water, biomass–waste, and coal and heavy oil for processes like combustion, gasification, liquefaction, pyrolysis, and anaerobic digestion can (a) reduce the emission of carbon dioxide, (b) allow biomass and waste to be used on a larger scale, and (c) reduce the use of fossil fuels, thus allowing those sources to last for a longer period. As shown in the book, in many cases, the multifuel strategy will overcome the shortcomings of renewables (such as storage, transportation, and unsustainable large-scale operation) and improve the overall performance due to synergy in the reaction process and more efficient use of the reactor vessel.

LEVEL 2: USE OF HYBRID NUCLEAR AND RENEWABLE ENERGY SYSTEMS WITH COGENERATION OF POWER AND HEAT

This approach involves implementations of advanced hybrid energy systems with or without cogeneration (simultaneous use of heat and power). Hybrid combinations of nuclear and/or various renewable sources of energy with backup biofuels and energy storage will allow the use of time- and location-dependent solar and wind energy to be used on a larger scale and in a sustainable manner. The use of cogeneration on the product side will enhance the thermal efficiency of the nuclear and solar power production by the efficient use of excess heat for various nonelectrical applications. This level of integration thus involves (a) the use of nuclear energy for both power and excess heat that can be applied to nonelectrical applications; (b) various renewable, nuclear, and fossil energy systems to allow penetrations of renewable sources of energy on a larger and sustainable scale for both power and heat productions; and (c) renewable energy systems with energy storage to satisfy the electricity need in remote locations where a connection to smart grid is not possible or economical. This type of integration will also allow more efficient use of nuclear and solar energy on smaller and intermediate scales.

LEVEL 3: HOLISTIC INTEGRATION OF ENERGY AND FUEL SYSTEMS USING GAS, HEAT, AND ELECTRICITY GRIDS

This approach involves the application of the holistic energy and fuel system integration concept through the use of gas (oil), heat, and electricity grids, which combine gas, heat, and power. These grids can also integrate various renewable, nuclear, and fossil energy power plants with the smart grids, which are also connected to the energy storage. The use of smart grids will allow the penetrations of renewable and nuclear energy at a smaller scale and at distributed levels. It will also allow a smart control of supply and demand of gas, heat, and electricity anywhere in the grid region with the advanced use of modern information and communication technologies and with the bidirectional flow of energy and fuel.

With further development of natural gas grid, the supply and demand of different sources of gas, that is, natural gas, biogas, shale and other unconventional gas, synthetic gas, and hydrogen, can be efficiently controlled and managed. Both gas and electrical grids can have not only horizontal integrations but also vertical integrations with other sources of fuels. In the future, the energy and fuel industry will heavily depend on the operations of smart, gas, heat, and electricity grids and interactions among them and other sources of fuels and energy. These grids will control the two-way transfer of gas, heat, and electricity with dynamic and demand response objectives. As grids are further developed, they will be the most useful tools to control supply and demand for both static and mobile use of energy and fuel.

These three levels of fuel and energy system integration will allow us to achieve the desired energy portfolio for the future. This book describes in detail these three methods of energy system integration with detailed examples. Chapters 2 through 6 address the use of different mixtures of fuels for combustion, gasification, liquefaction, pyrolysis, and anaerobic digestion processes. Chapters 7 and 8 examine our present state of knowledge on the use of hybrid nuclear energy and renewable energy systems for heat and power generations with numerous nonelectrical applications. Chapter 9 examines the holistic integration of renewable, nuclear, and fossil energy systems by gas, heat, and smart electrical grids. The importance of energy storage systems on the workings of gas, heat, and electrical grids is also illustrated.

The best example for the use of these three levels of integrations is the vehicle industry, which is one of the largest consumers of fuel and energy. This industry clearly demonstrates that vehicles in the future will predominantly use these types of fuel and energy system integration. Chapter 10 illustrates this point with numerous practical examples and recent developments of the industry in this direction.

Future energy and fuel demands and the preservation of the environment require more integration among the ten sources of fuels and energy. It is imperative that in future industries, different sources of energy and fuel cooperate and integrate with each other than simply compete.

Author

Dr. Yatish T. Shah received his BSc in chemical engineering from the University of Michigan, Ann Arbor, Michigan and MS and ScD in chemical engineering from Massachusetts Institute of Technology, Cambridge, Massachusetts. He has more than 40 years of academic and industrial experience in energy-related areas. He was chairman of the Department of Chemical and Petroleum Engineering at the University of Pittsburgh, dean of the College of Engineering at the University of Tulsa and Drexel University, chief research officer at Clemson University, and provost at Missouri University of Science and Technology, the University of Central Missouri, and Norfolk State University. He was also a visiting scholar at the University of Cambridge in the United Kingdom and a visiting professor at the University of California, Berkley, and Institut für Technische Chemie I der Universität Erlangen, Nürnberg, Germany. Currently, he is a professor of engineering and codirector of the Hodge Institute of Entrepreneurship at Norfolk State University. Dr. Shah has written five books related to energy: *Gas-Liquid-Solid Reactor Design* (McGraw-Hill, 1979), *Reaction Engineering in Direct Coal Liquefaction* (Addison-Wesley, 1981), *Cavitation Reaction Engineering* (Plenum Press, 1999), *Biofuels and Bioenergy—Processes and Technologies* (CRC Press, 2012), and *Water for Energy and Fuel Production* (CRC Press, 2014). He has also published more than 250 refereed reviews, book chapters, and research technical publications in the areas of energy, environment, and reaction engineering. He is an active consultant to numerous industries and government organizations in the energy areas.

1 Introduction

1.1 INTRODUCTION

There are 10 sources of energy and fuel in this world: oil, gas, coal, biomass, waste, nuclear, solar, wind, geothermal, and water. Except for waste all of them are natural sources. Waste can be natural or man-made. Oil, gas, coal, biomass, and waste are raw fuel sources, while solar, wind, nuclear, and geothermal are sources of energy. As indicated in my previous book [1], water can be either fuel or a direct source of energy. Waste is the most heterogeneous source of fuel, which largely contains different types of biomass but can also include glass, metals, polymers, and medical and nuclear waste. It is for this reason that it is classified here as a different source of fuel. It is also the only source of fuel, which heavily depends on the level of the population and the nature of our living style. These 10 sources can also be divided into three categories: (a) fossil (oil, gas, and coal), (b) nuclear, and (c) renewable (biomass, waste, solar, wind, geothermal, and water) energies.

As shown in my previous book [1], the use of different sources of fuel such as oil, gas, coal, biomass, waste, and nuclear to generate heat, power, and synthetic fuels as well as useful materials and chemicals over the last more than two centuries has been largely single dimensional; an industry for each source of fuel and energy has been developed independently. We have oil, natural gas, coal, nuclear (uranium), and, more recently, biomass and waste industries, each of which has developed its own technology and infrastructure base to supply our power, heat, fuel, chemicals, and materials needed. There have been very little interactions among these industries. This has led to the developments of oil refineries, natural gas infrastructure, and coal processing industries. Similarly, nuclear industry has also taken its own independent life. More recently, renewable sources like biomass, waste, solar, wind, geothermal, and hydroenergy have similarly been developed independently to address the environmental issues and generate other options for heat, power, and useful fuels and materials.

There was and still there is a need for the independent technological development of various sources of fuel and energy. Each source brought about its independent issues such as supply and demand, pricing, infrastructure, environmental impacts, social and political acceptance, product delivery market, and processing issues to generate the desired products. There was also a need to develop small- and large-scale technologies for each fuel source. There was also a need to understand the diversities within each source of fuel such as the different ranks of coal and different chemical and physical compositions of crude oils. The effects of processing conditions for the production of different types of gas such as natural gas, synthetic gas of different heat content, propane gas, landfill gas, shale gas etc. needed to be understood. Finally the effects of different types of waste and biomass on fuel and energy production processes needed to be evaluated. This led to the independent commercial development for each source and these sources competed with each other for their share of energy and fuel market. Nuclear energy also competed with the fossil energy for its share of power and electricity market. In recent years, the development of renewable sources of energy such as solar, wind, geothermal, and water have followed the same trends for their needs in the required technological development as well as their penetrations in the energy and fuel market.

While in the past 10 sources of fuels and energies have developed in a "rainbow" fashion and improved our understanding of each source of fuel and energy, the future requires more interactions and integrations among these sources of fuel and energy. The need for energy and fuel is growing worldwide, and this need competes with environmental regulations and market economics now more than ever. Furthermore, the available sources for fuel and energy vary widely in different parts of the world.

The central theme of this book is to demonstrate that safe, secure, efficient, economical, sustainable, and environmentally acceptable strategy to meet future global energy needs requires multiple levels of integration among different sources of fuel and energy. Although the knowledge base for the technological development of each source of energy and fuel is still needed, a long-term sustainable solution to the energy need lies in the set of meaningful integration of various sources of fuel and energy. The best way to transition from an energy industry, which is largely fossil fuel based, to the one with balanced fossil, nuclear, and renewable energy based will be to carry out a set of integrated energy and fuel production and dissemination processes. This book lays out the arguments why future development in the energy and fuel industry requires more prominent strategy of building multifuel, hybrid, and grid-integrated energy and fuel systems.

We have more than 1.5 billion people in the world without any electricity and associated infrastructure to receive it. Climate change and other environmental needs will put more and more pressure on fossil energy sources, and they will not be able to meet the desired regulations while maintaining cheaper fuel price. Nuclear energy will continue to suffer public image unless it becomes a part of carbon-free energy solution and it is used for other nonelectrical applications. While the growth in renewable energy sources is desirable, they face many challenges for their implementations. We need a paradigm shift where the strategic and pragmatic solution to the desired energy landscape may require more hybrid and integrated energy sources and multiple-fuel systems for power, heat, and synthetic fuel generations.

We need more creative and effective integrations of various forms of fuel and energy industries. In particular, we need more integrations among three major sources of energy: fossil, nuclear, and renewable both at local and holistic levels. This will provide an energy portfolio that has (a) less reliance on fossil energy, (b) more use of nuclear energy for both electrical and nonelectrical applications, and (c) a larger share of renewable sources of energy. Better and faster penetration of renewable sources of energy in the total energy portfolio will also require more integration among renewable, fossil, and nuclear energy sources. This book illustrates the theme with specific strategies, success stories, and potential challenges for its accomplishment.

1.2   WHY ENERGY AND FUEL SYSTEMS INTEGRATION?

Before we address the issue of why energy and fuel systems integration is needed, it would be worthwhile to briefly state the major issues facing the energy and fuel industry:

1. Social and political pressures on energy and fuel industry to reduce emissions of carbon dioxide and other greenhouse gases (GHGs) are very significant. It is generally believed that a significant reason for climate change is human-induced emissions of carbon in the environment. It is also factual that the most significant amount of GHG emissions is caused by the energy and fuel industries and vehicles using fossil fuels. Going forward, carbon emission to the environment will have to be reduced with less use of fossil energy.
2. Within fossil energy industry, coal-based power plants will be required to reduce emissions of carbon dioxide and other pollutants. The stringent environmental regulations and the cost of clean coal technology will force more and more aging coal power plants to be replaced by natural gas–driven power plants. Conventional crude oil supply will diminish. The recovery and use of unconventional oils such as heavy oils, tar sands, and bitumen will be even more environmentally harmful. The natural gas industry will rely more and more on unconventional sources such as shale gas, coal bed methane, and tight gas. Overall, fossil energy industry will struggle to maintain its stronghold on the energy and fuel industry due to social, political, and economic pressures. There will be internal realignment within components of fossil energy.
3. While nuclear energy industry may be favored in some parts of the world, overall it will continue to face social and political oppositions due to safety and environmental concerns. The nuclear industry will gain more public acceptance by its role in allowing renewable energies to penetrate deeper in the energy market and its use (both at large and small scales) for nonelectrical applications. The latter can be accomplished by the use of excess heat during nuclear power production process (cogeneration process).
4. While renewable energy sources will have favorable social and political acceptance, they will have to penetrate the market based on their competitive costs. High capital costs, lack of stable and sustainable large-scale productions, lack of infrastructure and strong competition from fossil energy industry will create strong headwind against their growth.
5. Besides price and environmental impacts, the availability of energy and fuel in poor and remote parts of the world will become more and more important. We have 1.5 billion people on this earth with no access to electricity. Social and political pressures on the energy industries and governments to address this issue will increase.
6. There are more pressures on transportation industries to produce vehicles, which are more fuel efficient and emit less GHG (particularly CO_2)

in the environment. The transportation industry is one of the largest consumer (more than 25%) of energy and likely to grow. The internal combustion engine currently used in most vehicles has fuel-efficiency limitations.

7. While energy and fuel are important necessities around the world, sources of energy and fuel vary widely in different regions. Each region will have to be creative in using homegrown sources of energy to reduce foreign dependence.

These outlined realities indicate that we face a gradual transition of fossil fuel–based energy economy to the evolvement of a more balanced energy industry that reduces our reliance on fossil fuels and justifies the use of nuclear energy more based on its environmental impact and its role in nonelectrical applications and a faster, cost-effective, and sustainable penetration of renewable sources of energy within the total energy portfolio. The strategies adapted to realize such a transition must factor (a) our existing investment in infrastructure for fossil energy, (b) large capital costs required for new renewable sources of energy, and (c) our inabilities to build sustainable and competitive large-scale renewable sources of energy and fuel supply base.

This book shows that all of these objectives can be achieved by combining old sources (fossil and nuclear) and new sources (renewables) for the present and new applications in the most meaningful way. These combinations can be implemented at three different levels of systems integration: multifuel, hybrid (with or without cogeneration), and grid-integrated energy systems. The book is divided into three parts based on these levels of systems integration:

1. Use of various raw fuel combinations to generate environmentally acceptable and economical power and synthetic gas, liquid, or solid fuels and heat by thermochemical and biochemical transformation processes. This approach is often characterized as dual-fuel (or dual-feedstock) approach to power and synthetic fuel generations. This is the most basic level of integration where thermodynamic efficiency, basic energy conversion efficiency, and synergy, among different component reactions, affect the overall effectiveness of various power and synthetic fuel generation processes. With right fuel combinations, power and synthetic fuel can be generated on a large and sustainable scale and with minimum impact on environment. This book outlines advantages, opportunities, and challenges associated with the multifuel combustion, gasification, liquefaction, pyrolysis, and anaerobic digestion processes.

2. Use of two or more energy systems in a hybrid combination with or without cogeneration of heat and power to produce one or more products. This is the second level of integration wherein nuclear and/or renewable energy systems are combined in hybrid manners to produce stable, sustainable, large-scale, economical, and environment friendly operations to generate heat, power, and synthetic fuel. Heat (generated from nuclear, solar, and geothermal energy sources) can also be used for a variety of nonelectrical applications

through the process of cogeneration. Since future requires expansions of renewable energy and nuclear energy (both of which are carbon-free sources), this book examines possibilities and potential challenges for various types of hybrid nuclear and renewable energy systems. The importance of energy storage in the workings of hybrid systems is also delineated.

3. Use of a grid approach to holistically integrate generation and demand of gas, heat, and electricity. To some extent, fuel (particularly gas) heat and electrical networks are connected and interdependent. Energy storage also plays an important role on reliable and efficient grid operations. In recent years, significant advances on the development of natural gas and electricity grids have been made. Recent expansions of shale gas (and other unconventional gas), biogas, synthetic gas, and hydrogen productions have led to further development of the natural gas grid. The old electrical grid has been transformed into "smart grid," which dynamically controls the supply and demand of electricity by highly optimized and centrally controlled operation. The smart grid allows two-way flow of electricity and actively responds to the varying demands of the electricity. New developments of plug-in hybrid and electric cars and combined heat and power operations for nuclear and solar energies will make the role of smart grid even more important and complex. In future, operations of natural gas grid, smart grid, and localized heat grids will be more intertwined. This book examines various aspects of this holistic form of integration and its future potential for energy and fuel industry.

This book also presents a case study of vehicle industry, which exemplifies these three levels of integration to meet the demands for fuel efficiency and flexibility and environment regulations. Vehicle industry is aggressively pursuing the development of vehicles, which are capable of using multifuels feedstock, vehicles that are hybrid in operation and vehicles that can be eventually connected to smart electrical grids for their requirements of power. The vehicle industry demonstrates the success of these three levels of integration. Their success would also lead to the similar success for other stationary applications.

The word "energy and fuel systems integration" that involves three levels of integrations, multifuel, hybrid, and grid-integrated energy systems as mentioned in this section, refers to the use of more than one fuel or energy sources of the "rainbow of 10" described earlier to achieve the end purpose. These 10 sources are divided into 3 major subcategories: fossil, nuclear, and renewable. In this book, while significant emphasis will be placed on multifuel, hybrid, and grid-integrated energy systems that cut across these three major subcategories, all important multifuel, hybrid, and grid-integrated energy systems combination will be assessed.

Often, the use of multifuels like coal and biomass or coal and water is considered as co-fuels or dual fuels. Sometimes, two fuels like gasoline and ethanol are mixed and they can be processed together. We will not consider the use of mixture of different types of coal, biomass, or oil as multifuels for the present purpose. Coal, biomass, gas, oil, waste, and water (which can be a fuel at high temperatures and pressures as a part of multifuel system) industries are separate, and a combined use of two or more of these fuels requires somewhat different strategies than the ones used for an individual fuel. Although water contains hydrogen (fuel), by itself it is not a fuel for combustion, gasification, liquefaction, or pyrolysis processes. In recent years, water has been found to provide either reactant or catalyst role in multifuel systems, which include coal, biomass, and waste. This will be further demonstrated in detail in the subsequent Chapters 3 through 6.

1.3   DETAILS ON METHODS OF INTEGRATION

1.3.1   INTEGRATION OF DIFFERENT SOURCES OF FUEL

The first method is to combine different sources of fuel to generate power, heat, and synthetic fuels. This book examines the effects of multifuel usage in five very important power and synthetic fuel generation technologies: combustion, gasification, liquefaction/extraction, pyrolysis, and anaerobic digestion. The use of multifuel such as coal and biomass, biomass and water, coal and natural gas, and coal and waste to generate heat and power (via combustion technology), synthetic gas (via gasification technology), synthetic liquid (via liquefaction and extraction technology), or synthetic fuel in general (gas, liquid, or solid) via pyrolysis technology is described in detail in Chapters 2 through 5 respectively. An anaerobic digestion of a mixture of different types of biomass and waste is also investigated in details in Chapter 6.

Combinations of various raw fuels offer some interesting opportunities and challenges. The processing of coal and biomass (or waste) mixture reduces carbon dioxide emission to the environment. There are also some synergies in the combustion and gasification of coal and biomass mixtures [14]. The mixture of coal and biomass/waste also allows a larger-scale operation for biomass and waste. If the mixture is properly chosen, the liquid production rate can be improved through synergies during co-liquefaction of coal and heavy oil (co-processing). As shown in my previous book [1], water can be an important fuel or a part of fuel mix under certain conditions. Combinations of water with coal, biomass, and waste under certain operating conditions can lead to improved synthetic fuel productions. Co-digestion of different types of wastes and biomass can significantly improve the efficiency of methane production by anaerobic digestion process. Such co-digestion of waste/biomass has gained such an acceptance in fermentation industry that it has now become a norm than an exception.

The concept of multifuel system is not new. It has been practiced since the last part of the twentieth century particularly as co-combustion, co-gasification, co-pyrolysis, co-liquefaction, and co-digestion. There were numerous different reasons for the development of multifuel systems. In the beginning, it started as a source of flexibility and convenience when a particular source of raw material was not easily available. The combustion of coal and oil mixture as well as coal and water mixture for boilers, diesel engines, and turbines gained some popularity due to their ease in providing long-distance transportation (through a pipeline). This also allowed a slurry injection instead of dry injection system to be used for boilers, combustors, and turbines. The combustion of coal and waste materials was also pursued as a means for making positive use of waste materials.

The advances in our understanding of unique properties of water at high temperature and pressure, particularly under supercritical conditions, led to significant exploration of liquefaction and gasification of coal, biomass, and waste in high-temperature and high-pressure water both under subcritical and supercritical conditions. In the last several decades, the use of water for the productions of synthetic fuels has expanded considerably. Some of these advances are outlined in Chapters 2 through 5. Simultaneously, the development of multifuel cars also made some progress largely boosted by the legislation dictating the use of renewable fuel like ethanol (10%) in gasoline. Today, most cars on the road are capable of using multiple fuels in some format.

In recent years, the combustion, gasification, liquefaction, and pyrolysis of the coal and biomass mixture gained significant attention because of the positive effects of biomass on the GHG emissions. As the use of fossil energy becomes more unattractive because of its harmful effects on environment, replacement of some part of fossil energy by renewable forms of energy such as biomass or waste is becoming more and more popular.

As demonstrated in a recent report by ExxonMobil [2], the global landscape of energy and fuel has changed continuously over the last more than two centuries. These changes have occurred due to a number of reasons including market demand, environmental regulations, availability of supply, pricing structure, storage and transportation infrastructure, and development of new technologies and applications. Different sources of fuels and energies have gained their prominence through competition and social and political acceptance.

While the dominant source of energy and fuel over the last several decades has been the fossil energy and fuel, in recent years, other forms of renewable energy have gained some grounds due to environmental concerns, social and political pressures, and success of new technologies in harnessing renewable forms of energy. The displacement of fossil energy and more use of nuclear and renewable sources of energy still face some headwinds. The use of multifuels for five very important synthetic fuel and power generation technologies provides a method for diverting the headwinds. Our present energy and fuel infrastructure is more geared toward the use of fossil energy, and the development of renewable energy sources will require large capital investment and new infrastructure. A sustainable short- and long-term strategy would be to build hybrid fuel market that can use both renewable and fossil energy sources and at least partially use the existing infrastructure. In the case of "co-digestion," the use of multiple waste and biomass significantly improves the efficiency of methane production. Similarly, in co-liquefaction or co-processing of coal and heavy oil, synergistic actions allow better production of synthetic liquid than what is possible for each individual source.

1.3.1.1 Combustion

This is one of the oldest technologies for power generation from fuels. It is well commercialized and is used all over the world. Combustion of coal to generate power has lately come under severe criticism because the process generates harmful chemicals and carbon dioxide, which are emitted to the environment. The process also generates ash, which needs to be either used or discarded. The environmental impacts of the coal-based power plants have forced new power plants to be more natural gas based. Unfortunately, while natural gas–based power plants generate less harmful chemicals, they still generate significant amounts of carbon dioxide. More aggressive push has been made on the so-called clean coal technologies, which handle emission issues. Unfortunately, these new processes are expensive and increase the cost of power for the customers.

The use of multiple fuels for combustion process is not new. In 1970 and the 1980s, coal–oil mixtures were used for the combustion. Similarly, the technology for coal–water slurry for the combustion was developed. In this book we will briefly assess these technologies. Unfortunately, while these technologies have some merits and are used in boilers, diesel engines, and turbines, they do not completely address the environmental issues.

Major recent efforts on the use of multifuel for the combustion have been focused on coal and biomass, coal and waste, and different types of waste (such as biomass, polymers, and glass). A multifuel feedstock that combines fossil energy and renewable form of energy (such as biomass) makes the most sense because a suitable mixture of coal and biomass is neutral toward the emission of carbon dioxide in the environment. Such a mixture has other advantages such as easy scale-up and more options for optimization. Co-combustion processes open up many possibilities and significantly expand the availability of raw materials for the combustion process. These and other issues related to co-combustion are discussed in this book.

1.3.1.2 Gasification

Another very important and versatile technology that is heavily commercialized is gasification. Just like combustion, this technology is also mature and is widely used to produce synthetic gas of different heat contents depending on the process conditions. Most gasification processes have used coal, oil, waste, and some other materials like oil shale and heavy oil. In recent years, the use of biomass for gasification has also been explored.

Just as for combustion, the use of a multifuel feedstock containing coal and biomass, coal and waste, or waste and polymers makes significant sense for the same reasons as those mentioned earlier for the combustion in Section 1.3.1.1. The difference between combustion and gasification is the amount of oxygen used in the process. In combustion, excess oxygen is used so that all materials are completely burned resulting in nonfuel products such as carbon dioxide, water, and oxygen. The process of combustion produces thermal energy that is used to drive turbine and generate electricity. The gasification process, on the other hand, uses substoichiometric amount of oxygen (partial oxidation) so the product contains fuel gases such as methane, hydrogen, carbon monoxide along with carbon dioxide, water, and oxygen. The products of gasification can be used either to generate power or to produce liquid fuel, chemicals, and materials with proper downstream catalytic conversion processes. This versatility of the gasification process has made it the most widely used thermochemical process in the world. A multifuel gasification process provides many benefits that include (a) more environment friendly gasification, (b) possibilities for a larger-scale operation for renewables, and (c) more options for process optimization, among others. This book critically assesses all the benefits and challenges of the multifuel gasification process.

One of the interesting multifuel gasification processes is the sub- and supercritical water gasification of coal and biomass. In recent years, supercritical gasification and reforming of numerous coals, biomass, and other substances are extensively examined because they easily produce either syngas (hydrogen and carbon monoxide) or hydrogen and carbon dioxide. Under certain conditions and in the presence of a catalyst, biomass in subcritical water can also produce gaseous fuel. Thus, the use of sub- and supercritical water for gasification and reforming of coal and biomass or coal and waste presents another exciting route for the production of hydrogen. This book also examines this novel and exciting approach for the gasification of multifuel (in this case three, coal, biomass/waste, and water) feedstock.

1.3.1.3 Liquefaction/Extraction

This is another important technology to produce oil from solid coal, biomass, waste, oil shale, or heavy oil. The production of oil produced by this method can be upgraded using standard refining technologies to produce naphtha, fuel oil, gasoline, diesel oil, or jet fuel. The direct liquefaction of coal to produce synthetic liquid is already a multifuel process because it involves two fuels: coal and oil (donor solvent). Often, liquid is produced by the extraction process. When liquefaction is carried out in the presence of a solvent, it becomes an analogous process to the solvent extraction. Most often liquefaction is carried in the presence of a donor solvent so that solvent extraction becomes a chemical dissolution process wherein solvent reacts with the coal, biomass, or oil shale molecules to produce a liquid containing a wide boiling point range of oil fractions.

In general, coal is liquefied in a donor solvent, although attempts have been made to liquefy coal in high-temperature, high-pressure water. A multifuel system that has become very popular is the hydrothermal liquefaction of biomass and waste in water under sub- and supercritical conditions. This multifuel process has resulted in very positive results. Bio-oil produced by this method can be upgraded to suitable oil for heating or transportation purposes. Coal is also liquefied in supercritical water to produce oils. Another interesting multifuel example is the production of oil from algae using supercritical water. This process has also commercial viability.

In the 1980s and 1990s, co-liquefaction or co-processing of coal and heavy oil in one- or two-stage processes became very popular. Numerous pilot-scale operations were proposed and tested. Co-processing of coal and waste and waste and polymers were also investigated. The studies showed that certain types of mixtures showed synergistic reactions and improved liquid yield. This book examines these reported studies.

In recent years, a hybrid liquefaction process, which combines both direct and indirect liquefaction processes, has also been examined. In indirect liquefaction, coal or biomass is first gasified and the synthetic gases are converted to liquids using the Fischer–Tropsch synthesis. This hybrid process has given some positive results.

1.3.1.4 Pyrolysis

Unlike combustion and gasification, pyrolysis is carried out in the absence of oxygen. This is also a very versatile and commercially adopted process because the process can generate refined solid, liquid, or gas depending on the operating conditions. Pyrolysis of coal, oil, and biomass has been extensively examined in the literature. Pyrolysis of naphtha fraction of oil is commercially used to produce ethylene and propylene and raw materials for polyethylene and polypropylene. Unlike in gasification, pyrolysis produces many chemicals in the gas phase and can produce oils that can be easily upgraded to various kinds of fuels. Pyrolysis of mixture of coal and oil has also been examined in the literature.

This book examines the benefits of pyrolysis of multifuel feedstock for the productions of refined solid, liquid (oil), and gas. Due to the absence of oxygen, the process of pyrolysis does not produce excessive amount of carbon dioxide like the ones produced in combustion and gasification processes. The pyrolysis of multifuel feedstock offers the possibilities of a variety of refined solids, liquid oils, and gases with a wide variety of compositions. Pyrolysis of mixed feedstock also produces improved liquid yield due to synergistic reactions. This book examines these alternate possibilities.

One of the interesting options in the pyrolysis is the wet pyrolysis (pyrolysis in the presence of water) of single or multifuel like coal, biomass, or waste. As shown in my previous book [1], wet pyrolysis offers many advantages. Wet pyrolysis of biomass produces many useful solid, liquid, and gaseous products that are useful for fuel and other material industries. This book examines wet pyrolysis under subcritical conditions for the water producing refined solids.

1.3.1.5 Anaerobic Digestion

An anaerobic digestion of aqueous waste is used worldwide to produce methane (and hydrogen under specific fermentation conditions), which can be used for the power production and for various other downstream operations [15]. This is one of the most energy-efficient processes. While the process is slow, it is carried out even in landfills. Different types of waste and biomass produce different amounts of methane with a varying degree of efficiency.

In recent years, anaerobic digestion of multiple feedstock (waste and/or biomass) has been extensively used. Such co-digestion process, particularly with wastewater treatment plant, has improved the digester efficiency of the methane production significantly. The co-digestion also allows the optimal use of open or close digester area and integrates multiple distributed operations. Thus, it also saves the areas required for the digestion process. Besides sewage sludge, co-digestion technology has also been applied to treat human and animal manure. Co-digestion can also result in the improved digestate that can be used as fertilizer depending on the nature of the multiple feedstock. This process has gained such acceptance that in many parts of the world, co-digestion has become a norm instead of exception in the anaerobic digestion industry.

1.3.2 Integration of Different Sources of Energy

While nuclear energy is carbon free, its usage suffers from public support due to safety and unavailability of suitable methods for waste disposal. While renewable sources of energy have a very favorable public support, some of the renewable sources such as solar and wind energies are time and location dependent and will require significant storage capacity or a backup system to be useful on a sustained and continuous basis. Same is true for the use of biomass and waste where a large-scale operation is not possible due to problems involved in the transportation and storage of the raw materials. Thus, the displacement of fossil energy by nuclear and renewable energy faces many challenges.

Forsberg [3,4] pointed out that a long-term viable strategy for the world is to rely less on environmentally harmful fossil energy and more on carbon neutral nuclear and renewable energy resources. For this to occur, we need more hybrid energy systems that involve combinations of fossil and renewable; nuclear and renewable; and nuclear, renewable, and fossil energy sources. Such hybrid combinations will allow a balanced use of each source of energy and fuel with proper environmental and long-term sustainability considerations. It will also make the penetrations of nuclear and renewable energy sources in the fossil energy–dominated world more easy and sustainable. According to Forsberg [3–10], coordination and smart synergistic hybrid combinations of various forms of energy and fuel are the best strategy to satisfy the energy and fuel need of the society on a sustainable and environmentally friendly basis. In choosing hybrid energy systems, synergies between different fuel and energy systems are very important. Randomly picked hybrid systems won't result in the improved, sustained, or optimum performance.

For example, Forsberg [3–10] points out that within fossil energy, natural gas and light crude oil produce less carbon dioxide than heavy crudes and coal. The future oil market will be more based on unconventional oils such as heavy oil, tar sands, bitumen, and oil shale. The recovery and refining of these oils will produce more carbon dioxide and will require more hydrogen than what is presently needed for the conventional crude oils. The processing of coal will also require more hydrogen and it will produce more carbon dioxide. Forsberg [3–10] suggests some novel hybrid methods of processing heavy and unconventional crude and coal underground using nuclear heat so that liberated carbon remains underground and not emitted to the environment. Forsberg and others at INL laboratory [11–13] point out that carefully designed hybrid energy systems accompanied by cogeneration (combined use of heat and power) will allow (a) less liberation of carbon dioxide to the environment, (b) more penetration of renewable energy sources into energy market, (c) better use of nuclear and solar energies for various nonelectrical applications besides its use in the generation of power, (d) more possibilities for power generations in the remote locations, and (e) more even supply of fuel and energy across the world even when local resources for fossil energy are scarce.

A hybrid energy system involving nuclear and/or renewable and/or fossil energy sources can lead to more efficient, economical, and environmentally acceptable processes to generate power, heat, or fuels. Here, more than one source of energy like nuclear and renewable (like solar, wind, geothermal, and hydro) are combined and backed up by renewable (biomass) or fossil fuels to manage both base level and peak power demand. They can also be designed to manage numerous nonelectrical applications. Such combinations are particularly important for time- and location-dependent solar and wind energies where proper hybridization allows both stability and a larger scale of power delivery operation. This book examines several cases of literature-reported hybrid energy systems which includes the following topics:

1. Nuclear power plant or solar energy cannot only generate power, but the excess heat generated in the process can be used to produce and refine both fossil and renewable fuels as well as reduce the need for fossil fuels in numerous nonelectrical applications.
2. Excess heat generated from nuclear and renewable energies (particularly solar) power plants can be used to generate fuel (particularly hydrogen), which can be used for energy storage. This stored energy can be used to meet the peak power demand.
3. Time and location dependencies of solar and wind energies can be handled though a hybrid system where either nuclear or fuel-operated power plant can provide steady base supply as well as meet the peak demand.
4. Fuel run turbines or other energy storage systems can be backups to the renewable energy (solar, wind, geothermal)–operated power plant. This is useful for remote locations where a connection to a power grid is not available.
5. Numerous hybrid combinations of renewable, nuclear, and fossil energy with each other or energy storage systems to provide more stable and sustainable power, heat, and synthetic fuel productions at distributed levels.

This book outlines numerous examples of successful hybrid combinations both at smaller and larger scales that are currently used or conceptualized across the world.

1.3.2.1   Hybrid Nuclear Energy Systems

The discussion on hybrid energy systems is divided into two chapters (Chapters 7 and 8) in this book. Chapter 7 illustrates various examples of hybrid nuclear energy systems. Recent works of Forsberg and coworkers [3–10] have shown that hybrid energy systems combining nuclear energy with either fossil energy or renewable energy have enormous potentials. They and others [11–13] have examined numerous hybrid combinations of nuclear and fossil as well as renewable energies. The literature has also pointed out that hybrid use of nuclear energy for heat and power (cogeneration) has some very valuable applications.

Forsberg and coworkers [3–13] examined the use of excess heat in nuclear reactors for various types of nonelectrical applications such as district and industrial process heating, water desalination, synthetic fuel production, enhanced oil recovery and refining, transportation needs including hydrogen production, ship propulsion, and space applications, among others. These types of usages also allow the successful implementation of small modular nuclear reactors for specific nonelectrical applications. Depending on the nature of nuclear reactor technology used, the heat generated at different temperatures can be used for different nonelectrical applications. Forsberg and coworkers [3–10] feel that the use of nuclear energy as an enabling mechanism for the penetration of renewable sources of energy and its application for various nonelectrical applications can substantially increase the use of carbon-free nuclear energy. All these hybrid nuclear energy systems are critically assessed in this book.

1.3.2.2 Hybrid Renewable Energy Systems

Renewable sources of energy include solar, wind, geothermal, and water. Solar and wind energies are important sources but they are time and location dependent. For example, in the nighttime there is no solar energy. There is no wind energy when wind is not blowing. For these reasons, building a backup or complementary hybrid energy system with these renewable sources make significant sense. This book examines various hybrid combinations of renewable energy with nuclear and fossil energy to supply power that is stable and sustainable. The hybrid systems handle both base and peak load demands.

Recently, a very large hybrid renewable energy plant that includes both geothermal and solar energies is built and being operated in Nevada (this is further described in Chapter 8). This book examines this and other hybrid renewable energy systems in distributed and isolated areas where connection to the smart electrical grid is not possible or not economical. Numerous real case studies for hybrid solar and wind energy systems are critically analyzed. The need for energy storage as a backup system is also discussed. For solar energy, use of excess heat to produce solar fuels along with solar power is also examined. This combined use of heat and power is similar to the one used for the aforementioned nuclear power.

Along with solar and wind, both hydro- and geothermal energies have significant potentials to operate in hybrid fashion. Heat pumps operated by geothermal energy can operate jointly with the conventional heating provided by natural gas or oil. Hydrokinetic energy can be combined with wind energy offshore to use common electric transmission lines under water. Hydrokinetic energy can also provide backup power when wind is not blowing. All of these and many other examples of hybrid systems with hydro- and geothermal energies are also discussed in this book.

1.3.3 HOLISTIC INTEGRATION OF DIFFERENT SOURCES OF FUELS AND ENERGIES (GRID APPROACH)

While the concept of energy and fuel systems integration can be carried out at local levels using multifuels and advanced hybrid energy systems with and without cogeneration, the ultimate desire for energy and fuel industry is to create holistic integration of distributed fuel and energy (power) systems by grids (or often called network in fuel system) where all subsystems of multifuel and hybrid processes can be connected to satisfy the demand of the customers on a dynamic basis. We have had grids (networks) for fossil fuels, particularly for gas that has been widely used to efficiently manage supply and demand for various usages of gas. Regional storage systems for natural gas facilitate management of such integrated network. In recent years, the transporttion of shale and other unconventional gas, biogas, synthetic gas, and hydrogen is also carried out with the use of natural gas grid.

In a natural gas grid, supply and demand of different types of gas are controlled by unified transportation and storage infrastructure. This infrastructure is often controlled at regional levels because of slower time response to supply and demand. The grid, however, connects the different types of gas productions and different nature and location of the demands.

In a similar way, the power production, transmission, distribution, and customer demand are also met by newly developed smart electrical grid. This grid will allow distributed renewable energy power plants to be connected for effective management of supply and demand of power. The smart grid can handle fluctuations in power generation supply and demand strategically and optimize and control various subenergy systems. Along with gas and electricity grids, local heat grids will make the maximum use of thermal energy particularly for the combined use of heat and power systems.

This book discusses the recent developments in the operation of a smart power grid. The operation of power grid (totally centralized or somewhat distributed) can also be facilitated by the integrations of backup fuel and energy storage devices. This book evaluates in details various methods of fuel and energy storage and the effects of the integration of energy storage to the smart grid. Unlike gas grid, smart grids connect various states because electricity travels much faster than gas flow. The smart electrical grids are capable of managing two ways flow of electricity (grid to customer and customer to grid). This allows a better and dynamic handling of demand response with optimization of the pricing structure. The grid is a complex system of management where dynamic control of supply and demand is exercised using modern information and communication systems. In recent years, the grid has also allowed a smart management of electrical flow in plug-in hybrid and electric cars.

There are three smart grid regions: west, east, and Texas region in the United States. There are plans to build a national smart grid to control and manage electrical network of the entire country in an efficient manner. The grid concept can also be extended to integrate natural gas, electrical power, and heat distribution grids (a system of grids). Such integration will allow any distributed renewable and small-scale nuclear energy power production to operate in a stable and sustainable manner by its connection to the common grid. It will also allow easy handling of local peak demand without upsetting the need of the other customers. The development of smart grids will further facilitate the use of small-scale and distributed renewable and nuclear energy power plants in an economical manner.

This book also briefly discusses the importance of energy systems integration concept for the development of right mix of renewable, nuclear, and fossil energies at local and global levels. The use of multifuels for synthetic fuel and power generations, use of advanced hybrid energy systems with or without cogeneration at regional level, and a holistic and intelligent integration of various fuel and energy systems through smart grids will allow optimization of systems at both local and national levels. The full adoption of these three levels of integration will thus allow us to achieve our overall objectives for future energy landscape in the most efficient manner.

1.4 MULTIFUEL, HYBRID, AND GRID-INTEGRATED VEHICLES: A CASE STUDY

One of the most important applications of the concept of multifuel, hybrid, and grid-integrated energy systems is the development of multifuel, hybrid, and grid-integrated vehicles. As a part of the developments of more flex fuel, hybrid, and plug-in vehicles, both large and small vehicles are more and more fuel and energy integrated. These development efforts are aimed at producing vehicles that (a) are capable of operating with multifuels and thereby more fuel flexible, (b) are more fuel efficient (deliver more mileage per gallon of fuel), and (c) emit less pollutants, particularly carbon dioxide. This book critically assesses the development efforts and their future. It is predicted that in the next several decades, multifuel, hybrid, and plug-in electric vehicles will become the norm rather than exceptions.

Multifuel cars can come in several different formats. Due to government regulations, all gasoline-driven cars now use 10% ethanol, which is mixed with gasoline. Multifuel cars can have two different fuel storage tanks depending on the nature of the fuels used. This could be diesel and natural gas, natural gas and gasoline, hydrogen and gasoline, etc. Some of these options with specific case studies are discussed in this book.

Hybrid vehicles using more than one source for power are increasing. There are many forms of hybrid cars, which use either electricity, fuel cell, hydrogen, conventional gasoline or diesel fuels, or biofuels like ethanol, natural gas, and propane. The designs of various types of hybrid cars on the road are briefly discussed in this book. The number of hybrid vehicles on road and that of hybrid boats on the water are rapidly increasing. Efforts are also being made to build hybrid planes.

Finally, new plug-in hybrid or electric cars will allow better management of electricity flow through use of a smart grid. The smart grid through its two-way flow of electricity will allow better management and control of extra electricity needs that will be created by hybrid and electric cars. Hybrid and electric cars and their management through grid are the ultimate example of why and how integrated systems will dominate the future energy industry. Future grid will combine static and mobile uses of energy.

The case study of multifuel, hybrid, and grid-integrated vehicles strongly supports the main theme of this book. The theme of fuel and energy systems integration at various levels is clearly exemplified by the vehicle industry. The successes of multilevel integrations in vehicle industry indicate that in future more integration of different sources of fuel and energy must be explored in static applications as well. Such a strategy will not prohibit the advancement in each fuel and energy technology. Instead, it will make the use of each source more robust and more sustainable. Multifuel and hybrid energy systems, which combine fossil and nuclear, fossil and renewable, and nuclear and renewable energy, appear to have long-lasting future.

1.5 SCALES OF ENERGY AND FUEL SYSTEMS

There are basically three main purposes of fuel and energy systems: (a) to produce appropriate solid, liquid, or gaseous fuels that can be used for a multiple downstream usages such as in productions of chemicals and materials, in mobile vehicles and stationary devices, and for energy storage; (b) to provide fuels and heat for residential, commercial, industrial, and other heating needs as well as their use for other non-electrical applications; and (c) to produce power and electricity. Out of the 10 aforementioned sources of fuel and energy, nuclear, solar, wind, and geothermal energy sources are predominantly useful for the generation of heat and power. Coal, oil, gas, biomass, waste, and water are generally used for all three purposes mentioned above in Section 1.3. Water can be a direct source for power or it can generate fuel in combination with coal, biomass, and waste or through its dissociation.

Energy and fuel systems can be divided into three categories based on their sizes. Each serves a different purpose.

1.5.1 LARGE-SCALE CENTRALIZED FUEL OR POWER SYSTEMS

Large-scale oil refineries, coal and natural gas processing plants, and large power plants are a part of this category. For these large-scale systems, an economy of scale works well. There are numerous examples of these systems all across the world. In order for these systems to work effectively, infrastructures for raw material transportation to the plant, delivery of the product from the plant, and storage of both raw materials and products must exist. In the United States, this is well established particularly through fuel and electricity grid structures. The smart grid structure for electricity developed since 2006 has considerably helped the transmission and distribution of power generated by these large systems.

The large-scale systems are dominated by fossil energy. Nuclear power plants are also most cost-effective at a larger scale. In nuclear power plants, excess heat at various temperatures (depending on the nature of the nuclear reactor) can be used for nonelectrical applications such as to produce synthetic fuels from coal, heavy oil, tar sand, and shale oil as well as from biomass and waste, recovery of unconventional oils, production of hydrogen from water, or reforming of carbonaceous feedstock, desalination of water, and district or industrial heating. This is illustrated in details in this book. The concept of combined heat and power use can also be applied to centralized solar power plants.

As shown in Chapters 2 through 5 and 8, unlike fossil energy and nuclear energy power plants, renewable energy fuel and power plants are difficult to build and sustain on large scale. Stand-alone biomass and waste fuel and power plants cannot be built at large scale due to difficulty in providing the sustained supply of raw materials needed for such plants. Similarly, large-scale solar and wind energy power plants are difficult to manage without backup storage systems or connections to electrical grids because of their time and location dependencies. Multifuel, hybrid, and grid-integrated energy and fuel systems will allow building of large-scale renewable fuel and energy systems in such cases.

1.5.2 SMALL- AND INTERMEDIATE-SCALE DISTRIBUTED FUEL AND POWER SYSTEMS

These types of fuel and power systems are developed to serve the needs of the local areas or regions that are relatively small or where the scope of the raw material and/or product delivery is limited. One example is the stand-alone biomass fuel generation or biomass or waste-driven power plant. This can serve local needs. Solar and wind energy–driven power plants can also be of this scale. These types of plants either can be connected to larger grids or can have its own independent grid serving local region. A connection to some types of energy storage system is also important for the stability and sustainability of the operations. If the power generated from these scale plants is not connected to a larger smarter power grid, independent hybrid systems need to be established to handle both base and peak power needs.

In recent years, smaller modular nuclear reactors are conceptualized for its use for nonelectrical applications. These reactors make the best use of heat and power for small-industrial-scale operations. The production of solar fuel from excess solar energy can also be carried out at this scale. The small-scale power generation systems are generally distributed systems.

1.5.3 MICROFUEL OR MOVING FUEL AND POWER SYSTEMS

These systems are generally located in remote areas or in moving vehicles. Besides its scale, the most important difference in these systems is that they may not be connected to any major fuel infrastructure or power grids. They are designed to be independently sustainable. Often, they are connected to microgrid. The plug-in electric or hybrid cars are, however, moving toward vehicle-integrated grid (VIG) systems. In remote areas, the availability of grid connections may not be possible. Strategies used in designing such systems are generally very different from the ones used in centralized large-scale or distributed intermediate-scale fuel or energy systems that are connected to the centralized infrastructure or smart power grids.

1.6 PRINCIPLES OF ENERGY AND FUEL SYSTEMS MANAGEMENT

Energy and fuel systems, large or small, are generally assessed, developed, and managed by the following 12 principles. We briefly assess each of this principle and wherever possible briefly outline the benefits of integrated, hybrid, and multifuel operations over single-fuel or single-energy source or distributed systems for each of this principle.

1.6.1 EFFICIENCY OF THE SYSTEM

Any fuel or energy system involves conversion processes wherein one form of fuel or energy is converted into another form. These conversion processes involve thermodynamic, process, and equipment-related efficiencies. Generally, overall efficiency is estimated in the form of how much energy is captured in the product compared to what is available in the feed. As shown in Chapters 2 through 10, multifuel, hybrid, and integrated energy systems, in general, allow more possibilities of improving the system efficiency than single-fuel, single-energy systems. For example, fuel efficiency in hybrid cars is better than a single-fuel internal combustion or diesel engines. Overall, they provide more mileage per gallon than single-fuel automobile. The hybrid energy system, if carefully chosen, can give better options for the improvement of overall energy efficiency. Often, hybrid processes help overall thermal efficiency by not requiring some of the steps in a single-fuel processing system. For example, the processing of coal–water or biomass–water for hydrothermal or co-combustion processes alleviates the need to remove water, which requires additional amount of heat. In principle, this applies to any mixture of coal, heavy oil, biomass, or waste and water.

Thermal efficiency of a nuclear reactor can be significantly improved by using waste thermal energy for numerous nonelectrical applications such as hydrogen production by water dissociation or for the recovery of unconventional oil by thermally enhanced oil recovery methods. Such integrated process will result in more efficient energy system than single-fuel, single-energy system. The same can be said for the generation of solar fuels by solar energy. A smart grid–integrated renewable energy power generation system can be more efficiently controlled than a single renewable energy system. We will illustrate many examples of improved energy efficiency by multifuel, hybrid, and grid-integrated energy systems in Chapters 2 through 10.

1.6.2 Safety and Security of the System

Any energy and fuel system should be safe and secure. For this purpose, the effects of extreme conditions to the safety and security of the system should be carefully analyzed. For example, the effect of hurricane or tornado on wind energy farm and safety and security of the wind power delivered needs to be analyzed. Similarly, the effect of flooding on the hydroelectricity and hydrokinetic energy generation needs to be carefully assessed. In these cases, a hybrid backup system will allow a continuous generation of power under the catastrophic situations.

Within a hybrid system if one part fails or needs to be shut down, the other part can take over the slack created by the failed part. When the causes for the extreme conditions for two or multiple parts in a hybrid system are significantly different, different parts can act as backup for other parts in unusual situations. The same principle can be applied to multifuel systems. For example, the processing of coal and biomass together can allow substituting biomass by coal if there are storage and transportation problems associated with biomass. In a bi-fuel car, the fuel for the car can be switched back and forth if one fuel is in short supply. In an integrated smart grid operation, a failure in one part of the grid can be easily handled by isolating the failed part and reorganizing the remaining circuit.

Generally, a multifuel or hybrid energy system offers more security to the overall system than a monotonic system. If a combustion or gasification process uses multifuels like coal and biomass, the temporary short supply in one of these fuels can be adjusted by more use of the other fuel without any disruption to the combustion or gasification process. Similarly, if a power plant is hybrid and operated by a combination of solar energy, wind energy, and combustion of diesel fuel, the temporary downturn in any one form of energy (such as solar during night) can be counterbalanced by the supply from the other forms of energy. In general, a hybrid or multifuel system can be designed to be more secure than a monotonic system.

Another example is the cyber security of an automated power plant. In a single-fuel or a single-energy power plant, if this security is compromised, the entire power plant may have to be shut down. In a hybrid system, however, while one system needs be shut down, the other system can possibly continue to operate. This principle can be even applied to home heating system. If only heat pump is used for heating and cooling, it will not work well in extremely cold environment. On the other hand, if a hybrid system of heat pump, coupled with conventional gas heating, is used, in the extremely cold environment, the heating can be switched to gas heating from heat pump assuring safe and secure heating operation. These and many other examples are further illustrated in this book.

1.6.3 Automation and Control of the System

Most energy and fuel systems need to be automated and properly controlled. An innovative system that performs well but cannot be automated or controlled will have a limiting practical value. In any control system, a multivariable control system always gives more secure operation than a single variable control. In this respect, a hybrid or multifuel energy system may give more options for control than a single-fuel or single-energy system. Grid-integrated systems can offer even more options than multifuel or hybrid systems for the control.

An excellent example of most sophisticated automation and control system is the smart electrical grid. The grid controls supply and demand of electrical power by a two-way flow of electricity generation, transmission, distribution, and customer demand. The grid handles the demand response with most sophisticated grid automation and control. For example, heat and power requirement of a home from the grid can be altered if a home has installed a solar panel to generate heat and electricity. In fact, in this situation a house can return back some heat and electricity to the grid. This type of two-way exchange of heat and electricity allows better automation and control of supply and demand by the smart grid. The U.S. power transmission is controlled by three large grids: east, west, and Texas and associated region. These sophisticated grids allow best automation and control of all power generation, transmission, distribution, and customer demand at both base and peak levels. As shown in Chapter 9, the integration of energy storage systems to the grid improves automation and control of power distribution by smart grid.

1.6.4 System Optimization

Just like efficiency, the optimization of any fuel and energy system requires options available in the processing conditions. While multiple options generally make the system more complex, it also allows more possibilities for the optimization. As shown in Chapter 3, the gasification of the multiple fuels of coal and biomass allows various options for feed preparation, nature and arrangement of gasifiers, and their processing conditions as well as various options for product purification processes. These types of multiple options allow more possibilities for optimization. In a hybrid car where more than one source of fuel or energy (e.g., gasoline and electricity) is used, the performance of the car can be better optimized than for the car using a single source of fuel or energy. The usage of the power generated by wind energy can be better optimized if it is either stored or supplemented by other source of fuel or energy. In general, a multifuel, hybrid, or grid-integrated energy system will allow more efficient use of its subcomponents with more possibilities for its optimization.

Any energy or fuel system must be optimized to operate in the most cost-effective way. Sometimes, a balance between an optimum performance and a safe and secure performance needs to be achieved. It is also true that some systems are better optimized than others. An optimum system is not necessarily the system that gives best performance. It could be the one that has largest degree of sustainability or least environment impact. Thus, the parameters for the optimization should be clearly delineated. The optimum system must generally consider all the principles outlined here and prioritize the importance of various principles. The system should then be optimized around these priorities. Multifuel, hybrid, and grid-integrated systems, in general, offer more options and flexibilities for the system optimization.

Many people say that a smart grid operation is the optimized operation. Most smart grids optimize the electricity supply and demand needs based on defined objectives. Generally, a smart grid is operated in such a manner so as to satisfy the needs of the customer at the lowest pricing in a sustained and safe manner.

1.6.5 SCALE-UP AND SCALE-DOWN OF THE SYSTEM

For most synthetic fuel productions and power generations, an economy of scale works well. Unfortunately, while we have built large refineries and coal processing plants, these sizes are not possible for biomass processing. Large-scale refineries can produce up to 500,000 bbl/day of oil. The large-scale biomass plant can only produce up to 50,000 bbl/day of oil. This is because of the difficulties in transporting and storing required biomass in a sustained manner. One method being explored is to use multiple-fuel system consisting of both biomass and coal, which can allow the large-scale processing with a lower demand on amount of biomass.

The large-scale operation often brings its own share of issues such as storage and transportation of raw materials and products, available knowledge of process performance as a function of equipment and process size, and issues related to start-up and shutdown conditions. In principle, a hybrid system of solar, nuclear, wind, or fossil fuel can allow a much larger-scale operation than an individual solar or wind energy system. Due to their intermittencies, large-scale solar and wind energy systems are not sustainable without the support of another nuclear or fossil energy system.

A hybrid energy system consisting of solar, wind, or geothermal energy can be scaled up to a higher level better with the support of energy storage device than a single renewable energy system. In recent years, the scale-up of wind energy has been expanded by the use of a combined wind and hydrokinetic energy generation unit, which can be installed offshore. This unit is described in details in Chapter 8 in this book. The largest-scale operation is the large smart grid operation that can include electricity generation, distribution, and storage in multiple states.

Just as large-scale operations, small-scale operations can also be a problem for some energy and fuel systems. The generation of sustainable small-scale power in remote locations can be problematic. This problem is often alleviated using hybrid systems. The use of hybrid systems for an automobile is another example where high efficiency is achieved using multiple sources of energy. Generally, small-scale operations are not as efficient and cost-effective as large-scale operations. The lower efficiency can be counterbalanced by the use of multifuels or hybrid energy systems. Nuclear reactors to produce power are generally large scale for economic reasons. For targeted nonelectrical applications, small modular hybrid nuclear reactors are, however, more suitable, and if they are properly designed and used, they can be economical.

1.6.6   LONG-TERM STABILITY AND SUSTAINABILITY

As shown by ExxonMobil report [2], in 1800, the dominant form of energy was biomass for heating and light because other forms of energy and fuel, electricity, and mobile vehicles were not discovered. Over the following years, as fossil energy, electricity, and mobile vehicles were discovered, the use of biomass significantly decreased. Nuclear energy was also discovered and used. The landscape of energy and fuel thus significantly changed. In recent years, as more advanced technologies for the use of biomass, waste, solar, wind, water, and geothermal energies are developed, these forms of energy are slowly replacing other forms of energy in some parts of the world. The basic reason behind the development of this rainbow of energies and fuels is the competition among different forms of energy based on the energy management principles that are being discussed here. The winners and losers of this competition determine the dominant colors of this rainbow. Unfortunately, such a competition does not guarantee a long-term sustainability of the particular form of energy and fuel. In the long term, it is predicted that hydrogen and electricity will be the dominant modes of energy usage irrespective of their sources.

When a particular energy and fuel system contains more than one source of energy and fuel, it inherently increases its long-term stability and sustainability because the survival of such a system does not depend on market favorability of only one form of energy. For example, the use of coal and biomass for combustion or gasification can reinforce the supply and demand of each type of fuel. The mixture also has less environmental impact than coal alone. The mixture can better handle a larger scale-up (and therefore lower cost per unit) than the use of biomass alone. The hybrid system can also find more variables for optimization and improved efficiency than a single-fuel system. In short, multifuel coal and biomass system has a better long-term stability and sustainability than coal or biomass alone.

Both solar and wind energies are time and location dependent. Their use depends on our ability to store these energies so they can supply the need on a continuous basis. A sustainable approach for the use of these forms of energy is to use a hybrid system, which includes multiple sources of energy (solar, wind, and nuclear) along with an energy storage system. This approach can ensure satisfaction of both base and peak demand of power. Even better, a connection of these intermittent power generating systems to a smart electrical grid can guarantee even higher rates of stability and sustainability of their operations.

1.6.7   FUEL AND ENERGY STORAGE

The storage of raw fuels and energies produced is very important for sustainable operation of any energy and fuel system. For example, the storage of biomass is very important because it requires large space due to its low mass and energy density. Also biomass can undergo biochemical decay during storage if it is exposed to oxygen. For large-scale operations, storage of required biomass is a much more difficult task than that of coal. This storage problem prohibits the sole usage of biomass in large-scale operation. One solution is to use multifuel like coal and biomass to satisfy the need of the large-scale processing. Coal is easy to store in large quantity compared to biomass.

For time-dependent energy sources like solar and wind energies, storage of energy generated during peak time is very important so that energy and power can be delivered during the time when sources of these types of energy are not available. One approach has been to convert solar or wind energy to fuel (like hydrogen), which can store the energy in the form of fuel. The fuel can then be converted to power when needed. Hybrid renewable energy systems combined with a backup energy storage system can generate power in a stable manner for both base and peak demands. Various methods for energy and fuel storage are described in detail in Chapter 9 of this book. Integrations of these storage systems to any type of gas or electricity grid are vitally important for meaningful control of fuel and energy supply and demand.

1.6.8 FUEL AND ENERGY TRANSPORTATION AND DELIVERY INFRASTRUCTURE

One of the results of the development of individual fuel and energy industries is that we have invested enormously in the infrastructure to support a particular fuel and/or energy industry. For example, we have several hundred thousand miles of pipelines created to supply natural gas all across the country. The railroad industry will heavily suffer if coal is not transported across the country. The same will be true for the barge industry if coal is not transported across the oceans. Both land and sea base transportations of oil have resulted in huge investments.

The recovery of oil and gas from underground has also resulted in the creation of underground transportation infrastructure. If the productions of conventional oil and gas diminish, these infrastructures will become a waste. In recent years, efforts are being made to use these infrastructures for advanced geothermal systems. It is important that as new sources of fuels and energies (individual or mixed) are developed, the available infrastructure is used to the extent possible.

While large investments in transportation infrastructures are necessary, they prevent rapid changes and adjustments for the development of new sources of fuel and energy. One of the reasons for the evolutionary nature of development for energy industry is the time and the level of investment required to build the necessary infrastructure to support the changes in industry. It is also clear that a long-term sustainability of these infrastructures requires their adaptability to different sources of fuel and energy. If possible, future infrastructure should be built or adaptable for mixed or hybrid systems than for individual source of fuel or energy.

The use of existing infrastructure for new sources of fuel and energy is being heavily investigated. For example, landfill gas (which contains about 55% methane) has been transported in natural gas lines by first processing the gas for purification by adsorption process. This results in the gas, which contains 95%–98% methane. Such a gas can then be transported in the conventional natural gas pipelines. The use of these gas lines for shale gas will also require purification of shale gas. The transportation lines built for coal water mixtures as well as coal oil mixtures can now be modified to use for the transportation of tar sands or heavy oil and water.

Effective and efficient transportation systems for raw materials and products for any fuel and energy system can make or break the workings of the fuel or energy systems. For example, transportation of large quantities of biomass at long distance is very cumbersome and expensive due to its low mass and energy density. This makes building a large-scale biomass processing plant away from its source problematic. This is not the case for coal, oil, or gas. This can be an argument for building a hybrid plant that uses coal and biomass or coal and water. Such a plant can secure large quantities of raw materials by appropriately adjusting the concentration of various components in the mixture. We currently use a mixture of gasoline and ethanol in the infrastructure that previously used gasoline alone.

The transmission of power captured by offshore wind energy requires underwater power transmission infrastructure. While this can be built, it would be a lot more efficient if this can also be used for the power generated by the underwater hydrokinetic energy. A hybrid infrastructure using both wind power and hydropower will have more long-term sustainability. The transportation of electricity requires cables that are spread high above the ground all over the country. These transmission and distribution lines form the basis for a smart grid. The supply and demand of electricity across the country solely depend on the reliability and control of a two-way electricity flow in these transmission and distribution cables. Thus, transportation infrastructure for the fuel and electricity via gas, heat, and electricity grids is essential for the management of fuel and energy supply and demand. The success of hydrogen as fuel or for power generation will depend on its transportation and delivery infrastructure. Current natural gas network can handle 5%–15% (by volume) concentration. New infrastructure will be needed if hydrogen is transported in the pure form.

1.6.9 SUPPLY AND DEMAND MARKET AND OVERALL ECONOMICS

In the past, supply and demand market and the overall process economics and pricing structure of the commodity made the individual sources of fuel and energy succeed or fail. The viability of each source of fuel and energy depended on its competitive advantage over other sources of fuel and energy. While in principle such competition is good for the industry, it has not always resulted in the most optimum solution for fuel and energy systems. Furthermore, supply and demand market and pricing of unit commodity of fuel or energy depend on local social and political acceptance, available natural resources, market demand, and government subsidy. For example, gasoline price varies significantly across the world. The same applies to other sources of fuel and energy. While individual markets for fuel and energy compete with each other, optimum solutions may lie in the hybrid use of various fuel and energy sources. The hybrid systems can allow different options to use multifuel and hybrid energy sources when the price of a given source may fluctuate significantly due to global or local conditions. If there's nothing else, hybrid systems provide more options for optimization of the cost and weather the price fluctuation of individual fuel and hybrid energy source better over a long haul for the customers. Multifuel and hybrid energy systems can be better optimized at local level than a single-fuel or single-energy source. These arguments also apply to the supply and demand and pricing market for the electricity. More controlled supply and demand and stable pricing are better achievable in an integrated system than in a single-source system.

1.6.10 ENVIRONMENTAL IMPACTS OF THE SYSTEM

This is perhaps the largest force for the changes in fuel and energy industry at the present time. There appears to be a constant conflict between the need for larger fossil energy resources and the need to protect environment. Global climate change and emission of GHGs have pushed many regulations that will change the landscape of energy industry. In general, reduction in carbon emission to the environment will demand more use of renewable and nuclear energy and less use of fossil energy.

The coal-based power plants will face a significant decline unless these power plants can come up with more efficient and economical ways to process coal with less emission of hazardous gases and carbon dioxide in the environment. At the present time, this is not possible without an increase in the price of the product. The problem can be eliminated by using a hybrid system of coal and biomass mixture, which can be carbon dioxide neutral to the environment. A study [1,15,16] has shown that a 70% coal and 30% biomass mixture can be carbon dioxide neutral to the environment.

New "fracking process" has allowed more production of unconventional gas such as "shale gas." This increased production will allow all new power plants use less harmful gas than coal. The "fracking process" is, however, not without some environmental concerns, and these will have to be addressed in the future. Similarly, a fossil fuel (coal, oil, or natural gas)–based power plant will be less environment friendly than a hybrid plant that uses both fossil energy and renewable forms of energy such as solar, wind, or hydroenergy. Such a hybrid plant will not only have less environment impact but better long-term sustainability due to multiple sources of energy and fuel. Such a hybrid process will also benefit by the newer technological developments both in fossil fuel processing and in renewable forms of energy.

In general, renewable and nuclear energy sources have lesser environment impact than fossil energy. The penetration of renewable energy in the fossil energy–dominated market faces few challenges. High capital costs along with smaller sustainable plants make them economically unattractive. A combination of renewable energy sources with fossil or nuclear sources will allow larger-scale and sustainable operation. This will also make more use of renewable and nuclear energies and less use of fossil energy. The use of heat from nuclear reactor for nonelectrical applications will allow more use of carbon-free nuclear energy and less use of fossil energy. This suggests that a hybrid system combining renewable energy sources with fossil energy sources and a combined use of heat and power from nuclear or solar reactors may be effective ways to reduce the carbon emissions to the environment. Nuclear energy can also be used in combination with renewable energy sources to provide stable and sustainable power production. Hybrid automobiles emit very low carbon dioxide and they already meet the 2030 standard laid out by EPA.

The penetration of renewable energy sources in the power industry will be significantly facilitated by the use of smart grids and energy storage systems. Larger use of renewable sources will help our environment. These and many other issues related to the benefits of multifuel, hybrid, and integrated energy systems to the environment are discussed in Chapters 2 through 10.

1.6.11   SOCIAL AND POLITICAL ACCEPTANCE OF THE SYSTEM

Social and political acceptance of the given fuel and energy system plays a very important role in its success. In Brazil, a political decision was made to partially or fully replace gasoline by ethanol to gain better energy independence of the country. They used sugarcanes as a raw material to produce ethanol. This effort was supported by the government subsidy. Thus, in this case, social and political acceptance of the idea of multifuel (gasoline and ethanol) made huge success.

Similarly, in this country, a decision was made to substitute 10% gasoline by ethanol from corn using government subsidy. This idea worked until the price of corn became too high for its use as food. The shortage of corn crop due to drought also made the idea unworkable. The nuclear energy will gain more social and political support if it is used more for nonelectrical applications such as desalination of water and district and process heating. In general, political subsidy and social acceptance can enormously help the success of the particular fuel or energy system. Multifuel and hybrid systems have a better chance for gaining these acceptances.

A multifuel or hybrid energy system often gains more social and political acceptance than a new monotonic system because it can be accomplished by retrofitting the existing infrastructure with lower cost or by building a more efficient multipurpose infrastructure that can be sustained over a longer period. For example, the new law of using 10% ethanol in gasoline did not require any changes in the internals of vehicles, thus making its implementation much smoother than the law requiring replacement of gasoline by ethanol. Such a replacement would have required significant alterations in the car engines. Similarly, a hybrid energy system of wind energy, hydroenergy, and energy generated by diesel turbine would get more social and political support because of its more favorable environment impact. For this matter, a combination of any form of fossil energy and renewable energy will get favorable political support due to its less impact on the environment.

Globally, hybrid cars are rapidly expanding because of their social and political acceptance. On the other hand, all electric cars are still working their way to get the desired social and political supports. Shorter travel distance and longer time for battery recharge work against the social acceptance of all electric cars. Hydrogen cars are attractive in California. However, a lack of infrastructure has stalled its social acceptance. The required infrastructure needs more political support. Plug-in electric car and VIG will increase social and political acceptance of electric cars. There appears to be general social and political acceptance of the idea of hybrid fuel cell and electric cars. Thus, political and social acceptance of new ideas is very important, and multifuel, hybrid, and grid-integrated fuel and energy systems appear to have a better chance for these acceptances than single-fuel, single-energy systems.

1.6.12   SYSTEM ACCESS

While the demand for energy and fuel is increasing worldwide, there are parts of the world where there is no electricity available. More than 1.5 billion people in this world live without access to electricity. In future, one of our priorities should be to make energies (both electricity and fuel) accessible to as many people as possible. Since fossil energy sources are unevenly distributed and biomass, waste, and other renewable forms of energy are available in all parts of the world, the use of latter forms of energy will provide better access to electricity and fuel in all parts of the world. As mentioned in Section 1.3.2, use of a hybrid system is a better method to tap energy from renewable resources.

Another method to provide the electricity in the remote area is to connect them to a regional or local electrical grid if possible. Grid can provide access to electricity that may not be possible otherwise. Holistic integrations of fuel and electricity through respective grids are the best methods to expand the access to energy and fuel. In remote areas where grid connections are not possible, hybrid renewable energy power systems are the only workable solutions for power, particularly when significant fossil energy sources are also not available.

The first six of the principles described in Section 1.6 relate to the internal factors of the system itself, while the last six principles relate to the external factors surrounding the system. Both internal and external factors are equally important in the overall success and effectiveness of the system. It is possible that sometimes, internal factors are all positive for a given system and the external factors change its effectiveness. A good recent example is the recent decline in coal industry in favor of surge in gas industry. This has occurred because of the environmental effects of coal industry and recent developments of the new technology of "fracking process" that has changed the gas supply market. The long-term stability of individual sources is always questionable. Access to multifuel and energy sources always provide better long-term stability and sustainability.

1.7 ORGANIZATION AND SCOPE OF THIS BOOK

The content of this book is divided into three parts. Chapters 2 through 6 are devoted to analyze the advantages and disadvantages as well as the opportunities and challenges for the use of various types of multifuels in five most important thermochemical and biochemical processes for synthetic fuel and heat and power productions. In Chapter 2, the use of multifuels in the production of heat and power by combustion process is examined. This is one of the oldest and most widely used processes. Chapter 3 examines the effects of multifuels in the production of synthetic gas via different types of gasification process. Synthetic gas can be used to either generate power or produce fuels and chemicals by various downstream processes. Chapter 4 evaluates the effects of multifuels on liquefaction/extraction processes to produce a variety of liquid fuels. Finally, Chapter 5 examines the behavior of the most versatile pyrolysis process for multifuel feedstock. In each chapter, based on the literature information, the values added and the new challenges by the use of multifuels for these thermochemical processes are critically reviewed. Wherever possible, suggestions for best synergies among fuels and the need for possible new investigations are identified.

Chapter 6 examines the use of multiple feedstock of biomass and waste for the production of methane in an anaerobic digestion process. This process of "co-digestion" significantly improves the efficiency of the fermentation process. In recent years, this has become so popular that it has become a norm rather than an exception in the fermentation industry.

Chapters 7 and 8 examine hybrid energy systems involving nuclear, solar, wind, geothermal, and hydroenergies. In Chapter 7, various types of hybrid energy systems for nuclear energy are evaluated. Methods to use excess heat coming out of various types of nuclear reactors for downstream nonelectrical applications are also examined. The hybrid use of nuclear reactors for power and useful other applications suggested by Forsberg [3–10] and others [11–13] is assessed. The nonelectrical applications (by cogeneration method) of nuclear reactor will make the operations of small modular nuclear reactors in future feasible.

Chapter 8 illustrates various successful examples of hybrid renewable energy systems, which can allow operations of renewable sources for stable and sustainable power productions at larger scale. Chapters 7 and 8 also discuss how nuclear and renewable energy hybrid systems can be further facilitated by connections to energy storage systems or smart gas, heat, and electrical grids. Small-level hybrid systems can also be created with the support of microgrids. These chapters outline some successful examples of hybrid systems. For example, nuclear energy power plant can be connected to solar energy and/or wind energy to obtain higher level of base power and handle peak electricity demand. Wind energy can be combined with hydroenergy to obtain more stable production of power. The hybrid use of nuclear, solar, wind, geothermal, and hydroenergies is particularly important for stable and peak load power production when the power plant is not connected to a centralized smart power grid. Such distributed systems are important in remote locations.

Chapter 9 examines the third and final method of energy systems integration. This chapter discusses gas, heat, and smart electrical grids and their roles in the holistic integration of fuel and energy systems. In recent years, gas and power productions, transmissions, distributions, and supply to the customers are managed and controlled by smart gas and electrical grids. The grids allow demand response management through a two-way communication. Local heat grids and their interactions with both gas and electricity grids further enhance the values of gas and electricity grids. These grids represent the final layer of integration, and in future, there may be another layer of grids that will provide more flexibility to control all fuel and energy systems. Chapter 9 also discusses the importance of energy storage systems and their role in the operation of the grids. In future, natural gas grid, heat grid, and smart electrical grid will be more intertwined and may also be connected to other sources of fuels. The proliferation of grid-integrated hybrid and electric cars will make the management of the grids more complex and more important. This issue is also discussed in Chapter 9.

Finally, in the last chapter, Chapter 10, one of the most important applications of the multifuel, hybrid, and grid-integrated energy systems, that is, vehicle industry, is examined in detail as a case study for the theme of this book. The vehicles operated by multiple fuels, hybrid power system, and vehicle integration to grid for plug-in hybrid and electric vehicles have become very important in recent years. This change in vehicle industry represents its future. Chapter 10 discusses this case study with numerous examples of multifuel, hybrid, and plug-in electric vehicles.

REFERENCES

1. Shah, Y.T. *Water for Energy and Fuel Production*, CRC Press, New York (2014).
2. ExxonMobil. The outlook for energy: A view of 2040, US Edition. ExxonMobil Report, ExxonMobil, Irving, TX (2012).
3. Forsberg, C. Sustainability of combining nuclear, fossil and renewable energy sources. *Prog. Nucl. Energy*, 50, 1–9 (2008).
4. Ridge, T.N. and Forsberg, C.W. High-temperature nuclear reactors for in-situ recovery of oil from oil shale. In *Proceedings of the 2006 International Congress on Advances in Nuclear Power Plants (ICAPP'06)*, Reno, NV, June 4–8, 2006. American Nuclear Society, La Grange Park, IL (2006).
5. Forsberg, C.W. Meeting U.S. liquid transport fuel needs with a nuclear hydrogen bio-mass system. In *Proceedings of the American Institute of Chemical Engineers Annual Meeting*, Salt Lake City, UT (November 4–9, 2007).
6. Forsberg, C.W. Economics of meeting peak electricity demand using nuclear hydro-gen and oxygen. In *Proceedings of the International Topical Meeting on the Safety and Technology of Nuclear Hydrogen Production, Control, and Management*, Boston, MA, June 24–28, 2007. American Nuclear Society, La Grange Park, IL (2007).
7. Forsberg, C.W. and Conklin, J.C. Hydrogen-or-fossil-combustion nuclear combined-cycle systems for base- and peak-load electricity production. ORNL-6980, Oak Ridge National Laboratory, Oak Ridge, TN (2007).
8. Forsberg, C.W., Rosenbloom, S., and Black, R. Fuel ethanol production using nuclear-plant steam. In *Proceedings of the International Conference on Non-Electrical Applications of Nuclear Power: Seawater Desalting, Hydrogen Production, and Other Industrial Applications*, Oarai, Japan, April 16–19, 2007. International Atomic Energy Agency, Vienna, Austria, IAEA-CN-152-47 (2007).
9. Forsberg, C.W., Gorensek, M.B., Herring, S., and Pickard, P. Next generation nuclear plant phenomena identification and ranking tables (PIRTs). In *Process Heat and Hydrogen Co-Generation PIRTs*, Vol. 6, U.S. Nuclear Regulatory Commission, Washington, DC, NUREG/CR-6944, ORNL/TM-2007/147.
10. Forsberg, C.W. High-temperature reactors for underground liquid-fuels production with direct carbon sequestration. In *Proceedings of the International Congress on Advanced Nuclear Power Plants*, Anaheim, CA (June 8–15, 2008).
11. Anderson, R.E., Doyle, S.E., and Pronske, K.L. Demonstration and commercializa-tion of zero-emission power plants. In *Proceedings of the 29th International Technical Conference on Coal Utilization & Fuel Systems*, Clearwater, FL (April 18–22, 2004).
12. Antkowiak, M., Boardman, R., Bragg-Sitton, S., Cherry, R., Ruth, M., and Shunn, L. Summary report of the INL-JISEA workshop on nuclear hybrid energy systems. INL/EXT-12-26551, NREL/TP-6A50-55650, INL Report, prepared for US DOE, Office of Nuclear Energy, Contract No. DE-AC07-051D14517, Idaho Falls, ID (July 2012).
13. Boardman, R. Advanced energy systems. Nuclear-fossil-renewable hybrid systems. A report to Nuclear Energy Agency, Committee for Technical and Economical Studies on Nuclear Energy Development and Fuel Cycle. INL, Idaho Falls, ID (April 4–5, 2013).
14. Tchapda, A. and Pisupati, S. A review of thermal co-conversion of coal and biomass/waste. *Energies*, 7, 1098–1148 (2014).
15. Lee, S. and Shah, Y. *Biofuels and Bioenergy: Technologies and Processes*, CRC Press, Taylor & Francis Group, New York (2012).
16. Ratafia-Brown, J., Skone, T., and Rutkowski, M., Assessment of technologies for co-converting coal and biomass to a clean syngas—Task 2 report (RDS), DOE/NETL-403.01.08 Activity 2 Report (May 10, 2007).

2 Heat and Power by Co-Combustion

2.1 INTRODUCTION

Combustion is one of the oldest and widely used processes in the world. This process carries out the complete burning of solid, liquid, or gaseous fuels in the presence of excess oxygen. The products of the process are inert gases such as air (or oxygen), water, and carbon dioxide along with any impurities produced from elements present in the fuels. The process generates a large amount of heat that can be used to drive turbine (by steam) and generate electricity. All of the fuel feedstock such as coal, oil, gas, biomass, and waste are used to generate electricity by this process. Most processes, however, use a single feedstock at a given time.

Multifuel combustion is probably the least complicated and most advantageous way of utilizing biomass, waste, oil shale, oil, and water to replace most widely used fossil fuels such as coal and natural gas for stationary energy conversion for heat and power [1–22]. Coal has been the most dominant fuel used for power generation. In 1997, U.S. electrical power utilities consumed about 87% of the nearly 1.1 billion tons of coal produced [23]. Coal was used to generate about 51% of total power generation in the United States. While natural gas has in recent years increasingly replaced coal, both coal and natural gas will continue to be the dominant forces for power generation. CO_2 emission and the resulting global warming effect will continue to be the most forceful objection against the use of coal and natural gas for the power production. The use of renewable sources of energy such as biomass, waste, and water along with other fossil energy sources such as shale oil and bitumen will continue to be pushed.

The different sources of statistics [1,2,4,19,22,23] on biomass/waste reserve indicate that there is an enormous potential of their use in energy and fuel production. Fundamentally, this can be achieved in two ways: (a) build separate biomass/waste power or fuel plants or (b) combine the use of biomass/waste in the existing fossil fuel industries. The first approach will require a significant new capital investment compared to the second one. Furthermore, the commercial biomass/waste plants will be smaller in size compared to large-scale existing coal and natural gas power plants. Smaller plants will not be able to utilize the economy of scale. From an economical point of view, the second approach makes the most sense for renewable bioenergy to penetrate and grow in overall energy and fuel market. Currently, the coal-fired power plants consume 50,000 PJ of coal each year. If they were to be co-fired at a rate of 10% (thermal) of the total need, it will require 5000 PJ of biomass/waste per year, a quantity that would make significant impact on energy industry [23]. The proximity of biomass and wastes to power stations and other potential co-utilization sites will influence the scope of the market size.

While the use of multiple fuels in combustion has been in existence since the second half of the last century, its commercial applications and research interests have been exponentially increasing since the 1990s. This deduction is supported by the rising number of publications and pilot and commercial scale processes produced during the last two decades [23,24]. The increase in activity also indicates the general belief that it is a very viable approach to decrease the use of fossil energy and increase the use of biomass, waste, and water in power production. As shown later, the use of shale oil has some limited advantages that may need further investigations.

As shown in Section 2.5, the co-firing of multiple fuels in the same boiler is one alternative to achieve emission reduction. This allows fuels with different origin and property to interact with each other. While such interactions present many opportunities, they also present several challenges in the common boiler design. Fuel characteristics significantly affect the boiler design, and these challenges must be met in order to gain advantages of co-combustion of multiple fuels. As shown in Table 2.1, the sulfur, chlorine, alkali, and other properties of coal and biomass are considerably different. Also, volatile matter (VM) in wood-based biomass is around 80%, whereas in coal, it is 30%. Wood is highly reactive compared to coal, which results in its complete combustion at a lower temperature. Sulfur and nitrogen content in the wood is considerably lower than that in coal. All these differences need to be factored in the design of the co-combustion process. Against all these issues, there are some synergies among various elements of fuels that can be advantageous during co-combustion of multifuels.

The multifuel combustion can be carried out in various ways depending on the desired end usages. There are three acceptable alternatives for the use of multifuel combustion [3,19,23,24,26]:

TABLE 2.1
Typical Differences between Fuels

Parameter	Coals	Biofuels	Wastes
Sulfur	Medium to high	Low	Low to medium
Chlorine	Medium	Low to high	Medium to high
Potassium	Bound	Medium to high	Low
Other alkali	Normal	Low	Low to normal
Alumina, silica	High	Low to high	High

Source: Leckner, B., Co-combustion: A summary of technology, A report from Department of Energy and Environment, Chalmers University of Technology, Alliance of Global Sustainability, Gotenberg, Sweden, 2007.

Note: Net calorific value, MJ/kg.

1. A small amount of biofuel or waste is combusted with coal in a boiler, which is originally designed for coal. Here, the main purpose is to get rid of biomass or waste or to replace coal (thus conserving environmentally harmful coal) by biomass or waste. This is the easiest way renewable fuel can penetrate fossil fuel market.
2. A small amount of fuel such as coal with high heating value (HHV) is burned with the fuel with low heating value such as sludge to attain the desired combustion temperature.
3. Spontaneous use of multifuels in any composition depending on price, availability, and local supply conditions.

The first method reduces emissions of CO_2 and assesses the reliability of the use of biomass and waste in a plant designed for coal. The second method also maintains low-CO_2 emissions and tests the reliability of multifuel operations. This approach can also test the synergy effects among various components of multifuels, thereby improving the operation and performance of a boiler. The third method can examine a number of fuels such as sawdust, wood chips, peat, petroleum coke (petcoke) along with coal depending on the social and political acceptance of various fuels. Transportation of raw materials plays an important role in the selection of fuel. Combustor technology also plays a significant role in processing different multifuel options.

Over the last several decades, due to concerns over public health and environment, federal regulations regarding the emissions of air pollutants, particularly CO_2, NO_x, and SO_2 from fossil fuel combustion have become particularly demanding. While a number of techniques and methods to control emissions have been proposed, many of them are expensive and increase power production costs. The use of multifuels (particularly a combination of fossil fuel and renewable fuel) is a less-expensive alternative and therefore has gained significant popularity. Besides its effect on environment, the use of multifuel has been accepted as the easiest method for biomass and waste to penetrate fossil fuel–controlled energy and power market.

This chapter mainly focuses on multifuel systems involving fossil fuels (mainly coal) and renewable fuels such as biomass, different kinds of waste, and water. The use of oil shale and oil as a component of multifuels will also be briefly examined. This chapter outlines the advantages and disadvantages, current progress, future opportunities, and problem areas for the combustion of these multifuels.

2.2 ADVANTAGES AND DISADVANTAGES OF MULTIFUEL COMBUSTION

The use of multifuels offers many advantages to the combustion process [23–26]. Here, we mentioned the most important ones:

1. The use of multifuels offers flexibility for the feedstock for combustion. This flexibility is very desirable when one source of fuel is not in abundant supply and for a sustainable operation of large power plants. The availability

of the substitute is not important because the main fuel remains coal or gas. A wide range of biomass and/or waste exists. Variations in quantity and quality of fuels can be compensated by proper adjustment of the co-firing rate.

2. The use of biomass allows the use of redundant agricultural and underutilized land, thereby improving the local economy and job creation. The use of waste can improve plant economics due to zero, low, or negative charge for the feedstock. In some cases, the discard of waste may allow the charge of "gate fee" for facilitating disposal. The combustion of coal/waste mixture reduces the waste compared to the plant, which only uses coal. This also eliminates the need for waste disposal.

3. If the waste or biomass feedstock is available locally, it reduces the overall transportation cost of the feedstock. The replacement of coal by biomass helps in long-term preservation of coal reserve.

4. Less use of coal and synergistic reactions between coal and biomass combustion can reduce the productions of pollutants. This also affects the overall economics of the process.

5. In many parts of the world, landfills are full. The use of waste in combustion allows disposal of waste without creating new landfill. The landfills across the world produce methane, which is harmful to the environment if it is emitted. Less use of landfill by this method helps the environment.

6. Co-combustion can be easily carried out in reactors such as pulverized fluidized bed, fluidized bed, and circulating fluidized bed (CFB). Thus, they alleviate the need for creating whole new combustion and gasification plants. As shown in Sections 2.5 and 2.7, many existing plants for coal combustion can be retrofitted (with some or little modifications) to handle multifuel feedstock. The additional investment in the co-combustion equipment is significantly lower than the investments needed for stand-alone biomass plants. (In case of pulverized coal [PC]-fired plants, it could be as low as 1/10th).

7. Due to higher electric efficiency and lower additional investment, the variations in feedstock prices are more tolerable for the fuel at the same electricity refund. This may allow use of more expensive biomass or reduce the cost for the utilization of the already available biomass.

8. The amount of net CO_2 emission will be reduced due to the fact that biomass or waste (cellulosic or organic) is CO_2 neutral. The overall electric efficiency is higher compared to the one obtained in the small stand-alone biomass plant. Due to higher efficiencies, the reduction of specific CO_2 emissions is significantly larger. This will allow additional CO_2 certificates and further lower the CO_2 reduction costs. CO_2 emissions can also be lowered by a lower use of fossil fuels.

9. Since nitrogen content in most biomass and waste are lower than that in coal, NO_x emission will also be lower during co-combustion. The use of biomass in larger plants will allow the use of deNO$_x$ installations. NO_x is better controlled in pulverized fuel (PF)-fired than in grate-fired boilers. Since most biomass contains lower sulfur than coal, in co-combustion, the

plant will generate lower SO_x. Higher ratio of sulfur to chlorine for the mixtures also reduces the risk of corrosion of process equipment compared to stand-alone biomass combustion plant.

10. Higher VM in the biomass will result in improved combustion resulting in better burn out and lower unburned carbon in the ash. In case of limited cocombustion, the fly ash can be valorized in cement or concrete. Also, due to high temperature in the combustion chamber during co-combustion, fly ash particles have an amorphous structure (glass) improving its pozzolanic properties.

Many other advantages can be feedstock specific. These will be further illustrated in Sections 2.4 and 2.5. The multifuel processing also poses some disadvantages [23–26]. Some of these are briefly illustrated in the following:

1. Biomass and waste in general carry lower heating value than coal, so the mixture will have lower heat and power generation capacity. High water and oxygen content of biomass and waste can make the overall process less energy efficient than the one using coal alone.

2. The economic advantages of using biomass and waste depend on local availability of these sources. If they are not available locally, they can restrict the economic advantages of the multifuel processes. The availabilities of biomass and waste are seasonally dependent, which may restrict its sustainable use in the process.

3. The compositions of different types of biomass and wastes can be highly variable. This can affect impurities in gaseous and solid products. This can also affect combustor performance and maintenance schedule.

4. Many types of biomass and waste will require feed preparation techniques, which are different than those for coal. These techniques can add costs to the process. Some biomass or waste may require an independent feeding system or even independent gasifier both adding significant costs to the overall process. For some biomass and waste, the reduction of particle size (such as straw) to the desirable range may become excessively expensive.

5. Low density and biodegradation of biomass can cause storage and transportation problems, which will have to be addressed and can add cost to the overall process. Low density causes several handling problems in storage and transportation.

6. Fuels such as municipal solid waste (MSW) may contain large amounts of chlorine, which can significantly increase the cost of gas cleanup requirement and plant component corrosion. Many biomass, such a straw contain large amount of alkali and inorganic materials, which can also cause more corrosion in plant equipment and also increase propensity of slagging and fouling phenomena.

7. Some biomass may generate excessive amount of particulates in flue gas, which can increase the cost of their removals. The quality of ash produced may decrease with multifuel processing, which may reduce its market potentials.

2.3 METHODS OF CO-COMBUSTION AND ASSOCIATED SYSTEM CONFIGURATIONS

In its simplicity, there are three major options for multifuel combustion: direct, indirect, and parallel [3,24]. Direct combustion involves combustion of multifuel in a single combustion chamber. Indirect combustion involves combustion of one fuel (often biomass or waste) in a separate vessel, and the heat (steam) or gas produced from this combustion is combined with the combustion of second fuel (generally coal) in a separate combustion chamber. In parallel combustion, the combustion of both fuels occurs in separate combustion chambers, and then gases from both chambers are combined for downstream purposes. These configurations can be further subdivided [3,24] depending on the nature of the feedstock, methods used to inject feedstock in the combustor, nature of combustor, and the methods used for product cleaning among other possibilities.

2.3.1 Direct Multifuel Combustion

In direct multifuel combustion, two or more fuels are directly injected in a combustor designed for the base fuel. The base fuel is often coal. This is the most convenient method and is often used for fluidized bed and pulverized fluidized bed boilers. Since fluidized bed can handle a larger variation in particle size and density, it is most commonly used for this type of multifuel combustion method. For pulverized fluidized bed, a requirement of well-defined fuel particle size distribution restricts its use only for small particle size and small concentration of second fuel [3,23,24]. For large concentrations of the second fuel, special mills and burners are desirable. In fluidized bed, except for the limit set by the heat balance, there is no restriction on the size of particles used for second fuel (generally biomass). Low ash content of biomass also helps flexibility in the particle size distribution. A grate-fired boiler can also have reasonable fuel flexibility as long as the additional fuels are mixed with the primary fuel outside the boiler before the injection of mixture in the boilers [3,23,24].

Depending on the nature of the combustor, other configurations for direct multifuel combustion are possible. The second fuel can be added on a grate, which is inserted in the bottom of a pulverized fluidized bed combustor (FBC) for coal. Mory and Tauschlitz [27] have shown a large-scale test for commercial PC boiler where 10 MW grate was installed at the bottom of the combustor. While the results were positive, this design restricts the available surface area for combustion of second fuel. Fuels can also be added in two or more stages, where the positions of the two or more inlets can vary. For example, the second fuel can also be added in the bottom fluidized bed combustion, while the coal is added in top PC bed section. This design can increase the production of flue gas particularly if the second fuel has high moisture content. The bottom fluidized bed can also have its own fuel and ash systems. This type of staged direct combustion is easy to implement if the main fuel is coal and not oil or gas [3,23,24].

Staging can also be accomplished by having separate feed lines and common burner in which (a) two inlets such that coal in the primary air and biomass in the swirling secondary air (or vice versa) or (b) three inlets in which two for primary air (central and annular), one for swirling secondary air [3,23,24]. This type of staging system is relatively inexpensive in the sense that a single swirl burner can be used to fire the blend. The coal and biomass are fed separately. At the burner entrance, the two fuels are unmixed. Near burner, the two streams mix due to the swirling section of the secondary air. When the good mixing occurs, higher combustion efficiencies and lower emission result. Sami et al. [24] showed the effectiveness of these configurations on emission of NO_x from boiler. However, if swirler is used, fuel fed into swirling secondary air can damage blades of swirler. This issue can be resolved by feeding secondary air at an angle [3,24].

2.3.2 INDIRECT MULTIFUEL COMBUSTION

As mentioned in Section 2.1, in an indirect multifuel combustion, the second fuel is combusted separately and either the heat or the heated products are merged with the combustion of the first fuel. Thus, two combustion processes have somewhat independent control although they are connected by the heat exchange or heated combustion products for the second fuel. This method can adopt a number of different configurations as well. For example, just like in direct combustion, feed preparations for the two (or more fuels) can be done separately. A combustion unit for the second fuel can be separated from the main coal-fired boiler on the fuel side but can be connected to it on the steam side [3]. In this way, the combustion of the second fuel can provide heat for the combustion of primary fuel. Furthermore, the ash coming out of primary fuel can be separated from the ash coming out of secondary fuel. This helps subsequent utilization of ash [1,3,23,24].

The configuration described earlier is used in the plant built in Denmark, which generates 600 MW of power. The secondary fuel—namely, straw with power of 100 MW fuel—is used [1,23]. The steam generated in the first boiler (which is similar in property with that of steam generated in the main boiler) joins the steam system of the main boiler before going through turbine. Since two combustors are combined only through steam, any harmful substances generated by the second fuel do not affect the performance of the primary boiler. The ashes and flue gases generated by both boilers can be treated separately. Any interruption in the second fuel boiler can be handled separately from the performance of the primary boiler. The feedstock used in the second fuel boiler can be chosen based on the steam properties needed for the success of the overall process. This arrangement is practiced in several plants worldwide.

Another configuration that is being used for indirect multifuel combustion is the one in which gas produced by the second fuel is mixed with the main fuel in the primary burner. In order to reduce volume flow of gas produced, size of the required steam system, and dimensions of hot gas connecting the additional furnace, the secondary boiler is operated as fluidized bed with low air flow sufficient to raise the bed temperature to about 800°C–900°C. At these temperatures, some of the chars may be gasified. This configuration has been used in some plants in Finland, Austria [28,29], and others, and they are described in an International Energy Agency report [30]. Combustible and low calorific value gas from secondary burner is subsequently burned in the primary burner. The ash from the secondary fuel is removed from the secondary burner, and it is separated from the ash generated in primary burner. The ammonia and alkalis from the secondary burner are transferred into the primary burner. The secondary burner could be a grate burner or rotary burner operating at lower temperatures (500°C–600°C). The char at those temperatures can be removed from the secondary burner and mixed with coal for primary burner or separately discarded if it contains significant alkali and sulfur impurities [3,23,24]. In this method, co-combustion of each fuel can be combusted at its own optimum temperature.

Another modification of indirect combustion that has been tried is to use some part of pulverized biomass and gas produced in the secondary burner for "reburning" in PC main boiler to provide additional fuel for special purpose or introduce biomass and gas as "afterburning" fuel in the cyclone in a CFB to reduce the production of N_2O [3,24,31,44]. The effects of reburning and other injection techniques to improve thermal efficiency and emission reduction have been extensively examined in the literature [3,24,32–44].

2.3.3 PARALLEL MULTIFUEL COMBUSTION

In this configuration, both primary and secondary combustors are operated independently, and the hot gases or steam generated from both combustors are combined before passing them to a common turbine and generator. The design and operation of feed system and combustor for both primary and secondary fluids are separate and independently controlled. While this gives maximum flexibility for the design of the combustor based on the feedstock properties, it involves the least integration for multifuel combustion. The capital cost for building this system would be the largest [3,23].

The parallel configuration allows independent control of combustion temperature, thereby an independent optimization of combustor temperature and other operating conditions can lead to corrosion and other fouling conditions. This approach is favored by Danish utilities. A new boiler unit using straw is incorporated into an existing coal-fired PC combustion unit by ELSAM at Sonderjylands Hojspoendingsoerk site [3,23]. ELKRAFT are retrofitting an existing coal-fired boiler at Asnaes with a 44 MW straw/wood boiler and are developing a new coal-fired plant at Avedore, which incorporates a 47 MW straw/wood boiler [1,23].

Nussbaumer [26] gave an excellent review of combustion and co-combustion of biomass. He points out that out of all thermal conversion technologies like combustion and gasification pyrolysis, only combustion has been proven for the size range of few kW to more than 100 MW. He points out that commercial power production is based on steam cycles whose economics and efficiency are more favorable at larger scales. Hence, large-scale co-combustion of coal and biomass has promising future. The emission control in the co-combustion is very important, and it depends on (a) incomplete combustion that is now well handled and (b) pollutants such as NO_x and particles that are formed as a result of fuel constituents such as N, K, Cl, Ca, Na, Mg, P, and S. In general, biomass furnaces exhibit relatively high NO_x and submicron particle emissions. NO_x emissions can be reduced by 50%–80% by air and fuel staging [3,23,24,45]. Particle reduction may require reduction in primary air and new furnace and burner designs. Combustion efficiency can also be improved by optimization of fuel to air ratio, heat distribution strategies among other parameters. The method of combustion, nature of feed preparation, and methods of fuel and air injections in the burner are very important in optimization of multifuel combustion system [3,23,24].

2.4 FEEDSTOCK PRETREATMENT

The multifuel combustion can use different types of combustors and system configurations as well as different combinations of fuel feedstock [46,82]. In all cases, however, a consistent supply of the feedstock in the appropriate grade and quality is the single most critical factor for good operational performance, which leads to the desired production and emissions targets. Since coal combustion is a well-established technology, the feed preparation of coal is also a proven and well-controlled technology. The major elements of feed preparation of coal involve particle sizing, sulfur and ash removals, and VMs (including water) removals irrespective of the nature of the coals. Biomass or waste on the other hand is a very nonhomogeneous material and harvested, stored, and transported differently depending on its nature and physical and chemical properties. Most biomass have higher water contents, have lower mass and energy densities, and are softer than coals. They also have higher VMs and ignite at higher temperature than coal. Transportation, storage, particle size reduction, and water removal can be significant issues for several types of biomass [1,3,23]. Constraints and role of biomass pretreatment in coal–biomass co-firing are well reviewed by Maciejewska et al. [47].

2.4.1 FEED PREPARATION AND HANDLING SYSTEM DESIGN ISSUES

As mentioned in Section 2.4, compared to coal, biomass and waste feedstock are more difficult to prepare and handle because of their (a) low heating value, (b) high moisture content, (c) low bulk density, (d) high fibrosity, and (e) great size variability. Several of these factors affect the transportation, storage and design, and operation of the feeding system adopted. For example, low bulk density can affect the transportation costs, size of conveyors, storage and feed bins, screw feeder size, etc. Often, a separate feeding system for biomass component of multifuel system is used. Both in Europe and the United States, most grain-producing countries or states produce substantial amount of surplus straw as a by-product. The straw available is a low-grade nonhomogeneous fuel source, characterized by high volatile content plus high chlorine and alkali levels. Because of its abundant availability, straw is still widely used in multifuel systems for several CFB power plants. Feeding systems adopted tend to be relatively complex and involve a number of separate stages. Sorting, blending, and preparing of fuels derived from wood and refuse-derived fuel (RDF) can also be equally complex.

Unlike in coal-fired power plants, the combustor using multifuel system of coal and straw results in excessive corrosion in downstream components of the plant due to high chlorine and alkali levels in straw. This issue requires a careful selection of construction materials for the plant components. Work carried out in Italy [23] with multifuel systems involving eucalyptus and poplar with coal required careful handling to achieve stable plant operating conditions. While both of these biomass are fast growing and in ample supply, the bark of their trees proved very difficult to handle, transport, and feed [23]. Furthermore, moisture content of the wood significantly affected its grinding characteristics and flow ability.

The burners are designed for certain particle size distribution of feed, dust, etc. For multifuel processing, research and development work is needed to economically prepare biomass/waste feed for the combustor. For example, for PC burners, biomass needs to be finely and evenly pulverized. This may be a very difficult task for softer and fibrous materials such as straw and miscanthus. Feeding of biomass/waste into a pressurized combustor at predictable feeding rates is problematic and requires further investigation. Straw has proven to be very difficult to process and handle, and its unpredictable behavior needs further assessment.

2.4.2 STORAGE AND TRANSPORTATION

Unlike coal, the transportation of biomass can be very difficult, time-consuming, and expensive. Low mass and energy densities, and fibrous nature of some biomass, make the long-distance transportation very expensive. Grassy materials need to be rolled into bales before transportation. For these reasons, multifuel plants are often situated at the locations of easy access to biomass and waste. Otherwise, the transportation costs can limit the size of the sustainable combustion operation. In most multifuel plant consisting of coal and biomass and waste, the concentration of biomass or waste ranges around 10% thereby requiring less materials for transportation.

Most biomass requires a combination of air drying before storage (in open environment) followed by further drying immediately prior to use. Biomass should be stored in an inert environment to avoid its biochemical degradation.

Bauer [48] described the storage issues in summer and winter for wood chips used in commercial-scale 100 MW plant that used both wood chips and lignite at Lubbenau power station at Brandenburg, Germany. While feedstock with 30% of moisture content was dried in this plant both in summer and winter within 5 weeks with no spontaneous combustion, this may not be always possible in humid summer weather. For storage of softer materials like straw, moisture control in the storage tank is very important for consistent quality of fuel supply to the plant. Very wet materials will be almost impossible to feed in the combustion plant. The ignition temperature for various biomass is generally higher than that for coal [23]. Biomass will have a higher tendency for spontaneous explosion than that for coal.

Unlike in coal, the nature of storage vessels also depends on the nature of the biomass. For example, dried sewage sludge requires storage in mass flow silos with stationary storage time being limited to 5 days in order to prevent sticking [23]. The discharge of this biomass from silos requires wide silo outlet openings, steep sides, and use of discharge aids such as air guns [23]. The storage of straw on the other hand requires an inert environment to avoid biological decay. The storage of soft biomass such as miscanthus can be problematic since this material tends to be unusable if stored at high moisture contents. The storage of coal on the other hand is relatively easy task.

2.4.3 Size Reduction

The reduction of coal particles and their handling are well-established technologies, and they are relatively easy and reliable. On the other hand, sizing of biomass is much more complex and may require a separate system. The degree of size reduction required depends on the nature of biomass/waste and the nature of the combustor technology. Many current size reduction and handling systems lack reliability [23,49–53]. Fuel variability, especially for softer biomass like straw poses particular difficulties for the size reduction and handling systems. In multifuel systems, two extreme options for feed handling are (a) separate reception, comminution, conveying, and combustion systems for biomass and coal and (b) separate reception followed by combined comminution, conveying, and combustion systems for biomass and coal. Numerous other alternatives in between these two extremes are also possible. In general, grinding equipment appropriate for a given feedstock must be selected carefully. Woody biomass can be easily grinded, while straw biomass needs chopping. More details on some techniques for feed preparation such as torrefaction and fast pyrolysis are covered in Chapter 3.

The required degree of size reduction depends on the choice of the combustor. FBC provides the most flexibility for particle size. Even for easily grindable fuel such as wood, significant fine grinding is required to achieve satisfactory burnout (<1 mm). The use of PF-fired boilers for multifuel requires expensive milling process. This also requires biomass to be dried and pelletized so that it can be milled to sufficient fine particles. As mentioned in Chapter 3, the power required for the grinding biomass increases exponentially below 1 mm particle size. Often, the process of torrefaction [52] is used to facilitate the grinding process. Chapter 3 also shows that power requirement for size reduction is significantly decreased for torrefied wood compared to untreated wood. The torrefied biomass also resists water absorption.

Hey et al. [5] outlined laboratory evaluations of fuel preparation methods for multifuel combustion of coal and biomass. They examined granular feeding in FBC and grinding sewage sludge for co-firing with coal in PC combustors. They also found that grinding of fibrous materials like straw and miscanthus in vibration or hammer mills was accompanied by stickiness or dust evolution and are therefore not suitable approaches. As long as conditions were carefully optimized, the grinding of coal and chopped barley and wood pieces in hammer mills was possible. However, such grinding in ball mill was problematic. For all of these cases, Hey et al. [5] concluded that grinding energies were of the order of 1%–2% of the feed calorific values.

For size reduction and handling purposes, biomass can be broken into four groups [23]:

1. Wood and wood waste, which are easy to grind and handle but grinding and handling of fines may be an issue
2. MSW and industrial waste, which are very nonhomogeneous and difficult to process uniformly

3. Sewage sludge, which can be broken into fine particles but its fibrous nature can cause handling problems
4. Straw and miscanthus whose particle size can only be reduced by chopping action and are difficult to feed in a mixture form particularly if they are wet

In the following paragraphs, we briefly examine size reduction and associated handling issues of these four distinct groups of biomass/waste for co-combustion.

2.4.3.1 Wood/Waste Wood

In Europe, several studies [8,54–56] using wood/waste wood and coal showed that with the use of secondary air, hogged wood of less than 12 mm can be blended with crushed coal on the feed belts. Boylan [55] successfully fed 12–25 mm size of tree trimming and garden waste with sawdust and coal (3:1 by volume) to a wall-fired PC fire plant. Some studies [8,23,54,56] have, however, suggested that 1 mm is the maximum acceptable size for wood combustion in PC combustion system. The study carried out for multifuel combustion at 125 MW Esbjerg plant in Denmark recommended hammer mill for grinding of wood. In VEAG' 100 MW Lubbenau plant in Germany, the wood chips were mixed with lignite before processing through fan mills. While the wood chips increased grinding energy by about 10%, the process worked very successfully for the mixture combustion.

The largest multifuel combustion process using waste wood and coal is Netherlands utility EPON process, which fires waste wood via separate wood burners retrofitted to a 635 MW wall-fired PC boiler. In this process, 60,000 tons of waste wood is co-fired annually [56]. The transported wood chips are grounded to a particle size of <1–8 mm in two hammer mills. The material is then further grounded in four pulverizers to less than 1 mm particle size and dried by preheated air. Sometimes, the uniform feeding of wood and sawdust has been found difficult due to their low density.

2.4.3.2 Municipal/Industrial Waste

The combustion of MSW and other industrial waste in different types of combustors, such as stoker, FBC, and circulating fluidized bed combustor (CFBC), is a well-established technology. Many of these combustors use RDF prepared from MSW and industrial wastes. A major drawback with the use of MSW as a fuel feedstock is its heterogeneous composition, which can present problems in a multifuel system. This can be partially handled by separating impurities like glass and metals and converting MSW into RDF pellets. This, however, is an expensive process. In order to maintain pellet structure integrity, water contamination to the pellets should be avoided. Often, the process of torrefaction precedes the pelletization process. The torrefied wood is more resistant to further water absorption.

As pointed out in References 23 and 57–61, in the United Kingdom, fiber fuel or packaging-derived fuel (PDF) is mixed with coal in fluidized bed boilers of Slough Heat and Power Ltd. Unlike RDF, PDF is made from segregated paper, packaging materials, plastic, and board and has a higher heating value. The multifuel feed can also include industrial waste such as tires, postconsumer carpets, auto shredder residues, and clinical wastes. Generally, tires are chipped to produce tire-derived fuel (TDF), carpet is shredded and premixed with coal, and auto shredder residues are usually conveyed pneumatically.

2.4.3.3   Sewage Sludge

Dried sewage sludge can be ground down more to smaller particle sizes than most biomass materials. The presence of fine particles, fibrous nature, and wide particle size distribution can, however, cause handling problems. The heat liberated during grinding can make pellets plastic, which in turn block the screens [62]. The dried sewage sludge can be well pulverized using a table mill. Probst and Wehland [9] successfully processed a mixture of PC (40%) and dried sewage sludge (60%) along with some limestone in a 10 MW CFB plant. The literature [23] also indicated that processing sludge (with less than 10% moisture) with coal in Weiher II power station at Saarland did not work well. The feed handling system suffered (a) excessive wear of the storage tank extraction systems, (b) leakage of the tanks, and (c) some clogging and flow problems of dried matter <0.5 mm. In this feed handling system, larger flow rates were easy to accommodate than smaller flow rates.

2.4.3.4   Straw and Miscanthus

The fibrous and soft nature of this type of waste requires the chopping action for size reduction. Fortunately, the high reactivity of these materials allows the use of large sizes. In Vestkraft coal-fired power station at Esbjerg, Denmark, some 15,000 tons of straw were combusted with wood and hard coal in 125 MW unit [23]. The feed was prepared by feeding baled straw to a shredder followed by a cutter. Straw feeding was difficult and below optimum. The straw moisture content above 25% caused failure in the process. The excessive moisture content caused problems in the straw feeding rates in Grenaa coal/straw dual fuel CFB plant in Denmark [23]. In 250 MW Amager power plant in Denmark, coal was mixed with 20% straw using cylindrical pellets of straw with 9% moisture content. The pellets were mixed with coal before feeding to the mills for cogrinding.

The chopping, handling, and feeding of straw have been found difficult [6]. Meschgbiz and Krumbeck [63] used a mobile straw mill designed for agricultural use, but the system was hazardous. A screw feeder was used to supply straw to coal feed; however, segregation between biomass and coal in the bunker resulted in uncontrolled feed rates. In general, however, feeding of straw in CFB combustion system by pneumatic feeding worked well due to longer residence time and technology being more tolerant of feed moisture or size. For pressurized fluidized bed combustion (PFBC) system, Andries et al. [64] fed up to 20% of crushed and pelletized straw and miscanthus and up to 20% of granular mixture of milled straw with PC. The results indicated that granular mixture was difficult to process through a lock hopper system.

2.5   EFFECTS OF COMBUSTOR DESIGN

Numerous different types of combustors are used to generate heat and power. The internal design of combustor is important in selecting appropriate combustor based on the components of multifuel systems. In some multifuel systems, the combustor design may need to be modified to efficiently operate multifuel systems.

2.5.1 PULVERIZED COAL COMBUSTORS

PC combustion is the most widely used technology for utility scale power generation in the world. With modern technologies, these types of combustors give electrical generation efficiency of about 38% with as high as 43% in some locations [1,3,23,24]. In this type of combustor, the fine powder coal is pneumatically blown with air into boiler plant. The combustion heat produces high-pressure superheated steam, which drives turbine. The power plants using this technology are very large with a capacity of 300–500 MWe. The product gas then goes through cleaning system to remove particulate matters, SO_2 and NO_x.

Leckner [3] identified three different methods to process mixture of coal and biomass in PC boilers. In the simplest case, blend coal and biomass are processed together in fuel handling system and PC coal burners. This works best for the small quantity of added fuel. When the concentration of biomass is increased, fuel handling and comminution systems for coal and biomass need to be separated. Once both fuels are properly prepared and appropriate particle sizes are achieved, they can be processed in a common burner. Finally in some cases, both coal and biomass require separate feed preparation and burning facility (like coal and straw) because of their distinctly different needs for feed preparation and distinctly different combustion characteristics. The problems mentioned earlier in this section are not important for the FBCs since they are most flexible in handling variations in particle size, VMs, and combustion characteristics. PC combustion plants are, however, very popular in processing multiple fuels. Some examples of large-scale PC plants using multiple fuels are listed in Table 2.2 [23].

The choice of the aforementioned three options significantly depends on the physical and chemical properties of the biomass along with its concentration in the feed. The lower energy density of most biomass and wastes compared to coal necessitates high volume flow in the boiler. The performance of the boiler and the associated equipment will depend on the properties of biomass/waste. These properties will dictate fuel supply, storage and handling, combustion-related behavior, boiler performance and efficiency, and emissions.

Since PC combustors require fine particles for second fuel as well, the cost of grinding is increased. A utility trial co-firing wood with coal at Georgia Power's Hammond Unit in the Unites States demonstrated that the mills power requirements were raised by 10%–15% by incorporation of the wood into the stream. The wood also required additional drying. Similar experience was found at EPON in Denmark [23].

Most biomass/wastes have combustion characteristics different to those of coal and their combustion in PC combustion chamber designed for single fuel will affect the overall combustion process and performance. This is particularly true if a separate set of burners is used. The combustion efficiency and heat release patterns directly affect boiler efficiency, emissions, and ash collection [3,23,24].

TABLE 2.2
Pulverized Coal Plants Co-utilizing Coal and Biomass/Wastes

Owner/Location	Feedstock/Capacity
Epon/Nijmegen, Holland	Coal and waste wood/602e
Lakeland Electric/Florida, United States	Coal and RDF/350e
VEAG/Magdeburg, Germany	Coal and wood/350e
Vasthamnsvert Uppsala Energi/Uppsala, Sweden	Coal, peat, and wood/320e
Elsam/Amager, Denmark	Coal, straw, and miscanthus/250e
Vasthamnsvert CHP/Halsingbourgi, Sweden	Coal and wood/180e
TVA/Tennessee, United States	Coal and waste wood/150e
Midkraft/Esbjerg, Denmark	Coal and Straw/150e
GPU GENCO/Illinois, United States	Coal and waste wood/130e
Bayernwerke Ag/Bavaria, Germany	Coal and straw/108e
Georgia Power/Hammond, United States	Coal and waste wood/100e
Ames municipal/Iowa, United States	Coal and RDF/75e
Saarbergwerke, AG/Saarberg, Germany	Coal and sewage sludge/75e
SEPCO/Savannah, United States	Coal and waste wood/54e
Stockholm Energy/Hasselbyvaerket, Sweden	Coal, wood, and olive waste/54e
Iowa light and power/Iowa, United States	Coal and agri waste/45e

Source: Opportunities and markets for co-utilization of biomass and waste with fossil fuels for power generation, A report from European Union, U.K., 2000, also found in website: http://ec.europa. eu/energy/renewables/studies/doc/bioenergy/2000_opportunities_and_markets.pdf.

While PC combustors have been successfully used for straw (up to 25%) and coal mixtures, both in the United States and Europe, this has required some changes in nature of feeding and burner design. While the straw is burned separately from coal, straw feeding is difficult. Also, airflow in combustor required adjustments in order to avoid the impingement of the straw on the boiler wall. While using straw, slag can accumulate around the straw burners, and some additional carbon can be present in fly ash. In general, straw lengths <50 mm were found necessary to achieve smooth feeding to the beater mill. Premixing of straw and brown coal was not feasible due to segregation of straw from coal [23].

The use of PC combustor for the mixed fuel containing coal and wood has been successful since wood can be ground to very fine particle size (such as sawdust). Within combustor, wood burners can be separated from coal burner, or both coal and wood are pulverized and combust together. These methods work well as long as moisture content in wood is less than 25% and wood concentration in the mixture is less than about 20%. Foster Wheeler [23] found that during co-firing of coal and wood in a tangentially fired PC unit, flame temperature decreased by about 40°C, and both SO_2 and NO_x in product decreased due to low levels of sulfur and nitrogen in the fuel blend. Boiler efficiency decreased by less than 1.5% during co-firing with 10%–15% wood. Besides wood, sewage sludge also worked well as long as it was pulverized and dry. Brouwer et al. [65] carried out co-firing of RDF and coal with common feeder and burner. Two methods of firing coal with biomass were used. In the first method, the biomass was premixed with coal and injected through the main line. In the second method, biomass was injected after the recirculation zone as a reburn fuel. The combined results of these two methods indicated that co-firing wood waste with coal may not lead to reduction in NO_x emissions in a low NO_x configuration unless large co-firing ratios are used [23].

Aerts et al. [66] studied the co-firing of switchgrass with coal in a 50 MW, radiant wall-fired PC boiler. Coal and switchgrass were fed and burned separately. The operation was normal with no slagging even with the use of larger switchgrass particles during co-combustion. During co-firing, NO_x emissions decreased by about 20% due to low level of nitrogen in the original switchgrass, a result also found by Christensen and Jespersen [67]. Higher VM content of the biomass did not affect combustor performance but resulted in the production of highly porous char thus accelerating its combustion [66]. Christensen and Jespersen [67] estimated that the corrosion rate with co-firing would be twice as high as that for coal alone. Dyjakon et al. [68,192] indicated that slagging and fouling risks occur on the heat exchange surfaces when biomass contains high amounts of K and Na. The increase in corrosion occurs when chlorine and sulfur contents in biomass and/or coal are high. The study concluded, however, that with many precautions up to 50%, biomass co-firing is possible.

Siegel et al. [69] co-fired straw and cereal with hard coal in 500 kW PF test unit. The tests were made for both common and separate feeders and burners. At higher biomass loading (about 60%), injecting coal in the annular pipe resulted in a decrease in NO_x, whereas using a central coal jet caused an increase in the emissions. These results are similar to the ones reported by Abbas et al. [62]. At lower biomass thermal loading (<40%), injection of coal in the central jet had lowered emissions. For a variety of cereals and straw tested, burnout was 99% for a co-firing ratio up to 60% [24]. Above this ratio, combustion efficiency dropped. The authors concluded that fuels with higher nitrogen content should be injected into the fuel-rich zone in order to reduce emissions [23,24,69].

Abbas et al. [62] evaluated a dual burner designed for the co-firing (with separate feed lines) of waste-derived solid fuel with PC in a PC combustion system. The biomass used was sawdust with some sewage sludge. The study showed that the combustion efficiency and NO_x emissions depended on the injection position of biomass and coal particles [3,23,24,62]. The highest efficiency and lowest NO_x emission were obtained when sawdust particles were injected in the primary air with a swirling annular stream of coal particles surrounding the biomass particles. The study also recommended the optimum co-firing ratio that resulted in the maximum particle burnout and minimum NO_x emissions [23,24,62]. Co-firing of coal and straw and coal and wood in PC combustor was also studied by Kierkegaard Petersen and Hansen [70].

2.5.2 STOKER COMBUSTOR

This type of combustor, which is used commercially for some years, refers to chain grate, traveling grate, and spreader stoker systems. Chain and traveling grates are similar in characteristics and are typically designed for boiler plant up to 80 MW [23,24]. In general, these types of combustion plants are not suitable for fuels with high moisture, volatile contents, and low calorific values. With such materials, ignition problems can occur as well as poor combustion performance resulting in unburned carbon loss and high CO, particulate, and hydrocarbon emissions. This type of combustor is also not suitable for fuels of varying consistency. The most acceptable method of co-firing would be to blend the fuel feed prior to boiler addition.

The use of waste in this type of combustor can require the addition of new gas cleanup system. If unstable combustion occurs, unacceptable intermittent or continuous high emissions of CO and organics could occur. If fuels in a multifuel system are not adequately blended and distributed on the grate, hot spots could be generated leading to grate damage and increased maintenance outages and costs, a situation not acceptable to small boiler operator who is looking for minimal cost and maintenance [23,24].

Stoker combustor is well suited for multifuel system of coal and rubber tire. In 1992, New York State Electric and Gas corporation co-fired coal and tire at its 74 MWe Jennison Station [23]. The plant using four stoker boilers worked well and processed more than 1 million tires with feed composition of (25% TDF and 75% coal). While Northern power Co. successfully co-fired wood and coal, other experiences indicated that this requires higher grate speed and the use of undergrate air [23]. The process can also generate acid gas and heavy metal emissions and perhaps dioxin emissions.

The co-firing of coal and wood chips (with premixed feed) at a stoker-fired steam plant was studied by Sampson et al. [21]. This study showed that there was a very little effect of fuel mix on the particulate emission. The feed system did not work well for wood concentration of 30 wt%. Both capacity problems and particulate emissions were reduced with lower moisture content of the feed. The study concluded that energy derived from wood be competitive with that from coal if more than 30 kton of wood chips per year were available and hauling distance was less than 60 miles.

Brouwer et al. [65] examined emissions reductions from co-firing of coal and RDF in a stoker system with common feed line and common burner. The RDF used was hard and softwood waste from manufacturing and chipped railroad ties. The NO_x concentration has lowered by 25% when co-firing occurred under clean conditions (excess air 50% and CO emission less than 20 ppm). Lower NO_x content was due to lower nitrogen content of railroad ties (0.22%). The NO_x emissions were further reduced with the injection of natural gas as a reburn fuel with 28% excess air and 15% gas injection; NO_x level decreased from 0.45 to 0.25 lb/MBtu.

2.5.3 FLUIDIZED BED COMBUSTOR

This type of combustor can also be characterized as atmospheric FBC and bubbling fluidized bed combustor (BFBC). Both types are extensively used in the combustion industries. BFBC plants are considered to be relatively easy to design for co-firing applications. A separate biomass or waste handling/metering plant will be required to deliver the co-fired fuel to the existing coal feed system. NO_x, CO, and organic control is built into the combustor units through the air staging systems [23].

The use of BFBC for a mixture of coal and waste has been in practice for a long time in the United States, Finland, and Sweden. Many of them are at paper and pulp sites. They are currently used in various sizes and different applications, which include steam and hot water boiling applications and some are connected to district heating schemes. BFBC has higher efficiency than pulverized flow combustor (PFC) and has the ability to burn a wide range of fuels. The multifuel units use gas and oil besides coal and biomass and wastes [1,3,23].

In Sweden and Finland, a majority of the BFBCs uses imported hard coal with local sources such as peat and wastes produced from paper and pulp industries. BFBC are very tolerant of inhomogeneous fuel particle size that can range from sawdust to 3 in. hog fuel [23]. They also can tolerate moisture content from 35% to 60%. Besides wood/wood waste and sewage, old animal meat has also been considered as a cofuel with coal for BFBC [23]. Thus, BFBC is the most versatile and useful combustor for multifuel feedstock. BFBC also meet the emissions standards for NO_x and SO_2 levels for most multifuel systems [3,23,24]. When further NO_x reductions are necessary, this can be achieved by ammonia injection into the furnace. Similarly, sulfur levels can be reduced by limestone injection in the furnace; sulfur capture levels of 70%–80% are generally achievable. Particulates can be controlled using conventional electrostatic precipitators or bag filters [1,23].

Van Doorn et al. [71,72] co-fired mixtures of coal and wood and straw and municipal sewage sludge in a FBC with common feed line and common burner. The coal–wood mixture gave reduced emissions of NO_x and SO_2 and ease of combustion. No agglomeration of fuel particles was observed. The emissions of CO_2, SO_2, and NO_x decreased with increasing wood to coal ratio. Similar results were obtained with straw except that HCl concentration increased with increased straw to coal ratio due to high chlorine content in straw. Co-firing sewage sludge with coal caused agglomeration of fuel particles and high emissions [1,23].

Li et al. [24,73] evaluated the effect of co-combustion of chicken litter (CL) and coal on emissions in a laboratory scale FBC. The experimental results showed that CL introduction increased CO emissions and reduced the levels of SO_2. The ratio of H_2S/SO_2 increased with increasing fraction of CL NO emissions increased or decreased depending on the percentage of CL in the mixed fuels. The temperature in the freeboard region increased with increasing the fraction of CL. While reverse was true for the bed temperature.

Fahlstedt et al. [24,74] carried out a series of tests on co-firing of wood chips with olive pits, palm nutshells, and coal at the ABB carbon 1 MW fluidized bed facility. The results showed that due to higher VM content of the biomass fuels, the blend combustion had a slightly higher efficiency than coal only combustion. Due to low nitrogen content of biomass, an increase in wood chip concentration from 20% to 40% resulted in a decrease in NO_x concentration by about 25%. While in general no fouling was detected, in case of palm nutshells, some oxide layer on the bed surface was observed.

Aho et al. [75–77] studied co-firing of biomass with coal and lignite in 20 kW FBC and 100 kW CFBC. The study showed that co-combustion of coal and lignite with biomass led to many significant benefits to both FBC and CFBC combustion processes. The components of both biomass and coal played important roles in the positive benefits. Wood ash strongly adsorbed sulfur from coal, thereby reducing the SO_2 emissions. Biomass generated high concentrations of chlorine in the product. Aluminum, silicon, and sulfur present in coal and lignite effectively removed chlorine from deposits and fine fly ash. Experiments with lignite and forest residue indicated that sulfation alone can markedly reduce chlorine concentration in the deposits. The chlorine concentration in the fly ash can be equally reduced by sulfation or alkali aluminum silicate formation. Mixing coal with wood residues can also lower the toxicity of fine fly ash. Wood-derived biomass strengthened SO_2 capture from the flue gas of coal combustion. The benefit was stronger for CFBC than FBC.

Gulyurtlu et al. [24,78,79] examined co-combustion of coal and meat and bone meal (MBM) in FBC. This technology was chosen to operate in an efficient way at lower temperatures (750°C–850°C) with lower pollutant emissions. The experimental work examined the coal/MBM ratio in the feed, freeboard temperature, excess air levels, and primary/secondary air ratios. The study resulted in the following conclusions:

1. The co-combustion of coal and MBM was feasible, with minor impact on the emissions and particularly the NO_x, N_2O, VOC, and CO_2 and also on the composition of the ashes produced.
2. Based on analysis of ash from the bed, cyclones, and stack, it was concluded that these ashes were suitable to be deposited in municipal landfills.

Other applications of the fluidized bed for co-combustion of coal and biomass are described by Kicherer et al. [80], Armesto et al. [81], and Hiltunen [82]. FBC plants are popular at large scale. A list of BFBC units supplied by Ahlstrom/Foster Wheeler is shown in Table 2.3. In the United States, a number of large-scale FBC facilities use mixtures of coal and biomass [23]. For example, a large-scale (50 MWe) FBC facility at Tacoma, United States, uses a mixture of coal, wood, and RDF [23].

2.5.4   CIRCULATING FLUIDIZED BED COMBUSTOR

Just like BFBC, CFBC is also very widely used for co-firing paper and pulp mill waste with coal in countries like the United Kingdom, the United States, Taiwan, Finland, Sweden, Norway, and Austria. The majority of these plants use Ahlstrom Pyroflow/Foster Wheeler technology, although some use Tampella/Kvaerner units as well. The advantages of CFBC are the same as those of BFBC. They include

1. Low NO_x emission and low SO_2 emissions with in situ sulfur retention
2. Good combustion control and long gas–solid residence time results in low CO and organic emissions as well as they allow use of low-grade coals and other fuels

CFBC is generally used for power generation less than 250 MWe, although some larger plants operate in the world [1,3,23,24]. CFBC are used both for process steam and electricity. Some co-fired units are operated in cogeneration form and comprises

TABLE 2.3

Examples of BFBC Units Supplied by Ahlstrom/Foster Wheeler

Location	Feedstock/Capacity (MWth)
PT Indah Kiat Pulp and Paper/Indonesia	Coal, peat, wood chips, bark, oil/218
Ocean Sky Co., Indonesia	Coal, peat, bark, oil/155
Soderenergi AB, Sweden	Coal, wood waste, peat, oil/120
Nykoping Energy, Sweden	Coal, wood, waste, peat, oil/100th, 35e
Rauma Paper Mill, Finland	Coal, bark, sludges, fiber, wastes/60
Lohja paper mill, Finland	Coal, wood waste, paper waste/36
Ostersund district heating plant	Coal, peat, bark, wood waste, oil/25
Skelleftea, Sweden	Coal, peat, wood waste, HFO/25
Seinajoki Energy, Finland	Coal, peat, wood waste, HFO/20
Pieksamaki district heating, Finland	Coal, peat, wood waste, HFO/20
Outokumpu Oyj, Finland	Coal, peat, wood waste/17.5 and 24

Source: Opportunities and markets for co-utilization of biomass and waste with fossil fuels for power generation, A report from European Union, U.K., 2000, also found in website: http://ec.europa. eu/energy/renewables/studies/doc/bioenergy/2000_opportunities_and_markets.pdf.

part of a district heating scheme. One successful example is Tampella CFBC that forms as part of Norrkoping Energi AB's Handeloverket power station, which uses a multifuel system of coal and wood. In this plant, the use of 60/40 (coal/wood) mixture resulted in less than 50% of SO_2 and NO_x compared to those produced by the sites existing PC boilers. The plant since has used 10/90 (coal/wood) mixture and is capable of running coal/wood/waste mixture due to low NO_x formation caused by low combustion temperature and staged air distribution system.

CFBC is used at various scales all over the world. A brief summary of some of the existing use of CFBC is given in Table 2.4. Besides biomass (more often wood), the second fuel along with coal has been TDF. Long solid residence time in such a combustor helps the combustion of difficult polymeric tire materials.

The use of CFBC for coal/tire multifuel combustion system has been significantly increased both in the United States and Europe. Traditionally, co-firing of tire and coal has been carried out in cyclone boilers or stoker/grate-fired units; however, in recent years, United Development Group's Goodyear plant at Niagara and Southern Electric Co. have expanded the use of CFBC for co-firing coal and tire. As shown in Table 2.4, CFBC is also widely used for co-firing of coal/biomass/waste mixtures due to its ability to manage diverse feedstock [23].

Armesto et al. [24,83] studied co-firing of coal and pine chips in a CFBC and FBC. The operations in both cases were normal. The CFB process had a higher combustion efficiency than the FBC process and consequently lower CO emission. The NO_x and SO_2 emissions were also lower in these two systems. In both cases, the authors found an increase in combustion efficiency with an increase in co-firing ratio. Rasmussen and Clausen [84] observed a sharp decrease in SO_2 emission when they co-fired straw with coal in an 80 MWth CFB. Lower SO_2 concentration was

TABLE 2.4
Circulating Fluidized Beds Co-utilizing Coal and Biomass/Wastes

Owner/Location	Feedstock/Capacity (th,e)
Rumford, Cogen Co./Rumford, United States	Coal, oil, wood/260th, 76e
Kainuun Voima Oy/Kajaani, Finland	Coal, peat, wood sludge/240th, 85e
Black River Partners/Fort Drum, United States	Coal, anthracite, wood/168th
Rauma mill/Rauma, Finland	Coal, peat, sludge, bark/160th
UDG Niagara, Goodyear/Niagara Falls, United States	Coal, tires/149th
P.H. Glatfelter Co./Spring Grove, United States	Coal, anthracite, wood, oil/132th
Norrkopings Kraft/Norrkoping, Sweden	Coal, wood/125th
Southeast paper/Dublin, GA, United States	Coal, sludge/125th, 65e
IVO/Kokkola, Finland	Coal, peat, RDF, wood/98th
Lenzing, AG/Lenzing, Austria	Coal, lignite, wood, sludge/94th
Karlstad Energiverken/Karlstad, Sweden	Coal, wood, waste/90th
Etela-Savon Energia/Mikkeli, Finland	Coal, lignite, wood waste, oil, gas/84th
Brista, Kraft AB/Sweden	Coal, wood, various wastes/80th, 40e
Nykoping Energiverk/Nykoping, Sweden	Coal, wood, peat/80th
Metsa-Sellu Oy/Aanekoski, Finland	Coal, wood waste, peat, oil/76th
Midkraft Power Co./Grenaa, Denmark	Coal, straw/60th, 17e
Papyrus Kopparfors AB/Fors, Sweden	Coal, peat, wood/56th
Patria Papier & Zellstoff/Frantschach, Austria	Coal, lignite, oil, wood/55th
Hunosa power station/La Pereda, Spain	Coal, coal wastes, wood wastes/50e
Caledonian paper plc./Scotland, United Kingdom	Coal, wood, oil/43th
Solvay Osterreich/Ebensee, Austria	Coal, lignite, gas, oil, wood/38th
Sande Paper Mill A/S, Norway	Coal, wood, RDF/26th
Ostersunds Fjarrvarme/Ostersund, Sweden	Coal, wood, peat/25th
Lieksa, Finland/Lieksa, Finland	Coal, peat, bark, sawdust/22th, 8e
Kuhmon Lampo Oy/Kuhmo, Finland	Coal, peat, wood waste/18th
Avesta Energiverk/Avesta, Sweden	Coal, peat, wood/15th
Ba Yu paper/Peikang, Taiwan	Coal, sludge/NA
Slough Estates/Slough, United Kingdom	Coal, wastepaper/NA

Source: Opportunities and markets for co-utilization of biomass and waste with fossil fuels for power generation, A report from European Union, U.K., 2000, also found in website: http://ec.europa.eu/energy/renewables/studies/doc/bioenergy/2000_opportunities_and_markets.pdf.

because of lower sulfur content of straw compared to that for coal. Due to decreasing temperature at higher co-firing ratios, the NO_x emissions remained essentially constant. Particulate concentration was very small.

Yilmazoglu et al. [24,53] examined direct and indirect co-firing of lignite and biomass in CFB combustor. The fuel system examined was Soma lignite with some dried agricultural residue such as corncobs, cotton gin, and olive pit in Soma B thermal power plant, using THERMOFLEX simulation software [85]. In direct co-combustion, biomass was mixed with lignite in the same mill and fed into the boiler furnace, and in indirect co-combustion, biomass was fired in a separate CFB boiler, and the produced steam was supplemented into the steam network of the power plant. In both cases, CO_2, SO_2, and dust emissions all slightly decreased and net efficiency slightly increased. Olive pit alternatively displayed the best effect on the performance of TPP. Indirect co-combustion was found to the best way to utilize biomass with decreasing emissions; however, the cost associated with this approach was higher than the one for direct co-combustion method.

Gayan et al. [23,24,86] examined co-firing of low-grade coals, biomass, and organic wastes in CFBC. Two kinds of coals (i.e., subbituminous and lignite) with pine bark forest residues in two CFBC pilot plants of capacity of 0.1 and 0.3 MWth were examined. The effects of main operating parameters such as combustion temperature, percentage of biomass in the feed, air velocity, excess air, coal type, percentage of secondary air, and particle size distribution of coal on the combustion efficiency were studied. The study also presented a mathematical model for carbon combustion efficiency in CFB for co-firing of both coals and biomass. The model fitted the data obtained from both pilot plants well. Some other applications of CFB for co-combustion of coal and biomass are described by Lind [87], Desroches-Ducarne et al. [88], and Tsai et al. [89].

2.5.5 PRESSURIZED FLUIDIZED BED COMBUSTION

This is one of the advanced clean coal technology concepts used widely in the United States and Europe. Many of the design consideration of this type of combustor are the same as those for BFBC and CFBC. The added complication is the requirement of feeding of biomass across the pressure boundary into PFBC reactor. For co-firing, this type of combustor is not as widely used as CFBC and BFBC. Overall, this technology is at the development stage, and while gaining popularity, it will not be as widely used as BFBC and CFBC.

Andries et al. [24,90] co-fired straw with coal in a 1.6 MWth pressurized FBC test rig with a common feed line and common burner. Due to high VM content of biomass, the temperature downstream of the freeboard (the surface of the fluidized bed) was higher than the coal-only case by 30°C. The location of high-temperature region corresponds to the location of volatile combustion. Co-firing reduced the CO, NO_x, and SO_2 concentrations in the freeboard.

2.5.6 CYCLONE FURNACES

The cyclone combustor is a form of slagging combustor primarily used in utility boilers. There are more than 1000 combustors of this type installed in the United States and Europe. A wide variety of fuels can be burned in cyclone combustors. This includes most ranks of coal, fuel oil, natural gas, wood, bark, refuse, TDF, and petcoke. There are, however, some criteria such as ash > 6%, VM > 15%, iron oxide/calcium, and magnesium oxide between 10 and 25 of the coal feedstock that best suits the operation of this furnace.

In the United States, cyclone furnace has been used to co-fire wastes and biomass with coal. Wood as a cofuel is used most extensively. The studies done at the Electric Power Research Institute (EPRI) and Tennessee Valley Authority indicate that the most appropriate level of wood is in 1%–15% (by heat content) region for cyclone boilers. In recent years, scrapped tires or TDF have been used more as a cofuel for heat and power generation. Some examples are Otter Tail's Power's 440 MWe Big Stone Plant at Milbank, South Dakota, which uses tire (and some RDF) with coal, and Illinois power, which use 2% tire with coal [3,23,24].

Ohlsson [50] examined co-firing of RDF (binder enhanced) and densified MSW in a 440 MW cyclone-fired combustor with common feed lines and burner. For 12% RDF and 88% coal mixture over a 10 h period, NO_x emission was 2%–3% less because of low nitrogen content of RDF. Sulfur dioxide emission was 17% less than that for coal alone. The reaction between SO_2 and calcium hydroxide in the binder in RDF lowered the SO_2 emission. Since RDF contained more ash than coal, the particulate concentration was higher (about 1.5 times) with co-firing than for coal alone.

2.5.7 SMALL-SCALE LABORATORY BURNER

Frazzitta et al. [33] evaluated the performance of a small-scale boiler burner facility for co-firing of coal and manure blends with a common feed line and common burner. The co-firing ratio was 20% manure by mass. Three types of manure, raw (RM), partially composted (PC), and full composted (FC), were used. PC and FC refer to the compost time of 30 and 120 days. The authors measured the temperature distributions in the burner, composition of flue gases, and combustion efficiencies. Fuel blends were injected in a preheated burner with a hot secondary air stream of about 200°C. When the burned fraction was about 95%, the temperature distribution with and without the manure was approximately the same implying that manure did not affect flame stability. NO_x levels decreased from 555 ppm for FC manure, 500 ppm for RM, and 480 ppm for PC manure at 95% burned fraction. The decrease may be related to original nitrogen content of three manures. Lower SO_2 emissions were observed for blended fuel. This may be due to SO_2 capture by the alkali ash of the manure feedstock.

2.5.8 ISSUES GOING FORWARD

The combustion of straw and sewage sludge with coal in a bubbling fluidized bed results in bed agglomeration under some operating conditions. For these feedstock and some cereals, increased level of chlorine (HCl) in the fuel gas increases the corrosion rates of downstream components. Many biomass fuels contain large amounts of alkalis especially potassium, which may aggravate the fouling problems. The use of biomass such as straw in multifuel systems can also lead to increased fouling of boiler internal surfaces. Both straw- and wood-fired plants have experienced high level of maintenance compared to units fired on coal alone. High levels of chlorides and alkalis in biomass can create long-term plant and combustor performance reliability problems. These types of biomass need to be carefully proportioned with coal to avoid fouling and corrosion.

Another issue is the inconsistent quality of the mixed feed to the combustor and its resulting effect on the combustor performance. Since biomass and waste are nonhomogeneous, collection, blending, and feeding system requires careful design. Unlike coal, biomass has low heating values, high moisture contents, low bulk densities, high fibrosity, and variable particle size, and the effects of these properties on the size and nature of feeding system need to be considered. Furthermore, the cost of these design considerations can be significant. For some multifuel system, these feed preparation systems are still under development.

While the processing of multifuels consisting of coal and biomass or waste in PC combustor has proven to be successful, both in Europe and the United States, they still face several operational problems. The issues include necessary preparation and feeding systems, the use of special burners, and corrosion and emission problems due to high chlorine content in biomass as well as in some coal. The milling of biomass to a very small particle size can be highly energy intensive and expensive.

Indirect multifuel technologies for biomass/waste/coal are still not a mature technology and further R&D and demonstration plants are required to make them commercially viable. The experience in first large demonstration coal-based integrated gasification combined cycle (IGCC) plant can be used for the development of multifuel indirect combustion technologies.

2.6   EMISSIONS AND PRODUCT TREATMENTS

The principal emissions from coal-fired plants are particulates, sulfur dioxide, and oxides of nitrogen. Emissions of sulfur dioxide are approximately in proportion to the fuel's sulfur content, that is, high sulfur eastern coal will generate more sulfur dioxide. The effects of incorporating biomass with coal on sulfur dioxide will depend on the fuel's sulfur content. For fuels such as straw and wood, the sulfur content is lower than that for coal, and so multifuel combustion with these species will generate less sulfur dioxide. The materials such as tires and cotton gins contain high amount of sulfur, and they will produce more sulfur dioxide. In PC systems, the retention of SO_2 by biomass ashes has not been observed due to high combustion temperatures. Sewage sludge contains high sulfur levels, and in this case, the overall SO_2 emission will depend on total fuel S content in multifuel system.

Emissions of NO_x cannot be predicted solely from coal or biomass nitrogen content since they are influenced by burner design and operation and the nature of the oxidant used during combustion process. Changes in boiler configuration to accommodate biomass co-firing may influence NO_x production. This is especially true if a dedicated set of burners is introduced in the boiler. In general, high volatile content (and often low nitrogen content) of biomass limits the production of NO_x during multifuel combustion. For example, literature has shown [3,23,24,45] that NO_x levels from down-fired PC co-combustion with straw were reduced by on average 30% during staged combustion compared with firing of brown coal alone. The NO_x emissions from higher rank coals are improved markedly by the use of multifuels. Addition of the coal is best placed in the substoichiometric zone with the biomass inserted into the air-rich zones. Biomass is also proved to be suitable for reburning [3,23,24] applications.

Pilot scale co-firing of sewage sludge with coal in various locations have revealed a much greater sensitivity of NO_x emissions to co-firing conditions than for biomass.

Sewage sludge has higher nitrogen content than other biomass. The preferred position for sewage sludge addition is therefore through the central, oxygen-deficient burner zone. Either higher or lower NO_x emissions than for 100% coal may be obtained depending on the co-firing rate: the first 15% of sewage sludge generally increases emissions. However, air and fuel staging can prevent this increase.

Generally chlorine content of coal is low. On the other hand, chlorine content of some of the biomass can be high. In the combustion of biomass and waste, dioxins and chlorine (as HCl) can be produced. The high temperatures in PC systems generally prevent any increase in the emissions of dioxins, despite the presence of chlorine in biomass feedstock such as miscanthus or straw. This has been demonstrated at 125 MWe scale at Esbjerg, Denmark [3,23]. During co-firing, the chlorine is mainly released as HCl. This can increase corrosion of superheated tubes and some downstream equipment.

High alkali and heavy metal contents in some biomass can affect emissions and quality of slag and ash during multifuel combustion. Slagging and fouling have been found to slightly increase with straw as a second fuel in 125 MWe boiler at Esbjerg [23]. Particulate emission increases when co-fired with wood. Coal ash can also be affected by biomass during co-firing. The inorganic, alkali, and heavy metals contained in biomass can affect ash properties and its usefulness for the construction industry. Sewage sludge has much higher ash content than other biomass, and therefore in co-firing, its concentration is limited to 10% to avoid problems with slagging and poor ash quality. In summary, both emissions and solids remain (ash and others) can be affected by the original properties of the components of multifuel systems. These emissions and solid properties need to be carefully managed by the method of combustion, detailed designs of burners and airflow system, and appropriate product treatments. Various issues dealing with particulates, sulfur, nitrogen, chlorine, alkali, and heavy metals during co-firing are discussed in details in the published literature [101–144].

Kalembkiewicz and Chmielarz [91] studied the effects of coal–biomass co-combustion on functional speciation and mobility of heavy metals in industrial ash. The study concluded that

1. The addition of biomass to coal (50/50 m/m) results in a decrease in the content of Cd from 5 to 1.5 mg/kg and an increase of Zn contents from 205–239 to 245 mg/kg: No major changes in total content of Pb (about 159 mg/kg) and Cu (about 40 mg/kg) occurred in ash
2. Mobile fractions of metals available for leaching changes from 42% to 53% for Pb, 0% to 43% for Cd, 9% to 17% for Zn, and 17% to 18% for Cu of their total content in the ash by coal–biomass co-combustion
3. 50/50 mixture of coal and biomass reduced the environmental mobility of Pb, Cd, and Zn, but did not cause major changes in the mobility of Cu

The treatments of impurities in the flue gas of combustor can be carried out in the same manner as those outlined in Chapter 3 on gasification. The major difference between combustion and gasification is that combustion produces more oxidative impurities like SO_2 and NO_x; on the other hand, gasification produces H_2S, COS,

HCN, and NH_3 due to the presence of the reducing environment. Some of the unit operations described in Chapter 3 can also be applied to NO_x and SO_2, and some additional methods are also available in several references [101–144]. Both SO_2 and NO_x can be removed by direct injections of ammonia and lime into the combustor. Ammonia will react with NO_x and limestone can absorb SO_2. These impurities can also be removed using selective catalytic reduction process. These and other available processes are also described in numerous reports by Research Triangle International (RTI) [92] and others [93–100] for both high- and low-temperature impurity removals.

There have been only limited tests with multifuel combustion of coal boilers fitted with flue gas desulfurization and selective catalyst reduction equipment. There are some evidences that the contamination from biomass affect limestone reactivity and catalyst activity of $deNO_x$ SCR catalysts. Both of these aspects require further research and testing. While, in general, NO_x and SO_2 emissions from multifuel coal and biomass systems are lower than the ones for coal alone, SO_2 level may increase while using sewage sludge. Optimum conditions for NO_x minimization for multifuel systems need additional information.

The fate of certain trace elements in biomass and waste are not fully established during multifuel combustion. Some confirmation of the fate of dioxins from straw during multifuel combustion in PFC boilers and nonleachability of residues that contain higher level of heavy metals is required. Generally, dioxins are destroyed at high temperatures and heavy metals are trapped in ash. The carbon content of ash can be controlled by optimizing burner systems. The ash can be used for the construction purposes if its composition is carefully controlled.

For the utilization of the ash in cement and concrete industry, the concentrations of alkali metals, P_2O_5, SO_3, Cl, and unburned carbon in the ash are the critical parameters [3,23,24,45]. The proportion of ash operating from biogenic fuels should not exceed 10% of the total ash. In case of substitute fuels, high ash content combined with a high multifuel combustion rate can create problems with fly ash quality. Because most biomass contains less S than coal, the fly ash conductivity will generally decrease during multifuel combustion [3,23,24,45,101–134].

2.7 COMMERCIAL PROCESSES AND LARGE-SCALE MULTIFUEL TESTING PROGRAMS

While there are a number of issues that still need to be addressed, multifuel combustion has been well accepted by the industry. As an indicator of its support by the government and industry, in this section, we briefly outline few commercial processes. Multifuel concept has been tested in numerous full-scale utility coal-fired boilers. As shown in Figure 2.1, the Dunkirk power station in New York will use willow grown by New York farmers along with coal to generate electricity [162]. Similarly, as shown in Figure 2.2, the boiler plant at Savannah River site by Department of Energy uses a mixture of coal and biomass [163]. Worldwide, the United States leads the commercialization of mixed feedstock followed by Europe. Some of these are outlined in Table 2.5 [1,3,23,24,162–164]. More descriptions of several commercial processes are also given in References 145–161.

FIGURE 2.1   The Dunkirk Power Station, hybrid power station, which uses willow grown by New York farmers to generate renewable electricity. (From Wikipedia, "Dunkirk," New York. http://en.wikipedia.org/wiki/Dunkirk,_New_York, 2015.)

FIGURE 2.2   The boiler plant at the Department of Energy's Savannah River site co-fires coal and biomass. (From Biomass co-firing in coal-fired boilers, Federal technology alert, a new technology demonstration publication, DOE/EE-0288, Federal Energy Management Program, Produced for the U.S. Department of Energy, Energy Efficiency and Renewable Energy, by the National Renewable Energy Laboratory, a DOE national laboratory, June 2004.)

TABLE 2.5
Tests of Co-Firing in Full-Scale Utility Coal-Fired Boilers

Company	Feedstock/Type of Combustor
Northern States Power	Coal, sand dust/cyclone combustor
Santee Cooper Electric Corp.	Co-firing forest debris/pulverized combustor
Tacoma public utilities	Coal and Biomass/FBC
Georgia Power and Southern Co.	Co-firing waste wood/pulverized combustor
City of Ames, Iowa State University	Co-firing TDF/wall-fired grate combustor
TVA	Co-firing of low% of wood/wall-fired grate combustor
TVA, Foster Wheeler	Co-firing low% wood/tangentially fired pulverized combustor
Madison G&E, University of Wisconsin	Co-firing switchgrass/wall-fired grate combustor
Duke Power	Co-firing plastic mill residue/cyclone boiler
GPU/Penelec, Foster Wheeler	Co-firing wood/cyclone boiler
TVA, Foster Wheeler	Co-firing wood up to 20%/cyclone boiler
TVA, Foster Wheeler	Co-firing tire up to 15%/tangentially fired pulverized combustor
NYSEG	Co-firing mid% wood/tangentially fired pulverized combustor
TVA, Foster Wheeler	Co-firing wood up to 20%/tangentially fired pulverized combustor
Savannah Electric and SCS	Co-firing high percentage wood/tangentially fired pulverized combustor
South Carolina E&G	Co-firing plastic fiber/tangentially fired pulverized combustor

Sources: Opportunities and markets for co-utilization of biomass and waste with fossil fuels for power generation, A report from European Union, U.K., 2000, also found in website: http://ec.europa. eu/energy/renewables/studies/doc/bioenergy/2000_opportunities_and_markets.pdf; Sami, M. et al., *Prog. Energy Combust. Sci.*, 27, 171, 2001.

The world energy council has forecasted that electricity requirement world-wide will double between 1990 and 2020 and further increase beyond 2020. This means that electrical generation capacity will significantly increase over next several decades. Moreover, new environmental regulations will dictate that new energy generation capacity will have to be environmentally acceptable, that is, less greenhouse gas (GHG) such as CO_2. This implies that multifuel combustion will have significant opportunities for growth over next several decades.

2.7.1 ALHOLMENS KRAFT CHP PLANT IN FINLAND

This giant-scale multifuel commercial plant consumes 12,600 TJ of fuel annually generating 240 MW of electric output and 100 MW of process steam output and 60 MW of district heat output [1,3,23,24]. The process uses multifuels consist of industrial wood and bark residues (35%), forest residues (10%), Peat (45%), and heavy fuel oil and coal (10%). The process generates process steam for adjacent paper mill and electricity for utility industry and district heat. UPM-Kymmene Pulp and Paper Mill nearby supplies the power plant with wood and bark residues. The Mill produces

159,000 tons of papers and 95,000 tons of packaging materials [1]. Since the plant is at the giant scale, special attention is given to the logistics and fuel procurement. The Alholmens Kraft plant in Finland is the largest bio-fuelled power plant in the world.

2.7.2 Fuel Flexible Avedore Plant with Ultrasupercritical Boiler

This is one of the world's most energy-efficient combined heat and power plants (CHPs), and it is located in Copenhagen, Denmark [1,3,23,24]. The heat produced by this plant serves 280,000 single-family houses. The multifuelled Avedore Unit 2 was started in 2001. The plant was divided into three parts: a gas turbine plant, a biomass plant, and an ultrasupercritical (USC) boiler plant. The plant uses multifuels and contains most advanced technologies, which include a USC boiler, the most advanced steam turbines, and the largest straw-fired biomass boiler. In the new design, the USC boiler uses natural gas, oil, and 300,000 tons/year wood pellets. Avedore 2 can utilize as much as 94% of the energy in the fuel and can reduce CO_2 emission by 10%.

Avedore 2 biomass plant burns 150,000 tons of straw annually (which constitutes 10% of Avedore 2's fuel consumption). The ash from the plant is used for fertilizer. The total plant capacity is 585 MW with gas turbines, and these turbines generate 505 MW of electricity and 565 MW of heat. Gas turbine uses natural gas [1,3,23,24].

2.7.3 Norrkoping 75 MW CHP Plant

The Handelo CHP in Norrkoping, Sweden, is a new design of waste-to-energy [1,3,23,24]. The plant contains four boilers; two CFB boilers for combustion of biomass and waste materials, one vibrating grate for demolition wood, and one traveling grate for coal. The CFB is designed to handle MSW, industrial waste, sewage sludge, rubber, and demolition wood. Along with flexible combustion technology, special fuel handling (aluminum removal) and gas cleaning systems are designed. The plant generates 11 MW of electricity (annual electricity production 88 GWh) and thermal heat output as district heat and process steam at 64 MW. The plant consumes 200,000 tons of waste with composition of 30%–50% combined household waste, 50%–70% classified industrial waste, and maximum sewage sludge of 20%.

2.7.4 Allen Fossil Plant of TVA

EPRI initiated a program in 1992 to commercialize co-firing in utility stations. Hughes and Tillman [19] provided a detailed account of co-firing tests carried out in full-scale power plants, pilot plants, and laboratory-scale facilities. At the Allen Fossil plant, a multifuel blend of wood and coal (wood 20%, coal 80%) was burned in a cyclone boiler [1,3,19,23,24]. The coals examined were Eastern high-sulfur coal and Utah bituminous coal. The collected data indicated that multifuel combustion of biomass and coal can (a) accomplish a trade-off between boiler efficiency and fuel cost, (b) reduce SO_2 emissions particularly when wood is fired with high-sulfur eastern coal, (c) reduce NO_x emissions, (d) reduce fossil-derived CO_2 emissions through fuel displacement, and (e) achieve fuel diversity and customer service goals.

2.7.5   Kingston and Colbert Fossil Plants

As a part of EPRI programs, these two PC technology plants [23,24], one tangentially fired (Kingston plant) and other wall fired (Colbert plant), were tested with small concentrations of wet sawdust (less than 5 wt%). The tests indicated that pulverized combustors can operate in a stable manner at this level of sawdust concentration. Five weight percent may be the approximate upper limit for transporting fine woods through ball mills or bowl mills. Other co-firing tests performed in utility boilers and pilot plants are summarized by Sami et al. [24] and others [162–164]. The results of these tests can be summarized as the following:

1. Co-firing can be performed at moderate and high percentages of second fuel in cyclone boilers.
2. Five to ten percentage (mass basis) co-firing in PC boilers may require separate feed lines depending on the capacity of existing pulverizers, type and condition of biomass fuels, and type of pulverizers used.
3. PC boilers can process 0%–5% co-firing. Pulverizer performance depends on type of biomass fuel and co-firing ratio. The fibrous structure in biomass (straw, miscanthus, wood, switchgrass) makes grinding difficult at the same level as that for coal. If coal pulverizers are used, grinding cost determines the extent of pulverization.
4. The potential for successful application of co-firing is site specific. It depends on the power plant being considered, the availability and price of biomass fuel within 50–100 miles of the plant, and the economic value of environmental benefits.

2.7.6   Shawville Generating Station

This station uses both wall-fired and tangentially fired boilers, which are also equipped with low NO_x burners [1,3,23,24]. The multifuel tests were performed with less than 3 wt% of sawdust, poplar, and tree trimmings in PC boilers. In these tests, stability and operability of the boilers and overall plant efficiency were not compromised by the use of multifuels. The difficulties in feeding fibrous materials to the pulverizers required reduction in feeder speeds and increase in mill outlet temperature [1,3,23,24].

2.7.7   Plant Hammond

The Southern company conducted extensive testing of multifuel combustion at plant Hammond using a wall-fired coal boiler [1,3,23,24]. The wood percentage in the mixture ranged from 9.7% to 13.5%. The tests indicated (a) only modest losses in boiler efficiency with multifuels, (b) unburned combustibles were higher during multifuel combustion, (c) mill energy requirement increased while cogrinding coal and wood, (d) mill fineness modestly decreased during multifuel combustion, and (e) no positive impacts on emissions during multifuel combustion. The boiler operated at full capacity during these tests.

2.7.8 Plant Kraft Station

Southern company also tested multifuel combustion at higher concentration of wood in a 55 MWe boiler at the Plant Kraft of Savannah Electric [1,3,23,24]. In a tangentially fired boiler, a separate wood feeding system that directed a flow of dry sawdust into the exhauster of the bowl mill was examined. Both wood and coal were co-fired in separate rows of burners. The tests indicated that high percentage of wood can be co-fired in this type of PC boiler using a separate dedicated wood feeding system.

2.7.9 APAS (European Commission Project, 1992)

This project looked at co-firing biomass in laboratory, pilot, and commercial units. The project had 25 partners from 8 European countries. Hein and Bemtgen [18] summarized the activities undertaken during this project. Two general types of biomass fuels were considered: woody (wood, straw, paper, miscanthus) and sewage sludge. The combustion facilities were either PC or FBC. The results derived from these studies are the following:

1. Both PC and FBC are well suited for co-firing provided a fuel-dependent feed and preparation system is installed.
2. Fuel conversion was not significantly affected by co-firing.
3. In order to avoid corrosion and slagging of the heat transfer surfaces, biomass fuel rich in chlorine and alkali metals should not be used in co-firing.
4. No increases in emissions of hazardous gases were observed. In some cases, substantial emission reductions were found, which were functions of biomass composition and fuel injection mode. The reduction in NO_x emission was significantly affected by the injection mode.

2.8 MARKETS AND ECONOMICS OF MULTIFUEL COMBUSTION

While the scientific interests in multifuel combustion has risen very rapidly over last several decades, its growth in commercial application will strongly depend on its economics and market potential. While generation of electricity by renewable sources of energy such as solar and wind will grow over next several decades, the largest provider for power generation will still be traditional combustion technology.

2.8.1 Market Potential

Two most important markets for co-firing multiple fuels are power sector and industrial sector. The power generation market is the largest single market potential for multifuels due to the fact that it is large and also because it is well suited for co-firing. Three types of co-firing combustors that are best suited for this sector are PC combustion, BFBC, and CFBC. In each case, the direct combustion technology will have greater application due to its relative simplicity. A market penetration of 500 PJ/year will be equivalent to retrofitting 10 GWe of coal-fired plant for 10% co-firing.

The potential for this market is huge because it is predicted that by 2040, 40% of our total energy consumption will be electrical [165]. The co-firing of biomass and waste with coal is best suited for larger systems where the flows are higher because of difficulty in their handling and resulting cost increase. This permits economy of scale and enables easier achievement of uniform feed rates.

Industry sector is a potential market because they will more likely than commercial sector be willing to examine new fuel options because of their expertise to examine them and put them into practice. Since BFBC and CFBC are the most widely used methods for burning biomass, these technologies can be easily used for the multifuel systems of coal and biomass. Many conventional stoker-fired boiler plants can also co-fire wood without problems. Similarly, many PC boilers can adopt some of the co-firing techniques currently being developed for power generation applications. RDF and straw can both be burned successfully on chain grate stokers, and co-firing would reduce the negative effects on output, corrosion, fouling, and emissions.

Another potential market is the expanded use of combustion for MSW and other waste materials. Globally, the production of MSW will soon surpass one billion tons per year, and it will continue to grow due to increased population. This MSW will have to be treated due to shrinking availability of landfills. This will thus provide another potential market for co-firing. For MSW, the selection of co-firing would be aided by the geographical spread in the availability of these materials: relatively small volumes of the material over wide areas would favor the less capital-intensive retrofitting of co-firing systems over new building of dedicated waste combustion plant. Large industrial boilers providing heat and power to industrial sites are the most likely candidates for new co-firing installations. The future scope for co-firing of sewage sludge is governed by the geographical spread of resources. Incineration schemes need to be near the sewage sludge production centers. This is generally accomplished by dedicated combustion facility for the sewage sludge, and a likelihood of the change in this behavior remains small.

In the United States, the markets for co-firing in commercial, industrial, and utility applications remain strong [151,153]. In recent years, a number of power generation facilities have adopted co-firing of multifuels. This reduces their environment impacts. The co-firing also helps the level of SO_2 and NO_x emissions. Multifuel approach also provides an alternative to coal fire utilities, should the need arise. The concept of multifuels combustion also helps the establishment of green pricing structures, the minimization of industrial waste–related problems, and the development of fuel crops as another market for U.S. agriculture industries [23].

The market for co-firing biomass with coal is best in large coal-fired boilers because they offer better economy of scale and efficiency over plant burning single biomass on its own. Due to its low heating value, high moisture content, and low density, it is only practicable to burn biomass at small scale to avoid transportation costs. Such plants, however, miss out the benefits of economy of scale. More penetration of biomass in co-firing processes will require more research and commercialization of biomass fuel handling equipment, fireside impacts (such as carbon burnout), ash fouling, ash disposal, and NO_x-related issues. Government and industries are working together to address these issues and enhance market share of multifuel combustion processes [1,3,23,24].

2.8.2 ECONOMICS

The economics of multifuel combustion, particularly coal and biomass/waste, are very site specific and depend on a variety of factors. These include availability of processed fuel and, for retrofit applications, the site layout and plant design. In existing power plants, it should be possible to install co-firing capacity at a lower investment cost than that of new biomass–waste-dedicated plant. This also increases the power generation using renewable energy sources. The impact on the operating and maintenance costs of existing plant must also be considered in the final decision. The economic environment in which the plant will operate also affects the viability of the project. For example, environmentally inspired tax breaks, power price support mechanism affects the overall economics. Finally, since most often co-firing is done on a larger-scale plant compared to stand-alone plant, the economy of scale will favor the cost per unit power generated to the co-firing option. Specifically, some of the technical economical issues for co-gasification of straw and sewage sludge are described by Exelby and Davison [166]. Krebs [167] gives an assessment on the role of renewable energies in response to CO_2 problem on commercial level.

There are three types of costs associated with the direct co-firing process: capital costs for retrofit, fuel costs, and other operating costs. For direct co-firing technologies, EPRI has quoted the capital cost of co-firing in the range $100–$200/kWe [3,23,24,19,168–172]. The lowest cost is for cyclone with up to 10%–15% wood and co-firing in pulverized combustors with 1%–3% wood, based on thermal input. Higher co-firing rates can increase the cost because that may involve separate biomass preparation system and cost in extreme case can range from $200 to $400/kWe [3,23,24].

Fuel costs depend on the nature of biomass and their availability. EPRI give the actual cost of biomass fuels from dedicated crops as $1.95–$3.50/GJ. In a recent EPRI member feasibility studies, projected fuel cost ranged from $1.35 to $2.65/GJ. Other operating costs include raw materials, maintenance, auxiliary power, labor costs, land taxes, disposal costs, and by-product credits [3,23,24]. Other contributions may come from tax breaks or environment credits due to minimization of emissions by the use of co-firing systems. EPRI recent calculations show that the cost of avoiding CO_2 emissions by using biomass in a PC boiler firing 15% biomass by heat would be <$5/tons of carbon. This number is considerably smaller than similar numbers for other methods of CO_2 reduction options. For indirect co-firing technologies, EPRI estimates for 100% wood-fired IGCC at $1765/kWe at 100 MWe scale [3,23,24].

In general, published cost data on processes and on the cost of biomass vary widely. Also, the economics of co-firing process depends on the assumptions. For example, sewage sludge can be obtained free and even with some gate fee. If, however, one needs to use specially grown energy crops, the cost can be substantially higher [1,3,23,24]. Most economic analyses have been carried out on wood or straw as cofuel for power generation applications. The commercial tests appear to be indicating that technical risks for these systems are not too high. Any new or retrofit co-firing plants should be prepared to operate fully on single fuel (coal) in case second fuel supply is exhausted and new supply possibility is not imminent.

Operating costs are typically higher for biomass than for coal. The most sensitive factor is the fuel cost that includes costs for transportation, preparation, and on-site handling. The favorable situation for multifuel combustion plant will require commitments of governments in terms of tax relief, subsidy, and guaranteed market for electricity to relieve environmental pressures such as pumping of sewage to sea, the excessive use of landfill, and the emission of fossil fuel-based CO_2.

2.9   OTHER CO-FIRING OPTIONS

Although in Sections 2.4 and 2.5, we have evaluated various issues dealing with biomass and waste as co-firing fuel for a variety of coal burner systems, these multifuel systems provide ways to substitute environmentally harmful coal by renewable sources of energy, which are less harmful to environment. The method also allows penetration of renewable source of energy for heat and power production at less capital investment than building an independent stand-alone biomass and waste combustion systems. Finally, this method can take advantage of economy of scale that is not possible for stand-alone biomass and waste combustion systems.

2.9.1   COAL–OIL CO-FIRING

One multifuel system that has been very widely examined in the past is combustion of coal–oil mixtures. This system offers some advantages: (a) slurry injection as opposed to dry injection in the combustor, (b) properties of oil and coal can be manipulated such that it will decrease emissions of pollutants, (c) the basic nature of solid ash produced will not significantly change from coal-alone system, (d) concentration of particulate materials can be reduced, (e) the transportation of such mixtures can be more easily carried by pipes and other means, (f) combustion efficiency can be maintained, and (g) burner designs can be easily adapted for the mixtures. Unfortunately in recent years and into the future, the decline in crude oil production and its high cost may not strategically favor the use of oil resources for this purpose. The use of unconventional oil will not be attractive due to its harmful effects on the environment.

2.9.2   COAL–WATER CO-FIRING

While coal–oil mixture may not be as attractive as it once was, due to a decrease in supply of oil, another option of coal–water fuel (CWF) may still find attractive usage in countries like Indonesia (even the United States) and others where oil supply is scarce but the country has a large amount of coal supply. In these countries, coal–water mixture can substitute fuel oil for various combustion purposes. Strictly speaking, coal–water is not a multifuel because water is not a fuel generally at the combustion temperature but a provider of oxygen and steam in the system. As shown in my previous book [165], CWF, however, provides an interesting combustion mixture that can be used in boiler, turbines, and diesel engines. The modern CWF is based on the use of ultraclean coal of very fine particles preferably less than 10 µm to have a variety of usages.

While the production of ultraclean coal with very fine coal particle size involve additional energy penalty for advanced coal processing, this can be offset by other benefits such as very low ash, avoidance of transmission, and distribution costs by the ability to achieve high-efficiency generation at smaller scale. While best available supercritical PF and IGCC plants can achieve fuel efficiency of 38%–40% (HHV) at a scale of around 700 MW, and a natural gas turbine 48%–50% at an intermediate scale of 250–500 MW, the highest overall cycle efficiency of 52% (HHV) is achieved by a low-speed diesel engine using fuel oil, at a scale around 30 MW. The literature [173–191] has shown that a properly prepared CWF can replace fuel oil with thermal efficiency only 2%–3% lower than that for fuel oil. The development of low-speed, high-capacity (up to 100 MW), and two-stroke diesel engines provides a useful technology platform for efficient distributed generation based on CWF.

The recent developments of sophisticated technologies to generate ultralow ash, ultrafine coal particles allow higher quality CWF than in the past. This high-quality CWF cannot only be a replacement of heavy fuel oil but a direct competitor of PC in diesel engines and gas turbines along with its traditional use in boilers. The use in diesel engine requires particle size to be less than 20–30 µm, whereas for gas turbines particle size of less than 10 µm is required. The concentration of around 50 wt% is used. The direct-fired coal is mostly applied to compression ignition (diesel) engine [173–191].

The injection and combustion characteristics of CWF in diesel engines are considerably different from those of diesel fuels due to poor atomization of CWF and time required to evaporate the water. Combustion and thermal efficiencies for CWF, however, compare well with that of diesel fuel, up to engine speed of 1900 rpm. While the development of an effective injector to the engine has been the most researched subject in recent years, electronic-controlled engine with its emission cleanup system, along with a number of other engine modifications to combat wear will allow faster penetration of CWF in the diesel engine industry. Currently, the use of a purged shuttle fuel pump plunger, diamond compact injector tip nozzles, electronically timed injection, cylinder with tungsten carbide covering, and top ring set and pilot injection of diesel are some of the technological requirements [173–191]. Other technical areas that need attention are the fate of mineral matter and its effect on the engine wear and how to minimize coal agglomeration during the evaporation of individual CWF droplets.

The use of CWF in diesel engine also allows cost-effective and efficient backup power and spinning reserve, which allows the efficient use of a range of biomass-derived liquid and gaseous fuels. The cost of electricity from CWF is very similar to the one for supercritical pf plants, and it gives 20% reduction in GHG emission (GGE). It also provides efficient, flexible, and adaptable distributed generation. In the long term, the use of CWF in large gas turbine combined cycle plants will give the highest fuel cycle efficiency and lowest GGE. The development of these machines is, however, very costly compared to diesel engine at the present time [173–191].

CWF thus offers another example of multifuel system that can be better than combustion of coal-alone system. It does have niche applications, and it will allow the reduced use of coal in more environmentally acceptable manner.

2.9.3 OIL SHALE–BIOMASS CO-FIRING

Other potential multifuel system involves the use of oil shale with coal or biomass/ waste. As pointed by Boardman et al. [195] and others [72,192,194,195], oil shale is an abundant, underdeveloped natural resource, which has natural sorbent and cementitious properties. It can be easily blended with coal, biomass, municipal wastes, waste tires, and other waste feedstock materials to provide joint benefit of adding energy content while adsorbing and removing sulfur, halides, and volatile metal pollutants and reducing the emission of NO_x.

Oil shale is abundantly available in some parts of the world like Turkey and Canada and can be used to substitute coal. The depolymerization–pyrolysis–devolatilization and sorption studies show that oil shale adsorption capabilities can be as good as limestone to capture SO_2 and SO_3. Furthermore, the kerogens released from oil shale have potential to remove NO_x through "reburning" chemistry similar to natural gas, fuel oil, and micronized coal. The residual fixed carbon in oil shale is also capable of adsorbing mercury and other heavy metals. All of these sorption properties are functions of particle heating rate, peak particle temperature, residence time, and gas phase stoichiometry. The reaction zones that exist in PC or FBCs can provide high calcium reactivity and the production of fixed/activated carbon that has high sorption capability for many pollutants.

Numerous literature studies [3,23,24,72,193–196] have shown that the addition of oil shale in a fluidized bed or pulverized bed combustors can enhance power generation by

1. Playing a role of highly effective, natural acid gas adsorbent for both coal and MSW
2. Adding heating value to the fuel input
3. Improving combustion options for power generation for MSW
4. Creating cementitious ash that does not allow hazardous materials to be leached and thereby serving as a viable building material
5. Relieving the strain on available landfills and perhaps lessening the risk of leaching from existing and new landfills

Boardman et al. [195] and Idaho National Laboratory (INL) illustrated a possible application of oil shale use to reduce pollutant emissions in a PC combustor. This process is graphically illustrated in Figure 2.3. The INL had licensed this application to Nalco Mobotec who is examining its potential for emission reductions. Oil shale can also be used as a resource for ex situ combustion by retorting the shale into the coal boiler at the optimum locations. This method allows the injection of the calcined shale into the combustor at the optimum temperature for SO_2/SO_3 and mercury adsorptions.

In a more recent study, Ozgur et al. [194] examined thermal analysis of co-firing of oil shale with biomass. The study performed the co-firing of biomass/oil shale

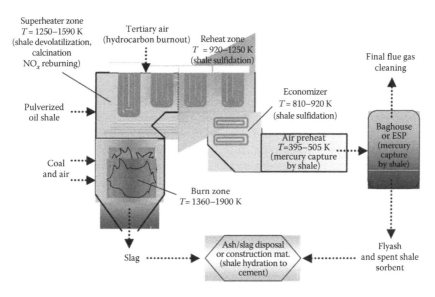

FIGURE 2.3 Injection of unreacted pulverized oil shale into a pulverized coal combustor. (From Boardman, R. et al., Co-firing oil shale with coal and other fuels for improved efficiency and multi-pollutant control, *Thirty-third International Technical Conference on Coal Utilization and Fuels Systems*, INL report INL/CON-08-14168, June 2008.)

blends using 10, 20, and 50 wt% biomass ratios. The study concluded that direct combustion of solid biomass and oil shale is a method of optimal use of these fuels. They simulated PF combustion using thermal gravimetric analysis (TGA) and differential scanning calorimetry (DSC) techniques. The study noted that blending the oil shale with biomass lowers the ignition temperature and improves the ignition process. High VM content of biomass allows easier ignition of blend than that for individual oil shale combustion. Activation energies for combustion blends were less than theoretical values calculated on the basis of arithmetic averages determined from individual fuels. This indicated positive interaction in the reduction of activation energy during combustion. The same level of synergism was not observed in the production of ash content of the blends. This may be due to short synergistic interaction period in the beginning followed by a long less-reactive char combustion period where ash is produced. The study concluded that blend of biomass and oil shale can be co-fired in the existing pulverized firing systems. In general, biomass fuels producing much VM and containing less cellulose are good candidates for co-firing with oil shale. Ten to twenty weight percent biomass blended with oil shale is a good option since it will provide sufficient amount of VM to maintain stability in ignition and combustion [194].

Wang et al. [196] examined the burnout characteristics of oil shale semicoke, corn stalks, and their blends using TGA. The maximum burning rate, temperature of the maximum burning rate, and burnout index were investigated. Batch charges of samples were burned in a laboratory-scale fluidized bed. The effects of particle size, initial bed temperature, airflow rate, and different semicoke percentages on the burnout time were investigated. The results showed that an increase in bed temperature or airflow rate was associated with a decrease in burnout time. On the other hand, an increase in mean particle size increased the burnout time. The mixture affected the burnout time in a significant way. Initially, it increased with the subsequent decline.

While commercial applications of multifuel systems involving oil shale and biomass are not developed as aggressively as the ones for coal and biomass, in January of 2011, Alstom power systems of France negotiated a full turnkey EPC contract in Estonia with Narva Elektrijaamad AS to supply 300 MW steam power plant [193]. The plant implements the CFB boiler technology at a large scale. For reduced environmental impact and increased operational flexibility, the boiler is designed to burn oil shale and peat and to co-fire up to 50% of wood chips. The main challenges of this plant are as follows: (1) oil shale exhibits heavy fouling and corrosion tendency, (2) wood chips have a higher moisture content and low density, which require high-volume storage and suitable handling equipment, and (3) peat shows a high potential for NO_x emission. In order to alleviate high-temperature corrosion problems, the heat exchangers operating at high temperatures are designed to be located outside the fluidized beds. Alstom advanced CFB technology will allow the required 150 mg/Nm³ emissions to be met without an additional denitrification system in spite of high nitrogen content of the fuel. This project is currently in progress with expected starting date of July 2015.

REFERENCES

1. Veijonen, K., Vainikka, P., Jarvinen, T., and Alakangas, E. Biomass co-firing—An efficient way to reduce greenhouse gas emissions. A report from European Bioenergy Networks (EUBIONET), VTT Processes, Altener Industries, Espoo, Finland (March 2003).
2. Al-Mansour, F. and Zuwala, J. An evaluation of biomass co-firing in Europe. *Biomass Bioenergy*, 34(5), 620–629 (May 2010).
3. Leckner, B. Co-combustion: A summary of technology. A report from Department of Energy and Environment, Chalmers University of Technology, Alliance of Global sustainability, Gotenberg, Sweden (2007).
4. Kelsall, G.J. and Laughlin, K. The development of co-gasification for coal/biomass and for other coal/waste mixtures. Final report: European Commission APAS Clean Coal Technology Program on Co-Utilization of Coal, Biomass and Waste, EC–Research project, APAS Contract COAL-CT92-0002, Vols. II and III, Lisbon, Portugal: Final reports (1995).
5. Hey, W., Pelz, W., Romey, I., and Sowa, F. Combined combustion of biomass/sewage sludge and coals of high and low rank in different systems of semi-technical scale. Final report: European Commission APAS Clean Coal Technology Program on Co-Utilization of Coal, Biomass and Waste, APAS Contract COAL-CT92-0002, Vol. II: Final reports (1995).
6. Morgan, D.J. and van de Kamp, W.L. The co-firing of pulverized bituminous coals with biomass and municipal sewage sludge for application to the power generation industry. Final report: European Commission APAS Clean Coal Technology Program on Co-Utilization of Coal, Biomass and Waste, APAS Contract COAL-CT92-0002, Vol. II: Final reports (1995).
7. Rasmussen, I. and Overgaard, P. Recent results concerning co-combustion of biomass and coal. In *Proceedings of the Ninth European Bioenergy Conference*, Vol. I, Copenhagen, Denmark, Pergamon Press, U.K., (June 1996).

8. Tumbull, J.H. Co-combustion of solid biomass. In Paper presented at *NOVEM/IA Symposium Co-Combustion of Biomass*, Nijmege, the Netherlands (November 8, 1995).

9. Probst, H.H. and Wehland, P. Combined combustion of biomass/sewage sludge and coals. Final report: European Commission APAS Clean Coal Technology Program on Co-Utilization of Coal, Biomass and Waste, EC–Research project, APAS Contract COAL-CT92-0002, Vol. II, Lisbon, Portugal: Final reports (1995).

10. Collins, S. Experience base grows for co-firing waste fuels. Power, 137, 45–47 (October 1993).

11. Swanekamp, R. Biomass co-firing technology debuts in recent test burn. *Power*, 139, 51–53 (February 1995).

12. Bestebroer, S.I., Smakman, G.J.J., and Kwant, K.W. Biomass to energy looks promising for the Netherlands. *Power Eng. Int.*, 30, 156 (November/December 1996).

13. Ekmann, J.M., Smouse, S.M., Winslow, J.C., Ramezan, M., and Harding, N.S. In *Co-Firing of Coal and Waste*. Morrison, G.F. (Ed.). IEA Coal Research, London, U.K. (August 1996).

14. EPON Project Summary Brochure. Co-combustion of pulverized wood in the Netherlands (1995).

15. van Loo, S. and Koppejan, J. *Handbook of Biomass Combustion and Co-Firing*, Twente University Press, Enschede, the Netherlands, 348pp., (2002).

16. Tillman, D.A. Biomass co-firing: The technology, the experience, the combustion consequences. *Biomass Bioenergy*, 19(6), 365–384 (December 2000).

17. Davidson, R.M. *Experience of Co-Firing Waste with Coal*, IEA Coal Research, London, U.K., 63pp. (February 1999).

18. Hein, K.R.G. and Bemtgen, J.M. EU clean coal technology—Co-combustion of coal and biomass. *Fuel Process. Technol.*, 54(1–3), 159–169 (March 1998).

19. Hughes, E.E. and Tillman, D.A. Biomass co-firing: Status and prospects 1996. *Fuel Process. Technol.*, 54, 127–142 (1998).

20. Easterly, J.L. and Burnham, M. Overview of biomass and waste fuel resources for power production. *Biomass Bioenergy*, 10(2–3), 79–92 (1996).

21. Sampson, G.R., Richmond, A.P., Brewster, G.A., and Gasbarro, A.F. Co-firing of wood chips with coal in interior Alaska. *Forest Prod. J.*, 41(5), 53–56 (1991).

22. Co-firing biomass with coal. A renewable and alternative energy fact sheet, From College of Agricultural Science, Agricultural Research and Cooperative Extension, Penn State University, College Park, PA (2011).

23. Opportunities and markets for co-utilization of biomass and waste with fossil fuels for power generation. A report from European Union, U.K. (2000); also found in website: http://ec.europa.eu/energy/renewables/studies/doc/bioenergy/2000_opportunities_and_markets.pdf. (accessed 2001)

24. Sami, M., Annamalai, K., and Wooldridge, M. Co-firing of coal and biomass fuel blends. *Prog. Energy Combust. Sci.*, 27, 171–214 (2001).

25. Advantages and limitations of biomass co-combustion in fossil fired power plants. Eight pages report from VGB Power Tech, Essex, Germany (March 2008).

26. Nussbaumer, T. Combustion and co-combustion of biomass: Fundamentals, technologies, and primary measures for emission reductions. *Energy Fuels*, 17, 1510–1521 (2003).

27. Mory, A. and Tauschlitz, J. Co-combustion of biomass in coal fired power plants in Austria. *VGB Power Tech*, 79, 50–55 (1999).

28. Anderl, H., Mory, A., and Zotter, T. BioCoComb-Vergasung von Biomasse und Mitverbrennung von Gas in einem Kohlenstaubkessel. *VGB-Kraftwerkstechnik* 80, 68–75 (2000).

29. Raskin, N., Palonen, J., and Nieminen, J. Power boiler fuel augmentation with a biomass fired atmospheric circulating fluid-bed gasifier. *Biomass Bioenergy* 20(6), 471–481 (2001).

30. Fernando, R. *Experience of Indirect Cofiring of Biomass and Coal*, IEA Clean Coal Centre, London, U.K. (2002).

31. Gustavsson, L. and Leckner, B. Abatement of N_2O emissions from circulating fluidized bed combustion through afterburning. *I&EC Res.*, 34, 1419–1427 (1995).

32. Berge, N., Carlsson, M., Kaliner, P., and Stromberg, B. Reburning of a pulverized coal flame with a LCV gas. Final report: European Commission APAS Clean Coal Technology Program on Co-Utilization of Coal, Biomass and Waste, APAS Contract COAL-CT92-0002, Vol. III, Lisbon, Portugal: Final reports (1995).

33. Frazzitta, S., Annamalai, K., and Sweeten J. Performance of a burner with coal and coal: Manure blends. *J. Propul. Power*, 15(2), 181–186 (1999).

34. Kaer, S.K., Rosendahl, L., and Overgaard, P. Numerical analysis of co-firing coal and straw. In *Proceedings of the Fourth European CFD Conference*, Athens, Greece, pp. 1194–1199 (September 7–11, 1998).

35. Sami, M., Annamalai, K., Dhanapalan, S., and Wooldridge, M. Numerical simulation of blend combustion of coal and feedlot waste in a swirl burner. In *ASME Conference*, Nashville, TN (November 1999).

36. Liu, H. and Gibbs, B.M. Hampartsoumian. The significance of re-burning coal re-burning coal rank on the reduction of NO in drop tube furnace. In *Eighth International Symposium on Transport Phenomena in Combustion*, San Francisco, CA (1995).

37. Dhanaplan, S., Annamalai, K., and Daripa, P. Turbulent combustion modeling coal: Bio-solid blends in a swirl burner. *Energy Week*, ETCE, ASME, IV, 415–423 (January 1997).

38. Smoot, L.D., Hill, S.C., and Xu, H., NOx control through reburning. *Prog. Energy Combust. Sci.*, 24(5), 385–408 (1998).

39. Harding, N.S. and Adams, B.R., Biomass as a re-burning fuel: A specialized co-firing application. *Biomass Bioenergy*, 19, 429–445 (2000).

40. Vilas, E., Skifter, U., Degn Jensen, A., López, C., Maier, J., and Glarborg, P. Experimental and modeling study of biomass re-burning. *Energy Fuels*, 18(5), 1442–1450 (2004).

41. Fan, Z., Zhang, J., Sheng, C., Lin, X., and Xu, Y. Experimental study of NO reduction through re-burning of biogas. *Energy Fuels*, 20(2), 579–582 (2006).

42. "Co-firing" *Wikipedia*, https://en.wikipedia.org/wiki/Cofiring (last modified on March 24, 2015).

43. Berge, N., Carlsson, M., Kallner, P., and Strömberg, B. Reburning of a pulverized coal flame with LVC gas. In *Co-Gasification of Coal/Biomass and Coal/Waste Mixtures* EC–Research project, *(APAS)*, Hein, K.R.G. (Ed.), Lisbon, Portugal. ISBN: 3-928123-15-7, pp. 1992–1994 (1993).

44. Maly, P.M., Zamansky, V.M., Ho, L., and Payne, R. Alternative fuel reburning. *Fuel*, 78(3), 327–334 (1999).

45. Tchapda, A. and Pisupati, S. A review of thermal co-conversion of coal and biomass/ waste. *Energies*, 7, 1098–1148 (2014).

46. Sipila, K. and Rossi, M. *Power Production from Waste and Biomass IV*, Advanced Concepts and Technologies Espoo, Finland, VTT Symposium 222, Organized by VTT, EC DG TREN, IEA Bioenergy, NOVEM, TEKES, KTM, pp.349. April 8–10 2002.

47. Maciejewska, A., Veringa, H., Sanders, J., and Peteves, S. Co-firing of biomass with coal: Constraints and role of biomass pretreatment. European Commission joint research report, EUR 22461 EN, DG JRC, Institute for Energy, Petten, the Netherlands (2006).

48. Bauer, F. Combined combustion of biomass/sewage sludge and coals of high and low rank in different systems of semi-industrial and industrial Scale. Final report: European Commission APAS Clean Coal Technology Program on Co-Utilization of Coal, Biomass and Waste. EC–Research project, APAS Contract COAL-CT92-0002, Vol. II, Lisbon, Portugal: Final reports (1995).

49. Beekes, M.L. Test facilities and trials for pre-treatment and co-combustion. In Paper presented at *NOVEM and TEA International Symposium on Co-Combustion of Bio-fuels*, Nijmegen, the Netherlands (November 1995).

50. Ohlsson, O. Results of combustion and emissions testing when co-firing blends of binder-enhanced densified refuse-derived fuel (b-dRDF) pellets and coal in a 440 MWe cyclone fired combustor, Vol. 1: Test methodology and results. Subcontract Report No. DE94000283, Argonne National Laboratory, Argonne, IL, p. 60 (1994).

51. Järvinen, T. and Alakangas, E. Co-firing of biomass-evaluation of fuel procurement and handling. Selected Existing Plants and Exchange of Information (COFIRING), Part 2, VTT-Energy Altener, Finland, p. 36 (January 2001).

52. Bergman, P.C.A., Boersma, A.R., Zwart, R.W.R., and Kiel, J.H.A. Torrefaction for biomass co-firing in existing coal-fired power stations. ECN Biomass, ECN-C-05-013, Petten, the Netherlands (2005).

53. Yilmazoglu, M., Calisir, T., and Amirabedin, E. Direct and indirect co-combustion of lignite and biomass. In *The Sixth PSU-UNS International Conference on Engineering and Technology* (*ICET-2013*), Nova Sad, Serbia (May 15–17, 2013).

54. Abbas, T., Costen, P., Kandamby, N.H., Lockwood, F.C., and Ou, J.J. The influence of burner injection mode on pulverized coal and bio-solid co-fired flames. *Combust. Flame*, 99, 617–625 (1994).

55. Boylan, D.M. Southern company tests of wood/coal co-firing in pulverised coal units. *Biomass Bioenergy*, 10(2–3), 139 (1996).

56. Gast, C.H. and Visser, J.G. The co-combustion of waste wood with coal. Effects on emissions and fly-ash composition. Final report: European Commission APAS Clean Coal Technology Programme on Co-Utilisation of Coal, Biomass and Waste, EC–Research project, APAS Contract COAL-CT92-0002, Vol. II, Lisbon, Portugal: Final reports (1995).

57. Lutge, C. et al. New applications for High Temperature Winkler Gasification (HTW) for thermal treatment of municipal solid waste and sewage sludge in combination with Catalytic Extraction Process (CEP). In *Gasification Technologies Conference,* Pittsburgh, PA (October 1996).

58. Tilman, D.A., Hughes, E., and Gold, B.A. Co-firing of biomass in coal fired boilers: Results of case study analysis. In Paper presented to *First Biomass Conference of the Americas*, Burlington, VT (1993).

59. McGowin, C.R. and Hughes, E.E. Efficient and economical energy recovery from waste by cofiring with coal. In *Clean Energy from Waste and Coal*, ACS Symposium Series No. 515, Khan, M.R. (Ed.) (1992) [Power/Internet, 1996]. (ACS Symposium Series pubs. acs.org/ISBN/97808441227736).
    (Developed from a symposium sponsored by the division of fuel chemistry at the 202nd national meeting of ACS, New york, [August 25–30, 1991]).

60. McGowin, C.R. and Wiltsee, G.A. Strategic analysis of biomass and waste fuels for electric power generation. *Biomass Bioenergy*, 10(2–3), 167 (1996).

61. Sundermann, B., Rubach, T., and Rensch, H.P. Feasibility study on the co-combustion of coal/biomass/sewage sludge and municipal solid waste for plants with 5 and 20 MW thermal power using fluidized bed technology. Final report: European Commission Research project APAS Clean Coal Technology Program on Co-Utilization of Coal, Biomass and Waste, APAS Contract COAL-CT92-0002, Vol. II, Lisbon, Portugal: Final reports (1995).

62. Abbas, T., Costen, P., Glaser, K., Hassan, S., Lockwood, F., and Ou, J.-J. Combined combustion of biomass, municipal sewage sludge and coal in a pulverised fuel plant. Final report: European Commission Research project APAS Clean Coal Technology Programme on Coutilisation of Coal, Biomass and Waste, APAS Contract COAL-CT92-0002, Vol. II, Lisbon, Portugal: Final reports (1995).

63. Meschgbiz, A. and Krumbeck, M. Combined combustion of biomass and brown coal in a pulverized fuel and fluidised bed combustion plant. Final report: European Commission Research project APAS Clean Coal Technology Program on Co-Utilization of Coal, Biomass and Waste, APAS Contract COAL-CT92-0002, Vol. II, Lisbon, Portugal: Final reports (1995).

64. Andries, J., Verloop, M., and Hein, K. Co-combustion of coal and biomass in a pressurized bubbling fluidized bed. In *Proceedings of the 14th International Conference on Fluidized Bed Combustion*, Vol. 1, Vancouver, British Columbia, Canada, pp. 313–320 (May 11–14, 1997).

65. Brouwer, J., Owens, W.D., Harding, S., Heap, M.P., and Pershing, D.W. Co-firing waste biofuels and coal for emissions reduction. In *Proceedings of the Second Biomass Conference of the Americas*, Portland, OR, pp. 390–399 (August 1995).

66. Aerts, D.J., Bryden, K.M., Hoerning, J.M., and Ragland, K.W. Co-firing switchgrass in a 50 MW pulverized coal boiler. In *Proceedings of the 1997 59th Annual American Power Conference*, Chicago, IL, Vol. 59(2), pp. 1180–1185 (1997).

67. Christensen, J. and Jespersen, P. Straw-firing tests at Amager and Kyndby power stations. Biomass for energy and the environment. In *Proceedings of the Ninth European Bioenergy Conference*, Copenhagen, Denmark, Vol. 2, pp. 1013–1018 (June 24–27, 1996).

68. Dyjakon, A., Cieplk, M., and van de Kamp, W. Deposition probe for on-line monitoring /investigation of slagging and fouling processes in the PF boilers. In *Proceedings of the International Technical Conference*, Vol. 35 (2, 138) (2010).

69. Siegel, V., Schweitzer, B., Spliethoff, H., and Hein, K.R.G. Preparation and co-combustion of cereals with hard coal in a 500 kW pulverized-fuel test unit. Biomass for energy and the environment. In *Proceedings of the Ninth European Bioenergy Conference*, Copenhagen, Denmark, Vol. 2, pp. 1027–1032 (June 24–27, 1996).

70. Kierkegaard Petersen, S.C.K. and Hansen, K.I. Co-firing of straw and design study and test co-firing of wood in pulverized coal fired boiler I/S Vestkdraft, Unit 1. Final report: European Commission Research project APAS Clean Coal Technology Program on Co-Utilization of Coal, Biomass and Waste, APAS Contract COAL-CT92-0002, Vol. II, Lisbon, Portugal: Final report 1995.

71. Van Doorn, J., Bruyn, P. and Vermeij, P. Combined combustion of biomass, municipal sewage sludge and coal in an atmospheric fluidized bed installation. In *Biomass for Energy and the Environment, Proceedings of the Ninth European Bioenergy Conference*, Copenhagen, Denmark, Vol. 2, pp. 1007–1012 (June 24–27, 1996).

72. Van Doorn, J., Bruyn, P., and Vermeij, P. Combined combustion of biomass, municipal sewage sludge and coal in an atmospheric fluidized bed installation. EC– Research project APAS Clean Coal Technology Program. v2. Lisbon, Portugal (1996) ISBN: 3-928123-16-5.

73. Li, S., Wu, A., Deng, S., and Pan, W. Effect of co-combustion of chicken litter and coal on emissions in a laboratory scale fluidized bed combustor. *Fuel Process. Technol.*, 89, 7–12 (2008).

74. Fahlstedt, I., Lindman, E., Lindberg, T., and Anderson, J. Co-firing of biomass and coal in a pressurized fluidized bed combined cycle. Results of pilot plant studies. In *Proceedings of the 14th International Conference on Fluidized Bed Combustion*, Vancouver, British Columbia, Canada, Vol. 1, pp. 295–299 (1997).

75. Aho, M., Skrifvars, B.-J., Yrjas, P., Veijonen, K., Taipale, R., Lybeck, E., Lauren, T., and Hupa, M. Benefits of biomass and coal co-combustion in fluidized bed boilers. In *Proceedings of the 12th European Biomass Conference*, Amsterdam, the Netherlands, Vol. I, pp. 456–460 (June 17–21, 2002).

76. Aho, M. and Silvennoinen, J. Preventing chlorine deposition on heat transfer surfaces with aluminium silicon rich biomass residue and additive. *Fuel*, 83(10), 1299–1305 (2005).

77. Aho, M. and Ferrer, E. Importance of coal ash composition in protecting the boiler against chlorine deposition during combustion of chlorine-rich biomass. *Fuel*, 84(2–3), 201–212 (2005).

78. Gulyurtlu, I., Boavida, D., Abelha, P., Lopes, M.H., and Cabrita, I. Co-combustion of coal and meat and bone meal. *Fuel*, 84, 2137–2148 (2005).

79. Gulyurtlu, I., Crujeira, A.T., Abelha, P., and Cabrita, I. Measurements of dioxin emissions during co-firing in a fluidised bed. *Fuel*, 86, 2090–2100 (2007).

80. Kicherer, A., Gerhardt, Th., Spliethoff, H., and Hein, K.R.G. Co-combustion of biomass/sewage sludge with hard coal in a pulverized fuel semi-industrial test rig. Final report: European Commission APAS Clean Coal Technology Program on Co-Utilization of Coal, Biomass and Waste, APAS Contract COAL-CT92-0002, p. 5, Vol. II: Final report: Lisbon, Portugal (1995).

81. Armesto, L., Veijonen, K., Bahillo, A., Cabanillas, A., Plumed, A., and Salvador, L. Co-combustion of coal and biomass wastes in fluidized bed. In *16th International Conference on Fluidized bed Combustion*, Reno, NV, ASME, 12pp. (May 13–16, 2001).

82. Hiltunen, M. EU waste incineration and LCP directives, co-firing and practical examples in fluidized-bed boilers/power plants. In *Power Production from Waste and Biomass IV*, Espoo, Finland, VTT Symposium 222, pp. 177–184 (1998).

83. Armesto, L., Cabanillas, A., Bahillo, A., Segovia, J.J., Escalada, R., Martinez, J.M., and Carrasco, J.E. Coal and biomass co-combustion on fluidized bed: Comparison of circulating and bubbling fluidized bed technologies. In *Proceedings of the 14th International Conference on Fluidized Bed Combustion*, Vancouver, British Columbia, Canada, Vol. 1, pp. 301–309 (1997).

84. Rasmussen, I. and Clausen, J.C. ELSAM strategy of firing biosolid in CFB power plants. In *Proceedings of the 13th International Conference on Fluidized Bed Combustion*, Orlando, FL, Vol. 1, pp. 557–563 (May 7–10, 1995).

85. Thermoflex, Thermoflow Inc., Sadbury, MA. (2013).

86. Gayan, P., Adanez, J., de Diego, L., Garcia-Labiano, F., Cabanillas, A., Bahillo, A., Aho, M., and Veijonen, K. Circulating fluidized bed co-combustion of coal and biomass. *Fuel* (Enviado para publication), 83, 277–286 (2004).

87. Lind, T. *Ash Formation in Circulating Fluidized Bed Combustion of Coal and Solid Biomass*, Technical Research Centre of Finland, VTT Publications 378, Espoo, Finland, 80pp. + app. 83pp. (1999).

88. Desroches-Ducarne, E., Marty, E., Martin, G., and Delfosse, L. Co-combustion of coal and municipal solid waste in a circulating fluidized bed. *Fuel*, 77(12), 1311–1315 (1998).

89. Tsai, M.-Y., Wu, K.-T., Huang, C.-C., and Lee, H.-T. Co-firing of paper sludge and coal in an industrial circulating fluidized bed boiler. *Waste Manage.*, 22(4), 439–442 (2002).

90. Andries, J., Vegelin, R.J., and Verloop, C.M. Co-combustion of biomass and coal in a pressurized fluidized bed combustor. Final report: European Commission Research project APAS Clean Coal Technology Program on Co-Utilization of Coal, Biomass and Waste, APAS Contract COAL-CT92-0002. Lisbon, Portugal: Final reports (1995).

91. Kalembkiewicz, J. and Chmielarz, U. Effects of biomass co-combustion with coal on functional speciation and mobility of heavy metals in industrial ash. *Polish J. Environ. Stud.*, 22(3), 741–747 (2013).

92. Dayton, D., Turk, B., and Gupta, R. Syngas cleanup, conditioning and utilization. In *Thermochemical Processing of Biomass: Conversion into Fuels, Chemicals and Power*, 1st edn., Brown, R.C. (Ed.), John Wiley & Sons, New York, pp. 78–123 (2011) (Chapter 4).

93. Ratafia-Brown, J., Skone, T., Rutkowski, M., and Cobb, J.T. Assessment of technologies for co-converting coal and biomass to a clean syngas. Task 2 report (RDS), Department of Energy, Washington DC (May 15, 2007).

94. Equipment design and cost estimation for small modular biomass systems, synthesis gas cleanup, and oxygen separation equipment, Task 2: Gas cleanup design and cost estimates: Wood feedstock. NREL/DOE Subcontract Report NREL/SR-510-39945, Nexant Inc., San Francisco, CA (May 2006).

95. Air Products Tees Valley Renewable Energy Facility, EA/EPR/JP3331HK/A001, Environmental Permit Application, Air Products PLC, Surrey, U.K. (May, 2011).

96. Bertocci, A. and Patterson, R. Wet scrubber technology for controlling biomass gasification emissions. In *IT3'07 Conference*, Phoenix, AZ (May 14–18, 2007).

97. Bertocci, A. Wet scrubbers for gasifier gas cleaning. In *IT3'07 Conference*, Phoenix, AZ, Paper No. 49 (May 14–18, 2007).

98. Energy evolved: Clean, sustainable energy recovery through plasma gasification. Westinghouse Plasma Corporation, A Division of Alter NRG Corp., Madison, PA; also Calgary, Alberta, Canada (February 2011). http://www.alternrg.com.

99. Krishnan, G., Wood, B., Tong, G., and Kothari, V. Removal of hydrogen chloride vapor from high temperature coal gases, Abstracts of the papers for American Chemical Society. *Fuel*, 195, 32 (1988).

100. Gil, J., Caballero, M., Martin, J. et al. Biomass gasification with air in a fluidized bed: Effect of the in-bed use of dolomite under different operating conditions. *I&EC Res.*, 38(11), 4226–4235 (1999).

101. Belén Folgueras, M., María Díaz, R., and Xiberta, J. Sulfur retention during co-combustion of coal and sewage sludge. *Fuel*, 83(10), 1315–1322 (2004).

102. Leckner, B. and Kilpinen, P. NO emission from co-combustion of coal and wood—A discussion on modeling. In *Opening Lecture of the Joint Meeting of the Scandinavian-Nordic and Italian Sections of the Combustion Institute*, Ischia, Italy (September 2003).

103. Åmand, L.-E., Leckner, B., Lücke, K., and Werther, J. Advanced air staging techniques to improve fuel flexibility, reliability and emissions in fluidized bed co-combustion. Värmeforsk Report 751, Swedish Energy Agency, Sweden (2001). http://www.varmeforsk.se/reports?action=show&id=4440.

104. Elled, A.L., Åmand, L.-E., Leckner, B., and Andersson, B.-Å. Influence of phosphorus on sulphur capture during co-firing of sewage sludge with wood or bark in a fluidized bed. *Fuel*, 85(12–13), 1671–1678 (2006).

105. Kling, Å., Myringer, Å., Eskilsson, D., Aurell, J., and Marklund, S. SCR at co-combustion of biofuels and waste fuels (in Swedish with English summary). Värmeforsk Report 932 (F4-220) Swedish Energy Agency, Sweden, (2005). http://www.varmeforsk.se/reports?action=show&id=4440.

106. Griffin, R.D. A new theory of dioxin formation in municipal solid waste combustion. *Chemosphere*, 15(9–12), 1987–1990 (1986).

107. Raghunathan, K. and Gullett, B.K. Role of sulfur in reducing PCDD and PCDF formation. *Environ. Sci. Technol.*, 30(6), 1827–1834 (1996).

108. Gullett, B.K., Bruce, K.R., and Beach, L.O. Effect of sulfur-dioxide on the formation mechanism of polychlorinated dibenzo-dioxin and dibenzo-furan in municipal waste combustors. *Environ. Sci. Technol.*, 26(10), 1938–1943 (1992).

109. Geiger, T., Hagenmaier, H., Hartmann, E., Römer, R., and Seifert, H. Effect of sulfur on the formation of dioxins and furans in sewage sludge incineration. *VGB Kraftwerkstechnik*, 72(2), 153–158 (1992).

110. Xie, Y., Xie, W., Liu, K., Pan, W.-P., and Riley, J.T. The effect of sulfur dioxide on the formation of molecular chlorine during co-combustion of fuels. *Energy Fuels*, 14(3), 597–602 (2000).

111. Gullet, B. and Raghunathan, K. Observations on the effect of process parameters on dioxin/furan yield in municipal waste and coal systems. *Chemosphere*, 34(4–5), 1027–1032 (1997).

112. Luthe, C., Strang, A., Uloth, V., Karidio, I., Prescot, B., and Wearing, J. Sulfur addition to control dioxins formation in salt-laden power boilers. *Pulp Paper Canada*, 99(11), 48–52 (1998).

113. Anthony, E.J., Jia, L., and Granatstein, D.L. Dioxin and furan formation in FBC boilers. *Environ. Sci. Technol.*, 35(14), 3002–3007 (2001).

114. Pedersen, L.S., Morgan, D.J., van de Kamp, W.L., Christensen, J., Jespersen, P., and Dam-Johansen, K. Effects on SOx and NOx emissions by co-firing straw and pulverized coal. *Energy Fuels*, 11(2), 439–446 (1997).

115. Leckner, B., Åmand, L.-E., Luecke, K., and Werther, J. Gaseous emissions from co-combustion of sewage sludge and coal/wood in fluidized bed. *Fuel*, 83(4–5), 477–486 (2004).

116. Dayton, D.C., Belle-Oudry, D., and Nordin, A. Effect of coal minerals on chlorine and alkali metals released during biomass/coal co-firing. *Energy Fuels*, 13(6), 1203–1211 (1999).

117. Åmand, L.E., Leckner, B., Eskilsson, D., and Tullin, C. Deposits on heat transfer tubes during co-combustion of biofuels and sewage sludge. *Fuel*, 85(10–11), 1313–1322 (2006).

118. Jiménez, S. and Ballester, J. Effect of co-firing on the properties of submicron aerosols from biomass combustion. *Proc. Combust. Inst.*, 30(2), 2965–2972 (2005).

119. Seames, W.S., Fernandez, A., and Wendt, J.O.L. A study of fine particulate emissions from combustion of treated pulverized municipal sewage sludge. *Environ. Sci. Technol.*, 36(12), 2772–2776 (2002).

120. Lundholm, K., Nordin, A., Öhman M., and Boström D. Reduced bed agglomeration by co-combustion biomass with peat fuels in a fluidized bed. *Energy Fuels*, 19(6), 2273–2278 (2005).

121. Yan, R., Gauthier, D., Flammant, G., and Badie, J.M. Thermodynamic study of the behavior of minor coal elements and their affinities to sulfur during coal combustion. *Fuel*, 78(15), 1817–1829 (1999).

122. Miettinen-Westberg, H., Byström, M., and Leckner, B. Distribution of potassium, chlorine and sulfur between solid and vapor phases during combustion of wood chips and coal. *Energy Fuels*, 17(1), 18–28 (2003).

123. Krause, H.H. High-temperature corrosion problems in waste incineration systems. *J. Mater. Energy Syst.*, 7(4), 322–332 (1986).

124. Robinson, A.L., Junker, H., and Baxter, L.L. Pilot scale investigation of the influence of coal–biomass co-firing on ash deposition. *Energy Fuels*, 16(2), 343–355 (2002).

125. Skrifvars, B.-J., Backman, R., Hupa, M., Sfiris, G., Åbyhammar, T., and Lyngfelt, A. Ash behavior in a CFB boiler during combustion of coal, peat or wood. *Fuel*, 77(1–2), 65–70 (1998).

126. Andersen, K.H., Frandsen, F.J., Hansen, P.F.B., Wieck-Hansen, K., Rasmussen, I., Overgaard, P., and Dam-Johansen, K. Deposit formation in a 150 MW utility PF-boiler during co-combustion of coal and straw. *Energy Fuels*, 14(4), 765–780 (2000).

127. Otsuka, N. Effects of fuel impurities on the fireside corrosion of boiler tubes in advanced power generating systems—A thermodynamic calculation of deposit chemistry. *Corros. Sci.*, 44(2), 265–283 (2002).

128. Wei, X., Lopez, C., von Puttkamer, T., Schnell, U., Unterberger, S., and Hein, K.R.G. Assessment of chlorine-alkali-mineral interactions during co-combustion of coal and straw. *Energy Fuels*, 16(5), 1095–1108 (2002).

129. Spliethoff, H. and Hein, K.R.G. Effect of co-combustion of biomass on emissions in pulverized fuel furnaces. *Fuel Process. Technol.*, 54(1–3), 189–205 (1998).

130. Ninomiya, Y., Zhang, L., Sakano, T., Kanaoka, C., and Masui, M. Transformation of mineral and emission of particulate matters during co-combustion of coal with sewage sludge. *Fuel*, 83(6), 751–764 (2004).

131. Salmenoja, K. Field and laboratory studies on chlorine-induced super heater corrosion in boilers fired with biofuels. Academic Dissertation. Report 00-1. Åbo Akademi, Faculty of Chemical Engineering, Åbo, Finland, 102pp. (February 2000).

132. Leckner, B. and Karlsson, M. Emissions from circulating fluidized bed combustion of mixtures of wood and coal. Department of Energy, Energy Information Administration, Annual Energy Outlook 1998, DOE/EIA-0383(98), Washington, DC (December 1997).

133. Biswas, B.K. and Essenhigh, R.H. The problem of smoke formation and its control, air pollution and its control. AIChE Symposium Series No. 126, Vol. 68, pp. 207–215 (1972).

134. Kuo, K. *Principles of Combustion*, Wiley, New York (1986).

135. Glassman, I. *Combustion*, 2nd edn., Academic Press, New York (1987).

136. Essenhigh, R.H. Fundamentals of coal combustion. In *Chemistry of Coal Utilization*, Vol. II, Elliott, M.A. (Ed.). Wiley/Interscience, New York (1979), pp. 1153–1312, Chapter 19.

137. Tillman, D.A., Rossi, A.J., and Kitto, W.D. *Wood Combustion: Principles, Processes and Economics*, Academic Press, New York (1981).

138. Pan, Y.G., Velo, E., and Puigjaner, L. Pyrolysis of blends of biomass with poor coals. *Fuel*, 75(4), 412–418 (1996).

139. Baxter, L.L. Ash deposition during bio-solid and coal combustion. A mechanistic approach. *Biomass Bioenergy*, 4(2), 85–102 (1993).

140. Hansen, L.A., Michelsen, H.P., and Dam-Johansen, K. Alkali metals in a coal and biosolid fired CFBC-measurements and thermodynamic modeling. In *Proceedings of the 13th International Conference on Fluidized Bed Combustion*, Orlando, FL, Vol. 1, pp. 39–48 (May 7–10, 1995).

141. Annamalai, K. and Ryan, W. Interactive processes in gasification and combustion-II. Isolated carbon, coal and porous char particles. *Prog. Energy Combust. Sci.*, 19, 383–446 (1993).

142. Åmand, L.-E. and Leckner, B. Reduction of Emissions of sulphur and chlorine from combustion of high volatile waste fuels (sludge) in fluidised bed. In *Proceedings of Third i-CIPEC, 2004*, K. Cen, J. Yang, and Y. Chi (Eds.), International Academic Publishers, Beijing, China, pp. 476–481 (2004).

143. Kim, M.-R., Jang, J.-G., Yoa, S.-J., Kim, I.-K., and Lee, J.-K. A mechanism of CaO-SO$_2$ reaction during fluidized bed sludge incineration. *J. Chem. Eng. Japan*, 38(11), 883–886 (2005).

144. Hein, K.R.G., Heinzel, T., Kicherer, A., and Spliethoff, H. Deposit formation during the co-combustion of coal–biomass blends. In *Proceedings of Applications of Advanced Technology to Ash-Related Problems in Boilers*, Baxter, L. and DeSollar, R. (Eds.). Plenum Press, New York (1996).

145. Adlhoch, W., Hoffinann, H., Klossek, K., and Schiffer, H.-P. Use of sewage sludge in the HTW process. Final report: European Commission APAS Clean Coal Technology Program on Co-Utilization of Coal, Biomass and Waste, EC– Research project APAS Contract COAL-CT92-0002, Vol. III, Lisbon, Portugal: Final reports (1995).

146. Wieck-Hansen, K., Overgaard, P., and Larsen, O. Co-firing coal and straw in a 150 MWe power boiler experiences. *Biomass Bioenergy*, 19, 395–409 (2000).

147. Hansen, P., Andersen K., Wieck-Hansen K. et al. Co-firing straw and coal in a 150 MWe utility boiler: In situ measurements. *Fuel Process. Technol.*, 54, (1–3), 207–225 (1996).

148. Gold, B.A. and Tillman, D.A. Wood co-firing evaluation at TVA power plants. *Biomass Bioenergy*, 10(2–3), 71–78 (1996).

149. Hansen, P.F.B., Andersen, K.H., Wieck-Hansen, K., Overgaard, P., Rasmussen, I., Frandsen, F.J., Hansen, L.A., and Dam-Johansen, K. Co-firing straw and coal in a 150 MWe utility boiler: In situ measurements. *Fuel Process. Technol.*, 54, 207–225 (1998).

150. Thorson, O. Modern boiler data for co-combustion (in Swedish with English summary), Värmeforsk 868, Project No. A4-217, Swedish Energy Agency, Sweden (2004).

151. Savolainen, K. Co-firing of biomass in coal-fired utility boilers. *Appl. Energy*, 74(3–4), 369–338 (2003).

152. Skoglund, B. Six years experience with Sweden's largest CFB boiler. In *Proceedings of 14th International Conference on FBC*, Preto, F.D.S. (Ed.), ASME, New York, pp. 47–56 (1997).

153. Kokko, A. and Nylund, M. Biomass and coal co-combustion in utility scale—Operating experience of Alholmens Kraft. In *Proceedings of 18th International Conference on FBC*, ASME, New York, ISBN: 0-7918-3755-6, Paper FBC2005-78035 (2005).

154. Koppejan, J. Introduction and overview of technologies applied worldwide. In *Second World Biomass Conference, Workshop 4: Co-firing*, Rome, Italy (2004).

155. Wieck-Hansen, K. and Sander, B. 10 Years experience with co-firing straw and coal as main fuels together with different types of biomasses in a CFB boiler in Grenå, Denmark. *VGB Power Tech*, 83(10), 64–67 (2003).

156. Gani, A., Morishita, K., Nishikawa, K., and Naruse, I. Characteristics of co-combustion of low-rank coal with biomass. *Energy Fuels*, 19(4), 1652–1659 (2005).

157. Ballester, J., Barroso, J., Cerecedo, L.M., and Ichaso, R. Comparative study of semi-industrial-scale flames of pulverized coals and biomass. *Combust. Flame*, 141(3), 204–215 (2005).

158. Buck, P. and Triebel, W. Operation experiences with co-combustion of municipal sewage sludge in the coal-fired power plant Heilbronn (in German). *VGB Kraftwerkstechnik*, 80(12), 82–87 (2000).

159. Luts, D., Devoldere, K., Laethem, B., Bartholomeeusen, W., and Ockier, P. Co-incineration of dried sewage sludge in coal-fired power plants: A case study. *Water Sci. Technol.*, 42(9), 259–268 (2000).

160. Sander, B. Full-scale investigation on co-firing of straw. In *Second World Biomass Conference, Workshop 4: Co-firing*, Rome, Italy (2004).

161. Mosbech, H. Biomass use in large power plants. Point of view of a utility. In Paper presented to *Final Conference of European Commission APAS Clean Coal Technology Program on Co-Utilization of Coal, Biomass and Waste*, Lisbon, Portugal (November 1994).

162. Biomass cofiring: A renewable alternative for utilities. Biopower Factsheet, a document produced for the U.S. Department of Energy (DOE) by the National Renewable Energy Laboratory, a DOE National Laboratory. DOE/GO-102000-1055, Golden, Colorado (June 2000).

163. Biomass cofiring in coal fired boilers. Federal technology alert, a new technology demonstration publication, DOE/EE-0288, Federal Energy Management Program, Produced for the U.S. Department of Energy, Energy Efficiency and Renewable Energy, by the National Renewable Energy Laboratory, a DOE National Laboratory, Golden, Colorado (June 2004).

164. Biomass co-firing. A final phase III report prepared by CCPC (Canadian clean power coalition) Technical Committee, Appendix C, pp. C01-C30 (November 2011). www.canadiancleanpowercoalition.com.

165. Shah, Y., *Water for Energy and Fuel Production*, CRC Press, Taylor & Francis Group, New York (May 2014).

166. Exelby, D.R. and Davison, J.E. A technical and economic assessment of co-gasification of straw or sewage sludge with coal in conceptual commercial plants. Confidential CRE Report. Assessment Report No. 94/10, 48 EPRI Journal CA (May/June, 1996).

167. Krebs, K.H. A prospective assessment on the role of renewables energies in response to the CO_2 problem. Report to the Forward Studies Unit of the CEC, Institute for Prospective Technological Studies, Seville, Spain (1995).

168. Tillman D.A., Hughes, E., and Plasynski, S. Commercializing biomass-coal co-firing: The process, status, and prospect. In *Proceedings of the Pittsburg Coal Conference*, Pittsburgh, PA (October 11–13, 1999).

169. Cuputo, J. and Macker, J. Biomass co-firing: A transition to low carbon future– Issue brief, Environment and Energy Study Institute, Werner C. and Vaugaan, E, (Eds.), Washington, D.C.20036 (March, 2009).

170. Patterson, W. *Power from Plants: The Global Implications of New Technologies for Electricity from Biomass*, Earthscan Publications and Royal Institute of International Affairs, London, U.K. (1994); Johansson, T.B., Kelly, H., Reddy, A.K.N., and Williams, R.H. *Renewable Energy: Sources for Fuel and Electricity*, Island Press/Earthscan, Washington, DC (1993).

171. Baxter, L. Biomass-coal co-combustion: Opportunity for affordable renewable energy. *Fuel*, 84(10), 1295–1302 (2005).

172. Goldberg, P. Impacts of co-firing biomass with fossil fuels. Report No. 2001-EERC-08-03; Final report (for the period April 1, 1999, through March 31, 2001) task 3.4, Cooperative agreement No DE-FC26-98FT40320UND4255 U.S. Department of Energy, National Energy Technology Laboratory, Pittsburgh, PA (2001).

173. Khodakov, G. Coal–water suspensions in power engineering. *Thermal Eng.*, 54(1), 36–47 (2007).

174. Wibberley, L., Palfreyman, D., and Scaife, P. Efficient use of coal water fuels. Technology Assessment Report 74, CSIRO Energy Technology, Mayfield West, New South Wales, Australia (April 2008).

175. Wibberley, L. Alternative pathways to low emission electricity. In *COAL21 Conference*, Crown Plaza, Hunter Valley, CA (September 18–19, 2007).

176. Schwartz, E. Process for the production of low-polluting CWF. In *BASF, Coal–Water Mixtures*, Sens, P. and Wilkinson, J. (Eds.), Elsevier, New York (1987).

177. Miller, B. Penn State's coal–water slurry fuel program. Earth and Mineral Science Energy Institute, College of Earth and Mineral Science, Penn State University, PA.

178. Fellner, C. and Hutson, N., Available and emerging technologies for reducing greenhouse gas emissions from coal-fired electric generating units. Prepared by Office of Air quality Planning and Standards, EPA, Research Triangle, NC (October 2010).

179. Cho, O. and Ku, S., Clean coal technologies in Japan—Technology innovation in the coal industry. Japan Coal Energy Center Report, NEDO, Tokyo, Japan, pp. 1–105 (January 2007).

180. Jaha, M. and Smit, F. Engineering development of advanced physical fine coal cleaning for premium fuel applications, AMAX Research Centre, Golden, Colorado for DOE Contract DE-AC22-92PC92208 (1993).

181. Cui, L., An, L., Gong, W., and Jiang, H. A novel process for preparation of ultra-clean micronized coal by high pressure water jet comminution technique. *J. Fuel*, 86(5–6), 750 (2007).

182. Mazurkiewicz, D. Investigation of ultra-fine coal disintegration effect by high pressure water jet. Rock Mechanics and Explosives Research Center, University of Missouri Rolla, Rolla, MO (2002). http://www.wjta.org/Book%203/4_1_Mazurkiewicz.pdf (accessed 2002).

183. Brooks, P., Clark, K., Langley, J., Lothringer, G., and Waugh, B. UCC as a gas turbine fuel (1999). http://www.australiancoal.csiro.au/pdfs/ucc.pdf (accessed 2000).

184. Steel, K. and Patrick, J. The production of ultra clean coal by sequential leaching with HF followed by HNO_3. *Fuel*, 82, 1917 (2003).

185. Steel, K. and Patrick, J. Regeneration of HF and selective separation of Si(IV) in a process for producing ultra clean coal. *Fuel Process. Technol.*, 86(2), 179 (November 2004).

186. Kashimura, N., Takanohashi, T. and Saito, I., Upgrading the solvent used for the thermal extraction of sub-bituminous coal, "Energy and Fuels, 20, 2063–2066 (2006).

187. Fu, Y.C., Bellas, G., and Joubert, J. Coal–water mixture combustion using oxygen-enriched air and staged firing. ACS preprint, personal communication (2012).

188. Siemon, J. Economic potential of coal–water mixtures. In *ICEASIE8*, IEA Coal Research, London, U.K. (September 1985).

189. Thambimuthu, K. Developments in coal–liquid mixtures. IEA Report, Paris, France (1994).

190. Khodakov, G., Coal water suspension in power engineering, Thermal Engineering, 54(1), 36–47 (December 2006). DOI: 101134/60040601507010027.

191. Ross, D.S., Green, T.K., Mansani, R., and Hum, G.P. Coal conversion in CO/water 2 oxygen loss and conversion Mechanism. *Energy Fuels*, 1, 292 (1987).

192. Dyjakon, A., Cieplk, M., Kallvodova, J., and van de Kamp, W. Laboratory- and full-scale investigations of slagging and fouling during energy production from co-combustion of coal and biomass. In *European Biomass Conference*, Mamburg, Germany, (June 29 – July 3, 209) CD-ROM-Edition, Vol. 17(2) (2009).

193. Le Tohic, F. and Joly, T. Narva power plant—The challenges of firing oil shale and biomass at a large scale. A report from Alston Power Systems, Levallois-Perret, France (2011).

194. Ozgur, E., Miller, S., Miller, B., and Kok, M. Thermal analysis of co-firing of oil shale and biomass fuels. *Oil Shale*, 29(2), 190–201 (2012).

195. Boardman, R., Carrington, R., Hecker, W., and Clayson, R. Co-firing oil shale with coal and other fuels for improved efficiency and multi-pollutant control. In *Thirty-Third International Technical Conference on Coal Utilization and Fuels Systems*, INL report INL/CON-08-14168, Idaho (June 2008).

196. Wang, Q., Xu, H., Liu, H., Jia, C., and Zhao, W. Co-combustion performance of oil shale semi-coke with corn stalk. *Energy Procedia*, 17(part A), 861–868 (2012).

197. "Dunkirk," *Wikipedia*, http://en.wikipedia.org/wiki/Dunkirk,_New_York (last modified on July 3, 2015).

3 Synthetic Gas Production by Co-Gasification

3.1 INTRODUCTION

Thermochemical transformation of coal, oil, and biomass to synthetic gases has been in commercial practice for a long time. The synthetic gases produced in this way have been the important part of energy portfolio of the world, and they are very important sources for heat, power, transportation fuels, and productions of chemicals and materials. While this type of transformation has so far been restricted to single raw material like coal, oil, or biomass, during the last two decades, more interests in using multiple feedstock like coal and biomass, coal and waste, and biomass and waste have evolved. This chapter assesses our state of knowledge of co-gasification of multifuels [1–13].

The gasification is fundamentally a partial oxidation process where solid or liquid fuels are oxidized in below stoichiometric requirements of oxygen. Like combustion, the process is exothermic and produces a mixture of synthetic fuel gas, which has significant heating value and can be used to generate heat and power in a downstream process if desired. The synthetic gas is also capable of producing liquid fuels and chemicals by processes such as Fischer–Tropsch synthesis, oxysynthesis, and isosynthesis [14]. While as shown in Chapter 2 the process of combustion (complete oxidation) produces heat and flue gas, which largely contains CO_2, water, and other gaseous impurities, the process of gasification produces gaseous fuel, which largely contains, methane, hydrogen, and carbon monoxide along with CO_2, water, and other impurities.

Coal gasification is a well-established industry. Large-scale coal gasification plants exist all over the world. In recent years, these plants have come under close scrutiny due to large emissions of greenhouse gases (GHGs). The modifications of plants to reduce GHG are expensive and difficult to implement. Biomass gasification, on the other hand, is a relatively young industry. Large-scale biomass and waste gasification plants face challenges of unsustainable raw material supply, expensive storage and transportation, and energy-intensive feed preparation. For these reasons, biomass and waste gasification plants are generally of smaller size than commercial coal gasification plants. This does not allow them to take full advantage of economy of scale.

Biomass in general has a high content of hydrogen (H), making it suitable as a blend to low H content coal. Biomass as gasification feedstock, although giving a high hydrogen yield, has the disadvantage of low energy density because of its high oxygen and moisture contents. This shortcoming is compensated for when blended with a higher-energy-content coal. Other challenges such as the seasonal limitation of biomass are somewhat mitigated through co-conversion with coal. The higher tar release (due to excessive volatile release and low gasification temperature from biomass gasification) is also reduced as blending with coal increases the temperature and enhances tar cracking. Blending biomass and coal as feedstock can reduce the shortcomings of each fuel and boost the efficacy of the overall system.

The most attractive benefit of coal and biomass co-gasification is the reduction of GHG emissions and environmental pollution. Biomass, compared to coal, has lower sulfur, N, and heavy metals. It is also carbon neutral if produced sustainably; thus, co-gasification of biomass and coal can make significant contributions in mitigating GHG and other emissions. Compared to carbon capture and storage (CCS), co-gasification of biomass with coal is relatively cost-effective, given the fact that CCS can incur high energy penalty, which can range from 15% to 40% of the energy output [12]. Even if CCS were to be implemented as a means for GHG mitigation, it would not, alone, be able to meet the 50% emission reduction target by 2050 suggested by the International Panel on Climate Change [13].

Just as in the combustion, the use of multifuels for synthetic gas production has gained significant momentum in the last two decades. Some of the arguments for this increased activity are the same as that for combustion, that is, reduced CO_2 emissions, more use of renewable feedstock, and diversion of waste from landfills to more constructive use. For the past two decades, the use of coal, biomass, and waste to generate synthetic gas has gained more popularity than simple combustion of these materials due to the versatility of the applications of synthetic gas in energy industry. The synthetic gas can be used for heat and power but it can also generate liquid fuels and chemicals that are needed for transportation and in industrial sectors. We also have an infrastructure for the natural gas, which can be used for the synthetic gas. Thus, gaseous fuel is one of the most important and versatile sources for heat, power, liquid fuels, and chemicals. It has also more sustainable future because it can be obtained from natural (conventional and unconventional) sources and it can be produced from coal, oil, biomass, and waste [1–13].

The use of coal and biomass or coal and waste can be assessed in a number of different ways. In co-gasification, biomass can be considered as a partial feed (co-feed) to large-scale coal-fed gasification process to avoid the key problems of stand-alone biomass gasification plants such as high cost, low efficiency, and shutdown risks if there is a biomass shortage. While large-scale refineries produce 500,000 bbl/day oil, biofuel refineries can only be 1/10th of this size. This size difference applies to both power and fuel productions. One way to avoid all the issues encountered with stand-alone biomass gasification plant is to co-feed biomass in large-scale coal gasification plants. Oxy-co-gasification allows increased efficiency and reduced environmental impact [1–13]. Also, while steam gasification of coal is not thermodynamically favorable, steam gasification of biomass is a very attractive way to produce hydrogen. Thus, co-gasification of coal and biomass in the presence of oxygen and steam appears to be an easier way to increase the role of renewable source of energy in the fossil-dominated industry.

Another way to look at the importance of co-gasification is that it is the method by which the use of coal in energy industry is reduced. This helps environment, particularly for CO_2 emission, and also preserves our valuable coal resources for a long-term purpose. The co-gasification will be more acceptable by the supporters of environment and politicians for the future use of coal. It will also gain more social acceptance for the use of coal.

Yet another positive argument for co-gasification of coal and waste or co-gasification of biomass and waste is that this allows less waste to be diverted to already overcrowded landfills and makes more constructive use of waste. While converting waste into fuel carries enormous acceptance by the proenvironment advocates and politicians, stand-alone waste gasification plants suffer from the same disadvantages that biomass plants do. Coal–waste gasification plants offer many positive benefits. For the past two decades, co-utilization of coal and waste at large scales has received considerably more attention than co-utilization of coal and biomass. One example is the large-scale Lurgi plant in the United Kingdom using coal and waste. The waste investigated has been municipal solid waste (MSW) that has had minimal presorting or refuse-derived fuel (RDF) that has had significant pretreatment such as mechanical screening, shredding, and torrefaction [15].

Just as for combustion of multifuels, the gasification of multifuels offers several advantages and some drawbacks. In this chapter, we focus on the multifuel systems that involve coal, biomass, wastes, and water.

3.1.1 Advantages

1. The use of biomass and/or waste with coal reduces the use of coal in the synthetic fuel industry. This reduction also produces less CO_2, which fits well for the green environment objectives.
2. The use of waste reduces the need for landfill space.
3. The use of coal/biomass mixture allows the use of biomass for the production of synthetic fuels without starting new stand-alone plants for biomass. In general, stand-alone plants will be more expensive than retrofitting existing coal gasification plants to accommodate biomass and waste.
4. The use of mixed fuel plant will allow the thermochemical conversion plants to be on the larger scale, thus using the economy of scale for the synfuel production. The stand-alone biomass and/or waste plants for synfuel production cannot be operated on large scale due to difficulty in providing feedstock in a sustainable manner. Unlike coal, biomass and waste are difficult to transport and store. Also, they require much more complex and expensive feed preparation systems for gasification. The long-distance transportation of biomass and waste can also be very expensive because of its low density, fibrous nature, and high moisture content.
5. In many cases, if properly chosen, mixture can provide synergy among its constituents, which may improve the conversion rate and quality of the product.
6. As shown in Section 3.2, the mixture of coal and biomass/waste can be processed in a number of different ways, which gives more flexibility to the overall operations. Due to its flexibility the process has higher long-term sustainability.

7. If mixture is carefully chosen, the quality of gaseous, liquid, or solid products can be significantly improved over single component feed. Hydrogen production can be optimized. The overall thermal efficiency can also be increased.

8. Co-gasification can help improving the public attitude displayed toward use of renewable feedstock and the development of multifuel supply network.

9. The project risk reduction by co-gasification process as opposed to stand-alone biomass process may provide enough security for project financiers [10].

10. Mixture provides more stable and reliable feed supply to the gasification process. Thus, mixture provides more security, less risk, and potential for large-scale operation.

3.1.2 DISADVANTAGES

1. The processing of mixtures will require different feed pretreatment processes (may be one for each fuel), different designs of gasifiers, or other processing equipment and downstream product cleaning processes. Separate equipment for all three cases may become expensive.

2. The feeding of biomass and waste to the gasifier may become difficult depending on the nature of biomass and waste. Depending on the type of gasification technology employed, pretreatment of biomass may be complex and energy intensive. For example, the particle size required by entrained-flow gasifier is less than 1 mm, which will require energy-intensive milling. The circulating fluidized bed, however, may not require such small-diameter particles.

3. An effective integration of biomass feeding with coal feeding may be difficult. The capacity of the biomass handling and feed system may be quite large and expensive compared to the energy content of biomass fed to the gasifier.

4. There may be impacts on the gasifier injection system relative to feeding a single mixture of coal–biomass versus utilizing separate injectors/burners. The dry feeding of biomass against the pressure (in case of pressurized thermochemical processes) may become challenging. Two separate feed injectors versus single feed injector may affect the gasifier performance.

5. Not much is known about gasification and pyrolysis of mixed fuel with different burning characteristics, particularly when physical and chemical properties of the biomaterials and waste are significantly different from that of coal.

6. Depending on the N, S, Cl, alkali, and heavy metal contents of the two or more feedstock, more treatments of gaseous emissions may be required. Generally, coal contains less chlorine and alkali and heavy metals compared to some biomass like straw and switchgrass. The high chlorine content in some biomass and waste may cause more corrosion and erosion problems in the process equipment. These would require more assessment

of the use of right materials of construction for process equipment during processing of mixtures.

7. Co-gasification of coal and different types of biomass may negatively impact the slagging behavior of the combined ash in the gasifier, depending on its design and type. The slagging behavior will also be affected by the nature and rank of coal. The ash composition of the mixed gasification process may be significantly different from the one for pure coal alone. The significant amount of inorganic metal content in the ash (which may be present in biomass) may limit its downstream usefulness.

8. Generally, biomass gasifies at a lower temperature than coal. When both are gasified together, selection of a temperature for the optimum reactor performance may be challenging.

9. For combined thermochemical transformation, it is good to have uniform particle size for all constituents of the feedstock. This may not be possible when using biomass like straw or miscanthus or shredded waste or the feed with different shape particles. Predictions and control of the reactor performance for nonuniform particle size and shape may become challenging.

10. Co-gasification of biomass may contribute to tar and oil formation in the raw syngas depending on the type of the gasification technology employed. This requires their separations from the syngas or their further conversion. Tar and oils can be avoided by operating gasification reactor at high temperatures (>1200°C). The tar and oils can also be converted to commodity products like phenols, an approach being taken by Dakota Gasification Company for lignite-based gasification plant.

3.2 CO-GASIFICATION OF COAL AND BIOMASS

3.2.1 GASIFICATION REACTORS AND THEIR SUITABILITY

Unlike in combustion based boiler operation where major emphasis in maximizing combustion efficiency, in co-gasification proper gasifier operation can affect both gasification efficiency as well as product distributions. The gasifier performance for gas composition and temperature must be maintained within appropriate ranges, and fuel/gas conversion must be optimized by adjusting two input variables, oxygen and steam, which are introduced in the gasifier. While gasification temperature cannot be constantly measured, it has to be kept within a certain range to determine cold gas efficiency and maintain safe operation. An error in oxygen feed can cause either very high temperatures that can damage the equipment (in slagging gasifiers) or low temperature that can stop slag flow resulting in reactor blocking. The values of all output variables to a certain extent depend on the control of input variables. These dependencies need to be understood and integrated in the reactor performance. Co-feeding of the biomass and coal depending on their relative quantities and nature can significantly alter the input basis and thus impact gasifier performance and output parameters. In general, high-temperature and high-pressure gasifier operations help to produce tar-free syngas, simplify downstream cleaning steps, and alleviate the need for intermediate compression before downstream usages [1].

TABLE 3.1

Advantages and Disadvantages of Different Gasifier Types

Reactor Type	Advantages	Disadvantages
Updraft	• Mature technology • Can handle high moisture • "No" carbon in ash • Small-scale applications	• High tar yields • Scale limitations • Slagging potential • Generates producer gas
Downdraft	• Small-scale applications • Low particulates and tar	• Scale limitations • Moisture sensitive • Generates producer gas
Fluidized bed/ circulating fluidized bed	• Large-scale applications • More flexibility on particle size • Can produce syngas • Better mixing • Direct or indirect heating	• Medium tar yield • Higher particle loading • Moderate temperatures • Higher residence time and lower throughput
Entrained bed	• Large scale • Short residence time • Oxidative environment • High throughput • High temperature • Negligible tar • Produces syngas	• Large carrier gas • Limited particle size

Sources: Ratafia-Brown, J. et al., Assessment of technologies for co-converting coal and biomass to a clean syngas—Task 2 report (RDS), DOE/NETL-403.01.08 Activity 2 Report, May 10, 2007; Bain, R., *Biomass Gasification Overview*, NREL, Golden, CO, January 28, 2004.

There are fundamentally three types of gasification reactors used for gasification of biomass or coal. These are moving bed reactors (either updraft or downdraft), fluidized bed reactors (either bubbling fluidized bed or recirculating fluidized bed), and entrained bed reactors. As shown in Table 3.1, each of these types of reactors has advantages and disadvantages [1,11].

3.2.1.1 Brief Overview of Biomass and Coal Gasification Reactor Technology

The reactor technology for coal gasification is much more mature than that for biomass gasification. Furthermore, coal gasification reactors operate at much larger scale than those for the biomass gasification. There are basically three types of reactors: moving bed (updraft or downdraft), fluidized bed (bubbling or recirculating), and entrained bed reactors used for coal and biomass gasification.

The moving bed biomass gasification reactors can be operated as updraft or downdraft reactors. In moving bed coal or biomass gasification reactors, large coal or biomass particles move slowly down the bed while reacting with gases moving up or down through the bed. Coal or biomass moisture content controls the product gas temperature. These reactors are generally operated at low (near atmospheric) pressures. All moving bed reactors have low oxidant requirement, need to handle caking coals and biomass with high mineral and volatile matters, and have limited ability to handle fines. They produce tar and oils and have high "cold gas" thermal efficiency when heating value of the hydrocarbon liquids is included.

There are two main commercial moving bed gasifier technologies. The Lurgi dry-ash gasifier was originally developed in the 1930s and has been extensively used for town gas production and in South Africa for chemicals from coal [1–13,16–20]. Slagging version of it called BG Lurgi (BGL) gasifier was developed in 1970 in which the temperature at the bottom is sufficient for the ash to melt. The most notable commercial application of this type is British Lurgi gasifier for coal–waste mixture. The moving bed reactors are used for the gasification of MSW where drying and feed particle size reduction is not carried out. These reactors can operate at temperatures as high as 1000°C, but the syngas produced from these reactors carries significant amount of tars that need to be removed as a part of syngas cleaning process. Many waste industries are making transformation of waste combustion to waste gasification using this type of reactors.

For the biomass gasification, fluidized bed reactors are most widely used because they provide the most flexibility in the range of operating conditions. Fluidized bed gasifiers can handle a wide range of feedstock including solid waste, wood, pulp sludge, MSW, RDF, corn stover, and high-ash coals with pressure range from 1 to 33 bar and average reactor temperature range from 725°C to 1400°C [1,16,17]. For complete tar and methane conversions, the reactor needs to be operated at temperatures greater than 1200°C–1300°C. Fluidized bed offers more flexibility in particle size variations. For most bubbling fluidized bed (BFB) gasifiers, biomass will have to be dried to increase operating temperatures. BFBs generally have uniform moderate temperature with good mixing and moderate oxygen and steam requirement and extensive char recycling [1–13]. Fluidized bed gasifiers may differ in ash conditions (dry or agglomerated) and in design configurations for improving char use. Commercial version of BFB includes the high-temperature Winkler and KRW designs. There are relatively few large fluidized bed gasifiers in operation. Directly heated circulating fluidized bed gasification of biomass is not as widely used as BFB. Very few have been operated at high pressures and most have operated at temperatures below 1000°C. Circulating fluidized bed (CFBs) have not demonstrated the use of pure oxygen and steam as reactants. The reported CO_2 content and H_2/CO ratios of the product coming from CFB are low [1–13].

While entrained bed reactors are normally not used for biomass gasification, more than 85% of commercial current coal gasification reactors are entrained bed reactors in which fine coal particles gasify with large amount of oxygen and steam [1–13]. Due to high gas flow rate, the residence time in these reactors is short. They operate at uniform high temperatures and with plug flow in the gas phase. Solid fuel must be fine and homogeneous and the reactors operate well above the ash-slagging conditions to assure high carbon conversion. They are able to gasify all coals regardless of coal rank, caking characteristics, or amount of coal fines. The coals with lower ash are preferred. Entrained bed reactors can use dry feed or wet slurry feed. The gasifier design can also vary in their internal designs to handle the very hot reaction mixtures and heat recovery configurations. Entrained bed gasifiers have been the choice for all coal- or oil-based Integrated gasification combined cycle (IGCC) plants under construction [1–13]. Commercial gasifiers include General electric (GE) gasifier, two variants of Shell gasifier, and the Prenflo gasifier. Both GE and Shell oil gasifier have more than 100 units in operation worldwide.

3.2.1.2 Best Option for Multifuel Coal–Biomass Gasification

As mentioned in Sections 3.2.1 and 3.2.1.1, BGL moving bed reactor is being used for coal–waste gasification in the United Kingdom. In a Royal Institute of Technology, Stockholm, Sweden, investigation of coal–biomass blend gasification in an oxygen-containing environment in a pressurized fluidized bed reactor was carried out. It was noted that char from woody biomass was very sensitive to thermal annealing effect, which occurred at relatively low temperature (around 650°C) and short 8 min of soak time. The study showed high gasification rate and lower yields of tar and ammonia in oxygen-rich environment for fuel mixture of birch and coal [1,16–20]. Also, higher oxygen content of biomass reduced gasifier oxygen consumption proportional to quantity of coal displaced.

While all three types of reactors can handle coal–biomass mixtures, the industry prefers entrained-flow reactor for coal–biomass mixture gasification. The main reasons are that (1) it provides high fuel flexibility; (2) high temperature allows it to operate under slagging conditions; (3) it mainly produces syngas, which can be easily used for a variety of downstream operations; and (4) it can be easily designed for large-scale, high-throughput operations [1–13,16–20].

A slagging entrained bed reactor is preferred for coal–biomass gasification because mineral matters from biomass can end up in the slag. Improved ash handling also occurs with entrained-flow gasifiers under slagging conditions. In order to obtain the proper slag properties, fluxing materials are often used. Sometimes, slag recycling may be necessary to obtain enough slag inside the reactor to ensure sufficient wall coverage [1–13].

At entrained bed reactor temperatures of 1300°C–1700°C, a gas with very low concentration of tar, methane, and carbon dioxide is obtained, which significantly decreases the gas cleaning cost. Modern entrained-flow gasifiers operate at a pressure of between 15 and 60 bar. These elevated pressures are advantageous for gas to liquid synthesis. High flow rate and small particle size requirement for the entrained bed will necessitate biomass feed pretreatment such as torrefaction or pyrolysis.

Since entrained bed reactors are widely used in coal industry and much is known about its commercial operation, its use for coal–biomass mixture (with small concentration of biomass) makes most sense. This choice will also allow easy retrofitting of existing large-scale entrained bed coal gasification reactors for coal–biomass mixtures. All indications are that future coal–biomass commercial gasification reactors would be entrained-flow reactors.

A co-gasification model for a large-scale entrained-flow gasifier was developed by the Center for Research of Energy Resources and Consumptions, University of Zaragoza, Spain. The model was validated with a large number of operating data from Elcogas IGCC power plant in Puertollano, Spain. The model has been successfully applied to a mixture of coal and coke and up to 10% biomass. The model predicts how oxygen and steam requirements should be changed and optimized with the change in the nature of the feedstock [1].

3.2.2 SYNERGY AND CATALYTIC REACTIONS DURING CO-GASIFICATION

3.2.2.1 Effects of Co-Gasification of Coal and Biomass/Waste on Tar Release

The literature pertaining pyrolysis under high temperature, pressure, and heating rate conditions that normally exist during gasification is sparse. The reported studies have also not examined the mechanism by which biomass components (cellulose, hemicellulose, and lignin) interact with each other under heating and the effect of mineral matter on these interactions. The knowledge of how these individual components and biomass as a whole would interact when blended with coal has also not been elucidated. Feedstock and their compositions can thus significantly influence the conversion behavior during the early stage of fuel decomposition. For an entrained bed gasification reactor operating at temperatures higher than 1200°C, the effects of feedstock components on char gasification are also not well understood [1,21].

Tar can be defined as the organics produced under thermal or partial oxidation regimes of any organic material and generally assumed to be largely aromatic [45]. While there are different tiers of tar components [1–3], saturated polynuclear aromatics such as benzene, toluene, and naphthalene are the hardest to decompose under gasification conditions. The decomposition of all tar components generally requires high temperature (>1200°C).

The tar content of the product gases from gasification of biomass is one of the major factors affecting the subsequent process stages. The type of feedstock used and gasification temperature are the main factors affecting tar yield during gasification [37]. Synergistic effects between coal and biomass particles are expected to lower the yield of tar compared to independent gasification of these feedstocks. Kumabe et al. [141] obtained a slight decrease in tar yield by varying Japanese cedar (biomass) concentration from 0 to 100 when co-gasifying with Mulia (Indonesia) coal with air and steam in a downdraft fixed bed gasifier at 900°C.

Pinto et al. [46] observed a decrease in tar yield for a mixture of 80% coal (high-ash coal from Puertollano, Spain) and 20% pine wood waste compared to coal alone in a fluidized bed gasifier operating at atmospheric pressure and temperatures of 850°C–900°C using a mixture of oxygen and steam as the gasifying agent. However, the addition of PE waste in the feed led to an increase in tar release. A probable reason for this could be that the polymeric structure of PE breaks into smaller fractions by thermal cracking, contributing to a greater amount of tars [46]. A ternary blend of coal, PE, and pine resulted in less tar release than when mixing coal and PE. They [46] and McKee et al. [35] also found that dolomite was an efficient catalyst for reducing the tar yield during co-gasification.

Collot et al. [144] found no synergetic effect during co-gasification of Polish coal and forest residue mixture. Aznar et al. [142] and Andre et al. [143] identified an increase in tar yield with an increase in biomass content of the feedstock, suggesting that synergetic effect during coal and biomass co-gasification might be highly dependent on biomass type as well as gasification conditions. Mettler et al. [217] point out that due to lack of understanding of intermediates formed during gasification, their effects on the co-gasification process are unknown.

Further investigations are still needed to clarify various issues of synergy during coal and biomass/waste co-gasification.

3.2.2.2 Roles of Mineral Matters and Slagging in Co-Gasification

Tchapda and Pisupati [21] analyzed the role of mineral matters on slagging during co-gasification. A summary of their analysis is briefly presented here. Slagging and fouling are caused by the relatively reactive alkali and alkaline earth compounds (K_2O, Na_2O, and CaO) found in biomass ash. The alkali and alkaline earth metals (AAEMs) present and dispersed in biomass fuels induce catalytic activity during co-gasification with coal. The catalytic activity is most noticeable when blended with high-rank coals. The presence of synergy during co-gasification is still controversial [21].

The merits of alkali (K^+ and Na^+) and alkaline earth (Ca^{2+}) metals as catalysts for coal gasification have been extensively investigated [22–25]. Due to high content of these elements in some biomass, co-gasification of coal and biomass is expected to exhibit some catalytic behavior, whose significance will depend on the type of biomass and coal being used. In fact, in lignite (low-rank coal) Na is the principal alkali metal, while the amount of K is low [26]. In bituminous coals (high rank), K is contained exclusively in illite or closely related clay structure [27], while Na is generally present as NaCl [28].

From low-rank coals to higher-rank coals, Ca is systematically changed from carboxyl bound to calcite [29], with decreasing catalytic activity [30]. The K present in the illite of bituminous coal is converted to K-aluminosilicate glass [27]. Therefore, mineral matters of high-rank coals have little catalytic activity during coal gasification. Any catalytically active ions such as Ca and Na in low-rank coals are highly dispersed [31].

As pointed out by Tchapda and Pisupati [21], the analysis presented in Sections 3.2.2.1 for coal is also applicable to biomass fuels. Ren et al. [32] compared co-gasification of meat and bone meal (MBM) blended with high-rank (anthracite) and low-rank coals (lignite). They found that for both rank coals, acid-based MBM samples gave a lower carbon conversion compared to the raw MBM blend. Thus, biomass mineral matter (Na, K, and Ca) demonstrated catalytic influence during co-gasification, as reported by other authors [33,34]. They also found that the catalytic effect of MBM minerals was not perceivable on lignite but was significant for anthracite. This is in accordance with McKee et al. [35] and Srivastava et al. [36] who demonstrated the increased catalytic activity of alkali metals on gasification with increasing coal rank.

The literature also reports [40] that the reaction rates of a blend of waste birch wood and Daw Mill coal was significantly increased, thereby reducing char in a pressurized fluidized bed gasifier under oxygen environment. Brown et al. [34] showed that the gasification of coal char from Illinois No. 6 coal increased by eightfold at 895°C in a mixture of 10:90 of coal char and switchgrass ash. In general, biomass is more reactive than any coal [37]. Its char is therefore continuously consumed during gasification, leaving very little remains at the end of the process, whereas coal char is less reactive and continuously accumulates in the bed during the course of the gasification. Blending biomass and coal takes advantage of both the high reactivity of biomass and its catalytic effect.

Habibi et al. [38] observed negative catalytic behavior between switchgrass and a subbituminous coal, while the mixture of switchgrass and fluid coke showed a synergy. They explain the negative behavior by a deactivation of mineral catalysts due to sequestration of the mobile alkali elements by reaction with aluminosilicate minerals in coal to form inactive alkali aluminosilicates [21].

The catalytic effect of biomass in co-gasification with coal is also expected to play an important role for the abatement of environmentally harmful species containing sulfur and nitrogen. Biomass species with a high content of K, Na, and Ca can form sulfate and capture sulfur from the gas phase when co-processed with coal [39]. During co-firing of straw and coal, Pedersen et al. [39] observed a net decrease of NO and SO_x emission, due to the decrease of fuel-nitrogen conversion to NO and due to retention of sulfur in the ash. Sjöström et al. [40] identified a lower ammonia yield during co-gasification of birch wood and Daw Mill coal. Cordero et al. [41] showed enhancement in desulfurization when blending coal with different types of biomass during co-pyrolysis. Haykiri-Acma and Yaman [42] showed that the addition of hazelnut shell to lignite contributed to the sulfur fixing potential of the resulting char in the form of CaS and $CaSO_4$ during co-pyrolysis of these feedstocks. For temperatures higher than 700°C, potassium is shown to be more trapped in Si in ash [43,44].

3.2.3 Feed Systems and Feedstock Preparation

As shown in Sections 3.2.5 and 3.2.6, while in past gasification technology for coal has used three different types of reactors, moving bed, fluidized bed (including circulating), and entrained bed, about 85% of existing and new coal gasification plants are based on the entrained bed technology. Biomass/waste gasification currently uses either moving or fluidized bed reactors. While fluidized bed reactors are suitable for smaller-scale biomass gasification, for large-scale coal–biomass gasification, industry prefers entrained bed reactors [1–3]. While we assess all three types of reactors, our major focus will be on the entrained bed reactors. Unfortunately, the existing coal feeding systems for this type of reactor are usually not suitable for biomass. A number of issues related to this matter need to be assessed and handled.

3.2.3.1 Issues Regarding Particle Size, Shape, and Density of Biomass

The entrained flow generally requires fine particle size and high gas flow. In coal-based entrained-flow gasification reactors, particle size used is around 90 μm. Milling biomass to the same size has a five times higher electricity consumption. Figure 3.1 shows the particle size versus the power (or energy) required for biomass. It is clear from this figure that power required for biomass increases exponentially as the particle size is reduced below 1 mm. Buggenum plant has shown [1] that for wood, particles of 1 mm is sufficient to get complete conversion. Thus, for woody biomass, downsizing to 1 mm is sufficient, and for this case, the electricity consumption is similar to the coal milling. For straw and miscanthus, on the other hand, reducing particle size to 1 mm may not be sufficient. Furthermore, in case of straw, its long narrow needle-type shape may require special type of chopping machines, and fibrous nature can create additional issues with feeding.

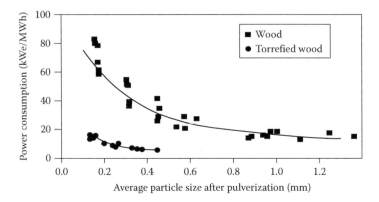

FIGURE 3.1 Power consumption for size reduction: untreated versus torrefied wood. (From Shah, Y. and Gardner, T., Biomass torrefaction: Applications in renewable energy and fuels, in: Lee, S. (ed.), *Encyclopedia of Chemical Processing*, Taylor & Francis Group, New York, pp. 1–18, April 24, 2012, published online.)

In GE gasifier, coal is fed into the gasifier as a slurry of coal and water. Milled biomass of 1 mm can be mixed with the coal–water slurry. As discovered by Polk plant [1–3], the way to make this work effective is to use automated method of milling, pneumatic transport, and feeding into the coal slurry system. It is absolutely critical that the larger particles are effectively screened and recycled for further milling. Often, the milling of particles to a smaller diameter will require the use of processes such as torrefaction [15,47–94] and briquetting or pelletizing to achieve small particles without excessive power consumption. Figure 3.1 also indicates the reduction in power requirement of torrified wood as a function of particle size.

The density difference between coal and biomass particles also creates some issues with effective feeding. Mixtures of coal and switchgrass have shown very poor flow characteristics from sloped storage bunkers due to difference in bulk density. This applies to mixtures of only 5% by volume of switchgrass and progressively gets worse as the switchgrass percentage increases [15,47–94]. The problem can be partially alleviated by preprocessing switchgrass into pellets before blending with coal. This, however, can be expensive. Other alternatives are to modify storage bunker design or separate coal and switchgrass feed systems. The recommended approach depends on the nature of switchgrass delivered to the plant and the desire to maintain single fuel processing train. The fibrous nature of the biomass like switchgrass, straw, and miscanthus and their low density also cause problems with the pneumatic feeding system with inert carrier gas. The fibrous and compressible nature of these biomass materials can aggregate and plug the feeding line. Larger 1 mm switchgrass or straw needles are not suitable for pneumatic feeding due to their shape [15,47–94].

3.2.3.2 Issues Related to Feeding Biomass to the Pressurized Reactor

As mentioned earlier, the best gasification reactor technology of the future is an entrained bed, oxygen-blown, high-pressure (10–50 bars) technology [1–3]. Unlike coal, the feeding of biomass in a pressurized reactor through an alternating pressurized and depressurized lock system poses several problems. While this periodic procedure is suited for high-bulk-density fuels and moderate pressures, at low bulk densities (e.g., biomass) and higher pressures, the amount of lock gas can exceed the fuel weight. Therefore, a well-designed feeding system to withstand back pressure from the gasifier is required, and this may be as expensive as the gasifier itself [1–3].

Pressurization of biomass in conventional lock hopper system requires much more inert gas than that for coal. Even when this system works for biomass, the consumption of inert gas on energy basis is approximately twice as high for biomass compared to coal due to its lower energy density. This will significantly lower the efficiency of the gasifier, as all the inert gas has to be heated to the gasification temperature. It also results in the dilution of the biosyngas. The extra inert gas has to be compressed, which results in additional electricity consumption.

For compression of biomass, a piston compressor has been developed in Europe in which approximately 50 times less inert gas is consumed [1–3]. A solid feed system developed jointly by DOE and Stamet appear to offer a solid feed alternative system to standard lock hopper system for feeding dry coal into pressurized gasifiers [98,99]. The Stamet system often called Stamet Posimetric pump was originally developed to feed oil shale into the gasifier systems. The system provides positive flow control. The device consists of a single rotating element that is made up of multiple disks and hub that are installed inside a stationary housing. The pump has been successfully used for lignite, bituminous, and PRB coal up to 560 psi pressure. The applicability of this and other devices for coal and biomass mixture dry feeding in pressurized gasifiers is still being improved.

The high reaction rates of entrained-flow gasifiers demand very small feedstock particle size (~100 µm), which is easily achievable for friable materials like coal, but more challenging and energy consuming for biomass due to its fibrous structure and hygroscopic nature. Therefore, the viability of entrained-flow gasification of biomass relies on its pretreatment. The fibrous nature of the biomass also prevents it from fluidizing and fluffs are formed that may plug piping. Among the pretreatment options available to alleviate these biomass shortcomings are pelletization, torrefaction, and fast pyrolysis [21]. Pelletization can be expensive and fast pyrolysis leads to liquids that can be mixed with coal for slurry feeding in the gasifier. For dry feeding, torrefaction has been found an attractive alternative. Because of their wide acceptance, we briefly discuss here torrefaction and fast pyrolysis alternatives for the feed pretreatment [1–3,15,47–94].

3.2.3.3 Pretreatment and Feeding Options for Biomass

Torrefaction is a process of mild pyrolysis at lower temperature (around 200°C–250°C) and longer contact time (between 30 and 60 min). This process dehydrates and depolymerizes the long polysaccharide hydrocarbon chains present in biomass. This results in a product that is hydrophobic and has a higher energy density and improved grinding and combustion capabilities [1–3,15,47–94]. These improved properties of the torrefied biomass allow for easy and more energy-efficient use in existing coal-fired power plants. The process of torrefaction is best illustrated through the Van Krevelen plot shown in Figure 3.2 [15,47–94]. This figure

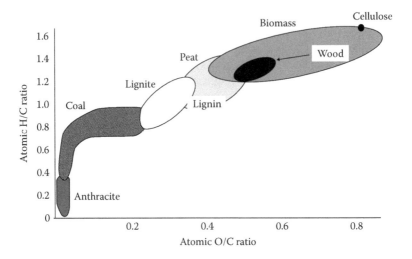

FIGURE 3.2 Van Krevelen plot illustrating the composition of various fuel sources. (Adapted from Hustad, J. and Barrio, M., Biomass, *IFRF Online Combustion Handbook*, Combustion File No. 23, Version No. 2, IFRF, 17-10-200, http://www.handbook.ifrf.net/).

illustrates that torrefaction results in the reduction of oxygen content and correspondingly, increased heating value of the treated biomass. Generally, during torrefaction, an increase in both mass and energy density occurs because approximately 30 wt% of the biomass is transformed into volatile gases. These gases carry 10% of the original biomass energy content [15,47–94]. This indicates that during torrefaction, a substantial amount of chemical energy is retained from the raw starting material to the product state. This differs from conventional pyrolysis where the energy yield varies from 20% to a maximum of 65%, even in very advanced pyrolysis technologies [15]. The torrefaction of biomass results in gases, such as H_2, CO, CO_2, CH_4, and C_xH_y; liquids such as toluene, benzene, H_2O, sugars, polysugars, acids, alcohol, furans, ketones, terpenes, phenols, fatty acids, waxes, and tannins, and solids comprise of char and ash.

Generally, the torrefaction process is operated under conditions that minimize the production of liquid products. As shown in Table 3.2 [15], torrefied biomass possesses very valuable properties. It has lower moisture content and therefore a higher heating value when compared to untreated biomass. The storage and the transportation capabilities of the torrefied biomass are superior over those of untreated and dried biomass. Torrefied biomass is hydrophobic and does not gain humidity in storage and transportation. It shows little water uptake on immersion, between 7 and 20 wt%, and it is more stable and resistant to fungal attack compared to charcoal and untreated biomass. Pelletization, by itself, produces biomass with higher mass density; however, the pellets are not hydrophobic and remain susceptible to fungal attack.

Torrefied biomass significantly conserves the chemical energy present in the biomass. The heating value of torrefied wood is approximately 11,000 Btu/lb and is nearly equal to that of a high-volatile bituminous coal that is 12,000 Btu/lb.

TABLE 3.2
Aspects of Torrefied Biomass for Gasification and Other Applications

Torrefied Product

1. Has lower moisture content and higher heating value
2. Is easy to store and transport
3. Is hydrophobic and does not gain humidity in storage and transportation
4. Is less susceptible to fungal attack
5. Is easy to burn, forms less smoke, and ignites faster
6. Significantly conserves the chemical energy in biomass
7. Has heating value (11,000 Btu/lb) that compares well with coal (12,000 Btu/lb)
8. Generates electricity with a similar efficiency to that of coal (35% fuel to electricity) and considerably higher than that of untreated biomass (23% fuel to electricity)
9. Has grindability similar to that of coal
10. Requires grinding energy 7.5–15 times less than that for untreated biomass for the same particle size
11. Has mill capacity 2–6.5 times higher compared to untreated biomass
12. Possess better fluidization properties in the gasifiers
13. Is suitable for various applications in heating, fuel, steel, and new materials manufacturing industries

Sources: Adapted from Bergman, P.C.A. and Kiel, J.H.A., Torrefaction for biomass upgrading, in: *Proceedings of the 14th European Biomass Conference and Exhibition*, Paris, France, October 2005; *ETA—Renewable Energies*, Florence, Italy, pp. 17–21, 2005; Bergman, P.C.A. et al., in: Van Swaaij, W.P.M., Fjällstrom, T., Helm, P.T., and Grassi, P. (eds.), *Proceedings of the Second World Biomass Conference on Biomass for Energy, Industry and Climate Protection*, Rome, Italy, May 10–14, 2004; Torrefaction for entrained flow gasification of biomass, Report No. ECN-RX-04-046, Energy Research Centre of the Netherlands (ECN), Petten, the Netherlands, pp. 679–682, 2004; Bergman, P.C.A., Combined torrefaction and pelletization: The TOP process, Report No. ECN-C-05-073, Energy Research Centre of the Netherlands (ECN), Petten, the Netherlands, 2005; Bergman, P.C.A. et al., Torrefaction for entrained flow gasification of biomass, Report No. ECN-C-05-067, Energy Research Centre of the Netherlands (ECN), Petten, the Netherlands, 2004; Bergman, P. et al., *Torrefaction for Biomass Co-Firing in Existing Coal-Fired Power Stations (BIOCOAL)*, ECN, Petten, the Netherlands, ECN-C-05-013, 72pp., 2005.

It generates electricity with an efficiency comparable to that of coal of approximately 35%, on a fuel to electricity basis [1–3,15,47–94], and much higher than that of untreated biomass that has an efficiency of 23%, on a fuel to electricity basis [15,47–94]. Bergman et al. [1,15] showed that torrefied biomass has better fluidization properties than that of untreated biomass, but similar to that of coal. Bergman et al. [1,15] also showed that torrefied willow ignites quicker and heats faster than refined coal particles.

Untreated biomass requires many times the grinding energy, by a factor of 7.5–15, to achieve a similar particle size compared to torrefied biomass (see Figure 3.1). This energy difference is significantly larger than the energy loss of biomass and energy supplied during torrefaction. The torrefied biomass is suitable for various applications such as working fuel, residential heating, and new materials for the manufacture of fuel pellets, as reducer in the steel smelting industry [1,15], in the manufacture of charcoal and active carbon, and in the gasification and co-firing with other fuels in gasifiers and boilers. Such a wide usefulness makes torrefied biomass a valuable and marketable product.

Fast pyrolysis is a rapid heating process at about 500°C in an inert atmosphere. Rapid condensation of the vapors in one to a few seconds occurs, and more than half of the biomass mass can be obtained as liquid pyrolysis oil at the expense of low char and gas yield. In these conditions, the pyrolysis gases contain less than 10% of the biomass energy, which is consumed on-site as energy input to the pyrolysis process. The brittle char obtained is ground to a fine powder and suspended in the liquid pyrolysis oil to form the bioslurry, with an overall energy content of up to 80%–90% of the original biomass energy [15], and considerable volume reduction.

The dispersed nature of the biomass feedstock requires the pretreated fuel to be transported to a central plant where it is co-converted with coal. Important parameters in deciding which of the two pretreatment options to consider are the energy density and postprocessing of the pretreated fuel.

Bioslurry is denser (and less bulky) than torrefied biomass. Thus, besides feeding bioslurry under pressure being easier, pneumatically feeding powdered torrefied wood requires pressurizing with an inert gas that reduces the calorific value and complicates the gas cleaning process. Another advantage of the fast pyrolysis process is that to avoid soil nutrient depletion, pyrolysis can be designed to optimize and obtain only an energy-rich biooil containing 60%–70% of the original biomass energy [15], while the remaining char may be immediately returned to the soil. This approach may well be more efficient since it avoids transporting the ash from the

TABLE 3.3
Options for Biomass Feed Preparation

Drying/Grinding	Torrefaction/Grinding	Pyrolysis
Physical process.	Physics/chemical process.	Chemical process.
Lowest energy.	Medium energy.	High energy.
Dry feed/high energy for grinding.	Dry feed/low energy for grinding.	Slurry feed.
Pelletization is possible.	Pelletization is possible.	Not applicable.
Storage and transportation are problematic because of moisture absorption and fungal attack.	Storage and transportation are easy.	Slurry storage and transportation require an inert atmosphere.
Grindability, fluidization properties, and combustibility for entrained-flow fluidization are superior.	Grindability, fluidization properties, and combustibility for entrained-flow fluidization are superior.	Pyrolysis can be considered as a first stage of two-stage gasification process.

Source: Shah, Y. and Gardner, T., Biomass torrefaction: Applications in renewable energy and fuels, in: Lee, S. (ed.), *Encyclopedia of Chemical Processing*, Taylor & Francis Group, New York, pp. 1–18, published online, April 24, 2012.

gasification plant to the forest. Moreover, it ensures that almost all the ash extracted by biomass is returned to the soil and prevents possible toxic elements that might be present in coal ash to be applied to the soil. Additional savings could result from the simplicity associated with handling biooil at the plant site. Handling biooil at the plant site will require fewer and comfortable metering and handling equipment compared to handling solid feedstock.

The method adopted for feeding biomass to entrained bed reactor depends on the nature of the feedstock. Woody biomass can be milled to 1 mm and then fed to the reactor by screw feeder with piston compressor. In general, all types of biomass can be subjected to the torrefaction followed by milling process and then feed to the entrained bed reactor either by pneumatic feeder and piston compressor or by screw feeder and piston compressor. All types of biomass can first be gasified in a pressurized fluidized bed gasifier, and the product gas and char can be fed to entrained bed gasifier for further gasification. Straw like biomass, which is difficult to mill, can be pyrolyzed to form bioslurry that can then be fed to entrained bed reactor by feeder and injector. These pretreatment and feeding options are not competitive but complementary or an alternative to each other. Table 3.3 describes various characteristics of the options for biomass feed preparation [1–3].

3.2.4 PRODUCTS CLEANING AND SEPARATION MECHANISMS

Syngas produced from coal–biomass mixture gasification will likely contain nitrogen, sulfur, chlorine, alkali, and heavy metal impurities since those elements are a part of original coal and biomass. These impurities must be removed from syngas before its downstream use. Here, we briefly examine the behavior of these impurities and processes required for their removals. The acceptable levels of these impurities in product gas are outlined in Table 3.4.

3.2.4.1 Behavior of Nitrogen, Chlorine, and Sulfur

Tchapda and Pisupati [21] gave a detailed accounting of nitrogen and their forms present in biomass and coal. They also pointed out that reaction pathways involving nitrogen during pyrolysis/gasification are complex due to the overlapping of reaction stages, the influence of reaction conditions (residence time, temperature, pressure, and gasification agent), and the differences between fuels. During pyrolysis, part of the fuel-bound nitrogen (FBN) is released with the devolatilizing gases, while the remainder is retained in the solid char to be released during subsequent gasification of the char [99,100]. The allocation of fuel N to volatiles and char is dictated by temperature [98,101–104], particle size [102], fuel type [98], residence time [102,104], heating rate [105,106,109,110], and pressure [105]. Temperature, fuel type, and particle size, are the dominant factors in this allocation while heating rate and pressure only have a minor effect.

Tchapda and Pisupati [21] also pointed out that any synergy during the decomposition of coal and biomass/waste blend influences the release of fuel nitrogen. Yuan et al. [107,108] found that synergy between coal and biomass enhanced the release of fuel N as volatile, reducing the amount of residual nitrogen left in the char during co-pyrolysis. Since synergy increases the volatile yields, it is expected that it will

TABLE 3.4
Allowable Concentrations of Contaminants in Syngas for Catalytic Synthesis

Syngas Contaminants	Contaminant Specification
$H_2S+COS+CS_2$	<1 ppmv
NH_3+HCN	<1 ppmv
$HCl+HBr+HF$	<10 ppbv
Alkali metals $(Na+K)$	<10 ppbv
Particles (soot, ash)	"Almost completely removed"
Organic components (viz. tar)	Removed to a level at which no condensation occurs upon compression to FT synthesis pressure (25–60 bar)
Heteroorganic components (incl. S, N, O)	<1 ppmv
CO_2, N_2, CH_4, and larger hydrocarbons	"Soft maximum" of 15 vol% that has been identified (the lower, the better)

Sources: Ratafia-Brown, J. et al., Assessment of technologies for co-converting coal and biomass to a clean syngas—Task 2 report (RDS), DOE/NETL-403.01.08 Activity 2 Report, May 10, 2007; Shah, Y.T., Biomass to liquid fuel via Fischer-Tropsch and related syntheses, in: Lee, J.W. (ed.), *Advanced Biofuels and Bioproducts*, Chapter 12, Springer Book Project, Springer Publ. Co., New York, pp. 185–207, September 2012.

also augment the yield of nitrogen species in the gas phase. At higher temperatures, N_2 was the dominant nitrogen species observed by Yuan et al. [107,108].

While the literature is not clear about the effects of experimental conditions on partitioning between HCN and NH_3, there are strong evidences [107,108,145] that slow heating rate produces NH_3, while fast heating rate produces HCN as the main N compound. The literature has also shown that at low temperature more N ends up in char than at high temperatures [21]. In co-gasification, the release of FBN in the volatile is desirable over its remaining in the char.

During co-gasification, chlorine is released in two steps: during pyrolysis at temperatures as low as 500°C in the form of HCl and during char burnout at temperatures higher than 700°C in the form of KCl and NaCl [111–114]. K will preferentially be released in the gas phase as KCl for Cl-rich biomass fuels.

Tchapda and Pisupati [21] pointed out that during co-gasification, sulfur is released mainly in the form of hydrogen sulfide (H_2S) and carbonyl sulfide (COS) [111]. In coal, sulfur exists in both inorganic and organic forms. Zhao et al. [115] observed that most of the organic sulfur in the temperature range of 400°C–900°C escaped as H_2S, a conclusion supported by other studies [232–236]. In co-gasification, since woody biomass and grasses contain much less sulfur than most coals, the replacement of coal by biomass will reduce H_2S and COS in the product syngas. Tchapda and Pisupati [21] point out that coal mineral sulfur could interact with the organic matrix during coal pyrolysis [116–125]. It has also been proven that H_2S evolved from decomposition of pyrite can be partially retained in the solid phase (char/ash) as alkali sulfides [123–125]. Si in coal can also have a beneficial effect on the sulfur retention at high temperature [43,111].

Organically associated Ca can recapture the H_2S to form sulfidic sulfur at temperatures above 600°C [125]. Given the high content of alkalis in some biomass fuels, sulfur capture by these metals could be an added advantage of co-gasification of coal- and biomass-based fuels.

3.2.4.2 Bottom and Fly Ash Characteristics and Role of Alkalis

Co-gasification results in the formation and deposition of ash, which can cause slagging and fouling of heat transfer surfaces. Inorganic matters and impurities in coal and biomass contribute to ash formation during co-gasification. The characteristics of ash formed depend on feedstock composition and the reaction conditions. The important factors contributing to ash characteristics and behavior are (1) the rank and type of coal and type of biomass [126] and (2) reaction conditions such as temperature, pressure, oxidizing or reducing atmosphere, and flow dynamics [126]. Generally, bark, straw, grasses, and grains have higher mineral contents compared to wood, and MSW and sewage sludge have higher heavy metals compared to coal and biomass [127–130].

In general, the inorganic matter of the feedstock has a major impact on the gasifier operation. This impact depends on the type of gasifier employed and whether or not gasifier is slagging or nonslagging. The nonslagging ash interferes with moving bed [21] if the temperature is not maintained below the melting temperature. If ash fuses together and forms clinker, it can stop or inhibit the downward flow of biomass. Ash fusion temperature also changes with sodium content. The clinker formation was observed in the moving bed Lurgi gasifiers at Dakota Gasification Great Plains plant, which forced shutdown of the plant [1,21,131]. Even if ash does not fuse, it can lower the fuel's reaction response. In a CFB, silica in the sand bed material can agglomerate with high sodium feedstock (like birch wood), and this agglomeration should be avoided by injecting additives.

Ash in some fresh wood does not melt even at 1300°C–1500°C due to the presence of CaO. This can cause problems with slagging entrained-flow gasifier. Despite the high-ash melting temperature in this case, slagging entrained-flow gasifier is preferred over nonslagging gasifier because (1) melt can never be avoided and (2) slagging entrained-flow gasifier is more fuel flexible. Slagging co-gasification often requires fluxing material in order to obtain proper slag properties [91,95–103,128–144].

Biomass mineral matter is usually rich in AAEMs, which melts at low temperatures. Aluminosilicates have higher melting temperatures and their concentrations are higher in coal than in biomass. Therefore, biomass ash tends to melt at lower temperatures than coal ash. It is therefore important to know the behavior of the ash mixture from coal and biomass in the design of co-gasification systems. The wide variation of mineral matter content and quality observed in biomasses as well as in coals causes the design of co-gasification systems to be carried out on a case-by-case basis. More quantitative assessment of characteristics of coal–biomass ash mixtures under different gasification conditions is needed [21].

Higman and van der Burgt [133] suggest that for gasification application, the ash melting characteristics should be determined under reducing conditions. The knowledge of possible chemical reactions among mineral matters present in coal and biomass during co-gasification can be very helpful for understanding ash characteristics. Mineral matter in solid fuels is generally grouped into two factions [21]:

1. The inherent inorganic material, which is most commonly associated with the oxygen-, sulfur-, and nitrogen-containing functional groups
2. The extraneous inorganic material coming from soil, or during harvesting, handling, storage, and processing of the fuel [134]

Tchapda and Pisupati [21] point out that inorganic materials in biomass can be further categorized in a number of ways. Vaporization, condensation, and coagulation/agglomeration are the main mechanisms involving biomass ash formation; alkali metals and chlorides contained in biomass vaporize to form $HCl(g)$, $KCl(g)$, $K_2SO_4(g)$, $Na_2SO_4(g)$, and $NaCl(g)$ [132]. Their condensation constitutes the main contributor to the fine ash fraction. Strand et al. [146] identified homogeneous nucleation and heterogeneous condensation as two principal routes for condensation of these vapors into particles.

The inherent and extraneous inorganic materials of coal have different pathways for ash formation. Inherent inorganic materials are close to each other in the char particle; therefore, they become molten and easily coalesce and agglomerate during char burnout. Extraneous minerals mainly form ash through fragmentation. The study of Wu et al. [136] partially agree with the conclusion that coarse ash particles are formed by coalescence and shedding and fine ash particles are formed through vaporization and condensation. The slagging and fouling indexes often used in the gasifier design are based on the fuel ash content and/or the ash chemical composition, and they give an indication of fuel inclination to form deposits in the gasifier. Chemical equilibrium calculation has been used in many studies to predict the behavior of ash at given pressure and temperatures [21,136,137].

There are no data available on the effects of alkalis on fly ash characteristics under co-gasification conditions. Most of the studies involving co-firing coal and straw have focused on alkali metals and the reactions of alkali metal with chlorine to determine the properties of fly ash deposits [137–140]. Significant amounts of K are released in the gas phase as KCl and KOH [137] as well as K_2SO_4 [137,147–149], and they may react with other species present in the gas phase. Zheng et al. [132] concluded that the total K in the fly ash can be predicted when the ash composition of the fuel and straw is known.

Based on observations during co-combustion, Na is likely to be released as NaCl, NaOH, $NaSO_4$, and Na [137] and reacts with the gas-phase compounds. These observations corroborate with the fuel characterization based on the reactive and nonreactive ash-forming matter. This prediction method segregates the ash-forming matter in the biomass fuels in the "reactive" and "unreactive" fraction. More detailed description of this method and other matters related to roles of alkalis in ash is given in the excellent review by Tchapda and Pisupati [21].

3.2.4.3 Ash and Slag Utilizations

Gasification ash may contain high concentrations of unburned carbon and harmful soluble organic compounds restricting its usefulness. If common gasifier is used for coal–biomass mixture, additional impurities like trace metal contaminants may make that ash even less useful. In biomass alone plant, dry ash (nonslagging) makes ash useful. They can be recycled and used for forestry and agricultural purposes. If ash contains heavy metals such as Cd, Pb, and Zn, it cannot be used for soil amendment. Volatile metals can also gather in the fly ash, making it nonuseful.

In co-gasification entrained-flow reactor, oxygen-blown slagging reactors are going to be used to avoid tar formation. Since biomass ash contains potash, phosphates, and calcium and iron oxides, these will melt at high slagging temperature and will not be useful for soil remediation. If fluxing agent with high Si and Al is added to facilitate slagging, slag will be the major solid product of co-gasification. The nonleachable slag can be used for structural landfill, blasting grit, roofing tiles, asphalt aggregate, portland cement aggregate, and other construction building products [1,21].

3.2.4.4 Syngas Cleaning Processes

Syngas coming from gasifier contain impurities such as particulate matters, nitrogen, sulfur, chlorine compounds, heavy metals, and tars (depending on gasifier operating conditions). These impurities can either be removed at high temperatures or low temperatures. We briefly outline some of the processes to remove these impurities [97,150–155].

3.2.4.4.1 Particulate Removal

The requirements for a particulate removal depend on the end use of the gas. Gasification coupled with gas engines used for stationary power applications requires particulate loading below 50 mg/N m^3 (N m^3 is the normal cubic meter). Particulate loading less than 15 mg/N m^3 with a maximum particle size of 5 μm is required to protect gas turbines in integrated gasification combined cycle processes. The most stringent requirements for particulate removal are for fuel synthesis applications that require particulate loadings of less than 0.02 mg/N m^3 to protect syngas compressors and minimize catalyst poisoning by alkali fumes and ash mineral matter.

Several technologies have been developed and are commercially available for particulate removal from high-temperature gas streams. Choosing an appropriate technology depends on the desired particle separation efficiency for expected particle size distribution to achieve the ultimate particulate loading based on the end use of the syngas. Pressure drop through the particle removal unit operation and thermal integration are also the key design parameters. Tars produced during coal–biomass gasification also have a significant impact on particulate removal strategies. Operating temperature of the most particulate removal devices should be above the tar dew point to avoid tar condensation and prevent particulate matter from becoming sticky and agglomerating.

There are primarily four types of devices that can be used to separate solid particles from a gas stream:

1. Cyclones
2. Filtration
3. Electrostatic separators (ESPs)
4. Wet scrubber

3.2.4.4.1.1 Cyclones Cyclones are used as an initial step in the gas cleanup process to remove the bulk of the char entrained in the syngas stream. This technology is standard in industry due to its low cost and high level of performance for removing particulates.

Commercial cyclones for particulate removal are well-known and proven technologies. A dirty gas stream enters a cyclone separator with high tangential gas velocity, and angular momentum forces particles close to the walls of the cyclones so that they no longer follow the gas stream lines. The cyclone efficiency is a function of gas flow, particle size, temperature, and pressure. Cyclones can be designed for optimum removal of particles of specific size distribution usually down to a lower size limit. Multiple cyclones of different designs can be used in series to achieve near submicrometer particle removal with high efficiency.

Cyclones are an effective device to remove solids when particles sizes are greater than 10 µm. A properly designed cyclone with significant pressure drop (and consequently greater power consumption) may separate out small particles down to 3–5 µm. Cyclone separation efficiency is expected to be low in this particle range. In general, the efficiency of cyclones drops considerably for particle size lower than 1 µm. The deposition of soot or high-melting-point metals from syngas on its surface may also lower its efficiency.

3.2.4.4.1.2 Filtration Operating temperature and product gas composition are the primary process parameters that need to be considered in selecting the appropriate filtering medium. Barrier filters are a technology option for high-temperature particulate removal. Filter housing design and filter media selection are key for optimizing particle capture efficiency within a manageable window of pressure drop across the filter. The initial pore size of the filter medium is the design basis for pressure drop and particulate removal; however, the filtration efficiency improves as particles collect on the surface to produce a filter cake, and pressure drop across the filter increases as the thickness of the filter cake increases. Pulsing inert or clean product gas back through the filters dislodges the filter cake, and in this way, the pressure drop across the filter can be restored to regain its original performance.

Generally, nonfabric filters (i.e., candle type) that can sustain high temperature are made of sintered metal, metal mesh, or ceramic. Problems with nonfabric filters operated at 450°C in dirty syngas environment have been reported; breakage and blockages have caused significant downtime. The resistance of metallic filters to corrosive and erosive nature of the raw syngas at high temperatures is also an issue.

1. *Ceramic filters*: Ceramic candles are being developed for high-temperature gas filtration (above 500°C) applications. Ceramic filters can be designed for any flow requirement and can remove 90% of particulates larger than 0.3 µm. In theory, the ceramic filter elements normally made of aluminosilicate or silicon carbide powder with a sodium aluminosilicate binder have exceptional physical and thermal properties and can withstand temperatures up to 1800°F. At temperatures below 850°F, ceramic filters have

demonstrated satisfactory operational reliability. The contact with sulfur, chlorine, and alkali metal salt impurities can lead to morphological changes and embrittlement of these filters.

2. *Metal candle filters*: Metal filters are used in high-temperature cleanup and it can achieve filtration level as low as 1 μm. Depending upon the material of construction, it can meet any flow requirements and it can operate over a wide range of temperatures. Metal filters made from stainless steel can be used in cleanup systems for temperatures below 650°F, while Inconel or alloy HR filters are suitable for operating temperatures up to 1100°F. Commercial operation of metal filters operating at maximum temperature of 915°F has been successful at a few gasification facilities in Europe. Corrosion and tar deposition on filter elements are the problem areas for these types of filters.

3. *Sintered metal filters*: Sintered metal filter elements are an alternative to ceramic candles. The operating temperature of sintered metal filters is typically lower than that for ceramic filters to minimize sintering. At the appropriate operating temperature, sintered metal filters are more robust than ceramic candles, as the risk for cracking and rupture is much lower.

4. *Low-temperature Baghouse filters*: Baghouse filters are not appropriate for high-temperature applications. They cannot replace cyclones as an effective solid removal option. Baghouse filters are made of a woven fabric or felted (nonwoven) material to remove particulate materials down to 2.5 μm. While for felted filter systems removal efficiency is constant, for other filters the efficiency can increase with the cake thickness. Baghouse filters are modular design and thus can accommodate a wide flow range from 1,500 to 150,000 CFM. The Baghouse filters made of polyester and acrylic are suitable for temperatures below 300°F, while Nomex, Teflon, Ryton, or fiberglass materials can take temperatures up to 500°F. The periodic bag replacement results in high maintenance cost. It also has a potential for bag fire or explosion.

3.2.4.4.1.3 Electrostatic Precipitators   In this device, electric charge is induced on the surface of the particles, which are removed from the gas stream as they follow the electric field lines to a grounded collector plate. ESPs can be applied to particulate removal in high-temperature, high-pressure gas streams. However, maintaining the stability of corona discharge for reliable steady-state operation is a technical challenge. Another technical challenge is ensuring materials' compatibility of the high-voltage discharge electrodes and other metal internal components with the syngas impurities. The overall size and capital cost of ESPs tend to make them best suited for large-scale operation.

ESPs are an effective particulate removal device for fine particles. However, ESPs are not suitable for high-temperature applications. Additionally, the soft and sticky nature of the particulates in high-temperature conditions would almost certainly cause widespread fouling and plugging of the ESP grids. The ESP is more suitable to remove the very fine particulates below 0.5 μm at near ambient temperatures. ESPs cannot replace cyclones for solid removal.

There are two types of ESP—dry and wet. Selection of the wet ESP as the final particulate removal unit is driven by several factors. With the specified particulate loading in the syngas, the saturation environment inherent in the WESP reduces the possibility of igniting the syngas with the electric charge and, consequently, an energy release. WESPs can achieve up to several times the typical corona power levels of dry precipitators, greatly enhancing collection of submicron particles. In wet precipitators, reentrainment in the last field is virtually nonexistent due to adhesion between the water and collected particulate. Also, the gas stream temperature is lowered to the saturation temperature, promoting condensation and enhancing particulate collection. WESPs are highly efficient collectors of submicron particulate, including condensable aerosols.

WESPs are known to also capture mercury and organic aerosols that may condense from syngas at saturation temperatures.

3.2.4.4.1.4   Wet Scrubbers   Wet scrubbers are useful for final cleanup with water quenching and scrubbing [153,154]. In addition to fine particulates, chloride, ammonia, and some H_2S and other trace contaminants are also removed from the syngas during scrubbing process. There are four major categories of wet particulate scrubbers: spray tower scrubber, packed bed or tray-type scrubbers, mechanically aided scrubbers, and Venturi scrubbers.

The separation principle of wet scrubbers is that particulates are knocked out with water droplets by various means. Effectiveness of different devices depends on particle size, water droplet size, density of water droplets, etc. Residence time can be an important factor in determining efficiency, depending on design principles. Out of these scrubbers, Venturi scrubbers are most suitable for low-temperature particulate removal from syngas. Packed bed, tray, and spray towers are useful for the removal of several gaseous impurities.

Wet scrubbing systems use liquid sprays either water or chilled condensate from the process to remove particulates that collide with liquid droplets. The droplets are then removed from the gas stream in a demister. Venturi scrubbers are the most common wet scrubbers but often require a relatively high-pressure drop to circulate quench liquid to be sprayed into the gas stream. Because wet scrubbing requires a liquid quench medium, operating temperatures need to be less than 100°C. This requires significant gas cooling for removing the sensible heat from the product gases. Heat losses from the wet scrubbing systems can adversely affect the energy efficiency of the overall process. On the other hand, for indirect gasification systems, the excess steam that is used as the gasifying agent needs to be quenched and recovered. Wet scrubbing systems are inevitable in indirect gasification systems to remove excess water vapor prior to compression and downstream syngas utilization.

Air Products design [152] for syngas cleaning used Venturi scrubbing in two stages with a downstream liquid/gas separator. The Venturi quench scrubber is designed to reduce syngas temperature from 850°C to ~85°C and solid loading from 30,000 to 500 mg/N m^3 (dry basis). In addition to cooling the syngas to near saturation temperature, the dual Venturi scrubber will also remove a high percentage (around 99%) of solid particulate over 5 µm in size that may be carried over from the gasifier, thereby minimizing downtime. The convergent section of the Venturi is covered with a water film to eliminate particles sticking to the wall. High gas velocity in the Venturi throat causes violent agitation in the water film creating numerous fine droplets to capture the particulates. Because of the close contact between the water and hot gas, a rapid temperature drop takes place over a very short distance and time and brings all chemical reactions to an abrupt stop. The addition of caustic (NaOH) to control the resultant water stream to a pH range of 6.5–7 will remove greater than 99% of HCl and HF and two-thirds of ammonia from the syngas stream. Separation efficiencies of 99% have been commercially demonstrated in a two-stage unit. Rapid quench of the syngas will prevent recombination of organic molecules into complex organics such as dioxins or furans.

3.2.4.4.2   Sulfur Removal

3.2.4.4.2.1   Low-Temperature Processes   For low H_2S concentration (<2000 ppm) and low sulfur quantities (<3 Tons per day [TPD]), an absorption process with chemical agents that are selective toward H_2S is most suitable. Some of the low-temperature sulfur removal processes are briefly described here:

1. THIOPAQ is a biotechnological process for removing H_2S from gaseous streams by absorption into a mild alkaline solution followed by the oxidation of the absorbed sulfide to elemental sulfur by naturally occurring microorganisms. The process was first used in 1993. Its application to syngas cleaning is not demonstrated.

   CrystalSulf is a new process to remove H_2S. It was developed for natural gas environment of high pressure and high CO_2/H_2S ratio containing heavier hydrocarbons. The process uses flammable and hazardous chemicals for absorption and sulfur oxidation. SO_2 is used as a reactant to undergo the modified Claus process for sulfur formation. The process is not well demonstrated.

2. Merichem's LO-CAT™ process for H_2S removal is based on an iron chelate redox system. This is an oxidation process that uses iron catalyst held in a chelating agent to oxidize H_2S to elemental sulfur. H_2S is absorbed in the solution and sulfur is oxidized by iron, which is reduced from +3 state to +2. When the gas stream comes into contact with the LO-CAT solution in the absorber, H_2S in the gas stream is converted into sulfur. The spent catalyst is regenerated by oxygen in the oxidizer, and an elemental sulfur is concentrated into a sulfur slurry. The sulfur recovered in the process contains some entrained residual catalyst, and this low-value sulfur is used for agricultural purposes.

   The process is most useful for small-scale applications where sulfur recovery is less than 20 TPD. The process can achieve 99.9% H_2S removal efficiency. The process can operate between atmospheric pressure to as high as 600 psi pressure. The temperature is normally maintained at about 110°F. Since the process only treats H_2S, a COS hydrolysis unit upstream of the process is needed to hydrolyze COS. Other acid gases such as HCN and mercaptans are removed by wet scrubbing.

3. For low temperature and high pressure, physical solvent processes such as Rectisol/Selexol are also used. The Rectisol process that uses methanol at temperatures less than 32°F can achieve a sulfur removal level as low as 0.1 ppm. The Selexol process that uses mixtures of dimethyl ethers of polyethylene glycol can achieve a sulfur level of 1 ppm.
4. Amine processes are proven technologies for the removal of H_2S and CO_2 from gas streams by absorption. This is a low-temperature process in which gas usually enters the absorber at about 110°F. Regeneration of rich amine is accomplished through the flash separator followed by a stripper column to recover H_2S and CO_2 from amine solutions. Amine systems normally operate in the low to medium pressure range of 70–360 psi. A sulfur removal level as low as 1 ppm is possible. Commercially available amines are as follows:

    a. *MEA*: Monoethanolamine removes both H_2S and CO_2 and is generally used at low pressures and in operations that require stringent sulfur removal.
    b. *DGA*: Diglycolamine is used when there is a need for COS and mercaptan removal in addition to H_2S. DGA can hydrolyze COS to H_2S; thus, COS hydrolysis unit is not needed in the cleanup system.
    c. *DEA*: Diethanolamine is used in medium- to high-pressure systems (above 500 psia) and is suitable for gas stream with high ratio of H_2S to CO_2.
    d. *MDEA*: Methyldiethanolamine has a higher affinity for H_2S than CO_2. MDEA is used when there is low ratio of H_2S to CO_2 in the gas stream so that H_2S can be concentrated in the acid gas effluent.

One problem with amine system is corrosion. In water, H_2S dissociates to form weak acid and CO_2 forms carbonic acid, both of which corrode metal.

Envitech syngas cleaning process [97,150,151,153,154] uses a media such as sponge iron. This method for H_2S removal is also used by Air Products [152] and Alter NRG processes [155]. Pyrogenesis (company in Canada) process (which uses plasma for tar removal) uses either a regenerative or a nonregenerative technology depending on H_2S concentration. In nonregenerative technology, the contaminated syngas is passed through a bed of iron oxide–impregnated wood chips where the hydrogen sulfide is converted to ferric and ferrous sulfide. These processes can achieve 90% H_2S removal. In regenerative process, a chelate iron solution is used to convert H_2S to elemental sulfur. Its environmentally safe catalyst does not use toxic chemicals and produces no hazardous by-products. The process unit can be designed for better than 99.9% H_2S removal efficiency. The decision to select either nonregenerative or regenerative technology is based on trade-off between higher capital cost with the latter technology and higher operating costs with the former technology.

3.2.4.4.2.2 High-Temperature Removal   Oxides of many metals (particularly transition metals) will react with H_2S as $MeO + H_2S \rightarrow MeS + H_2O$. This reaction can be effectively used to reduce H_2S concentration in the syngas [97]. ZnO possesses one of the highest thermodynamic efficiencies for H_2S removal and the most favorable reaction kinetics of all the active oxide materials within the 149°C and 371°C temperature range.

The economics of ZnO process become more attractive if the ZnO-based materials can be regenerated and used for multiple cycles. The process uses transport reactor that allows much higher throughput and smaller footprints, resulting in lower cost and excellent control of the highly exothermic regeneration reaction. Research Triangle International (RTI's) experience [97] with ZnO-based materials has demonstrated that these materials can effectively remove large amount of H_2S ranging from 500 ppmv to as high as 30,000 ppmv to effluent concentrations typically below 10 ppmv. ZnO-based materials are typically regenerated using a mixture of oxygen and nitrogen. RTI has demonstrated that ZnO materials also react with COS and CS_2. This reaction requires temperature to be above 232°C.

3.2.4.4.2.3 COS Hydrolysis While COS can be removed along with H_2S in some processes, in chemical absorption processes, the degree of COS removal is dependent upon the reactivity of the solvent solution with COS. While DGA can remove all COS, MDEA has little reactivity with COS. In physical absorption processes, the solubility of COS in the physical solvent and the COS partial pressure determine the level of its removal. A COS level of 0.1 ppm is attainable in Rectisol process [97,150,151], while the Selexol process [97,150,151] can achieve 10 ppm of COS. In the ZnO process [97,150,151], approximately 80% of COS can be removed by hydrolysis. The hydrolysis of COS follows the reaction

$$COS + H_2O \rightarrow H_2S + CO_2 \qquad (3.1)$$

The scrubbed syngas feed is normally reheated to 30°F–50°F above saturation before entering the hydrolysis reactor to avoid catalyst damage by liquid water. COS hydrolysis uses an activated alumina-based catalyst and is normally designed to operate at 350°F–400°F. The reaction is largely independent of pressure. Due to exothermic nature of reaction, equilibrium is favored at low temperature.

3.2.4.4.2.4 Sulfur Recovery Unit In sulfur recovery unit, the acid gas stream from amine or physical solvent is recovered to elemental sulfur. For sulfur recovery of more than 20 TPD, Claus sulfur recovery unit (SRU) is an economical approach. For a low sulfur recovery LO-CAT SRU is a more suitable process.

3.2.4.4.3 Nitrogen Removals

Dayton et al. [97] provides an excellent review of the methods for nitrogen removals. Although NH_3 is not a highly stable molecule, its dissociation requires a very high temperature due to its high activation energy. Krishnan et al. [156] indicated that Haldor Topsoe A/S catalyst showed excellent activity even in the presence of 2000 ppm H_2S. G-65, an SRI catalyst, showed a superior activity in the temperature range of 550°C–650°C [97,150,151]. A wide variety of metal oxides/carbides/nitrides can catalyze decomposition of ammonia. Group VIII metals such as Fe, Co, Ni, Ru, Rh, Pd, Os, Ir, and Pt are also active in the metal state. Groups Va (V, Nb) and VIa (Cr, Mo, W) are especially active for ammonia decomposition. CaO, MgO, and dolomite are also very active at temperatures as low as 300°C. Besides catalytic processes, two methods for removing ammonia include catalytic tar reforming and wet scrubbing [97,150,151].

3.2.4.4.4   Chlorides Removals

For bulk removal of HCl, different sorbent materials and processes have been studied and different technologies are being developed. In the two-stage "ultra clean process," synthetic dawsonite, nahcolite, and trona ($Na_2CO_3 \cdot NaHCO_3 \cdot 2H_2O$) have been tested. Trona was found to be the best sorbent for HCl. Krishnan et al. [156] showed that alkali minerals reduced HCl concentration from 300 to 1 ppm in the temperature range of 550°C–650°C. Nahcolite was the best absorbent, which contained 54% of chloride by weight. SRI, RTI, and GE showed that nahcolite can reduce HCl concentration below 1 ppm in fixed- and fluidized bed reactors in the temperature range of 400°C–650°C [97,156].

SRI observed that 0.3 ppm of HCl in the product is achievable using Katalco chloride guard 59-3 at 550°C. Most HCl treatment materials are once-through that cannot be regenerated. The naturally occurring alkali-based materials are less reactive than commercial products due to their lower surface area.

The halides can also be removed by wet scrubbing or purification by hydrogenation and ZnO absorption. The HCl can also be absorbed in a HCl scrubber, which uses a caustic/water recirculating process liquid (maintained at pH 7.5). This guarantees the HCl outlet concentration to be 5 ppm.

In several plasma processes, a packed bed scrubber is used to remove acid gases (mainly HCl) from the syngas stream. In order to efficiently absorb HCl, a large surface area of contact is provided. The scrubber solution pH is continuously adjusted using controlled amounts of caustic soda solution.

3.2.4.4.5   Alkali and Heavy Metal Removals

Compared to sulfur, chlorides, ammonia, and particulate removals, technologies for alkalis and heavy metals such as mercury, arsenic, selenium, and cadmium are not as well developed. Commercial technologies are heavily focused on mercury and arsenic. Adsorbents for mercury are developed by UOP LLC based on type X and Y zeolites that have been coated with elemental silver. Various forms of activated carbon are also used effectively to remove mercury. A mercury removal unit based on a fixed bed of sulfur-impregnated activated carbon has been designed to reduce mercury concentration in water-saturated natural gas from 1000 to 5 ppm. Synetix also describes a process for removing mercury from natural gas using metal sulfides on inorganic supports [97,151].

Alkali removal is normally accomplished by cooling the syngas stream below 1100°F to allow condensation of alkali species followed by barrier filtration or wet scrubbing. Sintered metal filters work better than ceramic filters in filtration process. Baghouse filters were used in Lahti's low-pressure gasification system and the FERCO facility in Vermont [1,97,151].

Alkali can easily be removed by wet scrubbing; thus, it is often the preferred method for alkali removal. While some mercury is removed in the quench, scrubber, and direct contact cooler, generally, a mercury removing guard is provided.

Mercury removal from gas is normally achieved by adsorption. There are three types of adsorbent available commercially:

1. Activated carbon
2. Sulfur-impregnated activated carbon
3. Silver-impregnated molecular sieve

Activated carbon whether sulfur impregnated or not is used in nonregenerative fashion to remove mercury below 0.1 ppm. The silver-impregnated molecular sieve can remove mercury to equally low level and is regenerable. The sulfur-impregnated activated carbon has more mercury adsorption capacity than conventional activated carbon. It can also absorb other trace components such as VOCs and dioxins/furans with an estimated efficiency of 95%. The carbon bed can reduce mercury concentration in the exit gas to as low as 0.1 ppm.

Besides the aforementioned materials, Baghouse captures fly ash and heavy metals from the gas stream using bags that are precoated with the mixture of feldspar and activated carbon. In several gas cleaning processes, after removal of fine particles, acid gases, and hydrogen sulfide, the syngas is passed through a deep bed scrubber to remove additional fine particles, lead, cadmium, mercury, and total reduced sulfur.

3.2.4.4.6   Tar Removal

Tars contain polyaromatic hydrocarbons and they are the most difficult to remove during gasification process. Gasification plants that do not use thermal tar destruction generally have lower outlet temperatures but higher energy content in the syngas. Numerous approaches and physical, chemical, and catalytic processes have been attempted both commercially as well as in pilot-scale units. Here, we briefly summarize our state of the art in each of these approaches.

3.2.4.4.6.1   Physical   In this method the objective of the tar removal unit is twofold: (1) to get as much of the tars as possible in the condensed phase and (2) to maximize tar collection and removal, particularly for tar droplets larger than 1–3 μm. The physical methods involve wet scrubbing and filtration. Wet scrubbing to condense tars out of product gas is an effective gas conditioning technology that is commercially available and can be optimized for the tar removal. Wet scrubbing technique does not eliminate tars but transfers it to oil or water phase, which will then have to be appropriately treated.

1. *Oil loop tar removal*: Envitech process [153,154] uses two-stage physical process for tar removal. In the first stage, it uses an oil-based scrubber to condense large particles and heavier tars. Oil loop makes sense when (a) tar concentration is high, (b) tar condenses as sticky solid rather than an oily liquid, and (c) there is a high degree of condensed tars that do not mix well with the water. Under these conditions, oil scrubber provides a medium where tars will be more soluble and less prone to foul the recirculation lines.

2. *Water loop tar removal*: While some syngas compositions will require an organic recirculation liquid, others will be suitable for a water loop configuration. The syngas is first cooled to saturation in a vertical downflow quencher using water. This is followed by a vertical Venturi scrubber to collect the condensed tar and particulate. In Envitech process [153,154], oil loop and water loop are operated as two stages for tar removal.

3. *Venturi wet scrubbers*: Wet scrubbing is generally used to remove water-soluble contaminants from the syngas by absorption into a solvent. It can be used for water-soluble tar components, particulates, alkali species, halides, soluble gases, and condensable liquids. A Venturi scrubber and mist eliminator is used to collect particulate and condensed tars in the recirculation liquid in both oil and water loop configurations. The literature data show that Venturi scrubber is effective at removing roughly 90%–99% of condensed tars and particulate larger than 1 µm.

3.2.4.4.6.2 Chemical/Thermal/Catalytic Treatments   A number of chemical and catalytic treatments can also be used to remove tars from syngas:

1. *Thermal cracking*: Thermal cracking of tars using solid acid catalysts such as silica–alumina and zeolites is well known [97]. The catalyst with strong acid sites can crack hydrocarbons above 200°C. The cracking of aromatics results in the formations of hydrocarbons and coke. Gil et al. [157] showed that cracking of tars with both spent FCC catalyst and dolomite was highly successful.

2. *Hydrogenation*: While hydrogenation of polyaromatic tar compounds is possible, it is thermodynamically not very favorable at higher temperatures. It is however possible to have ring opening activity at temperatures as low as 200°C for Rh, Pt, Ir, and Ru catalysts.

3. *Plasma conversion*: In pyrogenesis gas cleaning process, the soot and complex hydrocarbons coming out of cyclones and gasifiers' tars are converted to syngas by a secondary plasma gasifier. The main gasification reactions occurring in the secondary gasifier are

$$C + O_2 \rightarrow CO_2 \text{ (exothermic)} \tag{3.2}$$

$$C + H_2O \rightarrow CO + H_2 \text{ (endothermic)} \tag{3.3}$$

$$C + CO_2 \rightarrow 2CO \text{ (endothermic)} \tag{3.4}$$

$$CO + H_2O \rightarrow CO_2 + H_2 \text{ (exothermic)} \tag{3.5}$$

4. *Catalytic steam reforming*: Catalytic steam reforming of tars is a developing technology with significant research interests during last three decades. In this technology, tars are reformed by steam and a catalyst in a fluidized bed reactor to produce syngas. This technique offers several advantages: (a) catalyst reactor temperatures can be thermally integrated

with the gasifier exit temperature, (b) the composition of the product gas can be catalytically adjusted, and (c) steam can be added to the catalytic reactor to ensure complete reforming of tars. Among others, the catalysts that are well tested are (a) disposable catalysts such as dolomite and olivine; (b) conventional Ni and other catalysts that include ruthenium, iron, manganese, potassium, and barium; (c) catalysts with novel supports such as perovskites and hexaaluminate; and (d) more bi- and even trifunctional catalysts. Of all the catalysts tested, disposable catalysts or commercial Ni catalysts are preferred due to their low costs, durability, and long life. A proprietary nickel monolith catalyst has also shown considerable promise for destruction of biomass gasification tar [97,150,151]. A tar cracker known as the reverse-flow tar cracking (RFTC) reactor developed by BTG uses steam reforming process with a commercial Ni catalyst [151].

The typical conversion for the RFTC reactor is shown in Table 3.5.

5. *Partial oxidation*: The partial oxidation reaction was also investigated as the possible process for tar removal [97,150,151]. In this process, syngas enters the reactor with oxygen at 300°F and leaves at about 2500°F due to highly exothermic oxidation reactions. Tar, methane, light hydrocarbons, and benzene are converted to syngas. The disadvantage of this process is the low product heating value due to combustion of CO and H_2 as well. Also it is very difficult to selectively react methane by this process. The composition of the product is also shifted to lower H_2/CO ratio.

An autothermal reactor that includes both steam reforming and partial oxidation has also been successfully applied. Such an application would only apply to a particulate-free gas since any particulate in the gas could shortly blind catalyst surface.

TABLE 3.5
Typical Conversion for RFTC Reactor

Components	Conversion (%)
Benzene	82
Naphthalene	99
Phenol	96
Total aromatic	94
Total phenols	98
Total tar	96
Ammonia	99

Source: Equipment design and cost estimation for small modular biomass systems, synthesis gas cleanup, and oxygen separation equipment—Task 2: Gas cleanup design and cost estimates-wood feedstock, NREL/DOE Subcontract Report NREL/SR-510-39945, Nexant Inc., San Francisco, CA, May 2006.

3.2.5 POTENTIAL ISSUES AND OPTIONS FOR THE PLANT CONFIGURATIONS

Coal–biomass gasification complete plant can have a number of configurations, which are influenced by [1,2,150]

1. Overall scale of the plant and composition of feedstock
2. Required biomass capacity and its geographical dispersion relative to central gasification facility
3. Biomass pretreatment requirement
4. Levels of integration of various equipment
    a. At pretreatment stage
    b. Within the gasification stage
    c. Within the product cleaning and separation stage

Various configurations can provide different degrees of flexibility, operability, and reliability of the entire plant operations.

In order to take the maximum advantage of the economy of scale, the size of the co-gasification plant should be as large as possible. This will allow the most efficient gasification and syngas cleanup operations. Once the total plant size is fixed, the most important parameters that affect plant configuration are (1) the biomass concentration in the feed, (2) nature of biomass (e.g., wood vs. straw), (3) size requirement of biomass based on the chosen gasification reactor technology, (4) the availability of biomass locally or the transport distance for the biomass feedstock, and (5) required flexibility for processing of coal and biomass for stable and sustainable operation. An analysis of the last item sets up the degree of common versus segregated pretreatment, gasification production, and syngas cleanup systems.

Another important issue is the location of the pretreatment configurations: on-site or off-site. This issue is very important when biomass is not available locally and it has to travel long distance to the gasification plant. Low-density biomass is expensive to move to long distance. Often, it makes sense to carry out pretreatment at the site of the biomass and then transport the prepared biomass to the plant. This saves money if the density, particle size, shape, and structure are significantly altered during the pretreatment process like, torrefaction, pelletization, or flash pyrolysis to generate feedstock in a slurry form. The original nature of biomass plays important role in the pretreatment process. The pretreatment enhances both mass and energy densities of biomass, which significantly reduces biomass transportation costs. If the plant is designed for multiple types of biomass, each coming at significant distance from the central gasification facility, geographically dispersed biomass pretreatment facilities make significant sense. The transportation of coal at any distance is not a significant issue.

Process configuration is more important and more complex for gasification than combustion because the products are used for downstream upgrading as well as power generation. Process configuration has at least the following five options:

1. Gasify both coal and biomass together in the same gasifier and have a unified downstream operation. In this case, it is also possible to feed biomass

and coal separately in the same gasifier but at different locations. This is to take advantage of different reactivities of coal and biomass.

2. Gasify coal and biomass in separate gasifiers with separate pretreatment systems and then combine the product gas for a unified downstream operation. This allows the optimum gasifier design for each type of feedstock.
3. Same as (2) in pretreatment and gasifier systems but have separate cleaning steps and then combine the cleaned syngas coming from coal and biomass.
4. Any options from (2) or (3) but have different forms of feed for biomass (dry or slurry).
5. Any options from (2), (3), or (4) but share a common or separate air separation unit (ASU) to produce oxygen for the biomass or coal gasifiers.

These different options, and others, lead to different process configurations. Ratafia-Brown et al. [1,2,150] considered the following six possible configurations and their advantages and disadvantages. This analysis will allow others to consider other possible configurations:

- *Configuration 1*: Co-feeding coal and biomass to the gasifier as a mixture in dry or slurry form.
- *Configuration 2*: Co-feeding biomass and coal to the gasifier using separate gasifier feed systems, either in dry or slurry form.
- *Configuration 3*: Pyrolyzing as received biomass followed by co-feeding pyrolysis char and coal to the gasifier and separately feeding pyrolysis gas to the syngas cleanup system.
- *Configuration 4*: Biomass and coal are co-processed in separate gasifiers followed by a combined syngas cleanup.
- *Configuration 5*: Biomass and coal are co-processed in separate gasifiers followed by separate syngas cleanup trains, and the syngas feeds are combined prior to sulfur and CO_2 removal unit operations.
- *Configuration 6*: Same as (4) and (5) but share common ASU for oxygen feed to the separate gasifiers.

Ratafia-Brown et al. [1] gave an extensive assessment of these six process configurations. A summary assessment and advantages and disadvantages of each option is described in my previous books [2,3]. The analysis can also apply to other mixtures such as coal and waste and biomass and waste of widely differing properties.

The important thing here is that process options for co-gasification are many more and significantly more complex than the ones discussed earlier for co-combustion. While co-gasification is more complex than co-combustion, it also provides more flexible product distribution.

3.2.6   BRIEF REVIEW OF LITERATURE

While co-gasification is not as advanced and as much investigated as co-combustion, during last two decades, significant studies at various scales have been reported. Here, we examine some of them to illustrate new findings on the subject.

3.2.6.1   Co-Gasification of Coal and Biomass in Intermittent Fluidized Bed

Wang and Chen [172] examined co-gasification of coal and biomass in an intermittent fluidized bed reactor to investigate the effects of temperature (T), steam-
to-biomass ratio (SBMR), and biomass-to-coal ratio (BMCR) on hydrogen-rich gas
production. The results showed that H_2-rich gas, free of N_2 dilution, is produced, and
the H_2 yield was in the range of 18.25–68.13 g/kg. The increase of T, SBMR, and
BMCR is all favorable to the production of hydrogen. Both hydrogen and carbon
monoxide contents and hydrogen yield are increased with temperature. While gas
composition was not strongly affected by BMCR, the yield and content of hydrogen
increased with BMCR, reaching a maximum at BMCR = 4. While hydrogen content
and yield increased with SBMR, carbon monoxide showed a maximum with respect
to SBMR. The study showed that the order of the influence of the operation parameters on hydrogen production efficiency is T > SBMR > BMCR.

3.2.6.2   Co-Gasification of Coal and Biomass in
High-Pressure Fluidized Bed

McLendon et al. [173] examined co-gasification of coal and biomass in a jetting,
ash-agglomerating, fluidized bed pilot-scale-sized gasifier. Biomass used was sand-
ing waste from furniture manufacturer. Powder River Basin subbituminous and
Pittsburgh No. 8 bituminous coals were mixed with sawdust. Feed mixture ranged
up to 35 wt% biomass. The results with subbituminous coal/sawdust mixture showed
few differences in operations compared to only subbituminous coal tests. The bitdus
coal/sawdust mixture showed marked difference from bituminous coal-alone data.
Transport properties of coal/biomass mixtures were greatly improved compared to
coal only.

3.2.6.3   Co-Gasification of Coal and Polyethylene Mixture

Yasuda et al. [158,159] examined hydrogasification of coal/polyethylene (PE) mixtures at 1073 K under 7.1 MPa of hydrogen. The reaction time varied from 1 to 80 s.
Coal/PE mixtures in the ratio 90:10 and 75:25 were used in the study. Both product
distribution and temperature profiles were analyzed. For PE alone tests, yield (carbon basis) of methane reached 90%. A significant synergistic effect (even with 10%
PE in the mixture) was found when coal and PE were mixed and used in hydrogasification. The study indicated that the compensation of endothermic coal pyrolysis
process by heat evolved from hydrogenation of PE may be the reason for synergy.
The early drop in coal hydrogasification was prevented during co-gasification, and
the gasification of the mixture resulted in increased carbon conversion to methane.
The study recommended this mixture for future hydrogasification processes.

3.2.6.4   High-Pressure Co-Gasification of Coal with
Biomass and Petroleum Coke

Fermoso et al. [160] examined the effects of temperature, pressure, and gas composition on gas production, carbon conversion, cold gas efficiency, and high heating value
during the steam–oxygen gasification of a bituminous coal. The temperature and
oxygen concentration were the most important variables during gasification process.

Co-gasification was studied with biomass concentration up to 10% and petroleum coke up to 60%. The ternary mixture of coal–petcoke–biomass with 45:45:10 was also studied to evaluate the effect on gas production and carbon conversion. Co-gasification produced very positive results.

3.2.6.5 Co-Gasification of Woody Biomass and Coal with Air and Steam

Kumambe et al. [141] examined co-gasification of woody biomass and coal with air and steam to produce syngas for light fuels. The experiments were carried out in a downdraft fixed bed reactor at 1173 K. The study varied BMCR from 0 to 1 on carbon basis. The gas production increased with increased concentration of biomass, whereas the production of char and tar decreased. With the increase in biomass concentration, hydrogen concentration decreased and carbon dioxide concentration increased. However, CO concentration in the gas phase was independent of biomass concentration in the feed. A low biomass ratio produced gas-phase composition more suitable for methanol and hydrocarbon fuel synthesis, and a high biomass concentration produced gas favorable for DME synthesis. The synergy due to co-gasification may be observed in the extent of the water gas shift reaction. The gasification conditions in the study provided a cold gas efficiency ranging from 65% to 85%.

3.2.6.6 Co-Gasification of Coal, Biomass, and Plastic Wastes with Air/Steam Mixtures in Fluidized Bed

Pinto et al. [161] studied optimum conditions for co-gasification of coal and waste with respect to gas using only air, only steam, and mixtures of them. An increase in temperature increased hydrogen and decreased tars. Increasing temperature from 750°C to 890°C for a mixture of 60:20:20 coal/pine/PE waste (w/w) led to a decrease in methane and other hydrocarbon concentration of about 30% and 63%, respectively, while hydrogen concentration increased around 70%. An increase in air flow rate decreased hydrocarbon and tar production. The presence of air also decreased higher heating value of the product gas. The increase in steam increased reforming reaction and thereby increased hydrogen production.

3.2.6.7 Co-Utilization of Biomass and Natural Gas in Combined Cycles

Power production from biomass can occur through external combustion (e.g., steam cycle, organic Rankine cycles, Stirling engines) or internal combustion after gasification or pyrolysis (e.g., gas engines, IGCC). External combustion has the disadvantage of low efficiency (30%–35%). Internal combustion, on the other hand, has the potential of high efficiencies, but it always needs a more severe and mostly problematic gas cleaning.

De Ruyck et al. [162] examined an alternate route where advantages of external firing are combined with the potential high efficiency of the combine cycles through co-utilization of natural gas and biomass. Biomass is burned to provide heat for partial reforming of the natural gas feed. In this way, biomass energy is converted into chemical energy contained in the produced syngas. Waste heats from the reformer and from biomass combustor are recovered through a waste heat recovery system. The study showed that in this way, biomass can replace up to 5% of the energy in the natural gas feed. The study also showed that in the case of combined cycles, this alternative route allowed for external firing of biomass without important drop in cycle efficiency.

3.2.6.8 Steam Gasification of Coal–Biomass Briquettes

Yamada et al. [163] examined a biobriquette made by mixing low-grade Chinese coal and larch bark with $Ca(OH)_2$ as desulfurizing agent to measure sulfur evolution in mixture of nitrogen and steam at 1173 K. The briquettes were more effective in early pyrolysis condition than subsequent steam gasification condition for H_2S removal.

3.2.6.9 Syngas Production by Co-Conversion of Methane and Coal in a Fluidized Bed Reactor

Wu et al. [164] examined the concept of co-converting methane reforming and steam gasification of coal at 1000°C in a fluidized bed reactor without a catalyst. This concept is applicable where coal bed methane can be used to reform methane and gasify coal by steam simultaneously in situ. In addition, the integrated coal gasification and gas reforming offers an advantage that the H_2/CO ratio in the produced syngas can be adjusted between 1 and 3 by varying the ratio of coal/gas in the feedstock. The study showed some initial success of the concept in which over 90% conversion of natural gas in a laboratory fluidized bed reactor was obtained with favorable quality of produced syngas (i.e., high H_2 and CO and low CH_4 and CO_2).

3.2.6.10 Co-Gasification of Coal and Biomass in a Dual Circulating Fluidized Bed Reactor

Seo et al. [176] and others [177,178] examined the effects of temperature (750°C–900°C), steam to fuel ratio (0.5–0.8), and biomass ratio (0, 0.25, 0.5, 0.75, 1.0) on co-gasification of coal and biomass in dual circulating fluidized reactor (combustor/gasifier). Indonesian Tinto subbituminous coal and *Quercus acutissima* sawdust were used as coal and biomass, respectively. With increasing temperature and steam/fuel ratio, the product gas yield, carbon conversion, and cold gas efficiency of the mixtures were higher than the ones for coal alone. After pyrolysis, surface area, pore volume, and micropores of coal/biomass blend char increased. The maximum increase in gas yield can be obtained with a biomass ratio of 0.5 at the given reaction temperature. Calorific values of the product gas were 9.89–11.15 MJ/m³ with the coal, 12.10–13.19 MJ/m³ with biomass, and 13.77–14.39 MJ/m³ with coal/biomass blends at 800°C. The synergistic effects on the basis of calorific value and cold gas efficiency were pronounced with the coal/biomass blends. This study also presents a summary of previous co-gasification studies.

3.2.6.11 Co-Gasification of Coal and Chicken Litter

Priyadarsan et al. [166] examined co-gasification of coal and chicken litter in a 10 kW capacity fixed bed countercurrent atmospheric pressure gasifier at air flow rate of 1.3 and 1.7 m³/h under batch mode of operation. An increase in air flow rate decreased heating value of product gas from 5.2 to 4.9 MJ/m³, which resulted in a minor decrease in gasification efficiency. For both air flow rates, the product gas composition was CO at 30% dry basis and H_2 at 10% dry basis. The presence of coal completely inhibited ash agglomeration in the bed, which was observed during gasification of pure chicken litter. This can be attributed to reduced amount of Na and K in the blended fuel as compared to pure chicken litter biomass ash.

3.2.6.12   Co-Gasification of Biomass and Waste Filter Carbon

Sun et al. [165] examined co-gasification of waste filter carbon and char of wood chips with steam at atmospheric pressure. The effects of temperature 600°C–850°C and partial pressure of steam from 0.3 to 0.9 atm on the gasification rate were examined. The modified volumetric reaction model was used to evaluate kinetic data. The gasification rate of waste filter carbon was compared with that of co-gasification rate. The activation energies of filter carbon and wood chips were determined to be 89.1 and 171.4 kJ/mol, respectively.

3.2.6.13   Co-Gasification of Low-Rank Fuel–Biomass, Coal, and Sludge Mixture in a Fluidized Bed in the Presence of Steam

Ji et al. [167] examined co-gasification of low-rank fuels in a fluidized bed reactor. Within the range of experimental conditions examined, the highest amount of hydrogen and carbon monoxide was observed at 900°C and steam partial pressure of 0.95 atm. Temperature and steam were the most important variables in the system. High temperature favored hydrogen production and gas yield but did not always favor heating value. Sludge, oil, and coal mixture showed great potential for gas production. As reaction temperature and steam partial pressure increased, the heating value of product gas increased. Kurkela et al. [246] examined gasification of peat and biosolid.

3.2.6.14   Co-Gasification of Residual Biomass/Poor Coal in a Fluidized Bed

Pan et al. [170] examined co-gasification of residual biomass/poor coal blends and gasification of individual feedstock used in the blends in a bench-scale, continuous fluidized bed reactor working at atmospheric pressure. Two types of blends were prepared, mixing pine chips (from Valcabadillo, Spain) with black coal, a low-grade coal from Escatron, Spain, and Sabero coal, a refuse coal from Sabero, Spain, in the ratio range of 0/100–100/0. Experiments were carried out using mixtures of air and steam at gasification temperatures of 840°C–910°C and superficial fluidized gas velocities of 0.7–1.4 m/s. Feasibility studies were very positive, showing that blending effectively improved the performance of fluidized bed co-gasification of the low-grade coal and the possibility of converting the refuse coal to a low-Btu fuel gas. This study indicated that a blend ratio with no less than 20% pine chips for the low-grade coal and 40% pine chips for the refuse coal is the most appropriate. The dry product gas low heating value augmented with increasing blend ratio from 3700 to 4560 kJ/N m^3 for pine chips/low-grade coal and from 4000 to 4750 kJ/N m^3 for pine chips/refuse coal. Dry product gas yield rose with the increase of the blend ratio from 1.80 to 3.20 N m^3/kg (pine chips/low-grade coal) and from 0.75 to 1.75 N m^3/kg (pine chips/refuse coal), respectively. The study indicated that about 50% co-gasification overall process thermal efficiency can be achieved for the two types of blend.

3.2.6.15   Co-Gasification of Biomass and Coal for Methanol Synthesis

Chmielniak and Sciazko [169] examined the economy of methanol production through coal–biomass gasification by linking it with modern gas–steam power systems. The essence of linking is the full utilization of the capacity of coal–biomass gasification installations. The paper describes the up-to-date experience of coal–biomass gasification including processing toward syngas production and methanol production. A conceptual flow diagram of pressurized and oxygen-fed co-gasification of coal and biomass integrated with combined cycle and parallel methanol production is evaluated. The effect of methanol production rate on the economy of power production is assessed.

3.2.6.16   Underground Co-Gasification of Coal and Oil Shale

Zhao et al. [181] tested the feasibility of in situ co-gasification of coal and oil shale. Based on the specification analysis of coal and oil shale through simulating the occurrence state and characteristics of coal and oil shale, the underground coal gasification model test was carried out. The experiments were carried out for different oxygen-to-steam ratios of 0.3, 0.35, 0.4, 0.45, and 0.5. The effect of temperature on quality of gas was studied. The results showed that for oxygen/steam ratio of 0.4–0.45, temperature rising rate was 7°C/min and extended rate of gasification was 0.036 m/h, the extend of temperature field was continuous and stable, and both oil shale and coal temperature changes were uniform. The high temperature of 1000°C achieved here satisfied the requirement for oil gas production. The heating value of syngas improved 26.37% by co-gasification.

3.2.6.17   Co-Gasification of Petcoke and Coal/Biomass Blend

Khosravi and Khadse [178] presented an excellent review of petcoke and coal/biomass blend gasification. The following discussion is a brief summary of their review.

A number of studies have investigated synergetic effects of co-gasification of petcoke and coal blend. These include co-gasifications of anthracite and petcoke in thermogravimetric analyzer (TGA), which exhibited positive synergy due to catalytic effects of AAEMs [21,178]. Similarly, due to catalytic effects, CO_2 gasification of petcoke was found to be less than that of several coals [21,178]. Addition of a catalyst lowered the gasification temperature in the petcoke. Numerous studies [21,175–178] examined the effects of coal residue that had AAEMs and transition metals [21,178], the effects of transition metal and iron species [21,178], and the effects of calcium-promoted potassium carbonate catalyst on gasification. In all cases, gasification rate was significantly enhanced due to the presence of a catalyst.

Biomass was also found to have a positive effect of gasification if blended with petcoke [180]. Vera et al. [180] studied co-gasification of biomass and petcoke in a TGA and fluidized bed reactor. They found that higher biomass content led to shorter gasification times. Fermoso et al. [160] performed co-gasification of binary and ternary mixtures of biomass, coal, and petcoke in a highly pressurized fixed bed reactor and found that addition of biomass to coal up to 10% raised the cold gas efficiency and carbon conversion, whereas addition of more than 10% biomass into the blend did not have any effect on gas composition. The gas composition was highly altered in all blends by temperature and concentrations of oxygen and steam. Higher pressure slightly decreased CO and H_2 productions. Fermoso et al. [160] also found interactions among different components of the blends during co-gasification.

Nemanova et al. [247] studied co-gasification of petroleum coke and biomass in an atmospheric bubbling fluidized bed reactor and a TGA at KTH Royal University of Technology. Biomass ash in the blends was found to have a catalytic effect on the reactivity of petroleum coke during co-gasification. Furthermore, this synergetic effect between biomass and petcoke was observed in the kinetic data. The activation energy E_a determined from the Arrhenius law for pure petcoke steam gasification in the TGA was 121.5 kJ/mol, whereas for the 50/50 mixture, it was 96.3, and for the 20/80 blend, 83.5 kJ/mol.

Several other studies on co-gasification are also reviewed by Brar et al. [168] and others [141,158–182].

3.2.7 Commercial Co-Gasification Processes

In recent years, several large-scale demonstrations of coal–biomass co-gasification technology have taken place [1,181]. A combination of petroleum coke and coal has been studied on a large scale by the Elcogas project in Spain and the Global energy projects in Germany, the United States, and Scotland. There also seems to have been more interests in coal–waste than coal–biomass projects. For large-scale power generation (>50 MWe), the gasification field is dominated by plants based on the pressurized, oxygen-blown, entrained-flow, or fixed bed gasification of coal. Entrained-flow co-gasification operational experience to date has largely been with well-controlled fuel feedstock with short-term trial work, a somewhat narrow range of co-gasification ratios, and easily handled fuels [1,9,10,179].

A critical assessment of co-gasification of coal and waste or biomass is given by Rickets et al. [9]. Among other subjects, the paper gives an objective assessment of current status of co-gasification and supporting technologies. The paper also gives a critical assessment of commercial readiness, strength, and weakness of existing technologies for fuel handling, gasification, fuel gas cleanup, and conversion of fuel gas to electric power.

There are at least two commercial plants employing co-gasification of coal and biomass mixtures. These are briefly described here.

3.2.7.1 250 MWe IGCC Plant, Nuon Power, Buggenum, B.V.-Willem-Alexander Centrale

Due to Dutch government policy decision called "Dutch Coal Covenant," 253 MWe Nuon power plant at Buggenum, Netherlands, was converted from coal plant to coal–biomass plant to make CO_2 emission reduction of 200,000 metric tons/year. This amounted to about 30 wt% of biomass use with relative biomass and coal feeds as 185,000 and 392,000 metric tons/year, respectively. This plant was started in 1993 and uses Shell dry feed gasification technology. The gasifier is an oxygen-blown, continuous slagging, entrained-flow reactor. It is designed to accept a wide range of imported coals and contains several design features that differ from the

U.S. IGCC plants. The ASU and the gas turbine are tightly coupled where the gas turbine compressor supplies all the air to ASU. This increases the plant efficiency but also makes plant more complex and difficult to start. The plant efficiency based on lower heating value is about 43%.

The process involves pulverized and dried coal being pressurized in lock hoppers and fed into the gasifier with a transport gas via dense phase conveying. The carrier gas is nitrogen or product gas when nitrogen in the product gas is undesirable. The coal is oxidized by preheated, 95% pure oxygen mixed with steam. The coal is oxidized in the temperature range of 1500°C–1600°C and pressure range of 350–650 psi to produce a syngas principally composed of hydrogen and carbon monoxide with little carbon dioxide. Very-high-temperature gasification eliminates any hydrocarbon gases and liquids in the product gas [1,9,10,177,182]. The solid ash in the form of slag runs down the refractory-walled gasifier and is collected into a water bath as slurry. The gas leaving at about 1400°C–1650°C contains a small amount of char and about half of the molten ash. The hot gas is cooled by a couple of cooling stages where waste heat is recovered to generate steam that can be used in other parts of the processes. The solid particles from cooled gas are removed by cyclones, and the cooled gas then goes through a series of gas cleaning processes to remove sulfur (H_2S, COS) and chlorine before using the clean syngas for downstream purposes of heat, power, or liquid fuels and chemicals.

The plant processed mixed feedstock of coal and biomass first from 2001 to 2004 with about 18% by weight of pure and mixed biomass. More recently, biomass concentration has increased up to 30 wt%. In addition of demolition wood, tests were also performed with chicken litter and sewage sludge. The test program evaluated the effects of biomass on product gas and ash quality. The Nuon/IGCC plant uses the coal and biomass composition shown in Table 3.6. As shown in the table, the plant takes about 30 wt% biomass, most of which is waste wood to provide about 17% of energy input to the gasifier.

TABLE 3.6

Coal and Biomass Compositions of Nuon/IGCC Plant

Feedstock Type	Lower Heating Value (MJ/kg)	Feedstock Input (1000 Metric Tons/Year)	Feedstock (% by Weight)	Feedstock (% by Energy Input)
Waste wood	15.4	130	22.5	14
Dried sewage sludge	8.2	40	7	2.3
Other biomass	10.2	10	0.5	0.7
Total biomass	13.6	185	30	17
Coal	29	400	70	83
Total feed	24.4	577	100	100

Source: Ratafia-Brown, J. et al., Assessment of technologies for co-converting coal and biomass to a clean syngas—Task 2 report (RDS), DOE/NETL-403.01.08 Activity 2 Report, May 10, 2007.

3.2.7.2   250 MWe IGCC Plant, Tampa Electric's Polk Power Station

This is another co-gasification commercial plant that uses slurry-fed GE (formerly Texaco) gasification technology that also uses oxygen-blown, continuous entrained bed gasification reactor [1]. This process was brought online in 1996. In 2001/2002, the Department of Energy sponsored a project to demonstrate if Polk Unit # 1 can coprocess biomass as a fraction of its primary coal/coke feedstock (98% particle less than 12 mesh in size) without significant impact on its performance. The biomass used was a 5-year-old locally grown Eucalyptus grove with feed concentration up to 1.2% by weight for about 8.5 h. The original system was not designed to handle softer fibrous biomass. The results showed that biomass did not impede the performance of the plant and it yielded 860 kW (7700 kW h total) of electricity during the test period based on the relative heating value and flow rates of biomass and base fuel.

The Polk Power Station used old Chevron Texaco IGCC technology that is now owned by General Electric. In this process, 60%–70% of coal–water slurry is fed to the gasifier at the rate of 2200 tons (on dry basis) of coal per day. The normal feed is a blend of coal and petcoke, the solid residue from crude oil refining. The fresh feed is mixed with unconverted recycled solids and finely ground in rod mills until 98% of the particles are less than 12 mesh in size. The slurry passes through a series of screens before being pumped into the gasifier. The slurry and oxygen are mixed in the gasifier process injector. The gasifier is designed to convert 95% of carbon per pass, and it produces syngas of 250 Btu/scf heat content.

A schematic of the Polk Power Station plant is shown in my previous book [2]. In this process, the syngas coming out of gasifier is cooled in a series of steps, each recovering heat in the form of saturated high-pressure steam. The first syngas cooler, called the "radiant syngas cooler" (RSC), produces 1650 psig saturated steam. The gas from RSC is split into two streams, and they are sent to parallel convective gas coolers (CSC) where the process cooling and generating additional high-pressure steam (at lower temperature) is repeated. The gases then further go through a simultaneous cooling and impurity removal (particulates, hydrogen chloride, etc.) process. A final trim cooler reduces the syngas temperature to around 100°F for the cold gas cleanup (CGCU). The CGCU system is a traditional amine scrubbers system, and it removes sulfur, which is then converted to sulfuric acid and sold to the local phosphate industry.

The eucalyptus feedstock used in this power plant contained about 1/3 of heating value per pound at about half the density of coal. The characteristics of the mixed feedstock for the Polk power plant are shown in Table 3.7. These numbers indicate that even a modest concentration of this biomass will require a massive and expensive feed system. Although the combined characteristics of the mixed feedstock are not considerably different from the baseline, it increases hydrogen, oxygen, and ash content by 4.6%, 11%, and 3.4%, respectively. The CO_2 discharge is reduced by 0.87%. Biomass used in the Polk plant did not lend itself to size separation and screening, and it caused minor plugging of the suction to one of the pumps [1,2,6]. The results indicate that for a slurry system, feed preparation must be tailored to the nature of the biomass in order to prevent any malfunction by the slurry pump as well as downstream gas cleaning and turbine operation. Typical experimental results for

TABLE 3.7

Polk IGCC Plant Coal/Coke and Biomass Combined Feedstock

Feed Composition (wt%)	Coke + Coal	Biomass	Combined Feed	Recycle Solids to Gasifier
C	82.88	49.18	82.02	66.26
H	4.50	5.78	4.71	0.29
N	1.85	0.24	1.81	0.95
S	2.99	0.06	3.13	2.31
O	3.53	39.42	3.92	0.00
Ash	4.25	5.32	4.4	30.19
Total	100.00	100.00	100.00	100.00
HHV, Btu/lb dry	14,491	8,419	14,470	9,698
% of original feed recycled	—	—	—	48.6
lb carbon/million Btu	57.2	58.41	57.21	68.32
Effective lb carbon/million Btu[a]	57.2	1	56.71	

Sources: Ratafia-Brown, J. et al., Assessment of technologies for co-converting coal and biomass to a clean syngas—Task 2 report (RDS), DOE/NETL-403.01.08 Activity 2 Report, May 10, 2007; McDaniel, J., Biomass gasification at Polk Power Station—Final technical report, DOE award DE-FG26-01NT41365, May 2002.

[a] Accounts for biomass carbon recycle and carbon released during biomass preparation.

the Polk Power Station are described by Ratafia-Brown et al. [1,150]. The experience of the Polk Power Station can be extended to coal and other materials.

Based on the experiences obtained with these two commercial plants, the co-gasification of coal and biomass had a number of operational observations on both plants. A summary of those identified by Ratafia-Brown et al. [1,150] is described here:

1. While biomass size is not very important for fluidized and moving bed reactors, the preferred entrained bed reactor requires particle size to be 1 mm or less. In Polk and Buggenum plants [1,2], milling wood particles up to 1 mm was found difficult and energy intensive. Milling wood to 100 μm size required energy consumption of 0.08 kWe/kWth wood, a number highly uneconomical. Torrefaction can reduce this energy consumption by a factor of 4–8, but that process requires significant energy consumption.

2. While testing showed good performance with co-gasification of woody biomass and coal, the processing of other materials like straw, switchgrass, and miscanthus was more problematic. Transferring these materials to the plant in an economical manner and feeding into the pressurized gasifier in a suitable form were critical to performance and overall efficiency. While grass may cause agglomeration and it could be accompanied by fairly sturdy weed stalks, it is less likely to plug pumps and lines than woody fuels if fed in small quantities. If coal and switchgrass particles are stored together in a

bunker, they may create flow problems. Often, switchgrass is handled separately from coal. Due to fine particle size distribution of woody biomass, no increase of wear was measured in pulverized coal feed equipment at the Buggenum plant.

3. The short-rotation woody crop (SRWC) harvesting and preparation at the Polk Power Station was cumbersome and expensive due to its short test. Deliberate long-term strategy for harvesting and preparation was needed. Switchgrass may be prepared differently than SRWC feedstock. This depends on the form in which it arrives at the plant. It can be chopped down and mixed with coal, it can be torrefied and then pulverized with coal, or if arrived as pellets, it can be pulverized with the coal directly. Feed preparation will strongly depend on the nature and shape of biomass.

4. Fuel feeding and handling for Polk IGCC tests were also very labor intensive. Some large fragments got in the system and caused some problems. It was concluded that a dedicated automated feed system with better protection against oversize materials would be required in further commercial utilization of biomass at Polk. One other option is to use coal mill to further pulverize wood chips but this needs to be investigated.

5. Feeding biomass into high-pressure gasifier in dry form was very challenging. If lock hoppers are used, due to low density and heating value of biomass, (a) large amounts of inert gas are required and they must be compressed, and (b) gasification efficiency drops due to dilution of syngas. If pneumatic feeders are used, it consumes excessive energy in pressurizing and pneumatically feeding biomass fuel powder into a 40 bar reactor. If particles of 1 mm can be used in the entrained-flow gasifier, screw feeder can do the job with much less energy consumption. Furthermore, if piston feeder instead of lock hoppers is used, the inert gas consumption reduces even further. The total efficiency penalty can be reduced by more than 50% [1,2]. At Buggenum, dry feed lock hopper system was successfully demonstrated.

6. The biomass feedstock with high ash, chlorine, alkali, and heavy metals such as straw will affect the ash (with high alkali) and syngas with larger amount of chlorine and maybe heavy metals. While both Polk and Buggenum plants did not report change in slag quality due to small amount of biomass used, slag properties will change at high biomass concentration. This affects the industrial and commercial use of slag and needs to be studied. The nature of biomass such as grass (high ash) will be more problematic than low-ash woody biomass.

7. At the Buggenum's Nuon plant, fouling of the syngas cooler was caused by a high percentage of sewage sludge but not by woody biomass. The fouling is caused by the chlorine, alkali, and heavy metal contents of biomass feedstock. Different biomasses will therefore behave differently on the issue of fouling and corrosion to the downstream equipment.

8. In Nuon's WAC IGCC power plant, 50% co-gasification with biomass will yield an actual emission reduction of about 20% from the baseline CO_2 emission level.

3.2.8 SUMMARY, BARRIERS, AND POTENTIALS FOR FUTURE GROWTH

Co-gasification of coal and biomass allows for compensating the shortcomings of one fuel by another, since both fuels seem to be complementary in their drawbacks and advantages. In many specific cases, synergy among components enhances performance. GHG mitigation is one of the most attractive benefits of co-gasification of coal and biomass. Lifecycle assessment [1,4] on co-gasification of coal and biomass has shown that CO_2 emission declines proportionally to the amount of coal offset by biomass, considering biomass as a carbon-neutral source produced in a sustained manner [1–4]. The literature [1–4] has shown that 70/30 mixture of coal and biomass will be CO_2 neutral to the environment. With further reduction in CH_4 and N_2O emission, the overall outcome of co-gasification is that a percentage reduction of global warming potential is higher than the percentage of biomass in the blend.

Entrained-flow gasification produces a relatively clean gas compared to fixed bed and fluidized bed gasification. However, size reduction of biomass to the order of hundreds of microns, required in entrained-flow systems, may be expensive and difficult to achieve for some biomass feedstock. This may also require preprocessing steps like torrefaction or fast pyrolysis.

In general, biomass contains high concentrations of AAEMs and low concentrations of aluminosilicates. Coal contains high concentrations of aluminosilicates. Since AAEMs can be catalysts for coal gasification [83–86], some coal/biomass mixtures can exhibit improved performance during co-gasification. While there is no consensus on synergistic effects between coal and biomass during co-gasification, some believe that free radical formation and hydrogen transfer from biomass to coal can create synergy during co-gasification. This synergy is more pronounced at lower temperatures and not noticeable at higher temperatures [21,94,113]. Because of the low melting temperature of AAEMs and high melting temperature of aluminosilicate metals, biomass ash tends to melt at lower temperatures than coal ash. It is therefore important to know the behavior of ash mixture from coal and biomass in the design of co-gasification reactors. In the absence of this knowledge, the best strategy for the detailed design of co-gasification process is to handle the design in case-by-case basis because of a wide variation of mineral matter content and quality in both biomass and coals. The knowledge of formation rate and characteristics of coal–biomass ash mixtures under different gasification conditions is needed.

Based on thermodynamic equilibrium of co-gasification, the amount of gas produced, its gross calorific value, and cold gas efficiency increase with the concentration of biomass in the coal/biomass mixture. Higher temperature favors the extent of endothermic reaction and formation of hydrogen and carbon monoxide. However, the cold gas efficiency and the gross calorific value first increase and then decrease with an increase in temperature. Increasing the pressure has an opposite effect to that of temperature. At higher temperatures, the pressure effect is more dominant.

Coal gasification is a proven technology that has been operated successfully at commercial scale for heat, power generation, and production of synthetic fuels. On the other hand, based on the literature information, stand-alone biomass or cellulosic waste gasification and subsequent production of biofuels appear to be more complex and carried out at small scale. The share of biofuel in the transportation fuel market is likely to grow rapidly in the next decade due to numerous benefits, including sustainability, reduction of GHG emissions, regional development, reduction of rural poverty, and energy security [21]. The economics of scale is important for biomass gasification because only large biomass-to-liquid (BtL) plants with synfuel production capacities of at least 1,000,000 t/a would produce profit [21,183]. The desired scale is best achieved by processing coal and biomass mixtures.

Biomass, like most existing renewable energy resources, is both dispersed and variable over time. Regular and constant supply of a huge amount of biomass that is required for large-scale operation, coupled with processing and handling difficulties, makes a dedicated biomass power and heat plant or BtL plant highly improbable; but this limitation can be alleviated if coal and biomass are both used as feedstock. The backup storage capacity needed for sustainable power or fuel generation can also be enhanced using mixture as feedstock [1,21].

There are several barriers to future growth of coal/biomass or coal/waste co-gasification. Some of these relate to public image and perception. While gasification has a better public perception than combustion, waste is an area of public concern and low image. The large-scale coal/biomass and coal/waste plants will require transportation of biomass and waste at a significant level. Such transportation will need public support. Transport of fuel to a power station is always a contentious area. The transportation of waste on a larger scale is also problematic [1–3,21].

There is also a cultural issue of the acceptance of biomass and waste into an industry that has been single focused on coal and the promotion of its usage. We have developed a culture and know-how of single-dimensional industry like coal, oil, and gas. Co-gasification thinking at the industrial scale requires a paradigm shift to the energy and fuel system integration. Overall, due to its necessity and potential for success, the future of co-gasification is very bright.

3.3 HYDROTHERMAL GASIFICATION

During the last five decades, significant efforts have been made to use unusual property of water at high temperature and pressure for energy and fuel industry [184–191]. To that end, biomass/waste or their mixture gasification in the presence of high-temperature and high-pressure water becomes a multifuel system. The fast hydrolysis of organic molecules such as biomass at high temperature leads to a rapid degradation of the polymeric structure of biomass. A series of consecutive reactions lead to the formation of gas whose composition depends on the temperature and pressure of water, the contact time, and the catalyst if it is present. High solubility of intermediates in water, particularly at high temperature and pressure, allows further organic reactions to occur in aqueous media and prevents the formation of tar and coke. The reactive species originating from biomass (or other species) are diluted by solvation in water, thereby preventing polymerization to unwanted products. These conditions also lead to the formation of high gas yield at relatively low temperatures. Hydrothermal gasification (HTG) process is thus the process of gaseous fuel generation in an aqueous medium under sub- and supercritical conditions. This thus differs from "steam gasification" where solids react with gaseous steam to produce a set of gaseous products. Unlike steam in "steam gasification," the water in hydrogasification is not simply a provider of oxidation environment.

The goal of HTG is to obtain high quality and yield of fuel gas. The most important components of fuel gas are hydrogen, carbon monoxide, and methane (or lower volatile hydrocarbons). Just like in conventional gasification, temperature plays an important role in the formation of methane.

In principle, there are three types of HTG [194,195]:

1. *Low-temperature aqueous-phase reforming*: This occurs at low temperatures (215°C–265°C) and moderate pressures (23–65 bars) in the presence of a selective catalyst for carbohydrates with C:O ratio close to 1. This is a highly selective catalytic reaction involving sugar or sugar-derived molecules with water. The products of this reaction can be hydrogen, syngas, or lower alkanes depending on the nature of the catalyst and other operating conditions.

2. *High-temperature catalytic gasification under subcritical water conditions*: At higher temperatures up to supercritical temperature, in the presence of a catalyst, biomass/waste or organic compounds are gasified mainly to methane and carbon dioxide. In the absence of a catalyst, this region of temperature (250°C to critical temperature, 374°C) is also called hydrothermal liquefaction region wherein carbohydrates are liquefied to various organic products. In catalytic HTG process, the heat recovery is important for an efficient operation. The catalytic HTG process converts biomass/waste/water slurry into fuel gas. The gaseous fuel can be used for heat, power, or the generation of various chemicals. This process does not work well with coal.

3. *Gasification and reforming in supercritical water*: This can be carried out both in the presence and absence of a catalyst. The main products are hydrogen and carbon monoxide and carbon dioxide. This process works both for coal and biomass/waste.

The first type of reaction is a very selective catalytic reaction and mostly applied to very specific types of biomass molecules. The subject of aqueous-phase reforming is covered in my previous book [3] in great detail and will not be addressed here. This type of gaseous fuel production cannot be applied to coal. The second type of subcritical water gasification is enhanced by the use of catalyst. It is mostly effective for biomass and waste but not for coal. Since this type of reaction also has limitations on feedstock, it is only briefly covered here. The third type of gasification applies to both coal and biomass and it can be very effective for co-gasification. In the presence of a suitable catalyst, the third type of gasification is often accompanied by reforming reaction. We examine this type of reaction in some details.

3.3.1   Hydrothermal Gasification of Biomass/Waste in Subcritical Water

HTG of biomass/waste under subcritical conditions occurs when temperature is below the critical temperature of 374°C. In this case, biomass hydrolysis is slow and catalysts are required [184–200] for gas formation. Elliot et al. [188,196–199] examined subcritical gasification of biomass feedstock, which included cellulose, lignin, holocellulose (cellulose and hemicellulose), and a Douglas fir wood flour using nickel catalyst and added sodium carbonate cocatalyst. The results showed that at 350°C, the catalyst gave 42% of carbon fed compared to 15% of carbon fed in the absence of catalyst. Both hydrogen and methane concentrations were higher for the catalytic operations compared to the ones without catalyst. Carbon monoxide concentration was close to zero in the presence of catalyst. As regards the activity of alkali additions, the activity follows the order Cs > K > Na. The study of Elliot et al. [188,196–199] also indicated that conventional support for nickel—alumina (other than alpha-alumina), silica, various ceramic supports, minerals such as kieselguhr, and other silica–alumina—was unstable in hot liquid water environment due to mechanisms such as dissolution, phase transition, and hydrolysis. They reported useful supports as carbon, monoclinic zirconia or titania, and alpha-alumina.

Elliot [188] evaluated base metal catalysis, noble metal catalysis, and activated carbon catalysis for HTG. His important conclusions are summarized:

1. Of all the base metal catalysts examined [188,193–199], such as nickel, magnesium, tungsten, molybdenum, zinc, chromium, cobalt, rhenium, tin, and lead, nickel was found to be the most active and stable catalyst. Various supports such as kieselguhr, silica–alumina, alpha-alumina, alumina–magnesia on spinel form, and carbon examined in the literature [191–208] gave a varying degree of success. The most useful promoters were ruthenium, copper, silver, and tin impregnated at 1 wt%.
2. For noble metal catalysis, while some conflicting results are reported by various investigators [184–200], in general, platinum, palladium, and silver showed minor activities to HTG at 350°C; iridium had some activity but the best activities were shown by ruthenium and rhodium. Rutile form of titania and carbon supports were found to be effective. Vogel et al. [193,243] found ruthenium doping on nickel catalyst on carbon to be effective for HTG.
3. While activated carbon and charcoal were found to be effective catalysts by some investigators [184–190], these results were mostly obtained under supercritical conditions.

The study of Minowa and Ogi [200] indicated that the cellulose gasification depends on the nature of support and the size of metal particles on the support. They presented the following mechanism for the cellulose gasification:

$$\text{Cellulose} \xrightarrow{\text{Decompose}} \text{Water-soluble products} \xrightarrow{\text{Gasification/Ni}}$$

$$\text{Gases } (H_2 + CO_2) \xrightarrow{\text{Methanation/Ni}} \text{Gases } (CH_4 + CO_2) \qquad (3.6)$$

Vogel group [193,243] indicated that Raney nickel was more effective than nickel on alpha-alumina. They also studied nickel catalysts with ruthenium, copper, and molybdenum doping. Most effective results were obtained with ruthenium doping on nickel catalysts. Elliot [188,196–199] also reports that at 350°C, bimetallic Ru/Ni, Ru/C, and Cu/Ni gave favorable gas productions by HTG of a variety of biomass. Favorable yields were obtained for lignin gasification by Ru/TiO_2, Ru/Al_2O_3, Ru/C, and Rh/C catalysts.

Ro et al. [192] showed that subcritical HTG of hog manure feedstock can be net energy producer for the solid concentration greater than 0.8 wt%. While the costs for gasification are higher than the ones for anaerobic digestion lagoon system, the land requirement for the gasification process and costs of transportation and tipping fees are lower. In addition, catalytic gasification process would destroy pathogens and bioactive organic compounds and will produce relatively clean water for reuse. The ammonia and phosphate by-products generated in gasification have also potential value in the fertilizer market.

3.3.2 GASIFICATION AND REFORMING OF COAL AND BIOMASS/WASTE IN SUPERCRITICAL WATER

In Sections 3.1 and 3.2 we examined co-gasification of coal and biomass in an oxidative environment. Both coal and biomass can also be gasified in supercritical water. Li et al. [201] investigated coal gasification in the temperature range of 650°C–800°C and pressure of 23–27 MPa with K_2CO_3 and Raney Ni as catalysts and H_2O_2 as oxidant. Most experiments were performed with inlet slurry containing 16.5 wt% coal and 1.5 wt% CMC. The results showed that high temperature favors the gasification of coal in supercritical water, while pressure has a little effect on the gasification results. K_2CO_3 performed better than Raney Ni. Less char and tar were formed in the presence of catalysts. Supercritical water desulfurizes the coal and the solid particles remained had less carbon and hydrogen than original coal. The data of Li et al. [209] indicate that for the entire range they studied, 60% of the gas-phase concentration was hydrogen.

Vostrikov et al. [202] examined coal gasification in the temperature range of 500°C–750°C, pressure of 30 MPa, and reaction time of 60–720 s with and without CO_2. Once again, main gaseous products were CH_4, CO, CO_2, and H_2. Within the range of operating conditions examined, best carbon conversion was obtained at 750°C. The results showed a significant temperature dependence on product compositions for temperatures below 650°C. Benzene, toluene, and xylene and methane and carbon dioxide were the main products below 650°C. Similar results were obtained by Cheng et al. [203] who studied gasification of lignite coals in the temperature range 350°C–550°C and reaction time of 0–60 min in N_2 atmosphere.

Battelle Pacific Northwest Laboratory demonstrated that various alkali carbonate and Ni catalysts can convert wet biomass to methane-rich gas at temperatures between 400°C and 450°C and pressure as high as 34.5 MPa. Yu et al. [204] found that glucose at low concentration (0.1 M) can be completely gasified in 20 s at 600°C and 34.5 MPa with major products being hydrogen and carbon dioxide. Higher concentration of glucose, however, reduces the product concentration of hydrogen and carbon dioxide and increases the concentration of methane. Xu et al. [205] showed that a wide range of carbons effectively catalyze the gasification of glucose in supercritical water at 600°C and 34.5 MPa pressure with nearly 100% carbon gasification efficiency. Demirbas [206–208] examined decomposition of biomass such as bionutshell, olive husk, tea waste, crop straw, black liquor, MSW, crop grain residue, pulp and paper waste, petroleum-based plastic waste, and manure slurry waste in supercritical water and observed an increase in hydrogen production with temperature.

An extensive amount of work on supercritical water gasification of organic wastes has been reported in the literature [208,211,218,224,226–245]. The studies have shown that the gasification generally produces hydrogen and carbon dioxide mixture with simultaneous decontamination of wastes, particularly at higher temperatures. Guo et al. [209,210] presented an excellent review of supercritical water gasification of biomass and organic wastes. They as well as Lu et al. [211] showed the equilibrium effects of temperature, pressure, and feed concentration of wood sawdust on hydrogen, carbon dioxide, carbon monoxide, and methane concentrations in supercritical water. The gas yield in supercritical water was also affected by feedstock concentration, oxidant, reaction time, feedstock composition, inorganic impurities in the feedstock, and biomass particle size. Guo et al. [209,210] also concluded that alkali such as NaOH, KOH, Na_2CO_3, K_2CO_3, and Ca $(OH)_2$, activated carbon, metal oxides, and metals such as noble metal catalysts (Ru/a-alumina > Ru/carbon > Rh/carbon > Pt/a-alumina, Pd/carbon, Pd-a-alumina), as well as Ni catalysts and metal oxides such as CeO_2 particles, nano-CeO_2, and nano-$(CeZr)_xO_2$, enhanced the reactivity of biomass gasification in supercritical water. Xu and Antal [212,213,215] studied gasification of 7.69 wt% digested sewage sludge in supercritical water and obtained gas that largely contained H_2, CO_2, a smaller amount of CH_4, and a trace of CO. Other waste materials show similar behavior.

Kong et al. [189] briefly summarized the reported work for the catalytic HTG of various types of biomass in supercritical water. He showed that in the literature, catalytic HTG in supercritical water has been examined for glucose, organic wastewater, cellulose, soft and hard wood, grass, lignin, saw dust, rice straw, alkylphenols, corn, potato starch gels, potato waste, glycerol, cellobiose, bagasse, sewage sludge, catechol, vanillin, glycine, and many others. In all cases, the major products were hydrogen and some methane depending on operating conditions. The catalysts examined included Ni, Ru, Rh, Pd, and Pt on alumina and NaOH, KOH, Na_2CO_3, K_2CO_3, ZrO_2, activated carbon, and Ni on carbon. The preference was given to the disposable or cheap catalysts or to the reforming catalysts if the objective was to carry out reforming along with gasification. Carbon and base catalysts play important roles in the increased gas yields and hydrogen production. Tanksale et al. [186] and Azadi and Farnood [185] also provided an extensive review of catalytic supercritical water gasification.

Yamaguchi et al. [216,217] studied lignin gasification in supercritical water. They indicated that lignin gasification involves three steps: (1) lignin decomposition to alkylphenols and formaldehyde in supercritical water, (2) gasification of alkylphenols and formaldehyde over a catalyst, and (3) formation of char from formaldehyde. They showed that supercritical water gasification is a promising technique to reduce the lignin gasification temperature. They also studied lignin gasification with three different catalysts at 400°C, $RuCl_3$/C, $Ru(NO)(NO_3)_3$/C, and $RuCl_3$/C and found that the order of gasification activity was Ru/C = Ru(NO)(NO$_3$)$_3$/C > RuCl$_3$/C. EZXAFS analysis showed that during lignin gasification in supercritical water, ruthenium particle sizes in Ru(NO)(NO$_3$)$_3$/C and Ru/C catalysts were smaller than that in the RuCl$_3$/C catalyst. The study concluded that the ruthenium catalysts with smaller particle size of metal particles were more active for the lignin gasification. The results showed that for the lignin gasification, activity order followed ruthenium > rhodium > platinum > palladium > nickel, whereas hydrogen production rate followed the order palladium > ruthenium > platinum > rhodium > nickel. Both titania and activated carbon provided stable supports.

Byrd et al. [218] examined the supercritical gasification of biocrude from switchgrass at 600°C and 250 atm pressure. Nickel, cobalt, and ruthenium catalysts were prepared on titania, zirconia, and magnesium–aluminum spinel supports. Magnesium–aluminum spinel structure did not work. Over time zirconia-supported catalyst plugged the reactor, although Ni/ZrO$_2$ catalyst gave the best hydrogen production. Titania-supported catalysts gave lower hydrogen conversions but did not plug the reactor over time. All support materials suffered surface area loss due to sintering.

A number of investigators have looked at glucose as a model for biomass reforming under supercritical water. The pertinent reaction in this case is

$$C_6H_{12}O_6 + 6H_2O \rightarrow 6CO_2 + 12H_2 \qquad (3.7)$$

Generally, hydrogen yield is smaller than predicted from the aforementioned equation because varying amounts of methane are produced depending upon the reaction conditions. Kruse [187,219–221] gave a simplified reaction mechanism for cellulose reforming. Since glucose (and fructose) is the main product of hydrolysis of cellulose, their reaction mechanism also applies to glucose. The reforming of glucose was accelerated by alkali catalysts such as K$_2$CO$_3$ and KHCO$_3$. Both of these catalysts increased the hydrogen production and decreased coke formation. For biomass with low salt content and high protein content, these catalysts can increase the hydrogen yield.

Antal and Xu [215] and Antal et al. [214] showed the effectiveness of supercritical water reforming for the productions of hydrogen for numerous different types of biomass such as wood sawdust, corn starch gel, digested sewage sludge, glycerol, glycerol/methanol mixture, poplar wood sawdust, potato starch gels, and potato waste. Once again, higher temperature and catalysts gave better hydrogen productions. The final product distribution did depend on the nature of the feedstock. Similar results were obtained by Boukis et al. [223] for biomass slurries and sludges. Zhang et al. [224] examined the SCW reforming of glucose solution generated from a sludge hydrothermal liquefaction process. The experiments were performed using two different types of catalysts: 0.1 RuNi/gamma–Al$_2$O$_3$ or 0.1 RuNi/activated carbon catalysts (10 wt% Ni with a Ru to Ni molar ratio of 0.1). The first catalyst was very effective with glucose solutions and simulated aqueous organic waste giving hydrogen yield of 53.9 mol/kg dried feedstock at 750°C, 24 MPa, and WHSV of 6 h^{-1}. It was, however, not stable. The second catalyst exhibited higher stability.

The studies described earlier and many others [194,203,208,218,225–245] lead to some general conclusions. As the temperature increases above the critical temperature, more gases are generally produced from most carbonaceous materials. At lower temperature, for higher feedstock concentration, and in the absence of a catalyst, the gas production rate tends to be lower and contains more methane. At high temperature, for lower feedstock concentration, and in the presence of an effective catalyst, hydrogen production rate rapidly increases. Higher temperature and the presence of a catalyst promote reforming of gas and favor reverse water gas shift reaction producing more hydrogen and carbon dioxide. Pressure also affects the equilibrium of water gas shift reaction. Higher pressure favors methane formation as opposed to hydrogen production.

While very little literature is available for co-gasification of coal and biomass/waste in supercritical water, the aforementioned review clearly indicates high potentials for coal–biomass gasification in supercritical water. More details on this subject have been recently reviewed by Shah [3].

REFERENCES

1. Ratafia-Brown, J., Haslbeck, J., Skone, T., and Rutkowski, M. Assessment of technologies for co-converting coal and biomass to a clean syngas—Task 2 report (RDS). DOE/NETL-403.01.08 Activity 2 Report, Department of Energy, Washington, D.C. (May 10, 2007).
2. Lee, S. and Shah, Y. *Biofuels and Bioenergy: Technologies and Processes*, CRC Press, Taylor & Francis Group, New York (September 2012).
3. Shah, Y. *Water for Energy and Fuel Production*, CRC Press, Taylor & Francis Group, New York (May 2014).
4. Williams, R.H., Larson, E.D., and Haiming, J. Synthetic fuels in a world with high oil and carbon prices. In *Proceedings of the Eighth International Conference on Greenhouse Gas Control Technologies (GHGT-8)*, Trondheim, Norway (June 19–22, 2006).
5. Antal, M.J. Biomass pyrolysis: A review of the literature. Part II: Lignocellulose pyrolysis. In *Advances in Solar Energy*, Vol. 2, Boer, K.W. and Duffie, J.A. (Eds.). American Solar Energy Society, Boulder, CO, pp. 175–255 (1985).
6. Callis, H.P.A., Haan, H., Boerrigter, H., Van der Drift, A., Peppink, G., Van den Broek, R., Faaij, A., and Venderbosch, R.H. Preliminary techno-economic analysis of large-scale synthesis gas manufacturing from imported biomass. In *Proceedings of an Expert Meeting on Pyrolysis and Gasification of Biomass and Waste*, Strasbourg, France, pp. 403–417 (2003).
7. Kavalov, B. and Peteves, S.D. Status and perspectives of biomass-to liquid fuels in the European Union. Report No. EUR 21745 EN, European Commission, Joint Research Centre, Brussels, Belgium (2005).
8. Reed, T. and Gaur, S. *A Survey of Biomass Gasification*, 2nd edn., U.S. Department of Energy, National Renewable Energy Laboratory and the Biomass Energy Foundation, Golden, CO, p. 180 (2001).
9. Rickets, B., Hotchkiss, R., Livingston, B., and Hall, M. Technology status review of waste/biomass co-gasification with coal. In IChemE (Ed.). *Fifth European Gasification Conference*, Noordwijk, the Netherlands, p. 13 (April 2002).
10. Hotchkiss, R., Livingston, W., and Hall, M. Waste/biomass co-gasification with coal? Report No. Coal R216, DTI/Pub URN 02/867. Department of Trade and Industry, Cleaner Coal Technology Program, Great Britain (2002).
11. Bain, R. *Biomass Gasification Overview*, NREL, Golden, CO (January 28, 2004).

12. IEA State of art of biomass gasification, prepared by European Concerted action, analysis and coordination of the activities concerning gasification of biomass AIR3-CT94-2284 and IEA bioenergy, biomass utilization, task XIII, thermal gasification of biomass activity, Sweden and Canada country reports, IEA, Paris, France (1997).

13. Valero, S. and Uson, S. *Oxy-Co-Gasification of Coal and Biomass in an Integrated Gasification Combined Cycle (IGCC) Power Plant*, Center for Research of Energy Resources and Consumptions (CIRCE), University of Zaragoza, Zaragoza, Spain (April 2005).

14. Shah, Y.T. Biomass to liquid fuel via Fischer-Tropsch and related syntheses. In *Advanced Biofuels and Bioproducts*, Chapter 12, Lee, J.W. (Ed.). Springer Book Project, Springer Publ. Co., New York, pp. 185–207 (September 2012).

15. Shah, Y. and Gardner, T. Biomass torrefaction: Applications in renewable energy and fuels. In *Encyclopedia of Chemical Processing*, Lee, S. (Ed.). Taylor & Francis Group, New York, pp. 1–18, published online (April 24, 2012).

16. Drift, A., Boerrigter, H., Coda, B., Cieplik, M., and Hemmes, K. Entrained flow gasification of biomass; ash behavior, feeding issues, system analyses. Report C-04-039, ECN, Petten, the Netherlands, 58pp. (April 2004).

17. Bridgewater, A. *Fast Pyrolysis if Biomass: A Handbook*, Vol. 2, CPL Press (2002); Pels, J. et al., Utilization of ashes from biomass combustion and gasification. In *14th European Biomass Conference Exhibition*. Paris, France (October 2005).

18. Henrich, E. Clean syngas from biomass by pressurized entrained flow gasification of slurries from fast pyrolysis. In Paper presented at *SynBios, the Syngas Route to Automotive Biofuels, Conference* held at Stockholm, Sweden (May 18–20, 2005).

19. Cieplik, M., Coda, B., Boerrigter, A., Drift, V., and Kiel, J. Characterization of slagging behavior of wood as upon entrained flow gasification conditions. Report C-04-016 also Report RX-04-082, ECN, Petten, the Netherlands (July 2004).

20. Kurkela, E., Stahlberg, P., and Laatikainen, J. Pressurized fluidized bed gasification experiments with wood, peat and coal at VTT in 1991–1992. VTT publications no. 161, Finland (1993). http://www.vttresearch.com.

21. Tchapda, A. and Pisupati, S. A review of thermal co-conversion of coal and biomass/waste. *Energies*, 7, 1098–1148 (2014).

22. Radovic, L.R., Walker, P.L., Jr., and Jenkins, R.G. Catalytic coal gasification: Use of calcium versus potassium. *Fuel*, 63, 1028–1030 (1984).

23. Walker, P.L., Jr., Matsumoto, S., Hanzawa, T., Muira, T., and Ismail, I.M.K. Catalysis of gasification of coal derived cokes and chars. *Fuel*, 62, 140–149 (1983).

24. Freund, H. Kinetics of carbon gasification by CO_2. *Fuel*, 64, 657–660 (1985).

25. McKee, D.W., Spiro, C.L., Kosky, P.G., and Lamby, E.J. Eutectic salt catalysts for graphite and coal char gasification. *Fuel*, 64, 805–809 (1985).

26. Huffman, G.P., Huggins, F.E., Shah, N., and Shah, A. Behavior of basic elements during coal combustion. *Prog. Energy Combust. Sci.*, 16, 243–251 (1990).

27. Huffman, G.P., Huggins, F.E., Shoenberger, R.W., Walker, J.S., Lytle, F.W., and Greegor, R.B. Investigation of the structural forms of potassium in coke by electron microscopy and x-ray absorption spectroscopy. *Fuel*, 65, 621–632 (1986).

28. Huggins, F.E., Huffman, G.P., Lytle, F.W., and Greegor, R.B. The form of occurrence of chlorine in U.S. coals: An XAFS investigation. *ACS Div. Fuel Chem. Prep.*, 34, 551–558 (1989).

29. Huffman, G.P. and Huggins, F.E. Analysis of the inorganic constituents of low-rank coals. In *The Chemistry of Low Rank Coals*, Schobert, H.H. (Ed.). American Chemical Society, Washington, DC, pp. 159–174 (1984).

30. Lang, R.J. and Neavel, R.C. Behaviour of calcium as a steam gasification catalyst. *Fuel*, 61, 620–626 (1982).

31. Kreith, F. *The CRC Handbook of Mechanical Engineering*, CRC Press, Boca Raton, FL (1998).
32. Ren, H., Zhang, Y., Fang, Y., and Wang, Y. Co-gasification behavior of meat and bone meal char and coal char. *Fuel Process. Technol.*, 92, 298–307 (2011).
33. Zhu, W., Song, W., and Lin, W. Catalytic gasification of char from co-pyrolysis of coal and biomass. *Fuel Process. Technol.*, 89, 890–896 (2008).
34. Brown, R.C., Liu, Q., and Norton, G. Catalytic effects observed during the co-gasification of coal and switchgrass. *Biomass Bioenergy*, 18, 499–506 (2000).
35. McKee, D.W., Spiro, C.L., Kosky, P.G., and Lamby, E.J. Catalysis of coal char gasification by alkali metal salts. *Fuel*, 62, 217–220 (1983).
36. Srivastava, S.K., Saran, T., Sinha, J., Ramachandran, L.V., and Rao, S.K. Influence of alkali on pyrolysis of coals. *Fuel*, 67, 1683–1684 (1988).
37. Brage, C., Yu, Q., Chen, G., and Sjostrom, K. Tar evolution profiles obtained from gasification of biomass and coal. *Biomass Bioenergy*, 18, 87–91 (2000).
38. Habibi, R., Kopyscinski, J., Masnadi, M.S., Lam, J., Grace, J.R., Mims, C.A., and Hill, J.M. Co-gasification of biomass and non-biomass feedstocks: Synergistic and inhibition effects of switchgrass mixed with sub-bituminous coal and fluid coke during CO_2 gasification. *Energy Fuels*, 27, 494–500 (2012).
39. Pedersen, L.S., Nielsen, H.P., Kiil, S., Hansen, L.A., Dam-Johansen, K., Kildsig, F., Christensen, J., and Jespersen, P. Full-scale co-firing of straw and coal. *Fuel*, 75, 1584–1590 (1996).
40. Sjöström, K., Chen, G., Yu, Q., Brage, C., and Rosen, C. Promoted reactivity of char in co-gasification of biomass and coal: Synergies in the thermochemical process. *Fuel*, 78, 1189–1194 (1999).
41. Cordero, T., Rodriguez-Mirasol, J., Pastrana, J., and Rodriguez, J.J. Improved solid fuels from co-pyrolysis of a high-sulphur content coal and different lignocellulosic wastes. *Fuel*, 83, 1585–1590 (2004).
42. Haykiri-Acma, H. and Yaman, S. Synergy in devolatilization characteristics of lignite and hazelnut shell during co-pyrolysis. *Fuel*, 86, 373–380 (2007).
43. Khalil, R.A. Thermal conversion of biomass with emphasis on product distribution, reaction kinetics and sulfur abatement, PhD thesis. Norwegian University of Science and Technology (NTNU), Trondheim, Norway (2009).
44. Knudsen, J.N. Volatilization of inorganic matter during combustion of annual biomass, PhD thesis. Technical University of Denmark, Lyngby, Denmark (2004).
45. Milne, T.A., Evans, R.J., Abatzoglou, N. *Biomass Gasifier Tars: Their Nature, Formation, and Conversion.* NREL/TP-570-25357, National Renewable Energy Laboratory, Golden, CO, p. 204 (1998).
46. Pinto, F., Lopes, H., Andre, R.N., Gulyurtlu, I., and Cabrita, I. Effect of catalysts in the quality of syngas and by-products obtained by co-gasification of coal and wastes. 1. Tars and nitrogen compounds abatement. *Fuel*, 86, 2052–2063 (2007).
47. Mettler, M.S., Vlachos, D.G., and Dauenhauer, P.J. Top ten fundamental challenges of biomass pyrolysis for biofuels. *Energy Environ. Sci.*, 5, 7797–7809 (2012).
48. Prins, M.J., Ptasinski, K.J., and Janssen, F.J.J.G. Torrefaction of wood: Part 1. Weight loss kinetics. *J. Anal. Appl. Pyrol.*, 77(1), 28–34 (2006).
49. Shafizadeh, F. Pyrolytic reactions and products of biomass. In *Fundamentals of Biomass Thermochemical Conversion*, Overend, R.P., Mime, T.A., and Mudge, L.K. (Eds.). Elsevier, London, U.K., pp. 183–217 (1985).
50. Shafizadeh, F. Thermal conversion of cellulosic materials to fuels and chemicals. In *Wood and Agricultural Residues* Soltes, E.J. (Ed.). Academic Press, New York, pp. 183–217 (1983).

51. Bergman, P.C.A. and Kiel, J.H.A. Torrefaction for biomass upgrading. In *Proceedings of the 14th European Biomass Conference and Exhibition*, Paris, France, October 2005; *ETA—Renewable Energies*, Florence, Italy, pp. 17–21 (2005).

52. Bergman, P.C.A., Boersma, A.R., Kiel, J.H.A, Prins, M.J., Ptasinski, K.J., and Janssen, F.J.J.G. In Torrefaction for entrained flow gasification of biomass. Report No. ECN-RX-04-046; *Proceedings of the Second World Biomass Conference on Biomass for Energy, Industry and Climate Protection*, Rome, Italy, May 10–14, 2004, Van Swaaij, W.P.M., Fjällstrom, T., Helm, P.T., and Grassi, P. (Eds.). Energy Research Centre of the Netherlands (ECN), Petten, the Netherlands, pp. 679–682 (2004).

53. Bergman, P.C.A. Combined torrefaction and pelletization: The TOP process. Report No. ECN-C-05-073, Energy Research Centre of the Netherlands (ECN), Petten, the Netherlands (2005).

54. Bergman, P.C.A., Boersma, A.R., Kiel, J.H.A., Prins, M.J., Ptasinski, K.J., and Janssen, F.J.J.G. Torrefaction for entrained flow gasification of biomass. Report No. ECN-C-05-067, Energy Research Centre of the Netherlands (ECN), Petten, the Netherlands (2004).

55. Prins, M.J., Ptasinski, I.G., and Janssen, F.J.J.G. More efficient biomass gasification via torrefaction. In *Proceedings of the 17th Conference on Efficiency, Costs, Optimization, Simulation and Environmental Impact of Energy Systems* (*ECOS'04*), Guanajuato, Mexico, July 7–9, 2004, Rivero, R., Monroy, L., Pulido, R., and Tsatsaronis, G. (Eds.). Elsevier, Amsterdam, the Netherlands, pp. 3458–3470 (2004).

56. Prins, M.J. Thermodynamic analysis of biomass gasification and torrefaction, PhD thesis. Technische Universiteit Emdhoven, Eindhoven, the Netherlands (2005).

57. Williams, P.T. and Besler, S. The influence of temperature and heating rate on the slow pyrolysis of biomass. *Renew. Energy*, 7(3), 233–250 (1996).

58. Arcate, J.R. Torrefied wood, an enhanced wood fuel. *Bioenergy*, Boise, ID, Paper # 207 (September 22–26, 2002).

59. Arias, B., Pevida, C., Fermoso, J., Plaza, M.G., Reubiern, F., and Pis, J.J. Influence of torrefaction on the grindability and reactivity of wood biomass. *Fuel Process. Technol.*, 89(2), 169–175 (2008).

60. Bourgois, J. and Guyonnet, R. Characterization and analysis of torrefied wood. *Wood Sci. Technol.*, 22(2), 143–155 (1988).

61. Duijn, C. Torrefied wood uit resthout en andere biomassastromen. In *Proceedings of Praktijkdag Grootschalige Bioenergie Projecten*, SenterNovem, the Netherlands (June 2004).

62. Li, J. and Gifford, J. *Evaluation of Woody Biomass Torrefaction*, Forest Research, Rotorua, New Zealand (September 2001).

63. Pach, M., Zanzi, R., and Bjømbom, E. Torrefied biomass a substitute for wood and charcoal. In *Proceedings of the Sixth Asia-Pacific International Symposium on Combustion and Energy Utilization*, Kuala Lumpur, Malaysia (May 20–22, 2002).

64. Zwart, R.W.R., Boerrigter, H., and Van der Drift, A. The impact of biomass pre-treatment on the feasibility of overseas biomass conversion to Fischer-Tropsch products. *Energy Fuels*, 20(5), 2192–2197 (August 29, 2006).

65. Brooking, E. *Improving Energy Density in Biomass through Torrefaction*, National Renewable Energy Laboratory, Golden, CO (2002). http://www.nrel.gov/education/pdfs/e_brooking.pdf (accessed November 2002).

66. Reed, T.B. and Bryant, B. *Densified Biomass: A New Form of Solid Fuel*, U.S. Department of Energy, National Renewable Energy Laboratory, Golden, CO, p. 35 (1978).

67. Pentanunt, R., Mizanur Rahman, A.N.M., and Bhattacharya, S.C. Updating of biomass by means of torrefaction. *Energy*, 15(12), 1175–1179 (1990).

68. Kirk, J.T.O. *Light and Photosynthesis in Aquatic Ecosystems*, 2nd edn., Cambridge University Press, Cambridge, U.K. (1994).

69. Raven, P.H., Evert, R.F., and Eichhorn, S.E. *Biology of Plants*, 6th edn., W.H. Freeman/ Worth Publishers, New York (1999).

70. Weststeyn, A. and Essent Energie, B.V. First torrefied wood successfully co-fired, PyNe (biomass pyrolysis network), *Newsletter*, Issue 17 (April 2004).

71. Shafizadeh, F. and McGinnis, G.D. Chemical composition and thermal analysis of cottonwood. *Carbohydr. Res.*, 16(2), 273–277 (1971).

72. Alén, R., Kotilainen, R., and Zaman, A. Thermochemical behavior of Norway spruce (*Picea abies*) at 180–225°C. *Wood Sci. Technol.*, 36(2), 163–171 (2002).

73. Alves, S.S. and Figueiredo, J.L. A model for pyrolysis of wet wood. *Chem. Eng. Sci.*, 44(12), 2861–2869 (1989).

74. Di Blasi, C. and Lanzetta, M. Intrinsic kinetics of isothermal xylan degradation in inert atmosphere. *J. Anal. Appl. Pyrol. Species*, 40(41), 287–303 (1997).

75. Varhegyi, G., Antal, M.J., Jakab, E., and Szabó, P. Kinetic modeling of biomass pyrolysis. *J. Anal. Appl. Pyrol.*, 42(1), 73–87 (1997).

76. Yang, H., Yan, R., Chen, H., Lee, D.H., and Zheng, C. Characteristics of hemicelluloses, cellulose and lignin pyrolysis. *Fuel*, 86(12–13), 1781–1788 (2007).

77. Bradbury, A.G.W., Sakai, Y., and Shafizadeh, F.J. A kinetic model for pyrolysis of cellulose. *J. Appl. Polym. Sci.*, 23(11), 3271–3280 (1979).

78. Branca, C. and Di Blasi, C. Kinetics of the isothermal degradation of wood in the temperature range 528–707 K. *J. Anal. Appl. Pyrol.*, 67(2), 207–219 (2003).

79. Bridgeman, T.G., Jones, J.A., Shield, I., and Williams, P.T. Torrefaction of reed canary grass, wheat straw and willow to enhance solid fuel qualities and combustion properties. *Fuel*, 87(6), 844–856 (2008).

80. Broido, A. Kinetics of solid phase cellulose pyrolysis. In: *Thermal Uses and Properties of Carbohydrates and Lignins*, Shafizadeh, F., Sarkanen, K., and Tillman, D.A. (Eds.). Academic Press, New York, pp. 19–36 (1976).

81. Hakkou, M., Pétrissans, M., El Bakali, I., Gérardin, P., and Zoulalian, A. Wettability changes and mass loss during heat treatment of wood. *Holzforschung*, 59(1), 35–37 (2005).

82. Hakkou, M., Pétrissans, M., Gérardin, P., and Zoulalian, A. Investigation of wood wettability changes during heat treatment on the basis of chemical analysis. *Polym. Degrad. Stab.*, 89(1), 1–5 (2005).

83. Hakkou, M., Petrissans, M., Geradin, P., and Zoulalian, A. Investigations of the reasons for fungal durability of heat-treated beech wood. *Polym. Degrad. Stab.*, 91(2), 393–397 (2006).

84. Kamdem, D.P., Pizzi, A., and Jerrnannaud, A. Durability of heat-treated wood. *Holz Roh Werkstoff.*, 60(1), 1–6 (2002).

85. Weiland, J.J. and Guyonnet, R. Study of chemical modifications and fungi degradation of thermally modified wood using DRIFT spectroscopy. *Holz ala Roh und Werkstoff* (*Eur. J. Wood Prod.*), 61(3), 216–220 (2003).

86. White, R.H. and Dietenberger, M.A. Wood products: Thermal degradation and fire. In *The Encyclopedia of Materials: Science and Technology*, Buschow, K.H.J., Cahn, R.W., Flemings, M.C., Ilschner, B., Kramer, E.J., and Mahajan, S. (Eds.). Elsevier, Amsterdam, the Netherlands, pp. 9712–9716 (2001).

87. Voufo Panos, C.A., Maschio, G., and Lucehesi, A. Kinetic modeling of the pyrolysis of biomass and biomass components. *Can. J. Chem. Eng.*, 67(1), 75–84 (1989).

88. Bioenergy, a new process for torrefied wood manufacturing. *Gen. Bioenergy*, 2(4), 1–3 (2000).

89. Pétrissans, M., Gérardin, P., El Bakali, I., and Serraj, M. Wettability of heat-treated wood. *Holzforschung*, 57(3), 301–307 (2003).

90. Bourgois, J.P. and Doat, J. Torrefied wood from temperate an tropical species. In *Advantages and Prospects*, Vol. III, Egnéus, H. and Ellegârd, A. (Eds.). Elsevier, London, U.K.; *Bioenergy*, 84, 153–159 (1984).

91. Jannasch, R., Quan, Y., and Samson, R. A process and energy analysis of pelletizing switchgrass—Final report. Resource Efficient Agricultural Production (REAP-Canada) for Natural Resources, Canada, pp. 1–16 (2001).

92. Hustad, J. and Barrio, M. Biomass. *IFRF Online Combustion Handbook*, Combustion File No. 23, Version No. 2, IFRF, 17-10-200, (2001). http://www.handbook.ifrf.net/handbook/cf.html?id=2. (accessed 2001).

93. Bergman, P., Boersma, A., Zwart, R., and Kiel, J. *Torrefaction for Biomass Co-Firing in Existing Coal-Fired Power Stations* (*BIOCOAL*). ECN-C-05-013, ECN, Petten, the Netherlands, 72pp. (2005).

94. Maciejewska, A., Veringa, H., Sanders, J., and Peteves, S. *Co-firing of Biomass with Coal: Constraints and Role of Biomass Pre-Treatment*, EUR 22481–EN DGJRC Institute for Energy, Luxenbourg (2006). http://ie.jrc.cec.eu.int/publications/scientific-publications/2006/EUR22461EN.pdf. (accessed 2006).

95. Rutkowski, M.D., Schoff, R.L., and Keuhn, N.J. Analysis of Stamet pump for IGCC applications prepared for DOE/NETL. Parson Corporation, Pittsburgh, PA (July 2005).

96. Aldred, D., Saunders, T., and Rutkowski, M. Successful continuous injection of coal into gasification system operating pressures exceeding 500 psi. In Presented at the *Gasification Technology Conference*, Orlando, FL (October 9–12, 2005).

97. Dayton, D., Turk, B., and Gupta, R., Syngas cleanup, conditioning and utilization. In *Thermochemical Processing of Biomass: Conversion into Fuels, Chemicals and Power*, 1st edn., Chapter 4. Brown, R.C. (Ed.). John Wiley & Sons, New York, pp. 78–123 (2011).

98. Kambara, S., Takarada, T., Yamamoto, Y., and Kato, K. Relation between functional forms of coal nitrogen and formation of nitrogen oxide (NO_x) precursors during rapid pyrolysis. *Energy Fuels*, 7, 1013–1020 (1993).

99. Jiachun, Z., Masutani, S.M., Ishimura, D.M., Turn, S.Q., and Kinoshita, C.M. Release of fuel-bound nitrogen in biomass during high temperature pyrolysis and gasification. In *Proceedings of the 32nd Intersociety Energy Conversion Engineering Conference* (*IECEC*), Vol. 3, Honolulu, HI, pp. 1785–1790 (July 27–August 1, 1997); *Energies*, 7, 1141 (2014).

100. De Jong, W. Nitrogen compounds in pressurised fluidised bed gasification of biomass and fossil fuels, PhD thesis. Technische Universiteit Delft, Delft, the Netherlands (2005).

101. Blair, D.W., Wendt, J.O.L., and Bartok, W. Evolution of nitrogen and other species during controlled pyrolysis of coal. *Symp. Int. Combust.*, 16, 475–489 (1977).

102. Slaughter, D.M., Overmoe, B.J., and Pershing, D.W. Inert pyrolysis of stoker-coal fines. *Fuel*, 67, 482–489 (1988).

103. Solomon, P.R. and Colket, M.B. Evolution of fuel nitrogen in coal devolatilization. *Fuel*, 57, 749–755 (1978).

104. Pohl, J.H. and Sarofim, A.F. Devolatilization and oxidation of coal nitrogen. *Symp. Int. Combust.*, 16, 491–501 (1977).

105. Cai, H.Y., Güell, A.J., Dugwell, D.R., and Kandiyoti, R. Heteroatom distribution in pyrolysis products as a function of heating rate and pressure. *Fuel*, 72, 321–327 (1993).

106. Bassilakis, R., Zhao, Y., Solomon, P.R., and Serio, M.A. Sulfur and nitrogen evolution in the argonne coals. Experiment and modeling. *Energy Fuels*, 7, 710–720 (1993).

107. Yuan, S., Chen, X., Li, W., Liu, H., and Wang, F. Nitrogen conversion under rapid pyrolysis of two types of aquatic biomass and corresponding blends with coal. *Bioresour. Technol.*, 102, 10124–10130 (2012).

108. Yuan, S., Zhou, Z., Li, J., Chen, X., and Wang, F. HCN and NH_3 (NO_x precursors) released under rapid pyrolysis of biomass/coal blends. *J. Anal. Appl. Pyrol.*, 92, 463–469 (2011).

109. Di Nola, G., de Jong, W., and Spliethoff, H. The fate of main gaseous and nitrogen species during fast heating rate devolatilization of coal and secondary fuels using a heated wire mesh reactor. *Fuel Process. Technol.*, 90, 388–395 (2009).

110. Di Nola, G., de Jong, W., and Spliethoff, H. TG-FTIR characterization of coal and biomass single fuels and blends under slow heating rate conditions: Partitioning of the fuel-bound nitrogen. *Fuel Process. Technol.*, 91, 103–115 (2010); *Energies*, 7, 1144 (2014).

111. Knudsen, J.N., Jensen, P.A., Lin, W., Frandsen, F.J., and Dam-Johansen, K. Sulfur transformations during thermal conversion of herbaceous biomass. *Energy Fuels*, 18, 810–819 (2004).

112. Dayton, D.C., French, R.J., and Milne, T.A. Direct observation of alkali vapor release during biomass combustion and gasification. 1. Application of molecular beam/mass spectrometry to switchgrass combustion. *Energy Fuels*, 9, 855–865 (1995).

113. Björkman, E. and Strömberg, B. Release of chlorine from biomass at pyrolysis and gasification conditions. *Energy Fuels*, 11, 1026–1032 (1997).

114. Knudsen, J.N., Jensen, P.A., and Dam-Johansen, K. Transformation and release to the gas phase of Cl, K, and S during combustion of annual biomass. *Energy Fuels*, 18, 1385–1399 (2004).

115. Zhao, J., Hu, X., and Gao, J. Study on the variations of organic sulfur in coal by pyrolysis. *Coal Convers.*, 16, 77–81 (1993).

116. Calkins, W.H. Determination of organic sulfur-containing structures in coal by flash pyrolysis experiments. *Prep. Am. Chem. Soc. Fuel Chem. Div.*, 30, 450–465 (1985).

117. Cullis, C.F. and Norris, A.C. The pyrolysis of organic compounds under conditions of carbon formation. *Carbon*, 10, 525–537 (1972).

118. Winkler, J.K., Karow, W., and Rademacher, P. Gas-phase pyrolysis of heterocyclic compounds, part 1 and 2: Flow pyrolysis and annulation reactions of some sulfur heterocycles: Thiophene, benzo[b]thiophene, and dibenzothiophene. A product-oriented study. *J. Anal. Appl. Pyrol.*, 62, 123–141 (2002).

119. Ur Rahman Memon, H., Williams, A., and Williams, P.T. Shock tube pyrolysis of thiophene. *Int. J. Energy Res.*, 27, 225–239 (2003).

120. Huang, C., Zhang, J.Y., Chen, J., and Zheng, C.G. Quantum chemistry study on the pyrolysis of thiophene functionalities in coal. *Coal Convers.*, 28, 33–35 (2005).

121. Ibarra, J.V., Palacios, J.M., Gracia, M., and Gancedo, J.R. Influence of weathering on the sulphur removal from coal by pyrolysis. *Fuel Process. Technol.*, 21, 63–73 (1989).

122. Cleyle, P.J., Caley, W.F., Stewart, I., and Whiteway, S.G. Decomposition of pyrite and trapping of sulphur in a coal matrix during pyrolysis of coal. *Fuel*, 63, 1579–1582 (1984).

123. Gryglewicz, G. and Jasieńko, S. The behaviour of sulphur forms during pyrolysis of low-rank coal. *Fuel*, 71, 1225–1229 (1992).

124. Gryglewicz, G. Sulfur transformations during pyrolysis of a high sulfur polish coking coal. *Fuel*, 74, 356–361 (1995).

125. Telfer, M.A. and Zhang, D.K. Investigation of sulfur retention and the effect of inorganic matter during pyrolysis of south Australian low-rank coals. *Energy Fuels*, 12, 1135–1141 (1998).

126. Suárez-Ruiz, I. and Crelling, J.C. *Applied Coal Petrology: The Role of Petrology in Coal Utilization*, Elsevier Ltd., Burlington, MA (2008).

127. Biedermann, F. and Obernberger, I. Ash related problems during biomass combustion and possibilities for a sustainable ash utilization. In *Proceedings of the International Conference on World Renewable Energy Congress (WREC)*, Aberdeen, Scotland, U.K., p. 8 (May 22–27, 2005).

128. Baxter, L.L. Ash deposition during biomass and coal combustion: A mechanistic approach. *Biomass Bioenergy*, 4, 85–102 (1993).

129. Rushdi, A., Sharma, A., and Gupta, R. An experimental study of the effect of coal blending on ash deposition. *Fuel*, 83, 495–506 (2004).

130. Tumuluru, J.S., Hess, J.R., Boardman, R.D., Wright, C.T., and Westover, T.L. Formulation, pretreatment, and densification options to improve biomass specifications for co-firing high percentages with coal. *Ind. Biotechnol.*, 8, 113–132 (2012).

131. Li, S., Chen, X., Wang, L., Liu, A., and Yu, G. Co-pyrolysis behaviors of saw dust and Shenfu coal in drop tube furnace and fixed bed reactor. *Bioresour. Technol.*, 148, 24–29 (2013).

132. Zheng, Y., Jensen, P.A., Jensen, A.D., Sander, B., and Junker, H. Ash transformation during co-firing coal and straw. *Fuel*, 86, 1008–1020 (2007).

133. Higman, C. and van der Burgt, M. *Gasification*, Elsevier, New York (2008).

134. Livingston, W.R. *Biomass Ash Characteristics and Behaviour in Combustion, Gasification and Pyrolysis Systems*, Doosan Babcock Energy Limited, West Sussex, U.K. (2007).

135. Van Loo, S. and Koppejan, J. *Handbook of Biomass Combustion and Co-Firing*, Earthscan, London, U.K. (2008).

136. Wu, H., Wall, T., Liu, G., and Bryant, G. Ash liberation from included minerals during combustion of pulverized coal: The relationship with char structure and burnout. *Energy Fuels*, 13, 1197–1202 (1999).

137. Wei, X., Lopez, C., von Puttkamer, T., Schnell, U., Unterberger, S., and Hein, K.R.G. Assessment of chlorine-alkali-mineral interactions during co-combustion of coal and straw. *Energy Fuels*, 16, 1095–1108 (2002).

138. Dayton, D.C., Belle-Oudry, D., and Nordin, A. Effect of coal minerals on chlorine and alkali metals released during biomass/coal cofiring. *Energy Fuels*, 13, 1203–1211 (1999).

139. Andersen, K.H., Frandsen, F.J., Hansen, P.F.B., Wieck-Hansen, K., Rasmussen, I., Overgaard, P., and Dam-Johansen, K. Deposit formation in a 150 MWe utility PF-boiler during co-combustion of coal and straw. *Energy Fuels*, 14, 765–780 (2000).

140. Blevins, L.G. and Cauley, T.H. Fine particulate formation during switchgrass/coal cofiring. *J. Eng. Gas Turbines Power*, 127, 457–463 (2005).

141. Kumabe, K., Hanaoka, T., Fujimoto, S., Minowa, T., and Sakanishi, K. Co-gasification of woody biomass and coal with air and steam. *Fuel*, 86, 684–689 (2007).

142. Aznar, M.P., Caballero, M.A., Sancho, J.A., and Frances, E. Plastic waste elimination by co-gasification with coal and biomass in fluidized bed with air in pilot plant. *Fuel Process. Technol.*, 87, 409–420 (2006).

143. Andre, R.N., Pinto, F., Franco, C., Dias, M., Gulyurtlu, I., Matos, M.A.A., and Cabrita, I. Fluidised bed co-gasification of coal and olive oil industry wastes. *Fuel*, 84, 1635–1644 (2005); *Energies*, 7, 1134 (2014).

144. Collot, A.G., Zhuo, Y., Dugwell, D.R., and Kandiyoti, R. Co-pyrolysis and co-gasification of coal and biomass in bench-scale fixed-bed and fluidised bed reactors. *Fuel*, 78, 667–679 (1999).

145. Nelson, P.F., Buckley, A.N., and Kelly, M.D. Functional forms of nitrogen in coals and the release of coal nitrogen as NO_x precursors (HCN and NH_3). *Symp. Int. Combust.*, 24, 1259–1267 (1992).

146. Strand, M., Bohgard, M., Swietlicki, E., Gharibi, A., and Sanati, M. Laboratory and field test of a sampling method for characterization of combustion aerosols at high temperatures. *Aerosol Sci. Technol.*, 38, 757–765 (2004).

147. Aho, M. and Silvennoinen, J. Preventing chlorine deposition on heat transfer surfaces with aluminium-silicon rich biomass residue and additive. *Fuel*, 83, 1299–1305 (2004).

148. Kyi, S. and Chadwick, B.L. Screening of potential mineral additives for use as fouling preventatives in Victorian brown coal combustion. *Fuel*, 78, 845–855 (1999).

149. Raask, E. *Mineral Impurities in Coal Combustion: Behavior, Problems, and Remedial Measures*, Hemisphere Publishing Corporation, Washington, DC (1985).

150. Ratafia-Brown, J., Skone, T., Rutkowski, M., and Cobb, J.T. Assessment of technologies for co-converting coal and biomass to a clean syngas—Task 2 report (RDS), Department of Energy, Washington, DC (May 15, 2007).

151. Equipment design and cost estimation for small modular biomass systems, synthesis gas cleanup, and oxygen separation equipment—Task 2: Gas cleanup design and cost estimates-wood feedstock. NREL/DOE Subcontract Report NREL/SR-510-39945, Nexant Inc., San Francisco, CA (May 2006).

152. Air products tees valley renewable energy facility, EA/EPR/JP3331HK/A001, Environmental Permit Application, Air Products PLC, Hersham, Surrey, U.K. (May 2011).

153. Bertocci, A. and Patterson, R. Wet scrubber technology for controlling biomass gasification emissions. In *IT3'07 Conference*, Phoenix, AZ (May 14–18, 2007).

154. Bertocci, A. Wet scrubbers for gasifier gas cleaning, paper no. 49. In *IT3'07 Conference*, Phoenix, AZ (May 14–18, 2007).

155. Westinghouse Plasma Corporation, a division of Alter NRG Corp. Energy evolved: Clean, sustainable energy recovery through plasma gasification, Madison, AVA (February 2011).

156. Krishnan, G., Wood, B., Tong, G., and Kothari, V. Removal of hydrogen chloride vapor from high temperature coal gases. Abstracts of the Papers for American Chemical Society. *Fuel*, 195, 32 (1988).

157. Gil, J., Caballero, M., Martin, J. et al. Biomass gasification with air in a fluidized bed: Effect of the in-bed use of dolomite under different operating conditions. *I&EC Res.*, 38(11), 4226–4235 (1999).

158. Yasuda, H., Yamada, O., Zhang, A., Nakano, K., and Kaho, M. Hydrogasification of coal and polyethylene mixture. *Fuel*, 83(17–18), 2251–2254 (December 2004).

159. Yasuda, H., Yamada, O., Kaiho, M., and Nakagome, H. Effect of polyethylene addition to coal on hydrogasification enhancement. *J. Mater. Cycles Waste Manage.*, 16(1), 151–155 (February 2014).

160. Fermoso, J., Arias, B., Plaza, M., Pevida, C., Rubiera, F., Pis, J., Peria, G., and Casero, P., High pressure co-gasification of coal with biomass and petroleum coke. *Fuel Process. Technol.*, 90(7–8), 926–932 (July–August 2009).

161. Pinto, F., Franco, C., Andre, R., Tavares, C., Dias, M., Gulyurtlu, I., and Cabrita, I. Effect of experimental conditions on co-gasification of coal, biomass and plastic wastes with air/steam mixtures in a fluidized bed system. *Fuel*, 82(15–17), 1967–1976 (October–December 2003).

162. De Ruyck, J., Delattin, F., and Bram, S. Co-utilization of biomass and natural gas in combined cycles through primary steam reforming of the natural gas. *Energy*, 32, 371–377 (2007).

163. Yamada, T., Akano, M., Hashimoto, H., Suzuki, T., Maruyama, T., Wang, Q., and Kamide, M. Steam gasification of coal–biomass briquettes. *Nippon Enerugi Gakkai Sekitan Kagaku Kaigi Happyo Ronbunshu*, 39, 185–186 (2002).

164. Wu, J., Fang, Y., and Wang, Y. Production of syngas by methane and coal co-conversion in fluidized bed reactor, Institute of Coal Chemistry, Chinese Academy of Science, Taiyuan, China, personal communication (2013).

165. Sun, H., Song, B., Jang, Y., and Kim, S. The characteristics of steam gasification of biomass and waste filter carbon. A Report from Dept. of Chem. Eng., Kunsan National University, Gunsan, South Korea (2012).

166. Priyadarsan, S., Holtzapple, M., Annamalai, K., and Mukhtar, S. Co-gasification of coal and chicken litters. In *17th National Heat and Mass Transfer Conference and 6th ASME Conference*, College Station, TX (2011).

167. Ji, K., Song, B., Kim, Y., Kim, B., Yang, W., Choi, Y., and Kim, S. Steam gasification of low rank fuel-biomass, coal, and sludge mixture in a small scale fluidized bed. In *Proceedings of the Fourth European Combustion Meeting*, Vienna University of Technology, Vienna, Austria pp. 1–5, 14–17 (April 2009).

168. Lazar, M., Jasminska, N., and Lengyelova, M. Experiment of gasification of the synthetically mixed sample of waste in nitrogen atmosphere. A Report by Mechanical Engineering Dept., Technical University of Kostice, Kostice, Slovak Republic (2012).

169. Chmie lniak, T. and Sciazko, M. Co-gasification of biomass and coal for methanol synthesis. *Appl. Energy*, 74, 393–403 (2003).

170. Pan, Y., Velo, E., Roca, X., Manya, J., and Puigianer, L. Fluidized bed co-gasification of residual biomass/poor coal blends for fuel gas production. *Fuel*, 79, 1317–1326 (2000).

171. Li, M., Jie, L., and Lu, X. Model test study of underground co-gasification of coal and oil shale. *Appl. Mech. Mater.*, 295–298, 3129–3136 (2013).

172. Wang, L. and Chen, Z. Hydrogen rich gas production by co-gasification of coal and biomass in an intermittent fluidized bed. *Scient. World J.*, 276823 (2013), doi: 101155/2013/276823.

173. McLendon, T., Lui, A., Pineault, R., Beer, S., and Richardson, S. High pressure co-gasification of coal and biomass in a fluidized bed. *Biomass Bioenergy*, 26(4), 377–388 (April 2004).

174. Indrawati, V., Manaf, A., and Purwadi, G. Partial Replacement of non renewable fossil fuels energy by the use of waste materials as alternative fuels, CP1169. In *International Workshop on Advanced Material for New and Renewable Energy*, Handoko, L. and Siregar, M. (Eds.). American Institute of Physics, 978-0-7354-0706-0/07, Jakarta, Indonesia, pp. 179–184 (September 29, 2009).

175. Williams, R., Larson, E., and Jin, H. Comparing climate-change mitigating potentials of alternative synthetic liquid fuel technologies using biomass and coal. In *Fifth Annual Conference on Carbon Capture and Sequestration-DOE/NETL*, Pittsburgh, PA (May 8–11, 2006).

176. Seo, M., Goo, J., Kim, S., Lee, S., and Choi, Y. Gasification characteristics of coal/biomass blend in a dual circulating fluidized bed reactor. *Energy Fuels*, 24(5), 3108–3118 (2010).

177. Brar, J., Singh, K., Wang, J., and Kumar, S. Co-gasification of coal and biomass: A review. *Int. J. Forest. Res.*, 2012, 1–10 (2012).

178. Khosravi, M. and Khadse, A. Gasification of Petcoke and coal/biomass blend: A review. *Int. J. Emerg. Technol. Adv. Eng.*, 3(12) 167–173 (December 2013).

179. McDaniel, J. Biomass gasification at Polk Power Station—Final technical report. DOE award DE-FG26-01NT41365 Tampa Electric Co., Tampa, Florida (May 2002).

180. Vera, N., Abedini, A., Liliedahl, T., and Engvall, K. Co-gasification of petroleum coke and biomass. *Fuel*, 117, 870–875 (2014).

181. Zhao, L., Liang, J., and Qian, L., Coal and oil shale model test, underground co-gasification. *Appl. Mech. Mater.*, 295–298, 3129 (2013).

182. Krupp, U. Gasification technology: Shell gasification process, Gasification Brochure, ThyssenKrupp Technologies, Dortmund, Germany (1999). ThyssenKrupp.com.

183. Henrich, E., Dahmen, N., Dinjus, E. Cost estimate for biosynfuel production via biosyncrude gasification. *Biofuels Bioprod. Bioref.*, 3, 28–41 (2009).

184. Kersten, S.R.A., Potic, B., Prins, W., and VanSwaaij, W. Gasification of model compounds and wood in hot compressed water. *Ind. Eng. Chem. Res.*, 45, 4169–4177 (2006).

185. Azadi, P. and Farnood, R. Review of heterogeneous catalysts for sub and supercritical water gasification of biomass and wastes. *Int. J. Hydrogen Energy*, 36, 9529–9541 (2011).

186. Tanksale, A., Beltramini, J., and Lu, G. A review of catalytic hydrogen production processes from biomass. *Renew. Sustain. Energy Rev.*, 14, 166–182 (2010).

187. Kruse, A. Hydrothermal biomass gasification. *J. Supercrit. Fluids*, 47, 391–399 (2009).

188. Elliott, D.C. Catalytic hydrothermal gasification of biomass. *Biofuels Bioprod. Bioref.*, 2, 254–265 (May/June 2008).

189. Kong, L., Li, G., Zhang, B., He, W., and Wang, H. Hydrogen production from biomass wastes by hydrothermal gasification. *Energy Sources A*, 30, 1166–1178 (2008).

190. Akiya, N. and Savage, P.E. Roles of water for chemical reactions in high-temperature water. *Chem. Rev.*, 102(8), 2725–2750 (2002).

191. Brown, T., Duan, P., and Savage, P. Hydrothermal liquefaction and gasification of *Nannochloropsis* sp. *Energy Fuels*, 24, 3639–3648 (2010).

192. Ro, K., Cantrell, K., Elliot, D., and Hunt, G. Catalytic wet gasification of municipal and animal wastes. *Ind. Eng. Chem. Res.*, 46, 8839–8845 (2007).

193. Vogel, F. and Hildebrand, F. Catalytic hydrothermal gasification of woody biomass at high feed concentrations. *Chem. Eng. Trans.*, 2, 771–777 (2002).

194. Osada, M., Sato, T., Watanabe, M., Shirai, M., and Arai, K. Catalytic gasification of wood biomass in subcritical and supercritical water. *Combust. Sci. Technol.*, 178 (1–3) 537–552 (2006).

195. Osada, M., Hiyoshi, N., Sato, O., Arai, K., and Shirai, M. Subcritical water regeneration of supported ruthenium catalyst poisoned by sulfur. *Energy Fuels*, 22, 845–849 (2008).

196. Elliot, D., Sealock, L., and Baker, E. Chemical processing in high pressure aqueous environment. 3. Batch reactor process development experiments for organic destructions. *Ind. Eng. Chem.*, 33, 558–565 (1994).

197. Elliot, D., Peterson, K., Muzatko, D., Alderson, E., Hart, T., and Neuenschwander, G. Effects of trace contaminants on catalytic processing of biomass derived feedstocks. *Appl. Biochem. Biotechnol.*, 113–116, 807–825 (2004).

198. Elliot, D., Neuenschwander, G., Phelps M., Hart, T., Zacher, A., and Silva, L. Chemical processing in high pressure aqueous environment. 6. Demonstration of catalytic gasification for chemical manufacturing wastewater cleanup in industrial plants. *Ind. Eng. Chem. Res.*, 38, 879–883 (1999).

199. Elliot, D., Neuenschwander, G., Hart, T., Butner, R., Zacher, A., and Englelhard, M. Chemical processing in high pressure aqueous environments and process development for catalytic gasification of wet biomass feedstocks. *Ind. Eng. Chem. Res.*, 43, 1999–2004 (2004).

200. Minowa, T. and Ogi, T. Hydrogen production from cellulose using a reduced nickel catalyst. *Catal. Today*, 45, 411–416 (1998).

201. Li, Y., Guo, L., Zhang, X., Jin, H., and Lu, Y. Hydrogen production from coal gasification in supercritical water with a continuous flowing system. *Int. J. Hydrogen Energy*, 35, 3036–3045 (2010).

202. Vostrikov, A., Psarov, S., Dubov, D., Fedyaeva, O., and Sokol, M. Kinetics of coal conversion in supercritical water. *Energy Fuels*, 21, 2840–2845 (2007).

203. Cheng, L., Zhang, R., and Bi, J. Pyrolysis of a low rank coal in sub and supercritical water. *Fuel Process. Technol.*, 85(8–10), 921–932 (July 2004).

204. Yu, D., Aihara, M., and Antal, M. Hydrogen production by steam reforming glucose in supercritical water. *Energy Fuels*, 7(5), 574–577 (1993).

205. Xu, X.D., Yukihiko, M., Jonny, S. et al. Carbon-catalyzed gasification of organic feedstocks in supercritical water. *Ind. Eng. Chem. Res.*, 35, 2522–2530 (1996).

206. Demirbas, A. Biorefineries: Current activities and future developments. *Energy Convers. Manage.*, 50, 2782–2801 (2009).

207. Demirbas, A. Progress and recent trends in biofuels. *Progr. Energy Combust. Sci.*, 33, 1–18 (2007).

208. Demirbas, A. Hydrogen production from biomass via supercritical water gasification. *Energy Sources A*, 32, 1342–1354 (2010).

209. Guo, L., Cao, C., and Lu, Y. Supercritical water gasification of biomass and organic wastes. In *Biomass Book*, Ndombo, M. and Momba, B. (Eds.). In Tech Europe, Rijeka, Croatia pp. 165–182 (2010).

210. Guo, Y., Wang, S., Xu, D., Gong, Y., Ma, H., and Tang, X. Review of catalytic supercritical water gasification for hydrogen production from biomass. *Renew. Sustain. Energy Rev.*, 14, 334–343 (2010).

211. Lu, Y., Guo, L., Zhang, X., and Yan, Q. Thermodynamic modeling and analysis of biomass gasification for hydrogen production in supercritical water. *Chem. Eng. J.*, 131, 233–244 (2007).

212. Xu, X. and Antal, M., Kinetics and mechanism of isobutene formation from T– butanol in hot liquid water. *AIChE J.*, 40(9), 1524–1531 (1994).

213. Xu, X. and Antal, M. In mechanism and temperature-dependent kinetics of the dehydration of tert-butyl alcohol in hot compressed liquid water. *Ind. Eng. Chem. Res.*, 36(1), 23–41 (1997).

214. Antal, M., Allen, S., Schulman, D., and Xu, X. Biomass gasification in supercritical water. *Ind. Eng. Chem. Res.*, 39, 4040–4053 (2000).

215. Antal, M. and Xu, X. Hydrogen production from high moisture content biomass in supercritical water, a report for NREL/DOE under contact No. DE-FC36-94AL85804, University of Hawaii, Manoa, HI (2012).

216. Yamaguchi, A., Hiyoshi, N., Sato, O., Osada, M., and Shirai, M. EXAFS study on structural change of charcoal-supported ruthenium catalysts during lignin gasification in supercritical water. *Catal. Lett.*, 122, 188–195 (2008).

217. Yamaguchi, A., Hiyoshi, N., Sato, O., Bando, K., Osada, M., and Shirai, M. Hydrogen production from woody biomass over supported metal catalysts in supercritical water. *Catal. Today*, 146, 192–195 (2009).

218. Byrd, A., Kumar, S., Kong, L., Ramsurn, H., and Gupta, R. Hydrogen production from catalytic gasification of switchgrass biocrude in supercritical water. *Int. J. Hydrogen Energy*, 36, 3426–3433 (2011).

219. Kruse, A. Supercritical water gasification. *Biofuels Bioprod. Bioref.*, 2, 415–437, doi:10.1002/bbb.93 (September/October 2008).

220. Kruse, A. and Gawlik, A. Biomass conversion in water at 330°C–410°C and 30–50 MPa. Identification of key compounds for indicating different chemical reaction pathways. *Ind. Eng. Chem. Res.*, 42, 267–279 (2003).

221. Kruse, A. and Henningsen, T., Sinag, A. and Pfeiffer, J. Biomass gasification in supercritical water: Influence of the dry matter content and the formation of phenols. *Ind. Eng. Chem. Res.*, 42, 3711–3717 (2003).

222. Kruse A., Krupa, A., Schwarzkopf, V., Gamard, C. and Henningsen, T., Influence of proteins on hydrothermal gasification and liquefacation of biomass I. comparison of different feedstock. *Ind. Eng. Chem. Res.* 44, 3013–3020 (2005).

223. Boukis, N., Galla, V., D'Jesus, P., Muller, H., Dinjus, E. Gasification of wet biomass supercritical water. Results of pilot plant experiments. 14th European Biomass Conference, Paris, France, 964-967 (2005).

224. Zhang, L., Champagne, P., and Xu, C. Supercritical water gasification of an aqueous by-product from biomass hydrothermal liquefaction with novel Ru modified Ni catalysts. *Bioresour. Technol.*, 102(17), 8279–8287 (September 2011).
225. D'Jesus, P., Boukis, N., Kraushaar-Czarnetzki, B., and Dinjus, E. Gasification of corn and clover grass in supercritical water. *Fuel*, 85, 1032–1038 (2006).
226. Pinkwart, K., Bayha, T., Lutter, W., and Krausa, M. Gasification of diesel oil in super-critical water for fuel cells. *J. Power Sources*, 136, 211–214 (2004).
227. Takafumi, S., Mitsumasa, O., Masaru, W. et al. Gasification of alkylphenols with supported noble metal catalysts in supercritical water. *Ind. Eng. Chem. Res.*, 42, 4277–4282 (2003).
228. Takuya, Y. and Yukihiko, M. Gasification of cellulose, xylan, and lignin mixtures in supercritical water. *Ind. Eng. Chem. Res.*, 40, 5469–5474 (2001).
229. Takuya, Y., Yoshito, O., and Yukihiko, M. Gasification of biomass model compounds and real biomass in supercritical water. *Biomass Bioenergy*, 26, 71–78 (2004).
230. Tang, H.Q. and Kuniyuki, K. Supercritical water gasification of biomass: Thermodynamic analysis with direct Gibbs free energy minimization. *Chem. Eng. J.*, 106, 261–267 (2005).
231. Michael, J.A., Jr., Stephen, G.A., Schulman, D., Xu, X. and Divilio, R. Biomass gasification in supercritical water. *Ind. Eng. Chem. Res.*, 39, 4040–4053 (2000).
232. Paul, T.W. and Jude, O. Composition of products from the supercritical water gasification of glucose: A model biomass compound. *Ind. Eng. Chem. Res.*, 44, 8739–8749 (2005).
233. Peter, K. Corrosion in high-temperature and supercritical water and aqueous solutions: A review. *J. Supercrit. Fluids*, 29, 1–29 (2004).
234. Veriansyah, B., Kim, J., and Kim, J.D. Hydrogen production by gasification of gasoline in supercritical water. A Report from Supercritical Fluid Research Laboratory, Korean Institute of Science and Technology, Seoul, Korea, personal communication (2012).
235. Knoef, H. *Handbook Biomass Gasification*, Biomass Technology Group Press, Enschede, the Netherlands, pp. 22–23 (2005).
236. Lee, I.G., Kim, M.S., and Ihm, S.K. Gasification of glucose in supercritical water. *Ind. Eng. Chem. Res.*, 41, 1182–1188 (2002).
237. Andrea, K., Danny, M., Pia, R. et al. Gasification of pyrocatechol in supercritical water in the presence of potassium hydroxide. *Ind. Eng. Chem. Res.*, 39, 4842–4848 (2000).
238. Hao, X.H., Guo, L.J., Mao, X. et al. Hydrogen production from glucose used as a model compound of biomass gasified in supercritical water. *Int. J. Hydrogen Energy*, 28, 55–64 (2003).
239. Chakinala, A., Brilman, D., van Swaaij, W., and Kersten, S. Catalytic and non catalytic supercritical water gasification of microalgae and glycerol. *Ind. Eng. Chem. Res.*, 49(3), 1113–1122 (February 2010).
240. Susanti, R., Nugroho, A., Lee, J., Kim, Y., and Kim, J. Noncatalytic gasification of isooctane in supercritical water: A strategy for high-yield hydrogen production. *Int. J. Hydrogen Energy*, 36(6), 3895–3906 (March 2011).
241. Goodwin, A. and Rorrer, G. Conversion of glucose to hydrogen rich gas by supercritical water in a microchannel reactor. *Ind. Eng. Chem. Res.*, 47(12), 4106–4114 (May 2008).
242. Sinag, A., Kruse, A., and Rathert, J. Influence of the heating rate and the type of the catalyst on the formation of key intermediates and on the generation of gases during hydropyrolysis of glucose in supercritical water in a batch reactor. *Ind. Eng. Chem. Res.*, 43, 502–508 (2004).
243. Vogel, F., Waldner, M., Rouff, A., and Rabe, S. Synthetic natural gas from biomass by catalytic conversion in supercritical water. *Green Chem.*, 9, 616–619 (2007).
244. Osada, M., Hiyoshi, N., Sato, O., Arai, K., and Shirai, M. Reaction pathways for catalytic gasification of lignin in the presence of sulfur in supercritical water. *Energy Fuels*, 21, 1854–1858 (2007).

245. Osada, M., Sato, T., Watanabe, M., Adschiri, T., and Arai, K., Low temperature catalytic gasification of lignin and cellulose with a ruthenium catalyst in supercritical water. *Energy Fuels*, 18, 327–333 (2004).
246. Kurkela, E., Stahlberg, P., Simell, P., and Leppalahti, J. Updraft gasification of peat and biosolid. *Biosolid*, 19, 37–46 (1989).
247. Nemanova, V., Abedini, A., Liliedahl, T., and Engvall, K. Co-gasification of petroleum coke and biomass. *Fuel*, 117, 870–875 (2014).

4 Synthetic Liquid Production by Co-Processing

4.1 INTRODUCTION

Besides gasification, the direct liquefaction (and extraction) of coal, biomass, and waste to generate synthetic liquids is a very valuable and commercially viable thermochemical transformation technology. Unlike combustion, gasification, and pyrolysis, liquefaction (or extraction) generally involves multiple fuels, solid (coal, biomass, or waste), and liquid (oil or water). Coal, biomass, and waste materials can be liquefied (generally in the presence of a donor solvent) to produce a host of liquid products. The quality of liquid product depends on the nature of feedstock; nature of operating conditions such as temperature, pressure, and reaction time; and the presence of a catalyst. While coal, biomass, or waste can be liquefied on their own at high temperatures, the best results to produce desirable liquids involve liquefaction in the presence of a donor solvent, which actively interacts with solid feedstock in producing the liquid. The quality of the donor solvent is an important parameter in the production of both quantity and quality of liquid products as well as for the economics of the process. The liquid can also be produced by the extraction of various useful components of the solid feedstock (such as extraction of coal by supercritical water [SWC]).

Coal is generally liquefied in the presence of oil or another donor solvent, which takes an active role in the liquefaction process. During the last several decades, significant research and development activities on donor solvent coal liquefaction process have been carried out. The results of these activities are briefly summarized in Section 4.2. Unfortunately, this approach has not resulted in an economically competitive method for liquid oil production [1]. In more recent years, three other approaches have been investigated to produce oil from coal, biomass, and waste using different combinations of multifuels. These approaches are described in Sections 4.3 through 4.5. These approaches have found success in the laboratory-scale and pilot-scale operations. Some of them are also tested on a commercial scale. Thus, there are four basic multifuel methods for producing synthetic liquid directly from coal, biomass, and waste, which are as follows:

1. Direct liquefaction by a donor or reactive solvent. The direct liquefaction of coal requires the use of a donor solvent, which can actively react with defragmented coal at high temperature and pressure to produce synthetic liquids.

2. Hydrothermal liquefaction (HTL) of coal and biomass to produce oils. The conversion of biomass produces bio-oil.

3. Co-processing coal and heavy oil as well as other substances such as polymers, rubber tires, oil shale, and other types of wastes to produce liquid fuel. Synergy among various components of the mixture (for co-liquefaction or co-processing) plays an important role in gaining advantages for this approach.

4. Supercritical decomposition, extraction, and liquefaction of coal and biomass by water to produce oil.

It should be noted here that in methods 2 and 4, high-temperature and high-pressure water acts as a solvent, reactant, and catalyst for the liquefaction process. Thus, for all practical purposes, water acts as a fuel in these processes. As shown in my previous books [2,3], water possesses organic liquid–like properties at high temperature and pressure and becomes nonpolar allowing its active participation in the liquefaction process. Here, we examine each of these methods in some details.

4.2 DIRECT LIQUEFACTION OF COAL IN THE PRESENCE OF A DONOR SOLVENT

The direct liquefaction of coal by a donor solvent was practiced during the Second World War by Germans. The production of synthetic fuels using coal was carried out for military purposes where the cost of the production was not a factor. The technology of direct coal liquefaction became more important in this country during the 1970s due to oil embargo and the need for alternate sources for oil. Coal was also seen as feedstock for chemicals [4,5]. Unfortunately, the work done during the 1970s did not produce a very economically viable direct coal liquefaction process. The basic research carried out during this decade, however, did shed light on the basic mechanisms of coal liquefaction, which is briefly discussed in the succeeding text.

4.2.1 MECHANISM OF DIRECT COAL LIQUEFACTION

The starting point for the present state of knowledge on direct coal liquefaction is the mechanism of coal dissolution in the presence of a solvent. Basic structural units for coals of various ranks are shown in Figure 4.1. The reactive maceral components of coal are believed to be highly cross-linked, amorphous, macromolecular networks consisting of a number of stable cluster units or aggregates connected by relatively weak linkages. The most labile bonds under coal liquefaction conditions include [1]

$$Ar–CH_2–Ar \quad Ar–O–Ar \quad \text{sulfur analogs of oxygen} \qquad (4.1)$$

$$Ar–(CH_2)_n–Ar \quad R–O–Ar \quad \text{hydrogen bonds} \qquad (4.2)$$

$$R–O–R \quad \text{charge-transfer complexes} \qquad (4.3)$$

The weak bonds such as hydrogen bonding and charge-transfer complexes can be overcome to some extent at temperatures below 250°C, yielding extracts of up to

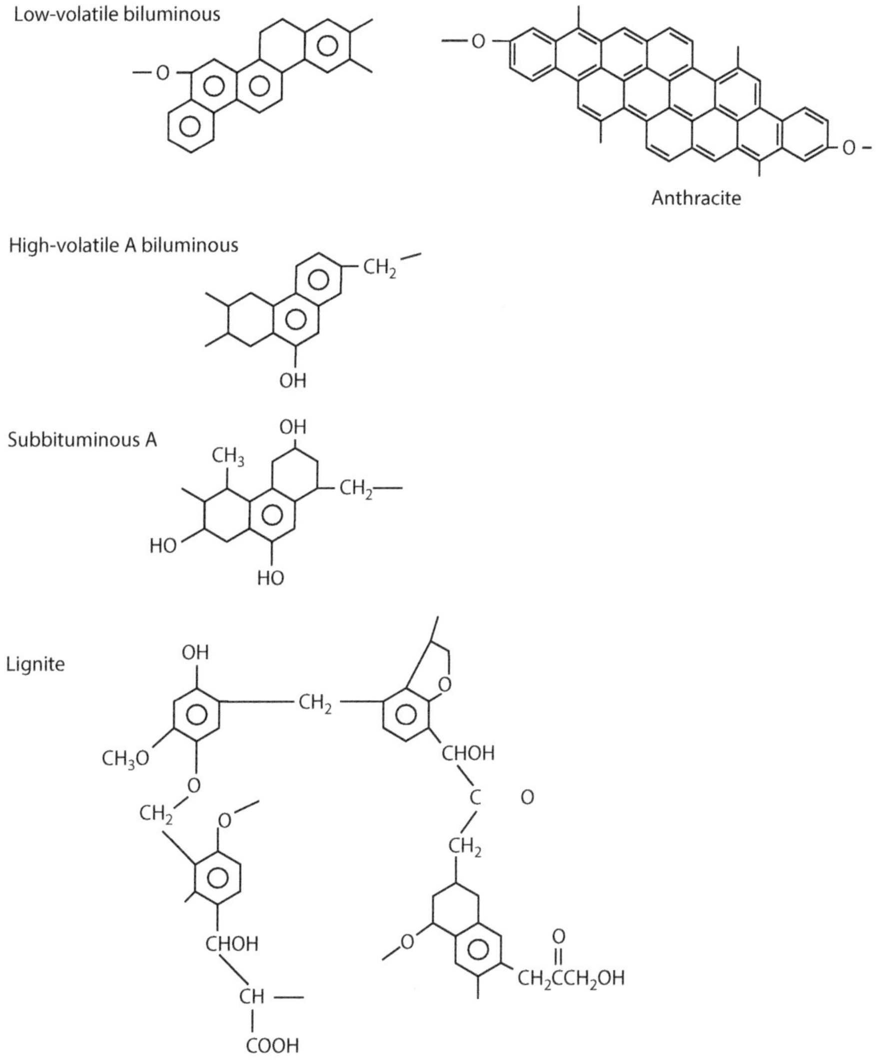

Low-volatile biluminous

Anthracite

High-volatile A biluminous

Subbituminous A

Lignite

FIGURE 4.1 Basic molecular structure of different rank coal. (From Shah, Y., *Reaction Engineering in Direct Coal Liquefaction*, Addison Wesley Publishing Co., Reading, MA, 1981; Wender, I., *ACS Div. Fuel Chem. Preprints*, 20(4), 16, 1975; Wender, I., *Cat. Rev. Sci. Eng.* (Marcel Dekker, New York), 14(1), 9, 1976.)

40%–50% of some bituminous coals [6–10]. As shown in Figure 4.2, this initial dissolution is accompanied by a change in viscosity. The swelling region and the extent of swelling can be lowered by the use of a donor solvent. As the temperature is raised above 250°C and it enters the regime of donor solvent coal liquefaction processes (400°C–460°C), thermal processes rapidly occur whereby the distributed activation energies are overcome for the labile bonds, yielding reactive fragments of free radicals [11–17]. If hydrogen is available to a radical from redistribution of hydrogen

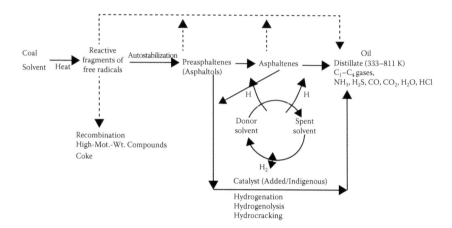

FIGURE 4.2 General reaction paths occurring during donor solvent coal liquefaction. (From Shah, Y., *Reaction Engineering in Direct Coal Liquefaction*, Addison Wesley Publishing Co., Reading, MA, 1981; Han, K. and Wen, C., *Fuel*, 58, 779, 1979.)

in the organic matrix (autostabilization) [18] or from donation by the solvent, the radical will be capped yielding a stable species with the molecular weight (MW) in the range of 300–1000. At conditions where hydrogen is unavailable or concentration of free radicals become excessively large, the fragments can recombine or polymerize to form very high-MW compounds or coke. The autostabilization process involves little in the way of net hydrogen transfer to the thermal fragments from the solvent but mostly redistribution of hydrogen within the coal matrix, with the solvent acting as net shuttler of hydrogen [11,12,15].

It has been suggested that the contact at the molecular level between the solvent and the fuel functional groups is essential in addition to the matching of the dehydrogenation rate of the solvent and the primary decomposition rate of coal. Miura et al. [19] ensured that the contact at the molecular level was achieved by swelling the coal with solvent at 100°C–250°C under a pressurized atmosphere. The observed increase of both the total volatile matter and the tar yield was brought about by not only the effective hydrogen transfer from tetralin to the coal but also by the effective utilization of small radicals such as OH· and H· from the coal for stabilizing the coal fragments. Mae et al. [20] used a similar technique by pyrolyzing a pyridine vapor swollen coal. The suppression of cross-linking reactions significantly increased the volatiles and tar yields. Awan et al. [21] conducted studies on tetralin-treated coals. They similarly obtained a significant increase in tar and volatile yields and explained this observation in terms of the suppression of the cross-linking reactions of coal fragments due to the penetration of tetralin into the micropores of coal, as well as the effective H· radicals transfer from tetralin to coal fragments during pyrolysis. Several experiments of this nature have been conducted to demonstrate the effectiveness of hydrogen transfer in the efficient release of gases during pyrolysis [19,22,23].

The dissolution of coal results in the formation of intermediate free radical species. As shown in Figure 4.2, these intermediate species are autostabilized and form the preasphaltenes, which are further reduced in MW to asphaltenes (somewhat stable aromatic nuclei) and then to distillable oils and hydrocarbon gases. The latter compounds are also generated at each step of the main reaction path as by-products [24]. As the preasphaltenes and asphaltenes are further decomposed, the aromaticity of the products can increase significantly due to dehydrogenation of hydroaromatic structures [16]. In fact, as asphaltene conversion increases and fewer polar functional groups are present, the remaining by-product asphaltenes are much more aromatic (condensed) and are hydrogenated at a relatively slow rate [25].

A generalized mechanism for solvent processing of coal is illustrated in more detail in Figure 4.3. From a more basic point of view, there are at least three pathways available for free radical stabilization: H-abstraction from the solvent or a hydrogen-rich portion of the coal, either elimination of part of the radical molecule with associated group migration or rearrangement to form another more stable structure or both, and a combination reaction such as condensation or alkylation with another molecule in the solvent, which could also be a radical. The presence and type of hydrogen donors will determine the preferred reaction path [1].

The physical solvation properties of the solvent largely determine whether or not converted coal dissolves. For example, solvent-refined coal (SRC) process solvent contains more phenols and polyaromatics than H-coal (Amoco process) and Exxon donor solvent process solvents, and therefore the former is a better physical solvent for converted coal [26]. Once the coal is in solution, the solvent controls the compositional changes of the intermediate products and the gas formation rate.

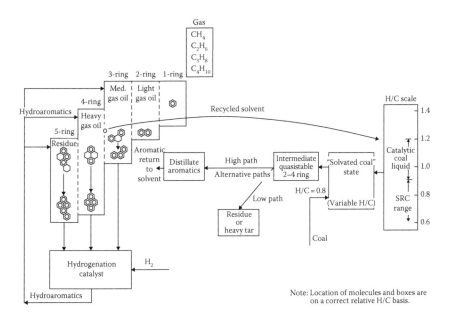

FIGURE 4.3 A generalized mechanism for solvent processing of coal. (From Shah, Y., *Reaction Engineering in Direct Coal Liquefaction*, Addison Wesley Publishing Co., Reading, MA, 1981.)

There is no doubt that hydrogen donation is a key factor in improving the yield of oil and increasing the conversion of the feedstock during liquefaction. The hydrogen transfer from a liquid carrier (tetralin) to coal was first described as a free radical (produced by thermal breakage of coal) reaction by Curran et al. [12]. They found that the degree of coal conversion was a function of the amount of hydrogen transferred and relatively independent of the solvent composition employed. Neavel [15] noted that the presence or lack thereof of a hydrogen donor vehicle had little effect on the thermal rupture of the coal fragment. This thermal cleavage results in the formation of highly reactive free radicals. In nondonor solvents, these free radicals react with surrounding molecules to form higher-MW compounds that become insoluble in subsequently employed extracting solvents. However, when a hydrogen-donating solvent is employed, hydrogen is abstracted from the solvent and stabilizes pyrolysis-formed free radicals. Products are then relatively low-MW substances capable of being dissolved in most solvents. Progressively, more donor hydrogen is abstracted for each increment of conversion as the reaction proceeds [15].

Vernon [27] conducted liquefaction of coal model compound dibenzyl and observed that when heated for 30 min at 450°C with an excess of a good donor solvent, tetralin, the conversion was 47% by mass, and the only major product was toluene; very little benzene and ethyl benzene were produced. When the same experiment was carried out in the presence of high-pressure hydrogen (11 MPa), the conversion increased to 58%. He then concluded that even molecular hydrogen could serve as a donor vehicle during liquefaction reactions. Li et al. [28] concluded that the primary factor inhibiting coal liquefaction is the consumption of hydrogen free radical (H$\cdot$) from solvent or H_2 and condensation of free radicals from coal pyrolysis after a period of reaction. So the essential approach for increasing oil yield and conversion of coal is to provide enough H$\cdot$ to stabilize the free radicals from coal pyrolysis. Unlike Neavel [15], McMillen et al. [29] offered that the direct transfer of hydrogen atoms from solvent-derived radicals to substituted positions in aromatic rings is substantial as preliminary steps in depolymerization of coal structures. Wei et al. [30] believed that molecular hydrogen enhances the thermolysis, hydrogenation, and hydrocracking of coal model compounds, as opposed to hydrogen-donating solvents, which inhibit these reactions. Although investigators' opinions regarding liquefaction process may vary, there is a consensus that free radicals released during cleavage of the coal molecule need to be stabilized by some hydrogen donor shuttle to suppress cross-linking reactions.

While the coal liquefaction behavior does depend on coal properties such as particles size (smaller the better) and coal surface properties as well as on physical processes that can occur during particle disintegration (which can be affected by local mixing, temperature, hydrogen partial pressure, etc.), it is now generally accepted that the quality of solvent plays a very important role on the mechanism of coal dissolution, extent of coal dissolution, and subsequent upgrading of the dissolved products. It has been pointed out that four quantities—H-donors, phenols, H-shuttlers, and H-abstractors—play important roles on the determination of the overall effectiveness of the solvent in the liquefaction process [1]. Solvents like tetralin and phenanthrenes are considered to be good donor solvents; cresol, on the other hand, is considered to be a poor solvent for the liquefaction.

Significant basic and developmental research carried both at laboratory and pilot scales on direct coal liquefaction carried out during the 1970s and 1980s indicated that the economics of the direct coal liquefaction process depend significantly on the extent of coal conversion and the quality of the products. Generally, feed coal concentration cannot exceed 40%–50%. The quality of the solvent and the degree to which it needs to be upgraded play important roles on the overall process economics. While tetralin and phenanthrenes are good donor solvents, their use in commercial practice is not economical. The processes were developed with the use of recycled solvent produced (with or without upgrading) by the process itself. While these processes were proven to be reasonable for certain types of boiler fuels, the solvent qualities were not good enough, in general, to be economical without its significant and expensive upgrading. Unlike crudes, coal liquids are highly aromatic, and the upgrading required significant amount of hydrogen [1]. The fuels produced from coal by this process, thus, could not compete well with the ones produced from crude oil (at its prevailing price).

4.2.2 EFFECT OF WATER PRETREATMENT OF COAL ON COAL LIQUEFACTION

The effect of water pretreatment of coal on the liquid oil produced by its liquefaction was studied by Serio et al. [31–33] and Ross and Hirschon [34]. Serio et al. [31–33] examined four different types of coals (Zap, lignite; Wyodak, subbituminous; Illinois No. 6, bituminous; and Pittsburgh, bituminous) pretreated by water at 4000 psig and 350°C and for the treatment times from 5 to 1200 min in a batch reactor. For each experiment, the yields of gases, water soluble materials, and residue were determined. The residues were subjected to analysis by a variety of techniques such as thermogravimetric analysis–Fourier transform infrared (FTIR), solvent extraction, donor solvent liquefaction, and FTIR. The study resulted in the following conclusions:

1. At short pretreatment times, the process loosened up the coal structure resulting in the increase of extractables and the yield. The oxygen content also decreased when coal was subjected to an accelerated aging process. However, the liquefaction yields appear to decrease relative to the raw coal.
2. At longer pretreatment times, the process partly recombines the structure resulting in a decline of extractable and tar yields. Oxygen continues to be removed, but ether groups go through a maximum. The liquefaction yields were closer to values for the raw coal.
3. The solvent adduction may be the reason for the decline in liquefaction yields for coals with short pretreatment times.
4. For Illinois coal, the yields were very sensitive to the amount of oxygen exposure. The participation by the oxidized form of pyrite in the liquefaction pretreatment chemistry appeared possible.

Bienkowski et al. [35,36] evaluated the effect of steam pretreatment on coal liquefaction. For a Wyodak coal stored under water (to avoid weathering), they pretreated the coal using 750 psig steam for 30 min at 200°C. Pretreatment of suction dried coal at 200°C increased production of extractables at 400°C from 30.5% to 38.5%. While the increase in pretreatment temperature to 240°C increased yield to 40.3%, an increase in pretreatment temperature to 320°C reduced the conversion to 33.8%. Bienkowski et al. [35,36] argued that an increase in pretreatment temperature increased coal matrix loosening and stabilization of some reactive components of the coal resulting in higher conversion. A further increase in temperature set up a higher rate of retrogressive reactions, which in turn decreased the conversion. Bienkowski et al. [36] also found that an addition of ammonia in both pretreatment and subsequent liquefaction stages gave even higher conversion due to the reactions between hydrogen and oxygen functional groups.

Graff and Brandes [37,38] and Brandes et al. [39] observed higher yields of liquid products from pyrolysis and solvent extraction of Illinois No. 6 coal, which was pretreated by steam at 320°C–360°C at 50 atm pressure. A similar pretreatment with helium had no effect, and the exposure to air of steam-pretreated coal lost the increase in yields. The study concluded that the pretreatment disrupts hydrogen bonds, reduces the number of covalent cross-links, and increases the hydroxyl groups in the coal [37–39]. The exposure to air weathers the coal with a negative effect on liquefaction yield. Khan et al. [40] showed that steam pretreatment at 1100–1300 psig and 300°C–320°C temperature for five coals of different ranks did not increase tar yields when pyrolyzed at a slow heating rate. The steam treatment reduced the concentration of oxygen functional groups for the low-rank coals and increased tar yields when pyrolyzed at a rapid heating rate.

Ross et al. [41–45] evaluated the effects of water pretreatment on Illinois No. 6 and Wyodak subbituminous coals and found no effects on toluene solubles in a subsequent donor solvent liquefaction process. Significant changes were, however, observed in the composition and MW distributions of the liquid products of the liquefaction process due to water pretreatment. The pretreatments were carried out at 250°C and 38 atm pressure. The coal liquefaction was carried out at 400°C, 500 psi H_2 pressure for 20 min in tetralin solvent.

4.3   HTL OF COAL AND BIOMASS

As shown in my previous book [2], water at higher temperature and pressure becomes nonpolar and behaves more like organic liquids. Water under these conditions becomes more reactive and takes active participation in the dissolution process of coal and biomass. While reaction chemistry between coal and water is not as strong as that between biomass and water, both are briefly described in this chapter.

4.3.1   COAL LIQUEFACTION IN HIGH-PRESSURE AND HIGH-TEMPERATURE WATER

A number of studies [46–52] have examined the coal liquefaction in water at high temperature and pressure. Mikita et al. [46] and Blaustein et al. [47] found tetrahydrofuran (THF) conversion of Illinois No. 6 coal in water to be about 67%, in water and SRC II solvent about 87%, and in water and 1000 ppm of Mo about 90% for reactions at 385°C, 1200 psig H_2 pressure, and 30 min residence time. A synergism was observed at low ratios (0.5 or less) of donor solvent to coal upon combination of

SRC II distillate and water. A similar effect was not observed when cyclododecane replaced water. The addition of Mo catalyst precursors to the water allowed complete elimination of donor solvent without loss in conversion.

Yoneyyama et al. [48] examined noncatalytic hydrogenation of several bituminous and subbituminous coals with or without water addition at 400°C. For comparison, similar experiments in nitrogen or undecane (n-C_{11}) were also carried out. In nitrogen or hydrogen atmosphere, water promoted coal conversion, but addition of undecane neither changed nor decreased the conversions. For higher-rank coal, undecane inhibited coal conversion in nitrogen. The conversion of coals using nitrogen and water increased with increasing carbon content of coals. On the other hand, when hydrogen and water were used, there existed no clear relationship between the coal conversion and carbon content of coals. Under pressurized hydrogen, coals containing pyrites gave significantly larger conversions implying its catalytic role in the conversion process. A synergistic effect (SE) existed between hydrogen and water on the conversion of coals, and the effect was more obvious for the coals containing larger amount of pyrite.

Ross and Blessing [49] and Ross et al. [50,51] found that for Illinois No. 6 coal in CO/H_2O system at 4000–5000 psig pressure and 400°C (under supercritical conditions), better toluene solubles were achieved than for tetralin under the same conditions. The CO/H_2O system was more effective than H_2/H_2O system and the latter system was not very effective for demineralized coal. The results were explained in terms of an ionic mechanism involving the initial formation of formate ion whereby hydrogen is donated to the coal.

Anderson [52] examined hydrothermal dissolution of coal and found that at high-temperature and high-pressure, coal dissolution is rapid and can be taken to completion. Breaking cross-linking structures will convert high-MW structures into low-MW products that can be processed and used as high-value chemical feedstock. Product is a pumpable liquid that can be further processed. Product recovery is high; up to 90% of the original carbon is recoverable as water-soluble product. Finally, inorganic components (pyrites, calcite) are readily converted to soluble products that can be recovered and/or treated in the liquid phase.

The studies described earlier clearly indicate that water plays an active role as a reactant for the coal liquefaction under high-temperature and high-pressure conditions. The reactive role of water is further increased near and above supercritical conditions. Thus, water should be evaluated as a possible solvent for the coal liquefaction process.

4.3.2   BIOMASS LIQUEFACTION IN WATER

Just like coal chemistry in organic solvents developed in the 1970s and 1980s, biomass chemistry with water has also been significantly developed in recent years due to expanding interests in renewable fuels. The use of water for steam gasification of coal or its use as a donor solvent found to have some possibilities [2,3]. However, biomass–water chemistry development showed that complex organic (particularly carbohydrate base) molecules from biowaste (manure and food processing waste), lignocellulose (crop residue), algae, and others can be converted to bio-oil (similar to crude oil) in the presence of water. As shown in Section 4.3.2.3, the extent of biomass conversion and quality of bio-oil depends on the nature of feedstock, temperature, pressure, reaction time, gaseous environment, and presence of a catalyst (if any) [53–61]. HTL of biomass to some extent mimics the natural geological process thought to be involved in the production of fossil fuels. HTL involves a direct liquefaction of biomass in water (and may be a catalyst) to liquid fuels under subcritical conditions. HTL process requires temperature range of about 250°C–400°C.

4.3.2.1 Reaction Mechanisms

HTL is a chemical transformation process of biomass in a heated and pressurized water environment where long-chain organic compounds break into short-chain hydrocarbons. It is believed that all fossil fuels found underground, petroleum, natural gas and coal, etc., based on biogenic hypothesis, are formed through HTL process from biomass buried beneath the ground and subjected to high pressure and temperature. The reported studies have shown that kerogens (which are a large part of oil shale) break down much easily in the presence of water [40–48]. HTL process reaction paths depend on the temperature, pressure, reaction time, water pH, solids particle size, and the nature of the catalysts (if present).

While the exact reaction pathways for HTL process is as yet not known, the study of Mikita et al. [46], Appell et al. [62,63] and Blaustein et al. [64] at Pittsburgh Energy Technology Center made some important points for the process. They studied liquefaction of wood particles with hydrogen and carbon monoxide at 370°C and 27 MPa pressure in the presence of sodium carbonate catalyst. Alkali salts such as sodium carbonate and potassium carbonate can initiate the hydrolysis of cellulose and hemicellulose into smaller fragments. The degradation of biomass into smaller products mainly proceeds by depolymerization and deoxygenation reactions. The amount of solid residue remaining depends upon the lignin content. Lignin contains alkyl phenols and free phenoxyl radicals formed by its thermal decomposition above 250°C, and they are likely to recombine and form solid residue through condensation or repolymerization reactions.

Appell et al. [46,62–64] Appell et al. [46,62–64] suggested that during the conversion of carbohydrates to oil, sodium carbonate reacts with carbon monoxide and water to form sodium formate as:

$$Na_2CO_3 + 2CO + H_2O \rightarrow 2HCO_2Na + CO_2 \tag{4.4}$$

which in turn reacts with cellulose in the wood wastes to form oil and regenerate sodium carbonate as:

$$2C_6H_{10}O_5 + 2HCO_2Na \rightarrow 2C_2H_{10}O_4 + H_2O + CO_2 + Na_2CO_3 \tag{4.5}$$

Vicinal hydroxyl groups in the carbohydrates undergo dehydration to form an enol followed by its isomerization to ketone. The newly formed carbonyl group is reduced to the corresponding alcohol with formate ion and water. The hydroxyl ion then reacts with additional carbon monoxide to regenerate the formate ion.

The aforementioned set of basic reactions is accompanied by a multitude of side reactions producing a whole host of intermediates. Some of the beneficial side reactions are facilitated by the alkaline conditions. When two carbonyl groups become vicinal, a benzylic type of rearrangement occurs, which results in the formation of a hydroxyl acid. The hydroxyl acid readily decarboxylates causing a net effect of reducing the remainder of the carbohydrate-derived molecule [46,62–64].

For HTL process, the segments produced by hydrolysis are further degraded by dehydration, dehydrogenation, decarboxylation, and deoxygenation. These types of reactions result in the formation of paraffin-type structures, which have less oxygen than the original compounds.

There are several other features in this reaction mechanism. Aldol condensation may be a part of the reaction. In the absence of a reducing agent such as CO or H_2, condensation reactions dominate, which lead to polymerization and the formation of solid-like products. The reducing agents keep the concentration of carbonyl groups low enough to produce liquid products instead of solid products.

Mikita et al. [46], Appell et al. [62,63] and Blaustein et al. [64], also pointed out that hydrogen radicals formed by the addition of CO and the presence of water gas shift reaction can react with various carbonyl and hydroxyl groups to form paraffins and water and thus avoid various condensation reactions. The addition of CO is thus more useful than the addition of molecular hydrogen [2,46,53,62–64].

In HTL process, other sets of complex reactions such as cracking and reduction of polymers such as lignin and lipids, hydrolysis of cellulose and hemicellulose to glucose and other simple sugars, hydrogenolysis in the presence of hydrogen, reduction of amino acids, dehydration, decarboxylation, C–O, C–C bond ruptures, and hydrogenation of various functional groups result in the production of liquids from biomass rather than solids as it occurs in the hydrothermal carbonization process [2].

4.3.2.2 Effects of Process Operating Conditions on HTL Process

The main objective of HTL process is to generate bio-oil of high quality. The process is designed to minimize the productions of solids and gas. Numerous process parameters affect the performance of HTL process, and these are well examined in the literature [53–61,65–86]. Major process operating parameters are (1) temperature, (2) residence time, (3) solids concentration, (4) pressure, (5) biomass heating rate, (6) biomass particle size, (7) presence of hydrogen donor solvent and reducing gas environment, (8) pH of slurry, and (9) nature of feedstock. The literature information on the effects of various operating parameters on the product distribution was well summarized in an excellent review article by Akhtar and Amin [53]. The following three sections briefly summarize their assessments.

4.3.2.2.1 Pressure, Temperature, and Residence Time

In any HTL process, high pressure allows a better manipulation of hydrolysis reaction and the reaction pathways that are thermodynamically favorable to produce liquids and gases. High pressure also increases solvent density resulting in better extraction capability of the solvent. For a catalytic operation, however, high solvent density can block the active catalyst sites and thereby reduce C–C bond breakage and the resulting degradation rate.

Generally, high temperature increases both the concentration of free radicals and the probability of repolymerization of fragmented species. The hydrolysis and fragmentation of free radicals dominate in the early stages of the reactions, while repolymerization occurs in the later stages of the reaction, which in turn forms char. Generally, at very high temperature, bio-oil production is reduced due to (1) secondary decompositions reactions that become active leading to high gas formation or (2) recombination of free radicals to form char. The overall process conditions and the presence of a catalyst generally dictate the dominant reaction mechanism. For most feedstock, however, the maximum bio-oil is obtained at temperature around 300°C–350°C [53]. Also, the literature results show that the largest shift in optimum temperature for bio-oil occurs for algae [66–78]. Both softwood and grass are generally more difficult to liquefy because of their higher lignin content and less reactive cellulose content.

Akhtar and Amin [53] showed that bio-oil production as a function of temperature reported by various studies exhibited an optimum, although the exact location of the maximum oil production depended on the nature of the feedstock. At temperatures above 300°C, the gas production can also increase particularly when a suitable catalyst is used. The typical temperature dependence of the product distributions of solid residue, oil, water solubles plus water, and gas are illustrated in Table 4.1 [53]. Once again, the exact phase composition will depend on the nature of the feedstock. Table 4.1 also shows that both oil/solid ratio and oil-to-gas ratio show maxima at a temperature around 300°C. Akhtar and Amin [53] suggested that a variation in the solid residue can be set as a reference point to measure optimum liquefaction temperature for bio-oil yield for a given feedstock. As shown in Table 4.1, an increase in temperature increases gas yield and decreases residue yield. The oil yield shows an optimum with respect to the temperature at around 300°C.

The effects of residence time on HTL process have been examined by numerous investigators, and these are well summarized by Akhtar and Amin [53]. Both biomass conversion and the nature of product distributions depend on the residence time. Since initial hydrolysis process is fast, normally short residence is preferred in HTL. Boocock and Sherman [87] showed that the bio-oil production was suppressed

TABLE 4.1
Product Distribution as Function Temperature

Temperature (°C)	Residue (%)	Oil (%)	WS/W (%)	Gas (%)	O/S	O/G
150	72	6	21	1	0.083	6
200	56	13	26	5	0.23	2.6
250	32	33	29	6	1.03	5.5
300	16	38	30	16	2.37	2.37
350	16	27	33	24	1.69	1.37

Source: Akhtar, A. and Amin, N., *Renew. Sustain. Energy Rev.*, 15, 1615, 2011. With permission.

Note: These are the best calculations/estimations from the graphical data.

O/G, oil/gas ratio; O/S, oil/solid ratio; W, water; WS, water soluble.

at high residence time except when biomass concentration in the feed was very high. The effect of residence time on the bio-oil yield also depended on the temperature. At low temperatures, an increase in residence gave higher bio-oil yield [53], whereas, at high temperatures (250°C–280°C), high residence time gave poorer bio-oil yields due to increase in gas yields. In general, higher residence time gave higher biomass conversion. Qu et al. [80] found decrease in heavy oil production at high residence time.

The effect of residence time on the product distribution in HTL process is complex. The intermediate products formed during the liquefaction process can form gas, liquid, or solid products by secondary and tertiary reactions depending on the nature of intermediates and prevailing local reaction environment. Karagoz et al. [81] showed that the decomposition products were not the same at low and high residence times for both low-temperature (180°C) and high-temperature (250°C) operations. In general, the composition and yield of bio-oil can be optimized by suitably adjusting temperature and residence time such that heavy residues containing asphaltenes and preasphaltenes are converted selectively to oil (and not gas). Often an addition of a reducing agent such as CO or H_2 prevents polymerization reactions and stabilizes the active free radicals.

4.3.2.2.2 Biomass Particle Size, Heating Rate, and Concentration

While the process of biomass particle size reduction (particularly for biomass like wheat straw, barley straw, and switchgrass) is energy intensive, in general, a smaller particle size results in higher degree of hydrolysis and fragmentation. However, literature data show that particle size has a secondary effect on biomass conversion and product distribution in an HTL process because of high solvation and extraction powers of water at high temperature and pressure. Zhang et al. [76] found no effect of particle size variation from 0.5 to 2 mm of grass perennials on the yield of bio-oils. Akhtar and Amin [53] recommended that the particle size between 4 and 10 mm should be suitable to overcome heat and mass transfer limitations at reasonable grinding cost.

While the aforementioned studies indicated marginal effects of particle size on the herbaceous biomass liquefaction process, Kobayashi et al. [79] showed a significant effect of particle size of woody biomass on HTL process. Wood powder pulverized by the vibration mill, cutter mill, and grinder was used as liquefaction materials. The wood powder was sieved between 212 and 500 µm. Based on the results of water solubles and specific surface areas for three different milling process obtained in this study, it was concluded that an increase in specific surface area increased the productions of water soluble (saccharine); however, the difference in water solubles between grinder and cutter mill was only marginal. This indicated that the crystallinity of the wood powder also affects the water solubles yield.

Bio-oil production during HTL process generally occurs at moderate heating rates. Slow heating rates usually lead to the formation of char residue due to secondary condensation and polymerization reactions. Very high heating rates also promote secondary reactions, which generally result in more gas productions. Compared to wet pyrolysis, heating rate is less important in HTL process because of the better dissolution and stabilization of fragmented species in hot compressed water medium.

Zhang [58] and Zhang et al. [76] observed that for heating rate range 5°C–140°C/min during HTL process for grassland perennials, bio-oil yield increased from 63% to 76% with an increase in heating rate.

The solid concentration also affects the bio-oil production [53]. In general, high amount of water favors productions of liquids largely due to enhanced extraction and higher degree of solvation of biomass. The solvent enhances the stability and solubility of fragmented components, thereby reducing the productions of solid residues and gases. At high biomass concentration, the interactions between fragmented biomass components and water decreases and the reactions among various fragments increase and thus influence of water on the product distributions diminishes. At high biomass concentration, HTL process more behaves like intermediate temperature dry pyrolysis resulting in less bio-oil production.

4.3.2.2.3 Gas and Liquid Properties

Generally, a reducing gas or a hydrogen donor stabilizes the fragmented products of liquefaction. Reducing environment inhibits condensation, cyclization, and repolymerization of free radicals, thereby reducing the char formation [53]. While H_2 is an effective reducing agent, it is also an expensive one. Often syngas (CO and H_2), steam, N_2 and argon, etc., are also used to provide a reducing environment [53]. The effects of various reducing agents on the maximum bio-oil production for cattle manure by HTL process is illustrated in Table 4.2. The results showed that CO was the most effective reducing agent, and it provided maximum bio-oil yield of 50% at 310°C [53]. This is in line with the assertion by Appell et al. [62,63]. The temperature (310°C) at which maximum occurred was independent of the nature of the gas. Air was ineffective because it led to combustion of biomass. The results also indicated that reactive gases gave better maximum oil yield than an inert gas like nitrogen.

The use of a hydrogen donor solvent such as tetralin and phenanthrene is also an effective way to stabilize free radicals and improve the bio-oil yield. A suitable catalyst can also induce or accelerate hydrogen transfer reactions to improve bio-oil yield. However, the cost of solvents and catalyst stability and cost can be limiting factors. In general, adsorption of reducing gas (like H_2) on the catalyst surface can increase the probability of hydrogen transfer reaction for the free radicals.

TABLE 4.2
Effect of Gas-Phase Condition on Oil Production

Reducing Gas	Maximum Yield of Liquid Oil (wt%)
Air	25
Nitrogen	36
Hydrogen	42
Carbon monoxide	50

Source: Akhtar, A. and Amin, N., *Renew. Sustain. Energy Rev.*, 15, 1615, 2011. With permission.
Note: These are the best estimations from the graphical data.

While most of the literature studies have focused on the neutral and alkaline water conditions for the HTL process, recently Yin et al. [71] examined the effect of water pH on reaction mechanism and product distribution of HTL of cellulose. The study examined water pH 3, 7, and 14 in the temperature range of 275°C–320°C and residence time of 0–30 min. Results showed that the composition of the products from HTL varied with pH. In acidic and neutral conditions, the main liquid product was 5-hydroxymethylfurfural (HMF). Under alkaline conditions, the main compound was C_{2-5} carboxylic acids. At all pH levels, high temperature and long residence times had negative effects on the bio-oil yields. The reaction mechanisms also depended on the pH level. In acidic conditions, polymerization of HMF to solids reduced the bio-oil production. In neutral conditions, HMF was converted to both solids and gases. Under alkaline conditions, bio-oil was converted to gases by the formation of short-chain acids and aldehydes. Different reaction mechanisms and different product characteristics mean different strategies required to improve quality and quantity of bio-oil under different pH conditions.

4.3.2.3 Role of Feedstock

The nature of feedstock is the most important variable affecting the quantity and quality of bio-oil by the HTL process [65–86,88–105]. HTL produces high-density liquid fuels, and it operates in the presence of water eliminating the need for feedstock drying that is important in dry pyrolysis process. HTL process uses agricultural biomass and biowaste including crop residues and wood, food processing waste, animal and human manure, and algae. HTL process can also be used as a pretreatment process for the subsequent fermentation for the feedstocks that are difficult to ferment.

While animal and food processing wastes contain lipids, proteins, and small amounts of lignocellulose, crop residues and wood primarily contain lignocellulose. The primary basic compounds in these feedstocks are various isomers of glucose (such as D-glucose, L-glucose), hemicellulose, cellulose, lignin, amino acids, proteins, and lipids including fatty acids such as stearic acid and palmitic acid. Over the years, more and more efforts have been made to examine the effectiveness of HTL process for a variety of lignocellulosic wastes containing different amounts of lignins and crystalline cellulose. Some of the materials examined are swine manure, garbage, Indonesian biomass residue, birch wood, sawdust, rice husk, phytomass, and chlorella [53–61,65–86,88–105]. These feedstocks have been tested in the temperature range of 280°C–375°C, pressure range of 5–50 MPa, and residence time of 5–180 min depending on the feedstock. In some cases, sodium carbonate was used as a catalyst. The final yield varied from 21% (in case of garbage) to as high as 61% (in case of swine manure) [53–61,65–86,88–105]. More lately, HTL has been more and more applied to algae to make biofuel. HTL can thus be applied to a variety of biomass with a varying degree of success.

The best raw materials for HTL process are perhaps biowastes such as various types of manures and wastewaters because they are mostly cellulosic (with very little lignin) and can be easily converted to bio-oil under hydrothermal conditions. In general, the presence of high cellulose and hemicellulose content in biomass yields more bio-oil. Appell et al. [46,62–64] were the first ones to examine various waste streams such as urban refuse, cellulosic wastes, and sewage sludge and found that at temperature around 380°C, pressure 1500 psig, and residence time of 20 min, even in the absence of a catalyst, oil yield of about 24.5% was obtained. Oil was largely paraffinic and cycloparaffinic at low temperatures with the presence of carboxyl and carbonyl groups, while at high temperatures some aromatics were present. Following this pioneering study, a significant number of additional studies were published [53–61,65–86,88–105] indicating the easy applicability of HTL process to cellulosic wastes. These studies are also extensively described in a number of recent reviews [53–61]. The studies included swine manure, bovine manure, cellulose, activated sludge, sewage sludge, artificial garbage, protein-contained biomass (such as food wastes), various types of wastewaters (including wastewater from paper and pulp industries), glucose, glycine, dairy manure, and poultry litter. Zhong and Wei [213] showed comparative experimental study for liquefaction of various types of wood.

The study done by Minowa et al. [82–85] using glucose and glycine as model compounds of carbohydrates and proteins indicates that a significant oil production started at temperatures >250°C and it increased with temperature. This and other studies have shown that fatty acids and lipid are the main reactants in HTL process. Below 300°C, aliphatic compounds are the major source of bio-oil. Protein is widely involved in HTL reaction possibly by peptide bond splitting and amino acid conversion dehydration. Within the range 300°C–450°C, the protein conversion reaction intensifies and peptide bond begins to react. Saccharide reaction mainly belongs to the splitting of branched chains and group transfer, while considerable dehydration and cyclization of the main chain still appears to be dominant. The decomposition of an individual cellulosic biomass differs based on their structure. Decomposition is easier in hemicelluloses due to amorphous structure. Cellulose has little crystalline to decompose due to beta [1,2,4,5] glycosidic linkages and relatively intermediate degree of polymerization (500–10,000). Major products of holocellulose degradation include cellohexaose, cellopentaose, cellotriose, cellobiose, fructose, glucose, erythrose, glycolaldehyde, glyceraldehyde, pyruvaldehyde, and furfurals [82–85].

The presence of liquid water as solvent is essential for HTL of lignocellulose feedstock; even more than that for cellulosic waste like manure or algae. Water in this case acts as a solvent and reactant along with its role as a vehicle for biomass and a carrier for the catalyst. Lignocellulose is the largest segment of the total biomass, and it contains a significant amount of lignin along with cellulose. It is the lignin component along with crystalline cellulose that is difficult to convert to bio-oil in HTL process. In the absence of a catalyst, lignin produces very little bio-oil, and it ends up as solid residue in the HTL process. While water is an excellent medium for the intermediate hydrolysis of cellulose and other higher-MW carbohydrates to water soluble sugars, it is not as effective for hydrolysis of heavily aromatic and multitiring aromatic structures. The breakdown of lignin requires high temperature or the presence of a catalyst. Within lignocellulosic substances, softwood gives much lower yield than hardwood because of the difference in their lignin contents. Zhang [58] and Akhtar and Amin [53] among others have given an extensive review of HTL of a variety of lignocellulosic biomass such as various energy crops, herbaceous products, forestry and other agricultural wastes, and various crop-oil wastes. Midgett [54] examined HTL process for materials such as tallow seed, switchgrass, and pine dust. The literature also indicates that bio-oil yield is affected by both temperature and the lignin content of the wood [53].

An interesting study was carried out by Sugano et al. [61] in which they examined the effectiveness of black liquor, paper regeneration wastewater, and the water on HTL of herbaceous eucalyptus biomass. Like softwood, this material contains lignin. The study showed that black liquor gave very low oil yield and high yield of water soluble components. The liquefaction of eucalyptus by water resulted in high residue production due to dehydration and polymerization, such as the formation of aliphatic ester bonds. The paper regeneration wastewater gave low residue and high yield for bio-oil compared to water. It appears that condensation reaction observed during the liquefaction in water was inhibited because carboxylic acid formed during the liquefaction of eucalyptus was neutralized with the cations in the wastewater. In the temperature range of 150°C–350°C, process wastewater gave the best oil yield compared to other two solvents. The optimum temperature for bio-oil yield was 300°C. This study indicated that solvent pH and other additives can affect the bio-oil productions of lignocellulosic materials.

Finally, high degree of polymerization (>10,000) and complex branching makes lignin difficult to decompose even at high temperatures. The aforementioned described studies among others showed that the conversion of lignin-containing lignocellulosic materials require catalysts to produce bio-oil instead of solids and gases. An extensive literature review of various catalyst studies for HTL process as applied to lignocellulosic biomass is given by Zhang [58].

Ever since Glen Meier of Renewable Energy Group introduced the concept of producing fuel from algae, it has caught everybody's attention [86,88–105]. Like biowaste, algae do not compete with food materials for fuel. The HTL of algae has been given some attention for the past two decades. Minowa et al. [101] converted *Dunaliella tertiolecta* with a moisture content of 78.4% directly into 37% oil by HTL process operated at 300°C and 10 MPa. The oil had a viscosity of 150–330 MPa s and a calorific value of 36 kJ/g, numbers comparable to that of fuel oil. Dote et al. [100] hydrothermally converted the artificially cultivated *Botryococcus braunii* Kiitzing Berkeley strain. The strain contained about 50% hexane solubles. The HTL process of this strain resulted in the production of 57% petroleum like bio-oil at 300°C. Similar work was carried out for *Microcystis viridis* harvested from a lake.

Brown et al. [104] reported hydrothermal conversion of marine microalga *Nannochloropsis* sp. into bio-oil in the temperature range 200°C–500°C and for residence time of 60 min. The highest bio-oil of 43 wt% was obtained at 350°C with a heating value of 39 MJ/kg, a number comparable to petroleum crude oil. The H/C and O/C ratios changed from 1.73 and 0.12 at 200°C to 1.04 and 0.05 at 500°C. Major components of bio-oil were phenol and its alkylated derivatives, heterocyclic N-containing compounds, long-chain fatty acids, alkanes, alkenes, derivatives of phytol, and cholesterol. Gases largely contained CO_2 and H_2.

Metal catalysts had been used in microalgae liquefaction. Matsui et al. [105] investigated liquefaction of *Spirulina*, a high-protein algae in water at 300°C–425°C using $Fe(CO)_5$–S catalyst. Other metal catalysts used were $Ru_3(CO)_{12}$ and $Mo(CO)_6$. Continuous culturing of the *B. braunii* Berkeley strain in secondary treated sewage was conducted and then liquefied by Sawayama et al. [102]. The liquefaction was carried out at 200°C, 300°C, and 340°C. The yield of the hexane soluble fraction was 97% compared with that in the feedstock algal cells. The heating value of the liquefied oil obtained from this reaction was 49 MJ/kg, and the viscosity was 64 MPa s at 50°C.

Different microalgae are not the same in producing oil through HTL process. Recent literature [86,88–105] have shown a wide variety of performance in the production rate of bio-oil and its quality. The performance of HTL of microalgae has been improved by the use of a catalyst such as Na_2CO_3 among others [86,88–105]. Zhou et al. [91] examined HTL of marine macroalgae *Enteromorpha prolifera* in the temperature range 220°C–320°C for 30 min and in the presence of 5 wt% Na_2CO_3 catalyst. The highest bio-oil yield of 23 wt% with higher heating value of 28–30 MJ/kg at 300°C was obtained. These numbers are smaller than what is reported for microalgae [92]. The bio-oil contained ketones, aldehydes, phenols, alkenes, fatty acids, esters, aromatics, and nitrogen-containing heterocyclic compounds. Acetic acid was the main component of water soluble components.

Vardon et al. [89,90] studied HTL of *Scenedesmus* (raw and defatted) and *Spirulina* algal biomass at 300°C and 10–12 MPa pressure and compared the performance with that of Illinois shale oil and bio-oil produced by dry pyrolysis (at 450°C). Both wet and dry pyrolysis gave energy dense bio-oil (35–37 MJ/kg) that approached shale oil (41 MJ/kg). Bio-oil yields (24%–45%) and physicochemical characteristics were highly influenced by conversion route and feedstock selection. Sharp differences were observed for mean bio-oil MW (dry pyrolysis: 280–360 Da; HTL: 700–1330 Da) and percentage of low boiling compounds (bp < 400°C) (dry pyrolysis: 62%–66%; HTL: 45%–54%). For wet algal biomass containing 80% moisture, energy consumption ratio for HTL (0.44–0.63) was more favorable than one for dry pyrolysis (0.92–1.24). In another study, Vardon et al. [89,90] also showed that *Spirulina* algal biomass gave 32.6% biocrude as opposed to 9.4% for digested sludge under the same reaction conditions as mentioned earlier and for 30 min residence time. While swine manure, digested sludge, and *Spirulina* algae gave biocrudes of similar heating value (32–34.7 MJ/kg), they differed substantially in their detailed chemistry. MW tracked with obdurate carbohydrate content followed the order *Spirulina* < swine manure < digested sludge.

Duan and Savage [92] were the first ones to evaluate the effects of various hydroprocessing catalysts on HTL of microalga *Nannochloropsis* sp. The experiments were performed at 350°C with Pd/C, Pt/C, Ru/C, Ni/SiO_2-alpha-Al_2O_3, CoMo/l_3-Al_2O_3 (sulfide), and zeolite catalysts. In the absence of hydrogen, all catalysts gave higher yields of bio-oil, but the elemental compositions and heating value of bio-oil (about 38 MJ/kg) were insensitive to the nature of the catalyst used. Gases contained H_2, CO_2, CH_4, and lesser amounts of C_2H_4 and C_2H_6. Ru and Ni catalysts produced nitrogen. H/C and O/C ratios of the products were about 1.7 and 0.09. While the presence of hydrogen and higher pressure suppressed the gas formation, the bio-oil yield and its characteristics did not significantly change.

Generally, high-lipid content of algal mass limits its conversion to bio-oil by HTL process. Yu et al. [94] examined the HTL process for low lipid microalgae *Chlorella pyrenoidosa* and found that at 280°C and 120 min reaction time, bio-oil yield of

39.4% was obtained. The bio-oil yield, water solubles, and gases were strongly dependent on the temperature and reaction time. Biller and Ross [95] correlated the performances of various types of algal biomass in HTL process by correlating the bio-oil yield with biochemical content of the biomass. They examined microalgae *Chlorella vulgaris*, *Nannochloropsis oculata*, *Porphyridium cruentum*, and cyanobacteria *Spirulina* and found that yields of biocrude from these species were 5–25 wt% higher than the lipid content of the algae depending upon biochemical composition. The yields of biocrudes followed the trend lipids > proteins > carbohydrates.

Ross et al. [103] examined the effects of alkalis and organic acids on HTL of low lipid content *C. vulgaris* and *Spirulina* algae at 300°C and 350°C. The effects of temperature and the catalyst types on the product yields and composition were examined. The catalysts used were alkali, potassium hydroxide, and sodium carbonate and the organic acids, acetic acid, and formic acid. The yields of biocrude were higher using an organic acid catalyst, and these crudes had a lower boiling point and improved flow properties. The higher heating value ranged from 33.4 to 39.9 MJ/kg. The biocrude contained 70%–75% carbon, 10%–16% oxygen, and 4%–6% nitrogen.

Biller et al. [98] examined a range of microalgae and lipids extracted from terrestrial oilseed for HTL process at 350°C and 150–200 atm pressure in the presence of a variety of heterogeneous catalysts. The results showed that the HTL process converted triglycerides to fatty acids and alkanes in the presence of certain heterogeneous catalysts.

The reviews presented in Section 4.3 for oil production by water for coal and biomass have not as yet extended to the mixture of coal and biomass/waste. The effectiveness of high-temperature, high-pressure water (and steam) on biomass/waste is much higher than that for coal. However, all evidences indicate that oil production from a mixture of coal and biomass and water under subcritical conditions may be worth pursuing.

4.4 CO-PROCESSING

The aforementioned discussion shows that coal liquefaction in a donor solvent and HTL of coal and biomass at elevated temperatures and pressures work very effectively to produce oil, although HTL of coal gives a somewhat poor-quality liquid. For the last four decades, several other co-processing combinations have been found very effective in generating large-scale commercial plants for synthetic liquids using coal, biomass, waste, and other fuels. These are

1. Co-processing of coal with heavy oil, bitumen, shale oil, vacuum gas oil (VGO), etc.
2. Co-processing of coal with biomass and different types of polymeric wastes to produce oil economically
3. Co-processing of some other unusual mixtures

As illustrated in this section by numerous examples [106–216], these different co-processing systems have shown some significant promises through synergies and are being pursued aggressively. Section 4.4.1 outlines numerous examples of the first category, which illustrate their usefulness. In the second category, noncatalytic SE observed in the release of volatiles and tar during coal/biomass/waste coconversion are examined. In the case of coal and biomass, the mechanism for synergy is a complex one, since the release of hydrogen from biomass should match the free radical release from coal and the two fuels should be spatially close for the transfer to happen. The discrepancy observed in the scientific community about this topic is probably due to these constraints. Section 4.4.2 also examines co-processing of coal and various types of polymeric wastes. The results show the presence of synergy as well. Finally, in Section 4.4.3, synthetic liquid productions by co-processing of several other multifuel combinations are examined.

4.4.1 Co-Processing of Coal and Various Types of Oils

The term co-processing became prominent during the 1980s and 1990s, when hydrocarbon fuels from coal are produced by coal/petroleum co-processing based upon the use of heavy oil, tar sand bitumen, and petroleum residue as solvents for the conversion of coal. Co processing is the simultaneous hydrogenation of coal and heavy oil fractions in specially designed reactors with coal content by weight ranging from as low as 1% to potentially as high as 50%–60% depending upon the technology employed [106–184].

While as shown in Section 4.2 the direct liquefaction of coal using a hydrogen donor solvent or recycle solvent has been intensely investigated since the 1970s, a parallel path to produce lighter crudes from unconventional heavy oils or crudes and tar sand bitumen has also been pursued over the last three decades. Upgrading and converting these heavy oils to distillate liquids using conventional petroleum thermal cracking, catalytic cracking, and/or hydrocracking technologies have required the installations of costly equipment to handle heavier oils. There is a sufficient evidence in the literature [106–184] to suggest that heavy oil converts more readily in the presence of coal and that significant enhancement in desulfurization, demetallization and denitrification, and conversion of asphaltene to oils also occur during co-processing of coal and heavy oil. Thus, significant interest arose in simultaneous conversion of coal and heavy oil to produce distillates. This process called "co-processing" of coal/oil was determined to be an effective method to achieve two separate objectives: (1) conversion of heavy oil to distillate and (2) simultaneously introducing coal liquids into the market place in a cost-effective evolutionary manner while greatly reducing the capital investment associated with the historical approach for establishing direct coal liquefaction industry.

Co-processing of *coal* with resids (1000°F+) is an attractive way of obtaining useful distillates from readily available cheap materials. Among the potential benefits for the implementation and utilization of the co-processing concept are [106–184]

1. Provision of a link or bridge between present day refining technology and a total coal-based synfuel industry. Improved economics compared to direct coal liquefaction due to smaller plant sizes and due to lower hydrogen requirements and the elimination of the use of process-derived solvent recycle

2. Improved residuum demetallization and improved product yields and mix. Minimization of the production of gases and undesirable by-products, such as high sulfur coke
3. Continued use of the U.S. hydrocarbon fuel infrastructure. A means of extending petroleum reserves by reducing crude utilization requirements

4.4.1.1 Types of Coal–Oil Systems

The co-processing schemes under consideration were generally an extension of two-stage coal liquefaction and application of residuum hydrocracking technology. It has been recognized that a possible synergism exists between coal-derived liquids and petroleum-derived residua. Co-processing improves the quality of synthetic liquid fuel products from coal by diluting them directly with petroleum-derived liquids. *Coal* is more aromatic with a lower H/C ratio and contains more inorganic material than found in heavy resids. Coal liquids contain a much higher proportion of aromatics compared to conventional petroleum-derived liquids, and the nonaromatic portion tends to be naphthenic rather than paraffinic. Coal liquids contain significant amounts of highly polar compounds and asphaltenes but a relatively low amount of sulfur-containing compounds. Further, petroleum-derived naphtha is low in nitrogen and oxygen. Coal-derived naphtha, on the other hand, has higher nitrogen and oxygen contents, is easier to reform, and has a higher octane number. Heavy oil contains significantly more sulfur and contains relatively small *amounts* of vanadium, nickel, and iron, some of which are trapped in porphyrin structures. Thus, combining coal-derived liquid with petroleum-derived liquid can provide some positive impacts on the overall product quality.

Broadly speaking, the co-processing processes can be divided into the following four categories [106–185]:

1. Hydrocatalytic processes
2. Extractive processes
3. Thermal processes (noncatalytic)
4. Hydrothermal processes

The first category includes H-Coal processes by Hydrocarbon Research Inc. (HRI), Lummus, Canada Centre for Mineral and Energy Technology (CANMET), Universal Oil Products (UOP), Chevron, and Kerr–McKee processes [185]. The second category includes processing variations incorporated for solid removal and deasphalting by Kerr–McGee, UOP, and Lummus [185]. The Cherry-P process falls into the thermal process category [185]. The process conditions are somewhat between those of visbreaking and delayed coking. The Pyrosol process falls into the aforementioned last category [185] and utilizes a mild hydrogenation of coal and heavy oil in the first stage. The second stage processes residuum under hydrogen pressure to produce more oil.

Often, co-processing of coal and heavy crude is carried out in two stages. In the first stage, both coal and heavy oil are liquefied, and in the second stage, they are both upgraded by hydrogenation. There are several pilot-scale operations that already exist demonstrating the success of this concept [106–185]. Co-processing is often carried out in the presence of a catalyst and/or another hydrogen donor solvent. Monnier [144] has presented an extensive review of catalytic co-processing. Curtis et al. [163] examined the effect of hydrogen donor addition in conjunction with catalytic hydrotreatment on the products obtained from upgrading residuum and co-processing coal with petroleum residuum. The use of hydrogen donor solvent gave improved products.

CANMET co-processing involves the simultaneous upgrading of coal and heavy oil or bitumen in a once-through mode of operation using a disposable iron catalyst. The CANMET additive (pulverized coal impregnated with iron sulfate, $FeSO_4$) has been identified as both hydrogenation and coke-reducing catalysts. Process feasibility has been investigated using a variety of coals and heavy oils/bitumen [151,170]. These studies and their references also demonstrated that in terms of product yields for subbituminous coals, CANMET processing is superior to liquefaction and comparable to hydrocracking. The effect of H_2S in hydrocracking of model compounds and in liquefaction is well documented [1]. The ability of H_2S to reduce coke formation and increase liquid yield during coal liquefaction has been patented by Exxon R&D Co. [1,151,170]. It has also been shown that H_2S has beneficial effects in noncatalytic crude oil hydrotreating processes [1]. Rahimi et al. [151,170] examined co-processing using H_2S as a promoter. The study showed hydrogen sulfide to be an effective promoter for achieving high coal conversions and distillate yields when co-processing subbituminous coal with bitumen vacuum bottom. Results also showed that at least at low and moderate severities of operation, H_2S performs as good as or better than $FeSO_4$ in terms of product yields as well as qualities. However, at higher severities, $FeSO_4$ is better than H_2S.

There are numerous other studies on various aspects of co-processing reported in the literature [106–185]. Ettinger et al. [153–156,160] examined various reaction pathways in co-processing. This study focused on the co-processing technology that was developed by Gatsis et al. [116,119,152] in which optimal reaction conditions (3000 psi, total pressure, 420°C, 2:1 resid to coal, and 1% molybdenum based UOP catalyst) enabled the conversion of about 90% of the coal to toluene soluble products and about 80% of the asphaltene to hexane soluble products.

Ignasiak et al. [150] examined two-stage co-processing of subbituminous coals and bitumen or heavy oil. They investigated the effects of pretreatment of coals with appropriately formulated carbon monoxide and water in the presence of bitumen or heavy oil and found fast reactions with high degree of coal solubilization and deoxygenation. The reaction was catalyzed by a mixture of alkali metal carbonates and proceeded readily at 380°C–400°C. The first stage reaction product appeared to be susceptible to further catalytic hydrogenation at 420°C–460°C with gaseous hydrogen yielding 65%–70% on dry-ash-free (daf) feed basis of hydrogen-rich distillate oil composed mainly of naphtha and middle oil. The study also presented a possible flow diagram for the two-stage process. The two-stage coal liquefaction/ resid hydrocracking integrated co-processing following Lummus technology was studied by Greene et al. [169]. They showed that at constant slurry concentration, reducing the ratio of solvent (524°C-boiling range) to coal from 1.5:1 to 1:1 had a minimal effect on observed coal conversions. However, this reduction in solvent/ coal ratio did adversely affect syncrude characteristics and net distillate yields.

Numerous other studies [106–185] have evaluated various aspects of two-stage co-processing (co-liquefaction) operations.

Besides these two-stage co-processing efforts, a single-stage slurry-catalyzed co-processing was examined by Gatsis et al. [152]. The study successfully demonstrated a single-stage slurry-catalyzed co-processing concept in laboratory batch experiments. The active UOP catalyst gave high conversion and high concentration of liquid product at relatively low temperature, and as a result, thermal degradation reactions and cracking of resid and coal-derived liquids to light gases were minimized. The liquid hydrocarbon product was of high quality and can be efficiently utilized as a feedstock in existing refineries. Similar data were obtained in a continuous bench-scale operation. The data however showed that the co-processing process is sensitive to high severity (temperature, residence time). High coal conversion and more conversion to high-quality liquid product can be achieved at relatively mild conditions where thermal degradation reactions are minimized.

Hajdu et al. [184] examined the pretreatment of the host oil in ways that would improve its performance and make it a better host oil. This can be carried out by (1) converting aromatic structures to hydroaromatics capable of donating hydrogen to coal, (2) cracking the heavy oil to lower-MW material that would be a better solvent, and (3) removing metals, sulfur, and nitrogen. Other investigators have found that co-processing performance *can* be improved by pretreating the host oil. Takeshita and Mochida [161] showed positive results by using a petroleum pitch that had been hydrotreated at high pressure over a $Ni/Mo/A1_2O_3$ catalyst. A similar approach was taken by Sato et al. [162] who obtained high *coal* conversion and high distillable oil yields after a *tar* sand bitumen that they were using was hydrotreated with a "Ni–Mo" catalyst. Curtis et al. [145,163,164] showed the importance of hydrogen donor compounds in host oil for achieving high *coal* conversion. Hydroaromatic compounds (hydrogen on a naphthenic *carbon* at the alpha or beta position from the aromatic ring) are one class of hydrogen donor compounds that are important in *coal* liquefaction.

Hajdu et al. [184] found that mild pretreatment of a Citgo resid (1000°F) using either Mo naphthenate or $Mo/Fe_2O_3/SO_4$, as well as pretreatment using the homogeneous catalyst $Co_2(CO)_8$ under synthesis gas can increase the available (donatable) hydrogen content of the resid. When these pretreated oils were thermally co-processed with an Illinois No. 6 coal, about 90 wt% of the coal (daf) was converted to soluble products. This high coal conversion was realized even at high coal loading (50 wt%). Pretreatment at 440°C that cracks the resid without adding much available hydrogen showed little promise for improving coal conversion above levels achieved with untreated resid. The products from co-processing coal and oil were equally split between high boiling material, mostly asphaltenes, and distillate. Distillate yields appeared to be affected by the concentration of coal in the feed, with maximum yields at coal loadings below 50 wt%.

Miller [111,113] and Miller and Baldwin [115] examined co-processing of coal and shale oil. Miller [111,113] examined co-processing of Wyodak subbituminous coal and shale oil derived from medium-grade Colorado shale at typical coal liquefaction conditions. Results indicated that pre-hydrotreatment of the shale oil, feed coal reactivity, and to some extent the use of a disposable catalyst all affected process performance. Distillate yields in excess of 60 wt% daf coal with corresponding hydrogen consumption values of less than 2.8 wt% daf coal were obtained in a once pass through process configuration. Encouraging results were also obtained at low severity conditions using CO/H_2O rather than H_2 as reducing agent. Miller and Baldwin [115] examined low severity liquefaction co-processing using Wyodak subbituminous coal and two shale oil samples from medium-grade Colorado oil shale. Results indicated that pre-hydrotreatment of the shale oil, lower reaction temperature, and higher initial CO pressure all contributed to enhance process performance. Distillate yields in excess of 85 wt% daf coal were obtained at 600°F and 1500 psig CO pressure and 60 min reaction time. Results from blank shale oil experiments suggested that overall distillate yield could be maximized by co-processing coal and shale oil rather than processing two feeds separately.

Barraza et al. [174] examined thermal and catalytic co-processing of subbituminous coal from La Yolanda colliery (southwest of Colombia) with lube oil waste in the presence of tetralin, hydrogen, as well as two catalysts $RuCl_2$ and $Ni–Mo/Al_2O_3$. Experiments were carried out at 380°C, 400°C, and 420°C with mass ratio of coal/lube oil/tetralin 1:5:3 and residence time of 30 min. Results showed that total conversion to liquid product was not a function of temperature in the thermal co-processing. However, total conversion to liquid product was improved when $RuCl_2$ was used as a catalyst. Total liquid distribution and maltenes concentration were dependent on thermal and catalytic co-processing. In the absence of catalyst and tetralin, co-processing of coal/lube oil produced high conversions and product yield.

4.4.2 CO-PROCESSING COAL AND WASTE POLYMERS

4.4.2.1 Coal and Waste Plastics

While the concept of co-processing started with the evidence that co-processing of coal and heavy oil shows an SE between coal conversion and heavy oil upgrading reactions, more recently synergy between coal liquefaction and waste conversion has also been examined [186–216]. About 97 million tons of waste plastics, paper, oils, and tires are generated in the United States every year. The most abundant waste is paper (with 73 millions) followed by plastic accounting for 16 million tons. United States generated about 300 million waste tires annually during early part of this century, which amounted to about 1.6 million tons of waste. The average annual production of waste oil was about 4.8 million tons in early part of this century. One solution to remove waste plastics or rubber was to co-process with coal or heavy oils. This co-processing plant can be placed near an existing refinery. Technical and economical feasibilities of this concept were initially examined by Gray and Tomlinson [130,205]. A preliminary design was developed to co-liquefy coal and plastic with 50:50 mixture. The plant products included hydrocarbon gases, naphtha, jet fuel, and diesel fuel. The results from the economic analysis identified profitability criteria for gross profit and return on investment based on variable conversion, yield, and tipping fee for plastic waste processed.

Liu et al. [136,180] screened catalyst for co-processing coal and waste plastic using thermogravimetry/gas chromatography/mass spectrometry analysis. At the high catalyst level used, the results revealed catalytic waste plastic cracking activity in the order

SiO_2/Al_2O_3 > HZSM-5 > $NiMo/Al_2O_3$ mixed with SiO_2/Al_2O_3 > solid superacids. On the solid superacids studied, the $ZrO_2/SiO_4{}^{2-}$ catalyst possesses the highest catalyst cracking activity, and approximate order of cracking activity is $ZrO_2/SO_4{}^{2-}$ > $Al_2O_3/SO_4{}^{2-}$ > $Pt/Al_2O_3/SO_4{}^{2-}$ > $Fe_2O_3/SO_4{}^{2-}$ > no catalysts. The HZSM-5 zeolite catalyst showed most promising results for co-processing coal with commingled waste plastic by increasing greatly the rate of decomposition reactions at the cost of low-MW products (high gas yields) and higher aromatic yields. Hydrocracking catalysts such as $NiMo/Al_2O_3$ mixed with SiO_2/Al_2O_3 showed potential promise for co-processing coal with commingled waste plastic due to their hydrogenation and cracking ability. Xiao et al. [208] showed that $Fe_2O_3/SO_4{}^{2-}$ and $ZrO_2/SO_4{}^{2-}$ showed good catalytic activity for depolymerization–liquefaction of plastics and rubbers like polyethylene (PE), polypropylene, and polybutadiene.

Comolli et al. [204] reported an initial study on co-processing of coal, oils, plastic, municipal solid waste (MSW), and biomass supported by Hydrocarbon Technology Inc. (HTI) and Department of Energy. The project was to examine the applicability of the concept of co-processing of waste (biomass, plastic, and MSW) with coal and oil. This combination was examined to see the effect of co-processing on lowering of CO_2 emissions. The initial set of data reported in this paper was very promising. Orr et al. [206] examined the thermal and catalytic co-processing of coal and waste tires and coal and waste plastics like PE, polystyrene (PS), and polypropylene. The study showed that co-processing at 430°C gave the highest total conversion yields for coal/rubber tire co-processing for 1 h under hydrogen atmosphere with a molybdenum catalyst. The zinc and carbon black contamination in the derived liquids was diminished when samples were co-processed at 430°C. Co-processing of coal with co-mingled plastic showed an SE on oil production but not on the total conversion. Pradhan et al. [167,175,190] showed that co-processing of waste plastics with either oil-only feedstock or coal/oil feedstock resulted in a significant improvement in process performance. Total feed conversion was enhanced as were 524°C + residuum conversion and C_4-524°C distillate yield. The addition of waste plastics to the feed increased hydrogen efficiency as both hydrogen consumption and C_1–C_3 light gas yield decreased. Co-processing of plastics with oil reduced the equivalent crude oil price required to have a 15% rate of return. The study concluded that commercialization of this concept is possible.

Dhaveji et al. [201] investigated co-processing of polyvinyl chloride (PVC) with coal. As mentioned earlier, while PVC decomposition produces HCl, it also produces chlorinated organics in the liquid phase at the liquefaction temperature 350°C–450°C. The literature [186–206] has also shown that vacuum pyrolysis of PVC at 500°C results in HCl gas yield of 53% with minor production of benzene. Dhaveji et al. [201] examined the effects of temperature, time, hydrogen pressure, and the catalyst on PVC liquefaction, coal liquefaction, and co-processing of coal and PVC. The catalyst used was a mixed metal ferric sulfide–based (Mg–Fe–S) catalyst. The results of the study indicated that liquefaction of PVC alone mostly produces HCl. The products of PVC and coal-alone liquefaction are mostly influenced by temperature with modest effects by time and hydrogen partial pressure. Addition of PVC to coal found to have SE for some of the products. A decrease in HCl yield during co-processing indicated interactions between coal and PVC during co-processing. The mixed metal catalyst did not significantly affect product yield during co-processing. The chlorine was mostly captured in THF soluble portion of the product during co-processing. More chlorine also went to hexane soluble portion of the product during co-processing than during PVC liquefaction alone. In the presence of the catalyst, more chlorine was observed in the THF soluble portion of the product.

Taghiei et al. [192] examined liquefaction of PE, polyethylene terephthalate (PET), and mixtures of PE and PET with different coals and actual plastic wastes from such items as milk jug and soft drink bottles. Both individual compounds and mixtures were liquefied for the temperature range of 400°C–450°C and residence time of 15–60 min and hydrogen pressure of 800 psi. The experiments used iron and HZSM-5 catalysts. Saturated oil, consisting mainly straight-chain alkanes and minor alkenes and some light hydrocarbon gases, was produced from both the plastics and plastic–coal mixtures. The conversions of PE and PET (from waste milk jugs and soda bottles) with HZSM-5 catalysts gave 100% conversion with 86%–92% oil yields. Using coal/plastic mixtures with iron catalysts gave lower conversion (53%–91%) and lower oil yields (26%–83%). The results indicated that while co-processing showed SE, the product distribution was strongly dependent on the nature of the catalyst.

Taghiei et al. [135,173,192] also conducted co-processing (co-liquefaction) of PE and mixed plastic waste with a bituminous coal, a subbituminous coal, and lignite in a tubing bomb reactor. Co-liquefaction of mixed waste plastics with a bituminous coal exhibited higher oil yields than obtained by liquefaction of either the waste plastics or the coal alone implying SE. Co-liquefaction of a 50:50 mixture of mixed waste plastic and an iron ion-exchanged subbituminous coal with the addition of 1.0 wt% of HZSM-5 zeolite catalyst gave a total conversion of over 90% and oil yield of approximately 70%. The oil yield of the mixture appeared to have been increased approximately 10% by SE. Co-liquefaction of iron ion-exchanged Beulah lignite with medium-density PE with no added solvent gave very good oil yields and total conversions at 450°C, indicating that the plastic plays the role of a hydrogen donor solvent for the coal under those conditions. It is interesting that the total conversion of lignite and plastic is possible at 450°C without a solvent.

Gimouthopoulos et al. [189] examined co-processing of lignite and post consumer polymers such as PS, polyisoprene (PISO), and PE. A series of silica–alumina acidic catalysts, which were prepared by sol–gel chemistry with different Si/Al ratio of Bronsted to Lewis acid sites, were tested. Thermal and catalytic co-processing was carried out with 1:1 mixtures of lignite and waste plastics. While the catalysts enhanced oil formation, lignite and polymeric materials demonstrated a variety of catalytic characteristics and resulting effects. While both PS and PISO were easily liquefied at 420°C, PE did not behave the same way in this condition. The co-processing behavior depended on the nature of the catalyst. Robbins et al. [140] studied characteristics of process oils from HTI coal/plastics (which included PS and high-density polyethylene [HDPE]) co-processing experiments. Unfortunately, due to lack of steady state behavior, the results of the study were inconclusive.

Kanno et al. [215] also observed synergy in the conversion of Yallourn coal (lignite) and PE under pressurized hydrogen using 1-methylnaphthalene and tetralin as solvent. The conversion and the oil yield were larger than the additive values of respective fuels processed separately, and the observed values of residues for both Yallourn coal and PE during the co-liquefaction were evidently smaller than the estimated values.

Anderson et al. [181] co-liquefied tires and a low-rank coal in a magnetically stirred autoclave under hydrogen pressures of 10, 5, and 1 MPa and nitrogen pressure of 0.1 MPa. They noted that the synergy was observed only at low hydrogen pressures. The oils obtained in the co-liquefaction showed a more aromatic nature than those obtained when each material was processed separately; a higher boiling point was also observed, suggesting that radicals from rubber and coal reacted between each other instead of reacting with hydrogen radicals. Although the existence of synergy has been agreed upon and the mechanism has been somehow well understood during co-liquefaction, it has not been the case for other thermal conversion processes.

Feng et al. [216] also noted increased conversion and oil yield during catalytic co-processing of coal and waste plastics. They used a mixture of subbituminous coal (Black Thunder), PE, and polypropylene using a number of catalysts: HZSM-5 zeolite catalyst, ferrihydrite treated with citric acid, co-precipitated Al_2O_3–SiO, and a ternary ferrihydrite—Al_2O_3–SiO containing 10% Al_2O_3–SiO.

4.4.2.2 Coal and Waste Rubber Tires

Besides the study of Orr et al. [206] mentioned in Section 4.4.2.1, several other studies have examined co-processing of coal and waste rubber. Farcasiu and Smith [141–143] showed that co-processing of coal and used rubber tire can provide an alternate use for waste rubber materials. The liquids formed in this co-processing were hydrogen rich and were a potential source for the aromatic oil component in new tires. These liquids can also be a source of transportation fuels. Farcasiu et al. [132,141–143,182,188] presented a review of the co-processing of coal and used rubber tire work carried out under the support of the Department of Energy.

Dadyburjor et al. [183,186,187,194,196,203] examined the effectiveness of mixed pyrrhotite/pyrite catalysts for co-processing of coal and waste rubber tires. The study concluded that the optimum processing conditions like temperature, pressure, residence time, and catalyst loading for co-processing depended upon the nature of coal and the tire used. The ferric sulfide–based in situ impregnated catalyst improved the coal conversion, but the effect was attenuated by the presence of the tire. The residue from the tire if suitably pretreated may increase the liquefaction conversion. The results for the effectiveness of co-processing were inconclusive. Sharma et al. [186] examined the effect of process conditions on co-processing kinetics of waste tire and coal. Thermal and catalytic liquefaction of waste (recycled) tire and coal were studied separately and together using different tire/coal ratio. Experiments were conducted in tube bomb reactor at 350°C–425°C. The effect of hydrogen pressure on the product slate was also studied. In the catalytic runs, a ferric sulfide catalyst impregnated in situ in the coal was used. Both the tire components and the entire tire exhibit the SE on coal conversion during co-processing. The extent of synergism depended on the temperature, hydrogen partial pressure, and tire/coal ratio. Experiments with coal and tire components indicated that the synergism was due to interaction between rubbery portion of the tire and not the carbon black. The synergism led to the increase in asphaltenes, which was nearly twice the one for coal-only run at 400°C. While the conversion of coal increased dramatically in the presence of catalyst, the catalytic effect was attenuated in the presence of tire, particularly at high tire/coal ratios. The data were analyzed using a second-order reaction mechanism between coal and asphaltenes and asphaltenes to oil and gas and an additional synergism reaction (first order) when both coal and tire are present. Parallel schemes were assumed for thermal and catalyzed reactions. The uncatalyzed liquefaction of coal had a lower activation energy compared to those of the synergism reaction and the catalytic coal liquefaction. The conversion of asphaltenes to oil and gas was relatively independent of temperature and the catalyst. The catalyst appeared to play a significant role in the conversion of coal to asphaltenes but a negligible role in the synergism reaction.

Mastral et al. [193,195] also examined co-processing of a low-rank coal (Samca coal, from Utrillas, Spain) and waste tires. They showed that the addition of tire to coal during co-processing increases total conversion by influencing the mechanism of stabilization of the generated coal radicals. This conclusion was drawn due to high yield of asphaltenes and low yield of C_1–C_3 hydrocarbons as compared to theoretical estimations. Mastral et al. [193,195] also revealed that the slight synergy found can be due to the small free radicals from vulcanized rubber decomposition, which were able to stabilize coal radicals to light products.

4.4.3　Other Co-Processing Systems

There are also some few other interesting studies carried out on co-processing. Baladincz et al. [202] studied co-processing of waste fat and gas oil mixtures on commercial hydrogenation catalysts to produce diesel fuel. Heterogeneous catalytic hydrogenation of mixtures (10%–30%–50%) fat content of rendered fat (from animal carcasses) and straight run gas oil on commercial $NiMo/Al_2O_3$ catalyst was pursued. The results showed that with increasing temperature, the yield of main product fraction decreased in the case of 10% and 30% rendered fat containing feedstock mainly because of the cracking reactions, while in the case of 50% rendered lard containing feedstock, the yield did not change significantly. The higher amount of normal and isoparaffins formed compensated the yield lost originating from the cracking reactions. The conversion of the triglyceride part of the feedstock was complete. Both sulfur and polyaromatic content of the products were decreased, although sulfur content did not meet the required standard. Both viscosity and density of the products met the required standards. The study concluded that commercial catalyst (whose activity did not significantly change during experiments) was suitable for the production of biogas oil containing gas oils with high yield (90.6%–97.3%) by co-processing and hydrocracking and hydrotreating of the gas oil and fat mixtures.

Xue et al. [106] examined co-processing of fluid catalytic cracking (FCC) slurry and coal to produce modifier from co-processing slurry and coal (MCSC) to be used as bitumen modifier. This heavy product can be used as bitumen modifier in asphalt concrete mix. The modification capability of MCSC was similar to that of commercial bitumen modifier, Trinidad lake asphalt (TLA), and met American Society for Testing and Materials (ASTM) standard specification of D5710-95. Compared to TLA, MCSC contained more aromatic, condensed, small-MW component.

Siddiqui and Ali [191] examined the co-processing of PVC and VGO at 430°C and 950 psi H_2 pressure for the duration of 2 h. PVC behaves differently from other plastic materials like PE, polypropylene, and PS. At temperatures in the range of 150°C–400°C, PVC decomposes into HCl and coke-/char-like residue. The residue can further decompose at higher temperatures like any other plastic materials. The effects of different catalysts on the product distribution for co-processing of PVC and VGO system were analyzed. The presence of VGO increased the overall conversion during co-processing compared to the single component reaction. Among various catalysts used, FCC and H-coal-1, hydrocracking catalyst (HC-1) catalysts were most effective in producing hexane soluble fraction. The rate of conversion in the co-processing system depended upon the chemistry and composition of PVC, VGO, and catalyst materials. The study showed that catalytic co-processing of VGO and PVC is a feasible method for producing transportation fuels.

Krasulina et al. [212] examined thermochemical co-processing of Estonian kukersite oil shale with peat and pine bark. Three mixtures of oil shale and bark, oil shale and peat, and oil shale, bark, and peat were studied in the temperature range of 340°C–420°C. Water and benzene were used as solvents. The influence of temperature, solvent and its type, and oil shale to peat and oil shale to biomass ratios on the gaseous, liquid, and solid yields was investigated. In co-processing experiments, several SE in product yields were observed. The most important SE was noticed for co-processing of oil shale with peat (10:4 by mass of the organic matter) at 360°C in the medium of water. For this case, the yield of the liquid product was 25% higher than the sum of the corresponding yields obtained in liquefaction of oil shale and peat separately under the same experimental conditions. The group composition of oil showed that various polar and high-polar oxygen compounds prevailed over hydrocarbon fractions. Data on the elemental and group composition demonstrated that partial substitution of biomass or peat for oil shale leads to the production of chemically modified shale oil.

Abdullah [211] studied co-processing of low-rank Mukah Balingian (MB) Malaysian coal with palm kernel shell (PKS) to investigate the synergism by comparing the conversion and product yields. Co-processing results indicated that there existed an obvious SE between MB and PKS, which depended on the liquefaction temperature. The results showed that the conversion and oil + gas yields during co-processing were considerably higher than those for the liquefaction of MB coal alone. The largest SE occurred at low temperature of 400°C and then decreased as temperature increased. The liquefaction conversion of 88.6% and oil + gas of 77.6% obtained during co-processing at the optimal temperature is higher than that of the corresponding value from the individual liquefaction of MB coal.

Ikenaga et al. [210] examined co-processing of microalgae with coal using coal liquefaction catalysts. Microalgae examined were *Chlorella*, *Spirulina*, and *Littorale*, and coals examined were Australian Yallourn brown coal and Illinois No. 6 coal. The experiments were carried out under pressurized hydrogen, in 1-methylnaphthalene at 350°C–400°C for 60 min with various catalysts. Co-processing of *Chlorella* with Yallourn coal was successfully achieved with excess sulfur to iron (S/Fe = 4), where sufficient amount of $Fe_{1-x}S$, which is believed to be active species, in the coal liquefaction was produced. The conversion and the yield of the hexane soluble fractions were close to the values calculated from the additivity of the product yields of the respective individual reactions. In the reaction with 1:1 mixture of *Chlorella* and Yallourn coal, 99.8% conversion and 65.5% of hexane soluble fraction were obtained at 400°C with $Fe(CO)_5$ at S/Fe = 4. When *Littorale* and *Spirulina* were used as microalgae, a similar tendency was observed with the iron catalyst. On the other hand, in co-processing with Illinois No. 6 coal, which is known to contain a large amount of sulfur in the form of catalytically active pyrite, the oil yield during co-processing was close to the additivity of respective reaction with $Fe(CO)_5$–S even at S/Fe = 2. $Ru_3(CO)_{12}$ was also effective for the co-processing of microalgae with coal.

Altieri and Coughlin [209] characterized the products formed during co-processing of lignin and bituminous coal at 400°C. The filterable solids from co-processing were about 30% hexane soluble compared to almost 10% soluble when the same amount of coal and lignin were reacted individually. In the case of co-processing, far more of the benzene soluble material was also pentane soluble oil. As a result of co-reaction, significant amount of nitrogen from the coal appeared in the liquid product in contrast to no observable nitrogen in the liquefaction of coal alone under comparable conditions. Gaseous products were significant and contained CO_2 as a major component. More carbon-14 from lignin was incorporated into the liquid product during co-processing of coal and lignin than in the liquefaction of lignin alone under comparable reaction conditions.

As shown by Tchapda and Pisupati [207], there appears to be clear evidence of the interaction during co-processing of some coal and biomass fuels. They indicated that during co-liquefaction, hydrogen transfer can occur between fragmented coal and biomass molecules. They illustrated mechanism of liquefaction process where (1) either H-donor solvent stabilizes thermally generated free radicals or (2) solvent mediated hydrogenolysis where H-donor engenders bond scission. The hydrogen transfer between coal and biomass can be very complex and depend on the nature of the coal and biomass. Coughlin and Davoudzadeh [214] co-liquefied lignin and Illinois No. 6 bituminous coal using a series of catalysts with tetralin and phenol as solvents. They concluded that when coal and lignin react together in a solvent, they depolymerize under mild conditions to produce a filterable liquid product with yields greater than would be predicted based on separate liquefaction of these species. They also noted that while the overall conversion of coal plus lignin increased as the fraction of lignin in the reaction mixture increased, the opposite was true for increased proportion of coal in the reaction mixture. However, they noted that the overall conversion was not improved further by increasing lignin ratio beyond 0.7.

4.5   DECOMPOSITION, EXTRACTION, AND CO-PROCESSING IN SUPERCRITICAL WATER

SWC carries unusual properties that allow it to decompose, extract, and easily liquefy coal, biomass (including algae), and polymeric materials [268]. SWC has very high solubility and diffusion coefficient for most materials. Various subcomponents of coal, biomass, and polymers can be extracted and stabilized to form liquid substances. This application generally produces useful liquids, which can be either a fuel or raw materials for various downstream chemicals. Feedstock normally used for the liquid productions are coal, polymeric materials, rubber tires, cellulose, etc., or mixtures of them [217–267].

4.5.1 EXTRACTION AND DECOMPOSITION OF SINGLE MATERIAL IN SUPERCRITICAL WATER

The extraction of coals with SWC is a promising route for the production of liquid fuels and chemical feedstock from coal. Deshpande et al. [217] obtained high conversion for extraction of a German brown coal and a Bruceton bituminous coal by SWC at 375°C and 23 MPa. They reported conversions of 70%–79% for the brown coal and about 58% for the bituminous coal. Scarrah [218], on the other hand, reported 35% conversion and only 10% liquid yield for North Dakota lignite at 400°C and 28 MPa pressure. Deshpande et al. [217] also obtained low liquid yield with lignite coal with high sodium content. Other studies [219–223,227–229] reported low conversion for bituminous coal particularly when solvent density was low. Kershaw and Bagnell [222] showed that at 380°C and 22 MPa, the conversions of Australian brown coals were considerably higher for supercritical extraction of water than with toluene. The reverse was, however, true for black coals. In general, they found SWC extraction was more effective for low-rank coals than high-rank bituminous coals. The extraction by water was also more dependent on pressure presumably due to solvent density effect. The hydroxyl concentration of liquid yield by SWC extraction was higher than the one obtained in the liquid produced by toluene extraction.

Swanson et al. [223] showed that for low-rank coals, the conversion and extract yields increased with increasing temperature and pressure. The conversion also decreased with increasing coal rank and correlated well with the percent volatile matter in the coals. The study also indicated that SWC extracts the volatile hydrogen-rich fraction of the coal. The extract was found to be highly polar in nature with significant quantities of phenols and long-chain aliphatic fatty acids.

Numerous other studies [219–268] have also addressed the behavior of coal, shale oil, and biomass under SWC conditions. Four typical studies illustrating the coal decomposition in SWC are reported by Nonaka et al. [235,243], Li and Eglebor [252], Vostrikov et al. [229], and Cheng et al. [228].

SWC has also been explored as a medium for the degradation of waste synthetic polymers [253–260]. Rubber tires were converted to a 44% oil yield by reaction in SWC at 400°C. When PS-based ion exchange resins were subjected to SCW at 380°C for 1 h [253–260], less than 5% of the polymer decomposed and the products included styrene and several oxygenated arenes such as acetophenone and benzaldehyde. SCW has also been used to extract oil and oil precursors from oil shale [244,246–251]. The process involved C–C bond cleavages, and in the presence of CO, higher hydrocarbons yields were obtained than the ones obtained in conventional pyrolytic treatment. Holliday et al. [261] showed that water near its critical point is a good medium for the hydrolysis of triglyceride-based vegetable oils into their fatty acid constituents.

Mitsubishi Materials Corp. [266] with the project support of Petroleum Energy Center, Japan, developed a thermal process that used SCW to crack vacuum distillation residue oil (VR) into clean lighter oil products. The final volume of solid waste generated was below 5%. The process was carried out in two stages in the same reactor. At the bottom of the reactor, heavy VR components (pitch) are decomposed into lighter components using 5% SCW at temperatures 400°C–450°C and pressures

200–250 atm. In the upper part of the reactor, lighter components are cracked at a slightly higher temperature with SCW and hydrogen to form lighter products. Untreated pitch was withdrawn at the bottom and sent to a reformer where it was partially oxidized by SCW at 1000°C to form hydrogen gas and soot. This hydrogen stream was passed onto the upper section of the cracking reactor. Overall, the process converted 70% of VR into lighter products, which included 15% gas, 7% liquefied petroleum gas, 11%, naphtha, 13% light oil, 24% VGO, 21% carbon dioxide, and 1% soot and 8% heavy oil. The process was proven in a test plant of size 1 bbl/day.

Glycerol is one of the important by-products of biodiesel generation by transesterification process. The conversion of glycerol in SCW was examined by May et al. [263]. They studied the conversion of glycerol in the temperature range of 510°C–550°C, 350 atm pressure in a bed of inert nonporous ZrO_2 particles as well as in a bed of 1% Ru/ZrO_2 catalyst for the residence time of 2–10 s. The feed solution contained 5 wt% glycerol. The experiments in the absence of a catalyst resulted in the formation of liquid products such as acetaldehyde, acetic acid, hydroxyacetone, allyl alcohol, propionaldehyde, acrolein, and acrylic acid, and gases such as H_2, CO and CO_2, and methane. The catalyst enhanced formation of acetic acid and inhibited formation of acrolein. In the catalytic experiments, the main products formed were hydrogen and carbon dioxide with little methane and ethylene. Complete glycerol conversion occurred at 510°C in 8.5 s and at 550°C in 5 s in the presence of the catalyst. This however did not result in complete gasification; some acetic acid and acetaldehyde were still present. At high residence times, methanol and acetaldehyde were formed. The hydrogen yield was only 50% of what is achievable by stoichiometry due to lack of high activity of the catalyst.

All the studies described earlier in this section indicate that near critical conditions and complex carbonaceous materials tend to decompose into a mixture of liquids and gases. The amount of each phase depends on the nature of feedstock, pressure, and reaction time. The use of a suitable catalyst increases both liquid and gas yields. An increase in temperature generally produces more gas. In the case of a mixture, the SE between the decompositions of two components depends on the nature of the components.

4.5.2 Co-Processing in Supercritical Water

There is a large body of literature that reports co-processing of coal and different types of biomass or waste or biomass and waste in SCW [232,236,249–268]. Here, we look at a few of them in detail. All studies indicate the possible SE during co-processing in SCW. One study reported here examines co-processing in supercritical ethanol. This study also supports the synergistic behavior between components of co-processing. One advantage of the studies reported here, compared to the ones reported in Sections 4.3 and 4.4, is that, co-liquefaction in SCW generally occurs at lower temperature, although pressure of the supercritical conditions can be higher than the conventional co-processing with a solvent.

A number of studies [235,241–245] examined the decomposition of mixed feedstock under SCW conditions. Veski et al. [244] examined the decomposition of a mixture of kukersite oil shale and pine wood in SCW and showed improved liquid and gas yields at 380°C temperature. The mixture indicated an SE and showed the product 1.5–2.0 times better than what would be predicted based on simple additive yields. The liquid product was richer in heterocompounds including polar ones compared to the one predicted from simple additive effects. Kim and Mitchell [234] examined decomposition of coal/biomass mixture. The results show that at 647.3 K and 220.9 atm pressure, small polar and nonpolar organic compounds released from the mixture were completely miscible with SCW. The hydrolysis of large organic molecules in SCW resulted in high concentrations of H_2, CO, CO_2, and low-MW hydrocarbons with very little tar, soot, and PAH formation. Sulfur, nitrogen, and many trace elements in coals were oxidized to form insoluble salts in SCW. There were no gaseous emissions and all products were dissolved in the SCW. The salts can be precipitated from fluid mixture and removed along with ash. Matsumura et al. [236] examined co-liquefaction of coal and cellulose in SCW at 673 K and 25 MPa. The coal used was Ishikari coal. Unlike the results of synergy reported by Veski et al. [244], in this study no synergy between coal and cellulose conversion was found. Simple additive method for each compound product distribution worked well for this system. There has been more discussion on SE in mixture decomposition recently by Lee and Shah [3].

Sunphorka et al. [242] examined co-liquefaction of coal and plastic mixtures containing high-density PE, low-density PE, PS, and polypropylene. The experiments were performed in the temperature range of 450°C–480°C, 40–70 wt% plastic mixtures, and water to feedstock ratio of 2:10. During co-liquefaction, all experimental variables had effects on liquid yield, but temperature did not have significant effect on the conversion. Long residue in the oil product decreased with increasing temperature, while it increased with increasing water/feedstock ratio. For the plastic mixture alone, only temperature had a significant effect on the oil yield. Maximum conversion and liquid yield of 99% and 66%, respectively, were obtained.

Onsari et al. [241] examined co-processing of coal and used tire in SCW. The experiments were performed for lignite coal and used tire in a 250 mL batch reactor under nitrogen atmosphere in the temperature range of 380°C–440°C, water/feedstock ratio 4:1–10:1 wt/wt, and percentage used tire content in the feedstock from 0 to 100 wt%. The results were obtained for conversion efficiency, liquid yield, and oil composition. The maximum conversion and oil yield were 67% and 50%, respectively, obtained at 400°C at 1 min with water to feedstock ratio 10:1 and 80% used tire content. The oil composition depended significantly on temperature. The co-processing of coal and used tire yielded a synergistically increased level of oil production. Moreover, the total conversion level obtained with co-processing alone was almost equal to those obtained in the presence of either Fe_2O_3 or NiMo catalysts, under the same conditions. The study concluded that SCW is a good medium for the dissolution of volatile matter from coal and used tire matrix.

Pei et al. [264] examined co-processing of microalgae and synthetic polymer mixture in sub- and supercritical ethanol. The study investigated *Spirulina* (S) microalgae with HDPE in a stainless steel 1000 mL reactor at different temperatures (T), different S/HDPE ratios (R1), (S + HDPE)/ethanol ratios (R2), and different solvent filling ratio (R3). Results showed that the addition of S to HDPE liquefaction could make the conversion conditions for HDPE milder. The yield of bio-oil obtained at 340°C with a 1/10R2 and 2/10R3 was increased by 44.81 wt% when R1 was raised from 0/10 to 4/6. Meanwhile, the SE between HDPE and S increased from 0 to 30.39 wt%. Further increase in R1 resulted in a decrease in SE. The yield of bio-oil increased with R2 first, and then declined at a higher R2. An opposite trend was observed for residue. The effect of R3 to the yields of the liquefaction products was similar to that of R2. The content of C and H in bio-oils reduced with increasing R1, while the content of O increased. The bio-oils from pure S liquefaction runs mainly consisted of oxygen containing compounds such as carboxylic acids, esters, and ketones. The major components of bio-oil, however, obtained from co-processing of S and HDPE mixtures were similar to those of pure HDPE-derived bio-oil in which aliphatic hydrocarbons dominated.

Cao et al. [266] examined co-processing of biomass (sawdust and rice straw) with plastic (HDPE) in subcritical water and SCW in a 500 mL stainless steel autoclave. The experiments were carried out individually as well as in mixtures. The effects of temperature, reaction time, ratio of water to biomass, and HDPE on degradation rates were investigated. The maximum yield was obtained at 380°C. At this temperature, the most important parameter for oil yield was biomass/plastic ratio in the feedstock. The yield of oil up to 60% was obtained using 1:4 weight ratio of biomass/PE mixture. Co-processing of biomass with plastic increased the yields of oils. Results showed that addition of biomass to plastic could decrease the degradation temperature of plastic to supply hydrogen, and the SE of biomass and plastic could enhance oil yield and ease of the requirement of reaction condition.

Wu et al. [267] examined co-processing of coal and PS in SCW. The experiments were carried out to investigate the effects of water to reactant ratio [10–29] (mass ratio); temperature (360°C–430°C) and addition of plastic (10%–40%) (mass ratio) on the coal conversion; and yields of asphaltenes, oils, and gas. The results showed that with the increasing water to reactant ratio, the total conversion increases first and then levels off, while the yields of oils and gas continues to increase. The yields of oil and gas increase dramatically with the temperature exceeding 420°C. The conversion reaches a maximum of 430°C at 31.2%, while the yields of oil and gas and asphaltenes were 12.6% and 18.6%, respectively. The yields of oils and gas for co-processing of coal and plastic are higher typically by 0.6%–2.7% than the weighted average oil and gas yields of coal and plastic individually. This implies an SE during co-processing.

Matsumura et al. [236] examined co-processing of coal and cellulose in SCW. A semibatch packed-bed reactor was employed for co-processing cellulose and Ishikari coal at 400°C and 25 MPa pressure. No interaction between coal and cellulose was observed for the production of residue and water insoluble products based on their yields and compositions. On the other hand, the yields of water soluble products increased during co-processing. Both hydrogen-to-carbon ratio and oxygen-to-carbon ratio of the water soluble product increased during co-processing. The mechanism of this interaction is proposed based on the addition reaction of compounds derived from cellulose with coal-derived compounds to increase recoverable yields of the water soluble product. Nonaka [235] also examined co-processing of coal and cellulose in SCW. The study showed that co-processing did not change the residue yield but led to production of compounds with lower MW and thus to a larger gas yield. Several other important studies for liquefaction and extraction in SCW are described in my previous book [2].

REFERENCES

1. Shah, Y. *Reaction Engineering in Direct Coal Liquefaction*, Addison Wesley Publishing Co., Reading, MA (1981).
2. Shah, Y. *Water for Energy and Fuel Production*, CRC Press, New York (2014).
3. Lee, S. and Shah, Y. *Biofuels and Bioenergy—Processes and Technologies*, CRC Press, New York (2012).
4. Wender, I. Catalytic synthesis of chemicals from coal. *ACS Div. Fuel Chem. Preprints*, 20(4), 16 (1975).
5. Wender, I. Catalytic synthesis of chemicals from coal. *Cat. Rev. Sci. Eng.* (Marcel Dekker, New York), 14(1), 9 (1976).
6. Halleux, A. and Tshamler, H. Extraction experiments on coal with various pyridine bases. *Fuel*, 38, 291 (1959).
7. Lowry, H. (Ed.). *The Chemistry of Coal Utilization*, Vols. I and II, Wiley, New York (1945).
8. Dryden, T., Chemical constituents and reactions of coal, in Lowery, H. (Ed.) *Chemistry of coal utilization*. Supplementary volume, Wiley, New York, pp. 223–295, Chapter 6 (1963).
9. Oele, A., Waterman, H., Goldkoop, M., and Van Krevelen, D. Extraction disintegration of bituminous coals. *Fuel*, 30, 169 (1951).
10. Van Krevelen, D. *Coal*, Elsevier, Amsterdam, the Netherlands, p. 393 (1971).
11. Curran, G., Struck, R., and Gorin, E. The mechanism of hydrogen transfer process of coal and coal extract. *ACS Div. Petrol. Chem. Preprints C*, 130 (1966).
12. Curran, G.P., Struck, R.T., and Gorin, E. Mechanism of hydrogen-transfer process to coal and coal extract. *Ind. Eng. Chem. Proc. Des. Dev.*, 6, 166–173 (1967).
13. Farcasiu, M. Short time reaction products of coal liquefaction and their relevance to structure of coal. *ACS Div. Fuel Chem. Preprints*, 24(1), 121 (1979).
14. Han, K. and Wen, C. Initial stage (short residence time) coal dissolution. *Fuel*, 58, 779 (1979).
15. Neavel, R. Liquefaction of coal in hydrogen donor and non donor vehicles. *Fuel*, 55, 237 (1976).
16. Whitehurst, D. and Mitchell, T. Short contact time coal liquefaction. *ACS Div. Fuel Chem. Preprints*, 21(5), 127 (1976).
17. Wiser, W., Anderson, L., Quader, S., and Hill, G. Kinetic relationship of coal hydrogenation, pyrolysis and dissolution. *J. Appl. Chem. Biotechnol.*, 21, 82 (1971).
18. Heredy, L. and Fugassi, P. Aspects of the chemistry of hydrogen donor solvent coal liquefaction. *ACS Adv. Chem. Series*, 55, 448 (1966).
19. Miura, K., Mae, K., Sakurada, K., and Hashimoto, K. Flash pyrolysis of coal following thermal pretreatment at low temperature. *Energy Fuels*, 6, 16–21 (1992).
20. Mae, K., Hoshika, N., Hashimoto, K., and Miura, K. A new coal flash pyrolysis method suppressing crosslinking through the swelling of coal by pyridine vapor. *Energy Fuels*, 8, 868–873 (1994).
21. Awan, I.A., Mahmood, T., and Nisar, J. Flash pyrolysis of Lakhara lignite utilizing effective radical transfer. *J. Chin. Chem. Soc.*, 51, 291–296 (2004).
22. Mae, K., Inoue, S., and Miura, K. Flash pyrolysis of coal modified through liquid phase oxidation and solvent swelling. *Energy Fuels*, 10, 364–370 (1996).
23. Miura, K., Mae, K., Sakurada, K., and Hashimoto, K. Flash pyrolysis of coal swollen by tetralin vapor. *Energy Fuels*, 7, 434–435 (1993).
24. Cronauer, D., Shah, Y., and Ruberto, R. Kinetics of thermal liquefaction of Belle Ayr coal sub-bituminous coal. *Ind. Eng. Chem. Proc. Des. Dev.*, 17, 281 (1978).
25. Bockrath, B. and Noceti, R. Coal derived asphaltenes-relationship between chemical structure and process history. *Fuel Process. Technol.*, 2, 143 (1979).

26. Whitehurst, D. The nature and origin of asphaltenes in processes coals. EPRI Final Report AF-1298, Project 410. University of Utah, Salt lake, Utah (December 1979).
27. Vernon, L.W. Free radical chemistry of coal liquefaction: Role of molecular hydrogen. *Fuel*, 59, 102–106 (1980).
28. Li, X., Hu, H., Jin, L., Hu, S., and Wu, B. Approach for promoting liquid yield in direct liquefaction of Shenhua coal. *Fuel Process. Technol.*, 89, 1090–1095 (2008).
29. McMillen, D.F., Malhotra, R., Chang, S., Ogier, W.C., Nigenda, S.E., and Fleming, R.H. Mechanisms of hydrogen transfer and bond scission of strongly bonded coal structures in donor-solvent systems. *Fuel*, 66, 1611–1620 (1987).
30. Wei, X., Ogata, E., Zong, Z., Zhou, S., Qin, Z., Liu, J., Shen, K., and Li, H. Advances in the study of hydrogen transfer to model compounds for coal liquefaction. *Fuel Process. Technol.*, 62, 103–107 (2000).
31. Serio, M., Kroo, E., and Solomon, P. Liquefaction of water pretreated coals. *ACS Preprints*, personal communication (2012).
32. Serio, M., Solomon, P., Kroo, E., and Charpenay, S. Enhanced coal liquefaction with steam pretreatment. Fundamental studies of water pretreatment of coal. A DOE Report under Contract No. DE-ACR22-89PC89878, September 5, 1989–March 5, 1992, Advanced Fuel Research Inc., East Hartford, Connecticut (1993). http://www.afrinc.com.
33. Serio, M.A., Solomon, P.R., Bassilakis, R., Woo, E., Malhotra, R., and McMillen, D. Fundamental studies of retrograde reactions in direct liquefaction. Quarterly Reports under DOE/PETC Contract No. DE-AC22-88PC88814. Advanced Fuel Research in East Hartford, Connecticut (1989).
34. Ross, D. and Hirschon, A. The effects of hydrothermal pretreatment on the liquefaction of coal. *ACS Div. Fuel Chem. Preprints*, 35(1), 37–45 (1990).
35. Bienkowski, P.R., Narayan, R., Greenkorn, R.A., and Chao, K.W. *Ind. Eng. Chem. Res.*, 26, 202 (1987).
36. Bienkowski, P.R., Narayan, R., Greenkorn, R.A., and Chao, K.C. Liquefaction of sub-bituminous coal with steam and ammonia. *Ind. Eng. Chem. Res.* 26, 206 (1987).
37. Graff, R.A. and Brandes, S.D. *Proceedings of the New Fuel Forms Workshop*, U.S. DOE Fossil Energy, p. 35 (1986).
38. Graff, R.A. and Brandes, S.D. Modification of coal by subcntical steam: Pyrolysis and extraction yields. *Energy Fuels*, 1, 84 (1987).
39. Brandes, S.D., Graff, R.A., Gorbaty, M.L., and Siskin, M. Steam pretreatment for coal-liquefaction. *Energy Fuels*, 3, 494 (1989).
40. Khan, M.R., Chen, W.-Y., and Suuberg, E.M. Influence of steam pretreatment on coal composition and devolatization. *Energy Fuels*, 3, 223 (1989).
41. Ross, D.S., Green, T.K., Mansani, R., and Hum, G.P. Water pretreatment of coal—Part 1. *Energy Fuels*, 1, 287 (1987).
42. Ross, D.S., Green, T.K., Mansani, R., and Hum, G.P. Water pretreatment of coal—Part 2. *Energy Fuels*, 1, 292 (1987).
43. Ross, D.S. and Hirschon, A. *ACS Div. Fuel Chem. Preprints*, 35(1), 37 (1990).
44. Ross, D.S., Hirschon, A., Tse, D.S., and Loo, B.H. *ACS Div. Fuel Chem. Preprints*, 35(2), 352 (1990).
45. Ross, D.S., McMillen, D., Ogier, W., Fleming, R., and Hum, G. Exploratory study of coal conversion chemistry. Quarterly Report No. 4, February 19–May 18, p. 26ff, DOE/PC/40785-4 (1982).
46. Mikita, M., Bockrath, B., Davis, H., Friedman, S., and Illig, E. Water and non donor vehicle assisted liquefaction of Illinois bituminous coal. *Energy Fuels*, 2, 534–538 (1988).
47. Blaustein, B., Bockrath, B., Davis, H., and Mikita, M. Water-assisted and nondonor vehicle assisted coal liquefaction. *ACS Preprints*, personal communication (2012).
48. Yoneyyama, Y., Okamura, M., Morinaga, K., and Tsubaki, N. Role of water in hydrogena-tion of coal without catalyst addition. *Energy Fuels*, 11, 1–8 (2001), doi:10.1021/ef010147r.

49. Ross, D.S. and Blessing, J.D. Hydro conversion of a bituminous coal with CO and H$_2$O. *Fuel*, 57, 379 (1978).
50. Ross, D.S., Blessing, J.E., Nguyen, Q.C., and Hum, G.P. Conversion of bituminous coal. *Fuel*, 63, 1206 (1984).
51. Ross, D.S., Nguyen, Q.C., and Hum, G.P. Conversion of bituminous coal in CO/H$_2$O. systems 3. Soluble metal catalysis. *Fuel*, 63, 1211 (1984).
52. Anderson, K. *Hydrothermal Dissolution of Coal and Other Organic Solids*, SIUC Technology Expo. (October 9, 2009).
53. Akhtar, A. and Amin, N. A review on process conditions for optimum bio oil yield in hydrothermal liquefaction of biomass. *Renew. Sustain. Energy Rev.*, 15, 1615–1624 (2011).
54. Midgett, J. Assessing hydrothermal liquefaction process using biomass feedstocks, MS thesis. Dept. of Biological and Agricultural Engineering, Louisiana State University, Baton Rouge, LA (May 2008).
55. Huber, G., Iborra, S., and Corma, A. Synthesis of transportation fuels from biomass: Chemistry, catalysts and engineering. *Chem. Rev.*, 1–51 (2006), also published on web (June 27, 2006).
56. Elliott, D. Process development for biomass liquefaction, personal communication (2012).
57. Demirbas, A. Progress and recent trends in biofuels. *Progr. Energy Combust. Sci.*, 33, 1–18 (2007).
58. Zhang, Y. Hydrothermal liquefaction to convert biomass into crude oil, Chapter 10. In *Biofuels from Agricultural Wastes and Byproducts*, Blaschek, H., Ezeji, T., and Scheffran, J. (Eds.). Blackwell Publishing, New York, pp. 201–232 (2010).
59. Behrendt, F., Neubauer, Y., Oevermann, M., Wilmes, B., and Zobel, N. Direct liquefaction of biomass. *Chem. Eng. Technol.*, 31(5), 667–677 (2008).
60. Chen, P., Min, M., Chen, Y. et al. Review of the biological and engineering aspects of algae to fuels approach. *Int. J. Agric. Biol. Eng.*, 2(4), 1–30 (December 2009).
61. Sugano, M., Takagi, H., Hirano, K., and Mashimo, K. Hydrothermal liquefaction of plantation biomass with two kinds of wastewater from paper industry. *J. Mater. Sci.*, 43, 2476–2486 (2008).
62. Appell, H.R., Wender, I., and Miller, R.D. Solubilization of low rank coal with carbon monoxide and water. *Chem. Ind.*, 1703 (1969).
63. Appell, H.R. Reactions of coal and coal model compounds with water. *Energy*, 1, 24 (1976).
64. Blaustein, B.C., Bockrath, B.C., Davis, H.M., Friedman, S., Illig, E.C., and Mikita, M.A. Water assisted and non-donor vehicle assisted coal liquefaction. *ACS Div. Fuel Chem. Preprints*, 30(2), 359 (1985).
65. Burkhard, K., Werner, H., and Friedhelm, B. Catalytic hydroliquefaction of biomass with red mud and cobalt monoxide molybdenum trioxide catalysts. *Fuel*, 69(4), 448–455 (1990).
66. Sudong, Y. and Zhongchao, T. Hydrothermal liquefaction of cellulose to bio-oil under acidic, neutral and alkaline conditions. *Appl. Energy*, 92, 234–239 (April 2012).
67. Theegala, C. and Midgett, J. Hydrothermal liquefaction of separated dairy manure for production of bio-oils with simultaneous waste treatment. *Bioresour. Technol.*, 107, 456–463 (March 2012).
68. Liu, H., Xie, X., Li, M., and Sun, R. Hydrothermal liquefaction of cypress: Effects of reaction conditions on 5-lump distribution and composition. *J. Anal. Appl. Pyrol.*, 94, 177–183 (March 2012).
69. Kang, S., Li, B., Chang, J., and Fan, J. Antioxidant abilities comparison of lignins with their hydrothermal liquefaction products. *Bioresources*, 6(1), 243–252 (2011).
70. Liu, Z. and Zhang, F. Removal of copper(II) and phenol from aqueous solution using porous carbons derived from hydrothermal chars. *Desalination*, 267(1), 101–106 (February 2011).
71. Yin, S., Dolan, R., Harris, M., and Tan, Z. Subcritical hydrothermal liquefaction of cattle manure to bio-oil: Effects of conversion parameters on bio-oil yield and characterization of bio oil. *Bioresour. Technol.*, 101(10), 3657–3664 (May 2010).

72. Xiu, S., Shahbazi, A., Shirley, V., and Cheng, D. Hydrothermal pyrolysis of swine manure to bio-oil: Effects of operating parameters on products yield and characterization of bio-oil. *J. Anal. Appl. Pyrol.*, 88(1), 73–79 (May 2010).

73. Liu, Z. and Zhang, F. Removal of lead from water using biochars prepared from hydrothermal liquefaction of biomass. *J. Hazard. Mater.*, 167(1–3), 933–939 (August 2009).

74. Kruse, A., Maniam, P., and Spieler, F. Influence of proteins on the hydrothermal gasification and liquefaction of biomass. 2. Model compounds. *Ind. Eng. Chem. Res.*, 46(1), 87–96 (January 2007).

75. Yanagida, T., Fujimoto, S., and Minowa, T. Application of the severity parameter for predicting viscosity during hydrothermal processing of dewatered sewage sludge for a commercial PFBC plant. *Bioresour. Technol.*, 101(6), 2043–2045 (March 2010).

76. Zhang, B., von Keitz, M., and Valentas, K. Thermochemical liquefaction of high diversity grassland perennials. *J. Anal. Appl. Pyrol.*, 84(1), 18–24 (January 2009).

77. Balan, V., Kumar, S., Bals, B., Chundawat, S., Jin, M., and Dale, B. Biochemical and thermochemical conversion of switchgrass to biofuels, Chapter 7. In *Switchgrass, Green Energy and Technology*, Monti, A. (Ed.). Springer-Verlag, New York, pp. 153–185 (2012).

78. Karagoz, S., Bhaskar, T., Muto, A., Sakata, Y., Oshiki, T., and Kishimoto, T. Low-temperature catalytic hydrothermal treatment of wood biomass: Analysis of liquid products. *Chem. Eng. J.*, 108(12), 127–137 (2005).

79. Kobayashi, N., Okada, N., Hirakawa, A., Sato, T., Kobayashi, J., Hatano, S., Itaya, Y., and Mori, S. Characteristics of solid residue obtained from hot compressed water treatment of woody biomass. *Ind. Eng. Chem. Res.*, 48, 373–379 (2009).

80. Qu, Y., Wei, X., and Zhong, C. Experimental study on the direct liquefaction of *Cunninghamia lanceolata* in water. *Energy*, 28, 597–606 (2003).

81. Karagoz, S., Bhaskar, T., Muto, A., Sakata, Y., and Azhar Uddin, Md. Low temperature hydrothermal treatment of biomass: Effects of reaction parameters on products and boiling point distributions. *Energy Fuels*, 18, 234–241 (2004).

82. Minowa, T. and Ogi, T. Hydrogen production from cellulose using a reduced nickel catalyst. *Catal. Today*, 45, 411–416 (1998).

83. Minowa, T. and Inoue, S. Hydrogen production from biomass by catalytic gasification in compressed water. *Renew. Energy*, 16, 1114–1117 (1999).

84. Minowa, T., Murakami, M., Dote, Y., Ogi, T., and Yokoyama, S. Oil production from garbage by thermochemical liquefaction. *Biomass Bioenergy*, 8, 117–120 (1995).

85. Minowa, T., Kondo, T., and Sudirjo, S. Thermochemical liquefaction of Indonesian biomass residues. *Biomass Bioenergy*, 14, 517–524 (1998).

86. Patil, V., Tran, K., and Giselrod, H. Towards sustainable production of biofuels from microalgae. *Int. J. Mol. Sci.*, 9, 1188–1195 (2008), doi:10.3390/ijms9071188.

87. Boocock, D.G.B. and Sherman, K.M. Further aspects of powdered poplar wood liquefaction by aqueous pyrolysis. *Can. J. Chem. Eng.*, 3, 627–633 (2009).

88. Zhou, D., Zhang, L., Zhang, S., Fu, H., and Chen, J. Hydrothermal liquefaction of macroalgae *Enteromorpha prolifera* to bio-oil. *Energy Fuels*, 24, 4054–4061 (2010).

89. Vardon, D., Sharma, B., Blazina, G., Rajagopalan, K., and Strathmann, T. Thermochemical conversion of raw and defatted algal biomass via hydrothermal liquefaction and slow pyrolysis. *Bioresour. Technol.*, 109, 178–187 (April 2012).

90. Vardon, D., Sharma, B., Scott, J., Yu, G., Wang, Z., Schideman, L., Zhang, Y., and Strathmann, T. Chemical properties of biocrude oil from the hydrothermal liquefaction of *Spirulina* algae, swine manure and digested anaerobic sludge. *J. Bioresour. Technol.*, 102(17), 8295–8303 (September 2011).

91. Zhou, S., Wu, Y., Yang, M., Imdad, K., Li, C., and Junmao, T. Production and characterization of bio oil from hydrothermal liquefaction of microalgae *Dunaliella tertiolecta* cake. *Energy*, 35(12), 5406–5411 (December 2010).

92. Duan, P. and Savage, P. Hydrothermal liquefaction of a microalga with heterogeneous catalysts. *Ind. Eng. Chem. Res.*, 50(1), 52–61 (January 2011).

93. Anastasakis, K. and Ross, A. Hydrothermal liquefaction of the brown macro-alga laminaria saccharina: Effect of reaction conditions on product distribution and composition. *Bioresour. Technol.*, 102(7), 4876–4883 (April 2011).

94. Yu, G., Zhang, Y., Schideman, L., Funk, T., and Wang, Z. Hydrothermal liquefaction of low lipid content microalgae into bio crude oil. *Trans. ASABE*, 54(1), 239–246 (January/February 2011).

95. Biller, P. and Ross, A. Potential yields and properties of oil from the hydrothermal liquefaction of microalgae with different biochemical content. *Bioresour. Technol.*, 102(1), 215–225 (January 2011).

96. Ross, A., Biller, P., Kubacki, M., Li, H., Lea-Langton, A., and Jones, J. Hydrothermal processing of microalgae using alkali and organic acids. *Fuel*, 89(9), 2234–2243 (September 2010).

97. Chakraborty, M., Miao, C., McDonald, A., and Chen, S. Concurrent extraction of bio oil and value added polysaccharides from chlorella sorokiniana using a unique sequential hydrothermal extraction technology. *Fuel*, 95, 63–70 (May 2012).

98. Biller, P., Riley, R., and Ross, A. Catalytic hydrothermal processing of microalgae: Decomposition and upgrading of lipids. *Bioresour. Technol.*, 102(7), 4841–4848 (April 2011).

99. Li, D., Chen, L., Xu, D., Zhang, X., Ye, N., Chen, F., and Chen, S. Preparation and characteristics of bio-oil from marine brown alga *Sargassum patens* C. Agardh. *Bioresour. Technol.*, 104, 737–742 (January 2012).

100. Dote, Y., Sawayama, S., Inoue, S., Minowa, T., and Yokoyama, S. Recovery of liquid fuel from hydrocarbon rich microalgae by thermochemical liquefaction. *Fuel*, 73, 1855–1857 (1994).

101. Minowa, T., Yokoyama, S., Kishimoto, M., and Okakurat, T. Oil production from algae cells of *Dunaliella tertiolecta* by direct thermochemical liquefaction. *Fuel*, 74, 1735–1738 (1995).

102. Sawayama, S., Minowa, T., and Yakoyama, S. Possibility of renewable energy production and CO$_2$ mitigation by thermochemical liquefaction of microalgae. *Biomass Bioenergy*, 17, 33–39 (1999).

103. Ross, A.B., Jones, J., Kubacki, M., and Bridgeman, T. Classification of microalgae as fuel and its thermochemical behavior. *Bioresour. Technol.*, 99, 6494–6504 (2008).

104. Brown, T., Duan, P., and Savage, P. Hydrothermal liquefaction and gasification of *Nannochloropsis* sp. *Energy Fuels*, 24, 3639–3648 (2010).

105. Matsui, T., Nishihara, A., Ueda, C., Ohtsuki, M., Ikenaga, N., and Suzuki, T. Liquefaction of microalgae with iron catalysts. *Fuel*, 76, 1043–1048 (1997).

106. Xue, Y., Yang, J., Liu, Z., Wang, Z., Liu, Z., Li, Y., and Zhang, Y. Heavy product from co-processing of FCC slurry and coal as bitumen modifier. *Preprint Paper ACS Div. Fuel Chem.*, 49(1), 24 (2004).

107. Miller, R.L. and Baldwin, B. Liquefaction co-processing of coal and shale oil at low severity conditions. *ACS Preprints*, 152–160 (1985).

108. Appell, H.R., FU, Y., Illrg, E., Steffgen, F. and Miller, R., *Conversion of cellulosic waste to oil*, Report of Investigations 8013, Pittsburgh, Energy Research Center, Pittsburgh, PA (1975).

109. Ross, D.S. et al. *ACS Div. Fuel Chem. Preprints*, 30(3), 94 (1985).

110. Ross, D.S. et al. *ACS Div. Fuel Chem. Preprints*, 30(4), 339 (1985).

111. Miller, R.L. Use of non-coal-derived heavy solvents in direct coal liquefaction. Interim Report for EPRI Project RP 2383-01. University of Wyoming, Laramie, WY (November 1985).

112. Shinn, J.H., Dahlberg, A.J., Kuehler, C.W., and Rosenthal, J.W. The Chevron co-refining process. In *Proceedings of the Ninth Annual EPRI Contractors' Conference on Coal Liquefaction*, Palo Alto, CA (May 1984).

113. Miller, R.L. *ACS Div. Fuel Chem. Preprints*, 31(1), 301 (1986); Miller, R.L. Effect of Wyodak (Wyoming) coal properties on direct liquefaction reactivity, Ph.D. dissertation. Colorado School of Mines, Golden, CO (1982).

114. Silver, H.F., Corry, R.G., and Miller, R.L. Coal liquefaction studies. Final Report for EPRI Projects RP 779-23 and RP 2210-1. Colorado School of Mines, Golden, CO (December 1982).

115. Miller, R.L. and Baldwin, R.M. The effect of Wyodak coal properties on liquefaction reactivity. *Fuel*, 64, 1235 (1985).

116. Gatsis, J. et al. Coal liquefaction co-processing. In *Proceedings of the DOE Direct Liquefaction Contractors' Review Meeting* (November 19–21, 1985).

117. Nafis, D. et al. Bench-scale co-processing. In *Proceedings of the DOE Direct Liquefaction Contractors' Review Meeting* (October 4–6, 1988).

118. Luebke, C. et al. Coal liquefaction co-processing topical report number 1. Prepared for the United States Department of Energy under Contract No. DE-AC22-84PC70002 (June 1, 1987).

119. Gatsis, J., Lea, C., and Miller, M. Coal liquefaction co-processing topical report number 2. Prepared for the United States Department of Energy under Contract No. DE-AC22-84PC70002. UOP Research Center, Des Plaines, IL (August 19, 1988).

120. Nafis, D. et al. UOP co-processing developments. In *Proceedings of the DOE Direct Liquefaction Contractors Review Meeting* (October 2–4, 1989).

121. Cugini, A., Lett, R., and Wender, I. Coal/oil co-processing mechanism studies. *Energy Fuels*, 3(2), 120–126 (1989).

122. Audeh, C. and Yan, T. Co-processing of petroleum residue and coal. *Ind. Eng. Chem. Res.*, 26(12), 2419–2423 (1987).

123. Fouda, S., Kelley, J., and Rahimi, P. Effects of coal concentration on co-processing performance. *Energy Fuels*, 3(2), 154–160 (1989).

124. Snjay, H., Tarrer, A., and Marks, C. Iron based catalysts for coal/waste oil co-processing. *Energy Fuels*, 8(1), 99–104 (1994).

125. Ibrahim, M. and Seehra, M. Free radical monitoring of the co-processing of coal with the chemical components of waste tires. *Fuel Process. Technol.*, 46(3), 213–219 (December 1995).

126. Cugini, A.V., Krastman, D., Martello, D.V., Frommell, E.F., Wells, A.W., and Holder, G.D. Effect of catalyst dispersion on coal liquefaction with iron catalysts. *Energy Fuels*, 8, 83 (1994).

127. Cugini, A.V., Rothenberger, K.S., Ciocco, M.V., and Veloski, G.V. Comparison of hydrogenation and dehydrogenation behavior and coal conversion activity upon the addition of coal for supported and unsupported molybdenum catalysts. *Energy Fuels*, 11, 213 (1997).

128. Cugini, A.V. PhD. thesis, University of Pittsburgh, Pittsburgh, PA (1993).

129. Rothenberger, K.S., Cugini, A.V., Thompson, R.L., and Ciocco, M.V. Comparison of coal liquefaction reactions conducted under different conditions of hydrogen delivery. In *Proceedings of the International Conference on Coal Science*, p. 1461 (1997).

130. Gray, D. and Tomlinson, G. Integration opportunities for coal-oil co processing with existing petroleum refineries. A report for DOE under Contract Number DE-AC22-95PC95054. Mitretek Systems, McLean, VA (1985).

131. Parker, R. and Clark, P. Coal-oil coprocessing using ebullated bed technology—Batch autoclave studies. In Paper presented at *Petroleum Conference of the South Saskatchewan Section*, Regina, Saskatchewan Canada (October 6–8, 1987).

132. Farcasiu, M. and Smith, C. Method of coprocessing waste rubber and carbonaceous materials, U.S. 5,061,363, to DOE (October 29, 1991).

133. Farcasiu, M. *Another use for old tires*, Chemtech, pp. 22–24, (January, 1993).

134. Anderson, L. and Tuntawiroon, W. Co-liquefaction of waste plastics with coal. *Preprints Paper ACS Div. Fuel Chem.*, 38(3), 816–822 (1993).

135. Taghiei, M., Huggins, F., and Huffman, G. *Preprints Paper ACS Div. Fuel Chem.*, 38, 810 (1993).

136. Liu, K. and Meuzelaar, H. Catalytic reactions in waste plastics, HDPE and coal studied by high-pressure thermogravimetry with on-line GC/MS. *Fuel Process. Technol.*, 49(1–3), 1–15 (October–December 1996).

137. Ibrahim, M. and Seehra, M. *Preprints Paper ACS Div. Fuel Chem.*, 38, 841 (1993).

138. Xue, Y., Wang, Z., Li, B., and Zhang, K. Co-processing of petroleum catalytic slurry with coal. *J. Coal Sci. Eng.* (China), 19(4), 554–559 (December 2013).

139. MacArthur, J., McLean, J., and Comolli, A. Two-stage co-processing of coal/oil feedstocks. U.S. Patent 4,853,111A (August 1989).

140. Robbins, G., Brandes, S., Winschel, R., and Burke, F. Characteristics of process oils from HTI coal/plastics co-liquefaction runs. In Paper presented at *1995 U.S./Japan Joint Technical Meeting* (*NEDO*), Sendai, Japan (October 16–19, 1995).

141. Farcasiu, M. and Smith, C.M. Modeling coal, liquefaction decomposition of A-(1-napthyl methyl) bibenzyl catalyzed by carbon black. *Energy Fuels*, 5, 83 (1991).

142. Farcasiu, M., Smith, C.M., Ladner, E.P., and Sylwester, A.P. *Preprints Paper ACS Div. Fuel Chem.*, 36, 1869 (1991).

143. Farcasiu, M., Smith, C.M., and Hunter, E.A. In *Proceedings of Conference on Coal Science*, IEA Coal Research (Ed.), University of New Castle-upon-Tyne, New Castle-upon-Tyne, U.K. Butterworth-Heinemann Ltd., Oxford, U.K., p. 166 (1991).

144. Monnier, J. CANMET Report 84-5E, Review of the Coprocessing of Coals and Heavy Oils of Petroleum Origin (March 1984).

145. Curtis, C.W., Tsai, K.J., and Guin, J.A. Evolution of process parameters for combined processing of coal with heavy crudes and residues. *Ind. Eng. Chem. Proc. Des. Dev.*, 24, 1259 (1985).

146. Kottensette, R.J. Sandia Report SANDB2-2495 (March 1983).

147. Bearden, R. and Aldridge, C.L. Hydroconversion of heavy hydrocarbons. U.S. Patent 4, 134825 (January 16, 1974).

148. Moody, T. Master's thesis, Auburn University, Auburn, AL (1985).

149. Liu, K., Jakab, E., Zmierczak, W., Shabtai, J., and Meuzelaar, H. *Preprint Paper ACS Div. Fuel Chem.*, 39(2), 576 (1994).

150. Ignasiak, B., Ohuchi, T., Clark, P., Aitchison, D., and Lee, T. Two-stage co-processing of sub-bituminous coals and bitumen or heavy oil. In Paper presented at the *Fuel Division Reactions of Coal in Novel Systems, ACS Meeting*, Anaheim, CA (September 7–12, 1986).

151. Rahimi, P., Fouda, S., and Kelley, J. Co-processing using H_2S as a promoter. *ACS Preprints*, 192 (1989).

152. Gatsis, J.G., Nelson, B.J., Lea, C.L., Nafsis, D.A., Humbach. M.J., and Davis, S.P. Continuous bench-scale single *stage* catalyzed *copmessing*. In *Contractor's Review Meeting*, Pittsburgh, PA (October 6–8, 1987).

153. Ettinger, M.D., Stock, L.M., and Gatsis, J.G. Coprocessing reactions of Illinois No.6 and Wyodak coals with Lloyd minister and Hondo Petroleum resids in the presence of dideuterium under severe conditions. *Energy Fuels*, 8, 960 (1994).

154. Ettinger, M.D. and Stock, L.M. Reactivity of phenolic compounds in coprocessing, *Energy Fuels*, 8, 808 (1994).

155. Ettinger, M.E. *Reaction Pathways in Coal/Petroleum Coprocessing*, University of Chicago Libraries (1993).

156. Stock, L.M. and Ettinger, M. and Gatsis J. G., Reaction pathways in co-processing. *ACS Preprints*, 1 (1993).

157. Sampson, C., Thomas, J.M., Vasadevan, S., and Wright, C.J. *Bull. Soc. Chem. Belg.*, 90, 1215 (1981).
158. Li, X.S., Xin, Q., Guo, X.X., Granges, P., and Delmin, B.I. Reversible hydrogen adsorption on MoS$_2$ studied by temperature-programmed desorption and temperature-programmed reduction. *J. Catal.*, 137, 385 (1992).
159. Blackbum, A. and Sermon, P.A.I. *Chem. Technol. Biotechnol.*, 33A, 120 (1983).
160. Stock, L.M. Hydrogen transfer reactions in coal conversion, Chapter 6. In *The Chemistry of Coal Conversion*, Schlosbexg, R. (Ed.). Plenum Publishing Co., New york, pp. 19–23 (1985).
161. Takeshita, K. and Mochida, I. Lap. Pat. 80-45703 (1980).
162. Sato, Y., Yamamolo, Y., Kamo, T., Inaba, A., Miki, K., and Saito, I. Effect of hydro treatment of various heavy oils as solvent for coal liquefaction. *Energy Fuels*, 5, 98–102 (1991).
163. Curtis, C.W., Tsai, K.-J., and Guin, J.A. *Fuel Process. Technol.*, 115, 71–87 (1987).
164. Bedell, M.W., Curtis, C.W., and Hool, J.L. *Fuel Process. Technol.*, 37, 1–18 (1994).
165. Owens, R.M. and Curtis, C.W. An investigation of hydrogen transfer from naphthalene during co-processing. *Energy Fuels*, 8, 823–829 (1994).
166. Vasireddy S., Morrcale, B., Cugini, A., Song, C. and Spirey, J. Clean liquid fuels from direct coal liquefaction: Chemistry, catalysis, technological status and challenges. *Energy Environ. Sci.*, 4(2), 311–345 (2011).
167. Pradhan, V.R., Hu, J., Tierney, J.W., and Wender, I. Activity and characterization of anion modified Iron (III) oxides as catalysts for direct liquefaction of low pyrite coals. *Energy Fuels*, 7, 446 (1993).
168. Friedman, S., Metlin, S., Svedi, A., and Wender, I. *J. Org. Chem.*, 24, 1287 (1959).
169. Greene, M., Gupta, A., and Moon, W. Coal liquefaction/resid hydrocracking via two-stage integrated co-processing. *ACS Preprints*, 208 (1993).
170. Rahimi, P.M., Fouda, S.A., Kelly, J.F., Malhotra, R., and McMillen, D.F. *Fuel*, 68, 422–429 (1989).
171. McMillen, D.F., Malhotra, R., and Tse, D.S. Interactive effects between solvent components: Possible chemical origin of energy in coal liquefaction and co-processing. *Energy Fuels*, 5, 179–187 (1991).
172. Brown, J.K. and Ladner, W.R. *Fuel*, 39, 87 (1960).
173. Taghiei, M.M., Feng, M., Huggins, F.E., and Huffman, G.P. Co-liquefaction of waste plastic with coal. *Energy Fuels*, 8, 1228–1232 (1994).
174. Barraza, J., Bolanos, A., Machuca, F., and Loiza, C. Thermal and catalytic co-liquefaction of a Columbian coal with lube oil waste. In *Second Mercosur Congress on Chemical Engineering and Fourth Mercosur Congress on Process Systems Engineering*, Bogota, Columbia (1998).
175. Pradhan, V.R. PhD thesis, University of Pittsburgh, Pittsburgh, PA (1993).
176. Rothenberger, K.S., Cugini, A.V., Schroeder, K.T., Veloski, G.A., and Ciocco, M.V. *Preprint Paper ACS Fuel Chem. Div.*, 39(3), 688–694 (1994).
177. Cugini, A.V., Krastman, D., Lett, R.G., and Balsone, V.D. *Catalysis Today*, 19(3) 395–408 (1994).
178. Hackett, J.P. and Gibbon, G.A. In *Automated Stream Analysis for Process Control*, Manka, D.P. (Ed.). Academic Press, pp. 95–117 (1982).
179. Ciocco, M.V., Cugini, A.V., Rothenberger, K.S., Veloski, G.A., and Schroeder, K.T. *Proceedings of the 11th Annual International Pittsburgh Coal Conference*, Vol. 1, pp. 500–505, Pittsburgh, PA (September 12–16, 1994).
180. Liu, K., Jakab, E., McClennen, W., and Meuzelaar, H. *Preprint Paper ACS Div. Fuel Chem.*, 38, 823 (1993).
181. Anderson, L.L. and Tuntawiroon, W. *ACS Div. Fuel Chem. Preprints*, 38(4), 810–815 (1993).
182. Farcasiu, M. and Smith, C.M. *Div. Fuel Chem. Preprints*, 37(1), 472–479 (1992).

183. Liu, Z., Zondlo, J.W., and Dadyburjor, D.B. Tire liquefaction and its effect on coal liquefaction. *Energy Fuels*, 8, 607–612 (1993).
184. Hajdu, P., Tierney, J., and Wender, I. Effect of modifying host oil on co-processing. *ACS Preprints*, 15 (1994).
185. Huber, D., Lee, Q., Thomas, R., Frye, K., and Rudins, G. An assessment of the potential for coal/residual oil co-processing. *ACS Preprints*, 221 (1991).
186. Sharma, R.K., Yang, J., Zondlo, J., and Dadyburjor, D. Effect of process conditions on coal liquefaction kinetics of waste tire and coal. *Catal. Today*, 40, 307–320 (1998).
187. Dayburjor, D., Zondlo, J., Sharma, R., Yang, J., Hu, F., and Bennett, B. The use of mixed pyrrhotite/Pyrite catalysts for coal liquefaction of coal and waste rubber tires. DOE Contract No. DE-FC22-93PC93053, Report for May 1993–April 1996, Department of Chemical Engineering, West Virginia University, Morguntown, WV (1996).
188. Farcasiu, M. Coprocessing of used rubber tires with coal. In Paper presented at the *144th Meeting of the Rubber Div.*, ACS, Orlando, FL (October 1993).
189. Gimouthopoulos, K., Doulla, D., Vlyssides, A., and Georglou, D. Waste plastics: Lignite mixtures co-liquefaction over Si/Al catalysts. *Waste Manage. Res.*, 17(3), 181–185 (June 1999).
190. Pradhan, V., Comolli, A., Lee, L., and Popper, G. Co-processing of waste plastics with coal and/or petroleum residuum. Report DOE Contract No. AC22-92PC2148, University of Pittsburgh, Pittsburgh, PA. (1999).
191. Siddiqui, M. and Ali, M. Catalytic co-liquefaction of model and waste plastics. In *Proceedings of the 15th Saudi-Japan Joint Symposium*, Dhahran, Saudi Arabia (November 27–28, 2005).
192. Taghiei, M.M., Huggins, F.E., and Huffman, G.P. Co-liquefaction of waste plastics with coal. In *Preprints of ACS Meeting*, Vol. 38(4), Chicago, IL, pp. 810–815 (August 1993).
193. Mastral, A., Murillo, R., Perez-Surio, M., and Callen, M. Coal hydrocoprocessing with tires and tire components. *Energy Fuels*, 10(4), 941–947 (1996).
194. Liu, Z., Zondolo, J., and Dadyburjor, D. Tire liquefaction and its effect on coal liquefaction. *Energy Fuels*, 8(3), 607–612 (1994).
195. Mastral, A., Murillo, R., Callen, M., and Garcia, T. Evidence of coal and tire interactions in coal-tire coprocessing for short residence times. *Fuel Process. Technol.*, 69, 127–140 (2000).
196. Dadyburjor, D.B., Stewart, W.E., Stiller, A.H., Stinespring, C.D., Wann, J.-P., and Zondlo, J.W. Disproportionated ferric sulfide catalysts for coal liquefaction. *Energy Fuels*, 8, 19 (1994).
197. Tang, Y. and Curtis, C.W. Thermal and catalytic coprocessing of waste tires with coal. *Fuel Process. Technol.*, 46(3), 195–215 (1995).
198. Sugano, M., Onda, D., and Mashimo, K. Additive effect of waste tire on the hydrogenolysis reaction of coal liquefaction residue. *Energy Fuels*, 20(6), 2713–2716 (2006).
199. Sunphorka, S., Prasassarakich, P., and Ngamprasertsith, S. Co-liquefaction of coal and plastic mixture in supercritical water. *J. Scient. Res. Chulalongkorn Univ.*, 32(2), 101–109 (2007).
200. Giray, E. and Sonmez, O. Supercritical extraction of scrap tire with different solvents and the effect of tire oil on the supercritical extraction of coal. *Fuel Process. Technol.*, 85(4), 251–265 (March 2004).
201. Dhaveji, S., Dadyburjor, D., and Zondlo, J. Co-liquefaction of coal and polyvinyl chloride (PVC). *ACS Preprints*, 1077 (1996).
202. Baladincz, P., Leveles, L., and Hancsok, J. Production of diesel fuel via co-processing of waste fat and gas oil mixtures on commercial hydrogenation catalyst. In *46th International Conference on Petroleum Processing*, Bratislava, Slovak Republic (June 7, 2013).

203. Stiller, H., Dadyburjor, D., Wann, J., Tian, D., and Zondlo, J. Co-processing of agricultural and biomass waste with coal. *Fuel Process. Technol.*, 49(1–3), 167–175 (October–December 1996).

204. Comolli, A.G. In *Proceedings of the Coal-Waste Coprocessing Workshop,* Pittsburgh, PA (September 9, 1994); also Comolli, A., Ganguli, P., Stalzer, R., Lee, T., and Zhou, P. The direct liquefaction of co-processing of coal, oil, plastics, MSW and biomass. *ACS Preprints*, Hydrocarbon Research Inc., Trenton, NJ, p. 300 (1994).

205. Gray, D. and Tomlinson, G. A techno-economic assessment of integrating a waste/coal coprocessing facility with an existing refinery. *ACS Preprints*, 20 (1996).

206. Orr, E., Tuntawiroon, W., Ding, W., Bolat, E., Rumpel, S., Eyring, K., and Anderson, L. Thermal and catalytic coprocessing of coal and waste materials. *ACS Preprints*, 44 (1995).

207. Tchapda, A. and Pisupati, S. A review of thermal co-conversion of coal and biomass/waste. *Energies*, 7(3), 1096–1148 (2014).

208. Xiao, X., Zmierczak, W., and Shabtai, J. Depolymerization-liquefaction of plastics and rubbers. 1. Polyethylene, polypropylene and polybutadiene. *ACS Preprints*, 4 (1995).

209. Altieri, P. and Coughlin, R. Characteristics of products formed during coliquefaction of lignin and bituminous coal at 400°C. *Energy Fuels*, 1(3), 253–256 (1987).

210. Ikenaga, N., Ueda, C., Matsul, T., Ohtsuki, M., and Suzuki, T. Co-liquefaction of microalgae with coal using coal liquefaction catalysts. *Energy Fuels*, 15(2), 350–355 (2001).

211. Abdullah, M. *Co-Liquefaction of Low Rank Malaysian Coal and Palm Kernel Shell—The Effect of Temperature*, IEEE Colloquium on Humanities, Science and Engineering (CHUSER), Kota, Kinabalu, pp. 627–630 (December 3–4, 2012). ISBN: 978-1-4673-4615-3.

212. Krasulina, J., Luik, H., Palu, V., and Tamvelius, H. Thermochemical co-liquefaction of Estonian Kukersite oil shale with peat and pine bark. *Oil Shale*, 29(3), 222–236 (2012).

213. Zhong, C.L. and Wei, X.M. A comparative experimental study on the liquefaction of wood. *Energy*, 29(11), 1731–1741 (2004).

214. Coughlin, R. and Davoudzadeh, F. Coliquefaction of lignin and bituminous coal. *Fuel*, 65, 95–106 (1986).

215. Kanno, T., Kimura, M., Ikenaga, M., and Suzuki, T. Coliquefaction of coal with polyethylene using Fe(CO)$_5$-S as catalyst. *Energy Fuels*, 14, 612–617 (2000).

216. Feng, Z., Zhao, J., Rockwell, J., Bailey, D., and Huffman, G. Direct liquefaction of waste plastics and coliquefaction of coal-plastic mixtures. *Fuel Process. Technol.*, 49, 17–30 (1996).

217. Deshpande, G., Holder, G., Bishop, A., Gopal, J., and Wender, I. Extraction of coals using supercritical water. *Fuel*, 63, 956 (1984).

218. Scarrah, W.P. In *Chemical Engineering at Supercritical Fluid Conditions*, Pauliatis, M., Penninger, R., Gray, D., and Davidson, P. (Eds.). Ann Arbor Sci., Ann Arbor, MI, pp. 385–407 (1983).

219. Modell, M., Reid, R., and Amin, S. U.S. Patent 4,113,446 (September 12, 1978).

220. Jezko, J., Gray, D., and Kershaw, J. The effect of solvent properties on the supercritical gas extraction of coal. *Fuel Process. Technol.*, 5, 229 (1982).

221. Vasilakos, N., Dobbs, J., and Parasi, A. *ACS Div. Fuel Chem. Preprints*, 28(4), 212 (1983).

222. Kershaw, J. and Bagnell, L. Extraction of Australian coals with supercritical water. *ACS Reprints*, 101–111 (1988).

223. Swanson, M., Olson, E., Diehi, J., and Farnum, S. Extraction of low-rank coals with supercritical water, personal communication (2012).

224. Minowa, T., Zhen, F., and Ogi, T. Cellulose decomposition in hot-compressed water with alkali or nickel catalyst. *J. Supercrit. Fluids*, 13, 253–259 (1998).

225. Mitsuru, S., Bernard, K., Roberto, M. et al. Cellulose hydrolysis in sub-critical and super-critical water. *J. Supercrit. Fluids*, 13, 261–268 (1998).

226. Mitsuru, S., Zhen, F., Yoshiko, F. et al. Dissolution and hydrolysis of cellulose in subcritical and supercritical water. *Ind. Eng. Chem. Res.*, 39, 2883–2890 (2000).

227. Wahyudiono, S., Shiraishi, T., Sasaki, M., and Goto, M. Non catalytic liquefaction of bitumen with hydrothermal/solvothermal process. *J. Supercrit. Fluids*, 60, 127–136 (December 2011).

228. Cheng, L., Zhang, R., and Bi, J. Pyrolysis of a low rank coal in sub and supercritical water. *Fuel Process. Technol.*, 85(8–10), 921–932 (July 2004).

229. Vostrikov, A., Psarov, S., Dubov, D., Fedyaeva, O., and Sokol, M. Kinetics of coal conversion in supercritical water. *Energy Fuels*, 21, 2840–2845 (2007).

230. Kumar, S. and Gupta, R. Biocrude production from switchgrass using supercritical water. *Energy Fuels*, 23, 5151–5159 (2009).

231. Kim, I.C., Park, S.D., and Kim, S. Effects of sulfates on the decomposition of cellobiose in supercritical water. *Chem. Eng. Process.*, 43, 997–1005 (2004).

232. *Supercritical Water Cracks Residue Oil*, Mitsubishi Material Corp., Chemical Engineering, p. 14 (September 2007).

233. Ayhan, D. Hydrogen-rich gas from fruit shells via supercritical water extraction. *Int. J. Hydrogen Energy*, 29, 1237–1243 (2004).

234. Kim, B. and Mitchell, R. Coal and biomass gasification under supercritical water conditions, personal communication, Mechanical Eng. Department, Stanford University, Stanford, CA (2012).

235. Nonaka, H. Development of liquefaction process of coal and biomass in supercritical water. *Fuel Energy Abstr.*, 39(1), 18 (January 1998).

236. Matsumara, Y., Nonaka, H., Yokura, H., Tsutsumi, A., and Yoshida, K. Co-liquefaction of coal and cellulose in supercritical water. *Fuel*, 78(9), 1049–1056 (July 1999).

237. Demirbas, A. Hydrogen production from biomass via supercritical water extraction. *Energy Sources*, 27, 1409–1417 (2005).

238. Phillip, E. Organic chemical reactions in supercritical water. *Chem. Rev.*, 99, 603–621 (1999).

239. Duan, P. and Savage, P. Upgrading of crude algal bio oil in supercritical water. *Bioresour. Technol.*, 102(2), 1899–1906 (January 2011).

240. Duan, P. and Savage, P. Catalytic hydrotreatment of crude algal bio-oil in supercritical water. *Appl. Catal. B: Environ.*, 104(1/2), 136–143 (April 2011).

241. Onsari, K., Prasassarakich, P., and Ngamprasertsith, S. Co-liquefaction of coal and used tire in supercritical water. *Energy Power Eng.*, 2, 95–102 (2010).

242. Sunphorka, S., Prasassarakich, P., and Ngamprasertsith, S. Co-liquefaction of coal and plastic mixture in supercritical water. In *The Fifth Mathematics and Physical Sciences Graduate Congress*, Organized by Faculty of Science, Chulalongkorn University, Bangkok, Thailand; in collaboration with National University of Singapore and University of Malaya, pp. 1–8 (2011).

243. Nonaka, H., Matsumura, Y., Tsutsumi, A., Yoshida, K., Matsuno, Y., and Inaba, A. Development of liquefaction process of coal and biomass in supercritical water. *Sekitan Kagaku Kaigi Happyo Ronbunshu*, 33, 73–76 (1996) [SciFinder].

244. Veski, R., Palu, V., and Kruusement, K. Co-liquefaction of kukersite oil shale and pine wood in supercritical water. *Oil Shale*, 23(3), 236–248 (2006).

245. Yokura, H., Nonaka, H., Matsumura, Y., Tsutsumi, A., and Yoshida, K. Effect of catalyst addition on co-liquefaction process of coal and biomass in supercritical water. *Sekitan Kagaku Kaigi Happyo Ronbunshu*, 34, 69–72 (1997).

246. Missal, P. and Hedden, K. Extraction of a Colorado oil shale by water in the sub- and supercritical phases. *Erdoel & Kohle, Erdgas, Petrochemie*, 42(9), 346–352 (1989).

247. Funazukuri, T., Yokoi, S., and Wasao, N. Supercritical fluid extraction of Chinese Maoming oil shale with water and toluene. *Fuel*, 67(1), 10–14 (1988).

248. Hu, H., Zhang, J., Guo, S., and Chen, G. Extraction of Huadian oil shale with water in sub- and supercritical states. *Fuel*, 78(6), 645–651 (1999).

249. Canel, M. and Missal, P. Extraction of solid fuels with sub- and supercritical water. *Fuel*, 73(11), 1776–1780 (1994).

250. Johnson, D.K., Chum, H.L., Anzick, R., and Baldwin, R.M. Lignin liquefaction in supercritical water. In *Thermochemical Biomass Conversion*, Bridgwater, A.V. and Kuester, J.L. (Eds.). Elsevier, London, U.K., pp. 485–496 (1988).

251. Palu, V., Kruusement, K., and Veski, R. Supercritical water extraction of biomass and oil shale. In *29th Estonian Chemistry Days*, Tallinn, Estonia, p. 77 (2005).

252. Li, L. and Eglebor, N. Oxygen removal from coal during supercritical water and toluene extraction. *Energy Fuels*, 6(1), 35–40 (1992).

253. Chen, D., Perman, C., Riechert, M., and Hoven, J. Depolymerization of tire and natural rubber using supercritical fluids. *J. Hazard. Mater.*, 44, 53–60 (1995).

254. Park, S. and Gloyna, E. Statistical study of the liquefaction of used rubber tyre in supercritical water. *Fuel*, 76(11), 999–1003 (1997).

255. Su, X., Zhao, Y., Zhang, R., and Bi, J. Investigation on degradation of polyethylene to oils in supercritical water. *Fuel Process. Technol.*, 85, 1249–1258 (2004).

256. Moriya, T. and Enomoto, H. Investigation of the basic hydrothermal cracking conditions of polyethylene in supercritical water. *Shigen to Sozai*, 115, 245 (1999) (Japanese).

257. Matsubara, W. et al. Development of liquefaction process of plastic waste in supercritical water. *Mitsubishi Juko Giho*, 34, 438 (1997) (Japanese).

258. Broll, D., Kaul, C., Kramer, A., Krammer, P., Richter, T., Jung, M., Vogel, H., and Zehner, P. Chemistry in supercritical water. *Angew. Chem. Int. Ed.*, 38, 2998 (1999).

259. Moriya, T. and Enomoto, H. Characteristics of polyethylene cracking in supercritical water compared to thermal cracking. *Polym. Degrad. Stab.*, 65, 373 (1999).

260. Moriya, T. and Enomoto, H. Role of water in conversion of polyethylene to oils through supercritical water cracking. *Kagaku kogaku Ronbunshu*, 25(6), 940 (1999) (Japanese).

261. Holliday, R., King, J., and List, G. Hydrolysis of vegetable oils in sub and supercritical water. *Ind. Eng. Chem. Res.*, 36(3), 932–935 (1997).

262. Li, Y., Guo, L., Zhang, X., Jin, H., and Lu, Y. Hydrogen production from coal gasification in supercritical water with a continuous flowing system. *Int. J. Hydrogen Energy*, 35, 3036–3045 (2010).

263. May, A., Salvado, J., Torras, C., and Montane, D. Catalytic gasification of glycerol in supercritical water. *Chem. Eng. J.*, 160(2), 751–759 (June 2010).

264. Pei, X., Yuan, X., Zeng, G., Huang, H., Wang, J., Li, H., and Zhu, H. Co-liquefaction of microalgae and synthetic polymer mixture in sub- and supercritical ethanol. *Fuel Process. Technol.*, 93(1), 35–44 (January 2012).

265. Shen, Y., Wu, H., and Pan, Z. Co-liquefaction of coal and polypropylene or polystyrene in hot compressed water at 360–430°C. *Fuel Process. Technol.*, 104, 281–286 (December 2012).

266. Cao, H., Yuan, X., Zeng, G., Tong, J., Li, H., and Wang, L. Co-liquefaction of biomass with plastic in sub- and super critical water. *Chem. Ind. Forest Prod.*, 29(1), 95–99 (2009).

267. Wu, H., Pan, Z., Jin, Z., Jin, Q., and Lin, C. Co-liquefaction of coal and polystyrene in supercritical water. *J. Fuel Chem. Technol.*, 39(4), 246–250 (2011).

268. Savage, P. Organic chemical reactions in supercritical water. *Chem. Rev.*, 99, 603–621 (1999).

5 Co-Pyrolysis

5.1 INTRODUCTION

While the removal of volatile matters and pyrolysis are the initial steps in the gasification, in general, pyrolysis of coal, oil, biomass, and waste is the most versatile thermochemical process. Unlike combustion, gasification, and liquefaction processes, the process of pyrolysis can produce solids, liquids, or gases depending on the operating conditions. The heating rate, residence time, and the temperature play very important roles in the composition of the final products. Several variations of the pyrolysis process are described in Table 5.1 [1,2]. For coal, biomass, and waste, often pyrolysis process is used to make oils, which can be upgraded for its downstream use in energy, fuel, and petrochemical industries.

While combustion is carried out with the excess of oxygen, and gasification is carried out with below stoichiometric requirement of oxygen (i.e., partial oxidation), pyrolysis is carried out in the absence of oxygen. Thus, pyrolysis is an endothermic reaction, while as shown in Chapters 2 and 3, gasification and combustion are exothermic reactions. Pyrolysis process thus requires the heat input in the system. While combustion process produces flue gas, which largely contains carbon dioxide and steam along with other contaminants depending on the impurities present in the feedstock, gasification produces synthetic gas, which contains hydrogen, methane, carbon monoxide, carbon dioxide, and steam along with other contaminants. Unlike flue gas, syngas has a significant heating value and it can be used for the productions of heat, power, liquid fuels, and chemicals.

The process of pyrolysis generates lower boiling organic chemicals along with the constituents of syngas. This is why ethylene and propylene, raw materials for polyethylene (PE) and polypropylene (PP), are produced by the pyrolysis of naphtha fraction of oil and other similar feedstock. Pyrolysis also produces liquids and solids. Thus, the process of pyrolysis is much more versatile in its product generations than gasification and combustion. Pyrolysis is also carried out at a lower temperature than gasification and combustion.

During the last three decades, the process of co-pyrolysis has gained significant momentum because researchers have observed that one can improve both quantity and quality of products of pyrolysis by mixing several feedstocks, which produce synergistic reactions. As shown in Section 5.3.1, the synergistic reactions are also useful in reducing the productions of harmful pollutants.

TABLE 5.1
Waste Pyrolysis Technologies, Operating Conditions, and Major Products

Technology	Residence Time	Temp. (°C)	Heating Rate	Major Products
Conventional carbonization	h–days	300–500	Very low	Charcoal
Pressurized carbonization	15 min–2 h	450	Medium	Charcoal
Slow pyrolysis	5–30 min	About 600	Low	Charcoal, oil, gas
Conventional pyrolysis	1 h or more	400–600	Low	Charcoal, oil, gas
Conventional pyrolysis	5–30 min	700–900	Medium	Charcoal, gases
Fast pyrolysis	0.5–5 s	About 650	Fairly high	Oil
Fast pyrolysis	0.1–2 s	400–650	High	Oil
Flash pyrolysis	<1 s	650–900	High	Oil, gases
Ultrapyrolysis	<0.5 s	1000–3000	Very high	Gases
Vacuum pyrolysis	2–30 s	350–450	Medium	Oil
Pressurized hydropyrolysis	<10 s	<500	High	Oil
Methanopyrolysis		<10 s	>700	High oil, chemicals

Sources: Lee, S. and Shah, Y., *Biofuels and Bioenergy: Technologies and Processes*, CRC Press, Taylor & Francis Group, New York, September, 2012; Shah, Y., *Water for Energy and Fuel Production*, CRC Press, Taylor & Francis Group, New York, May, 2014.

5.2 ADVANTAGES AND DISADVANTAGES

5.2.1 Co-Pyrolysis Offers Several Advantages

1. Just as combustion and gasification, co-pyrolysis of coal and biomass reduces the net production of CO_2 to the environment.
2. Co-pyrolysis of waste reduces the load to the landfill, thereby making constructive use of waste for energy, fuels, chemicals, and materials. This is not only true for biowastes but more importantly other more difficult wastes such as polymers, plastics, and waste tires. These waste materials do not disintegrate readily by anaerobic digestion in the landfills.
3. Co-pyrolysis makes the best use of synergy among various chemical transformations. Examples described in Sections 5.3, 5.4 and 5.5 indicate that one of the main purposes of co-pyrolysis is to use synergy to improve quantity and quality of the desired products. The proper choice of materials for co-pyrolysis is very important.

5.2.2 Disadvantages

1. Just as for combustion and gasification, the processing of multiple feedstock will require different types of experimental setup. Generally, fluidized-bed reactor is preferred to process co-pyrolysis feedstock.
2. The easy availability of feedstock is important for sustainable long-term operation of the co-pyrolysis project. When it uses biomass or waste mixtures, this may not be always possible.

3. Since pyrolysis generally produces multiple products (i.e., gas, liquid and solids) with more complex compositions, the use of co-pyrolysis will involve more changes in the downstream measurements and operations.

In this chapter, the effects of different types of mixture on the product distributions as well as other characteristics of pyrolysis are discussed in three parts: (1) co-pyrolysis of coal/ biomass/waste, (2) co-pyrolysis of other multifuel systems, and (3) wet pyrolysis (i.e., pyrolysis in aqueous-phase environment).

5.3 CO-PYROLYSIS OF COAL/BIOMASS/WASTE

5.3.1 ASSESSMENT OF SYNERGY

The understanding of synergy between pyrolysis of biomass and coal requires first understanding of reaction during individual pyrolysis [3]. Biomass contains three substances: cellulose, hemicellulose, and lignin. Pyrolysis at slow heating rate for these three components carried out by Yang et al. [4] showed that cellulose that is high in carbonyl groups releases CO, hemicellulose that is rich in carboxyl group releases CO_2, and lignin that is aromatic releases H_2 and CH_4 due to cracking and deformation of the aromatic C=C and C–H bonds as well as cracking of methyl groups.

For the temperatures less than 927°C, where most pyrolysis process occurs, the literature has presented conflicting reports on interactions among various components of biomass. Some [5] have claimed no interactions among biomass components, while others [6] have claimed interactions among various components of biomass during pyrolysis. In all studies, pyrolysis characteristics and product distributions were found to be significantly affected by the biomass mineral matters and biomass ash. The literature also indicates that during pyrolysis, volatiles come from cellulose and char from lignin and hemicellulose produces both char and volatiles [3].

Since inorganic matters (mineral matters) and ash from biomass have catalytic effects on the yields of volatile matters and char composition, these effects need to be separated while evaluating the synergy between pyrolysis of biomass and coal. It is known from the literature that pyrolysis conditions as well as nature of biomass feedstock [4,7–9] dictate the yield and composition of the volatile products formed during the early stage of the pyrolysis and they influence the char composition and the environment in which char decomposes during the later stage of the pyrolysis.

Tchapda and Pisupati [3] pointed out that there are two indicators for the synergy during co-pyrolysis of coal and biomass:

1. The extent of increase in the production of total volatiles (tars and light gases) and the decrease in char yield
2. Overall decrease in pollutants (oxides of nitrogen and their precursors as well as SO_x and their precursors)

The fate of nitrogen, sulfur, and chlorine compounds during pyrolysis of coal–biomass mixture was discussed in Section 3.2.3.1. The discussions in these sections show how synergy plays a role in the distributions of H_2S, COS, HCN, NH_3, HCl, and others in the gas phase and in the solid phase like char, ash, and slag. The discussions also indicate the roles of alkalis and mineral matters in these distributions. The readers are referred to those sections for this subject.

During co-pyrolysis of coal and biomass, biomass breaks down early and releases volatiles due to its weak covalent bonds and high oxygen content. This break down results in the production of free radicals, which facilitate the decomposition of coal. The hydrogen generated from cracking of heavy and light volatile molecules of biomass can react with the free radicals generated from coal. Thus, biomass can act as a hydrogen donor for coal pyrolysis. This type of hydrogen exchange during co-pyrolysis prevents recombination reactions and reduces the production of less reactive secondary chars [3].

The oxidation of high-heating-value volatile matters coming from coal further increases the pyrolysis temperature. This facilitates the cracking of tar molecules produced from coal and biomass as well as endothermic conversion of biomass and coal chars into gases. The net effect is the increase in volatiles yield, less char, and a higher fuel conversion rate. This type of synergy was reported by Ahmaruzzaman and Sharma [10] for the thermal decomposition of PP, petroleum vacuum residue, and biomass blend. As pointed out by Tchapda and Pisupati [3], the synergy occurred because free radicals generated from PP cracking reacted with the thermal decomposition products from petroleum vacuum residue and biomass.

The increase in total volatiles released (tars and gases) and the corresponding decrease in char production have received the most attention for synergistic evaluation during co-pyrolysis of coal and biomass [3]. As mentioned earlier in this section, synergy can also be manifested in overall decrease in pollutants. This type of synergy can occur through catalytic reactions by mineral matters or through other hydrogen transfer/free radical mechanisms. Tchapda and Pisupati [3] pointed out that the existence of noncatalytic synergistic effects remains controversial. Some found the evidence of synergy [11–35] and others found no synergy [36–48]. The way biomass and their components are intermeshed with coal can also affect pyrolysis behavior and the product distribution [3].

Since blending of coal and biomass/waste leads to an increase of the volatile products (CO, CH_4, and C_nH_m), the energy content of the gas is expected to be higher than when gasifying coal alone because of the high heating value of CH_4 and C_nH_m [49–51]. An increase in biomass to coal ratio in the mixture further increases conversion efficiency [50,52]. As shown by Franco et al. [52], an addition of polymeric waste like PE (in coal and pine waste mixture) also increases conversion and heating value due to its higher C and H contents. They, however, found that for mixtures of coal with equal amounts of pine and PE wastes, the gasification results were found to lie between those obtained for compositions with the same amount of only one waste.

Tchapda and Pisupati [3] pointed out that several investigators [36–48,53–57] did not find significant synergy during pyrolysis of coal and biomass mixtures. These results are summarized in Table 5.2. The reports include mixtures like bituminous coal (Gottelborn hard coal) and straw [52]; coal and sewage sludge [52]; coal and biomass [36]; subbituminous Collie coal with either waste wood or wheat straw [56]; Daw Mill coal (bituminous) and silver birch wood [53]; high-volatile bituminous

TABLE 5.2

Reported Studies on Synergy and No Synergy during Coal–Biomass–Waste Pyrolysis

Feedstock	Synergy	No Synergy	References
Bituminous coal/straw (Gottelborn hard coal)		×	Storm et al. [54]
Gottelborn hard coal/sewage sludge		×	Storm et al. [54]
Coal/biomass		×	Biagini et al. [36]
Subbituminous Collie coal/waste wood or wheat straw		×	Vuthaluru [56]
Daw Mill coal (bituminous)/silver birch wood		×	Collot et al. [53]
Drayton coal (high-volatile bituminous)/radiata pine sawdust		×	Meesri and Moghtaderi [37] and Moghtaderi et al. [38]
Malaysian subbituminous coal/empty fruit bunches, kernel shell, mesocarp fibers of palm tree		×	Idris et al. [57]
Lignite blends/olive kernels, forest residues, and cotton residues		×	Vamvuka et al. [39,55]
High-volatile bituminous or lignite coals/olive kernel and straw		×	Vamvuka et al. [39,55]
Polypropylene/petroleum vacuum residue/biomass	×		Ahmaruzzaman and Sharma [10]
(Wujek, bituminous; Kaltim Prima, bituminous; Turoszow, lignite) with a variety of biomass (pinewood, cellulose, lignin, xylan, and polywax model compounds)/batch reactor	×		Jones et al. [35]
(Wujek, bituminous; Kaltim Prima, bituminous; Turoszow, lignite) with a variety of biomass (pinewood, cellulose, lignin, xylan, and polywax model compounds)/py–GC–MS and TGA		×	Jones et al. [35]
Yallourn coal (lignite)/Taiheiyo coal (subbituminous)/fine powder of homopolypropylene, low-density PE, high-density PE	×		Hayashi et al. [58]
Seyitömer lignite/safflower seeds	×		Onay et al. [59]
Subbituminous coal/sawdust	×		Park et al. [19]
Dayan coal (lignite) and legume straw	×		Zhang et al. [14] and Gao et al. [60]
Subbituminous (Bitsch coal), Lemington bituminous coal/radiata pine sawdust	×		Ulloa et al. [61]
Coals of different ranks/straw	×		Kubacki [62]

coal (Drayton coal) and radiata pine sawdust [37]; Malaysian subbituminous coal and empty fruit bunches, kernel shell, and mesocarp fibers of palm tree [57]; lignite blends with olive kernels, forest residues, and cotton residues [39,55]; and high-volatile bituminous or lignite coals, blended with olive kernel and straw [39,55] where no synergy was found. More details on these studies are given in an excellent review by Tchapda and Pisupati [3].

Despite the lack of synergy in the co-pyrolysis of mixtures cited earlier, Tchapda and Pisupati [3] point out that many other investigators [11–35,58–62] found synergy during mixture pyrolysis. The mixtures include a variety of coals (Wujek, bituminous; Kaltim Prima, bituminous; Turoszow, lignite) with a variety of biomass (pinewood, cellulose, lignin, xylan, and polywax model compounds) [35]; Yallourn coal (lignite) and Taiheiyo coal (subbituminous), with fine powder of homo-PP, low-density PE (LDPE), or high-density PE [58]; Seyitömer lignite and safflower seeds [59]; coals of different ranks with straw [62]; sawdust and a subbituminous coal [19]; and Dayan coal (lignite) and legume straw [14]. For the last two mixtures and others [60], the synergy was pronounced at lower temperatures (around 600°C) but reduced as temperature increased and was less pronounced around 720°C. However, Ulloa et al. [61] noted interaction between subbituminous (Bitsch coal) and bituminous (Lemington coal) coals, with radiata pine sawdust at temperatures higher than 400°C and up to 1200°C. Park et al. [19], Onay et al. [59], and Gao et al. [60] noted that synergy was more pronounced at higher biomass blending ratio (BBR) (around 70%). All of these studies are also listed in Table 5.2. Once again, more details on these studies are outlined by Tchapda and Pisupati [3].

The aforementioned results indicate that in general, synergy seems to be more pronounced at intermediate temperatures (not too high nor too low), higher biomass concentrations, and higher concentrations of polymeric wastes. A thorough understanding of free radical formation and their evolution during pyrolysis seems to be the key to the possible interaction between coal and biomass. While extensive studies have been conducted on free radicals release during coal pyrolysis [10,59–67], such studies involving direct in situ observation of radicals released during biomass or biomass components (cellulose, hemicelluloses, and lignin) or polymeric wastes pyrolysis are lacking [3]. Understanding the mechanism of free radicals release during biomass/biomass components and polymeric wastes pyrolysis can pave the way for improvements in the modeling of coal/biomass/wastes blends decomposition during co-pyrolysis and the synergistic effect between coal, biomass, and waste.

5.3.2 Effects of Pressure on Volatiles, Tar, and Char

Tchapda and Pisupati [3] also gave an excellent review of the effect of pressure on volatile yield, tar, and char during pyrolysis of coal and biomass. The following paragraph briefly summarizes their analysis.

During the early devolatilization phase of pyrolysis, an increase in the pressure decreases the total volatile yield of coal. This was demonstrated by Lee et al. [68] for a bituminous coal at pressures up to 0.39 MPa and by Sun et al. [69] for two Chinese bituminous and anthracite coals. The latter work, however, indicated that the effect of pressure depends on the level of temperature, an assertion also supported by

Seebauer et al. [70] for a bituminous coal at pressure between 0.1 and 4 MPa and by Collot et al. [53] for Daw Mill coal in the pressure range of 1–25 bar. Gibbins et al. [71] obtained a decreased yield with an increase in pressure in a helium atmosphere.

The pressure effect during devolatilization of biomass is similar to that of coal. Pindoria et al. [72,73] observed a decrease in total volatile yield as pressure increased from 0.25 to 7 MPa during pyrolysis of eucalyptus sawdust. Collot et al. [53] obtained similar results for silver birch for pressure varying from 2 to 25 bar in a helium environment. All reported studies showed that the effect of pressure on total volatile yield is less pronounced at higher pressures.

High pressure also prohibits the release of larger tar molecules resulting in lower overall tar yields [74,75]. The secondary repolymerization of tar at higher temperature and the occurrence of autohydrogenation at higher pressure also increase production of lighter hydrocarbon gases like methane. A significant decrease in tar yield and increased yields of char and hydrocarbon gases during pyrolysis of five German hard coals in an inert gas (He) atmosphere as well as in hydrogen atmosphere, with pressure varying from 0.1 to 9 MPa and a temperature range of 950°C–1000°C are reported [3,76–78]. Similar results were obtained for pyrolysis of Pittsburgh No. 8 and Linby (UK) coals [72,73] and eucalyptus wood waste [70].

Pressure also significantly affects the reactivity of char and ash formation during char gasification [3]. Higher pyrolysis pressures lead to lower surface area of char and, therefore, lower reactivity during gasification. This assertion was verified for Illinois No. 6 bituminous coal [79] and for lignite, subbituminous, and bituminous coals [80,81]. Experiments with radiata pine, spotted gum (*Eucalyptus maculata*), and sugarcane bagasse showed comparable results [82,83].

Tchapda and Pisupati [3] pointed out that high pressure reduces the release of volatile matters, causing their large deposition on the pore surfaces, which creates secondary reactions and, therefore, deactivates active sites on the resulting char. They also suggest that another possibility may be that higher pyrolysis pressure enhances fluidity of coal, leading to a better mobility, alignment, and ordering of carbon layers with reduction of micropores and subsequent loss of gasification reactivity. Concurrent influences of these two phenomena are also possible.

5.3.3   ADDITIONAL CASES OF CO-PYROLYSIS OF INVOLVING COAL

Besides coal/biomass co-pyrolysis studies and their analysis by Tchapda and Pisupati [3] outlined in Sections 5.3.1 and 5.3.2, numerous other studies on co-pyrolysis that involve coal have been reported [84–92]. Several other studies are briefly summarized below.

5.3.3.1   Coal/Waste Tires/Waste Plastics/Waste Cotton

Kriz and Brozova [93] examined co-pyrolysis of noncaking hard coal (Lazy mine) and waste tires. The study was carried out on a laboratory scale. The experiments were carried out for 40/60 mixture of coal and tires and at a final temperature of 900°C. The study concluded that waste tires can be reused for co-pyrolysis. The addition of tires increased the yield of tar. The solid residue (coke) represented the main product in all the experiments. It can be used for production of sorbents or carbonizing agents. Hydrogen was the main component of pyrolytical gas obtained in all experiments. The study also examined waste plastics under the same condition as rubber tires. The results in this case were very similar to that for rubber tires. The study also showed that the use of waste cotton is not suitable for co-pyrolysis. The high oxygen content in the waste cotton predominantly produced reaction water.

5.3.3.2 Coal/Agricultural Waste

Aboyade et al. [94] examined coal/agricultural waste pyrolysis in an updraft fixed-bed pressurized gasifier. Coal/biomass blends of 100/0, 95/5, 50/50, and 0/100 wt% were examined. Sugarcane bagasse was selected as the biomass feedstock. The results showed significant synergistic effects during co-pyrolysis, which affected the yields of specific volatile species and functional groups such as H_2, CO, CO_2, CH_4, acids, furans, ketones, phenolics, polyaromatic hydrocarbons (PAHs), and other mono- and heterocyclic aromatics. The interactions among the species produced from individual feedstock gave significantly higher acids than what would be expected assuming no interactions. There was also a comparatively less drastic increase in the percentage of mono- and polycyclic aromatics. This observation agreed with the previous observations by Jones et al. [35] for their atmospheric slow pyrolysis of coal and pinewood. The study proposed a number of reaction pathways to produce PAH and phenols. The simple additive nature of prediction for the volatile products did not apply for this mixture. The possible synergistic effects during coal and biomass co-pyrolysis was also observed by Nyendu et al. [95].

5.3.3.3 Lignite/Corncob

Sonobe et al. [96] examined synergies that existed during co-pyrolysis of lignite and corncob. The experiments were performed in a TG-MS lab system and this was followed up by verification in a deep-fixed-bed reactor. The pyrolysis characteristics of this mixture gave good understanding of reaction mechanisms during co-combustion of lignite/corncob mixtures. The experiments were performed for pure corncob and lignite and various mixtures of lignite and corncobs. The devolatilization study showed that lignite showed a broad range of decomposition from low temperature to 600°C, while corncob shows a sharp decomposition at around 220°C–350°C. The devolatilization of lignite/corncob mixture indicated synergistic effects during co-pyrolysis. The co-pyrolysis of lignite and corncob was able to enhance the thermal decomposition of lignite through the secondary pyrolysis reaction of biomass. The CH_4 production yield was also increased during co-pyrolysis due to early decomposition of macromolecular chain of lignite accompanying with secondary tar cracking.

5.3.3.4 Lignite/Pistachio Seed

Onay [97] studied co-pyrolysis of lignite and pistachio seed in a Heinze retort under nitrogen environment. Heating rate of 10°C/min and 50/50 mixture of the two components were examined. Higher yield for pistachio seed was observed compared to that for lignite. The yield of the product was proportional to the temperature. A considerable synergistic effects during co-pyrolysis resulted in increased oil yield in a fixed-bed reactor. Maximum pyrolysis oil yield of 27.2% was obtained at pyrolysis temperature of 550°C. The co-pyrolysis also gave improved oil quality.

5.3.3.5   Coal/Capsicum Stalks

Niu et al. [98] examined the co-pyrolysis of capsicum stalks and Baoji coal in a TG-DSC. The thermal degradation of capsicum stalks ranged from 290°C to 387°C, while that for Baoji coal ranged from 416°C to 586°C. The study showed significant synergistic effects in the temperature range 314°C–369°C and 431°C–578°C. The synergistic effect can also be seen from the kinetic study performed according to the Friedman method. The rate of mass loss and the intrinsic kinetic constant during co-pyrolysis were higher than the calculated values from the pyrolysis of individual components in the temperature range of 314°C–368°C and lower in the temperature range of 413°C–578°C. The study showed that pyrolysis of capsicum stalks, Baoji coal, and their mixtures can be described by one, two, or four first-order reactions. Three thermal peaks were observed during co-pyrolysis of CS/BS blends, the first denoted the loss of water, the second one denoted the result of dominant CS pyrolysis, and the third corresponded to the BC pyrolysis. The chemical reaction constant was the highest at 40% CS in the mixture.

5.3.3.6   High-Sulfur Coal/Different Lignocellulosic Wastes

Cordero et al. [23] examined co-pyrolysis of high-sulfur coal and various lignocellulosic wastes in TG laboratory system and pilot-scale mobile bed furnace operating at 600°C. The raw materials were Spanish lignite from Teruel with high sulfur content and four residual biomasses: wheat straw, olive stone, almond shell, and pine wood sawdust. Initial particle size of all raw materials ranged from 0.8 to 1.5 mm. The study showed that the presence of biomass enhances coal desulfurization upon thermal treatment in significant relative amounts, giving rise to about as much as twice the percent of sulfur loss at high biomass–coal ratios in the starting blend in comparison with the sulfur loss occurring upon pyrolysis of coal alone. Combustion experiments with chars resulting from co-pyrolysis of these coal–biomass blends confirmed this significant improvement in desulfurization. Thus, this co-pyrolysis process provided a potential way to obtain improved solid fuels combining good heating values with environmentally acceptable sulfur contents. The chars resulting from co-pyrolysis showed heating values within the range of high-quality solid fuels, whereas the ash contents remained in the vicinity of that of the starting coal except in the case of coal–straw mixture. In the latter case, higher ash content of the biomass waste was reflected in the co-pyrolysis chars and this results in the lower heating value. More research is needed to examine the mechanisms involved in higher rate of sulfur removal during co-pyrolysis.

5.3.3.7   Coal/Raw and Torrefied Wood

Lu et al. [41] examined a mixture of anthracite coal and *Cryptomeria japonica* (both raw and torrefied) for co-pyrolysis using thermo-gravimetric analysis (TGA). The wood was torrefied at 250°C and 300°C. Five different blends, 100, 75, 50, 25, 0 wt%, were examined at 800°C. The results indicated that the thermal degradation processes of the single materials were characterized by a three-stage reaction, except at TW300, which demonstrated a four-stage reaction resulting from more lignin retained at higher torrefied temperature. While mildly torrefied (at 250°C) and severely torrefied (300°C) wood were mixed with anthracite coal in different blend composition, the pyrolysis process changed from a three-stage reaction to a four-stage reaction. The results indicated that pyrolysis characteristics of the blends were very close to those of individual materials, thereby implying that there was very little synergy between pyrolysis of wood/torrefied wood and that of anthracite coal during co-pyrolysis. Thus, irrespective of torrefaction, the behavior of mixture can be predicted from those of individual materials and known composition of the blends. An examination of the co-pyrolysis kinetics indicated that an increase in BBRs led to the increase in activation energy of the second stage and a decrease in the third stage. This was because cellulose in the biomass mainly affected the activation energy in the second stage. The activation energy in the third stage was mainly affected by the lignin in the biomass. These relative differences in the composition of cellulose and lignin also affected activation energies in second and third stages.

5.3.3.8   Coking Coals and Plastics

Sakurovs [99] examined the blends of three coking coals and PP, polystyrene (PS), polyacrylonitrile, and polyphenylene sulfide for co-pyrolysis. The heating was monitored using proton magnetic resonance thermal analysis. The results showed that the different plastics had different effects. PS strongly reduced the fluidity of all of the coals. PP did not affect the fluidity of two coking coals of lower rank. Polyphenylene sulfide reduced the fluidity of coals at temperatures near the solidification temperature of the coals, and polyacrylonitrile appeared to increase the fluidity of coals at temperatures near the softening temperature of the coals. These different effects exhibited by different plastics on different coals indicate that the interactions between plastics and coals must be carefully examined before plastics are added to coking coal blends.

5.3.3.9   Coal and Heavy Carbonaceous Residue

Onay [100] patented a process of co-pyrolysis of coal and heavy carbonaceous residue. The invention showed that a heavy hydrocarbonaceous residue can be upgraded to lower boiling fractions by the fluidized-bed co-pyrolysis of coal particles previously heated until about 1%–10% volatiles have been removed to a time sufficient to remove all the volatiles from the coal and residue condensable to oily liquids. The co-pyrolysis produced lower boiling oily liquids, char, and gas.

5.3.3.10   Coal Briquette/Spear Grass (*Imperata cylindrica*)/ Elephant Grass (*Pennisetum purpureum*)

Martin [101] examined the improvement in properties of coal briquette using spear grass and elephant grass. Briquettes of different compositions were produced by blending the plant material with the coal at various concentrations (0%, 10%, 20%, 30%, 40%, 50%, 100%). The physical, mechanical, and combustion properties of the briquettes were compared and it was found that ignition, burning rate, and reduction in smoke emission showed improvement with increase in biomass concentration. Compressive strength and cooking efficiency (water boiling time and specific fuel consumption) showed initial improvement and rendered to break with briquette containing 50% biomass concentration for elephant grass briquette. For spear grass, the compressive grass was at maximum at biomass concentration of 30%.

5.3.3.11   Low-Rank Coals/Biomass

Weiland [102] presented experimental and theoretical results for the effects of co-feeding lower-rank coal and biomass into a transport bed reactor. Both kinetic rates and product distributions were analyzed. The focus of this study was to evaluate complex pyrolysis reaction and to provide fundamental co-pyrolysis kinetic data under transport gasifier conditions. The experimental test matrix included subbituminous Powder River Basin (PRB) coal, Mississippi lignite, southern yellow pine, multiple PRB/pine blends, and 80/20 lignite/pine blend at temperatures ranging from 600°C to 975°C and pressure of 1, 4, and 16 atm. Co-pyrolysis tests were performed in an isothermal drop reactor, where rapid sampling via in situ mass spectrometry yields kinetic rates as well as quantitative product gas measurements of the primary syngas constituents: CO, CO_2, CH_4, H_2, H_2O, H_2S, and C_2H_4. These product distributions as well as char and tar were found to be approximately linear with respect to temperature and biomass blend ratio. The data were generated for all ranges of temperature, pressure, and blend ratios mentioned earlier. Both kinetic activation energies and pre-exponential factors were shown to be nonlinear with respect to biomass blend ratio, implying a significant reaction synergy in co-pyrolysis of low-rank coals and biomass that must be accounted for in accurate modeling of transport gasifier systems.

5.3.3.12   Coal/Woody Biomass

Kubacki [103] examined coal and biomass (wood waste, short rotation woody crops [e.g., willow coppice]), heterogeneous crops (miscanthus), refuse and waste-derived fuels, and wastes such as chicken litter and sewage sludge. The study showed that co-pyrolysis and co-combustion of the coal and these biomass materials result in the reductions of SO_2, NO_x, volatile matters, and PAH emissions in the product. The synergy during co-pyrolysis depended on the nature of coal and biomass (with different amount of catalytic components), heating rate, residence time, and the physical forms of the fuels. Co-pyrolysis study suggested that biomass type can lead to a small effect on the coal pyrolysis and on the total volatile matter released but there were no major changes in the nature of the volatiles. The study indicated that the synergy in emission reduction in the co-utilization of coal and biomass was not simply due to interactions of volatiles in the vapor phase; rather, the processes of pyrolysis and combustion were linked and as such needed to be studied together.

5.3.3.13   Coal/Oil Shale

Co-pyrolyzing coal and oil shale can have many advantages over pyrolysis of either material separately. First, preliminary Illinois State Geological Survey (ISGS) data indicate that the inert material in oil shale dilutes the coal–oil shale mixture, reducing the agglomeration of caking coals. Secondly, liquids and gases are produced from both coal and oil shale. Thirdly, the carbonate minerals found in oil shale (calcite, dolomite) and their corresponding oxides formed during pyrolysis can act as scavengers of hydrogen sulfide during pyrolysis and hydrodesulfurization. Johnson et al. [104] showed that under typical conditions for coal–oil shale pyrolysis, dolomite and calcium oxide are capable of maintaining the desired H_2S concentration in the reactor. Also, under typical conditions of char desulfurization, calcium oxide will be an effective sorbent. In a fluidized-bed reactor with few seconds residence time, the reaction rate needs to be fast enough to achieve thermodynamic equilibrium. These reaction rates need to be determined. Also, a thorough understanding of the influence of the gas environment and the effects of structural properties of the oil shale particles is essential.

5.3.3.14   Waste Plastic Co-Pyrolysis with Used Oil Followed by Co-Processing with Coal

Mulgaonkar et al. [105] examined a two-stage process to improve conversion and yield of oil for coal liquefaction. In the first stage, waste plastic was pyrolyzed with used oil. High-density PE (HDPE) was used with waste oil to co-pyrolyze at different temperatures (350°C–600°C) and residence time of 2–8 s. The product of the process was a liquid oil that has considerably reduced viscosity and that can be used directly as a feedstock to refineries compared to the liquid product produced from pyrolysis of used oil alone. In the second stage, the oil produced from the co-pyrolysis of waste plastics and used oil is co-processed with coal. The co-processing resulted in high conversion of coal (>80%) and a high selectivity of oil (>60%). The study also indicated that co-pyrolysis of waste plastic and used oil was also carried out in a pilot-scale unit where insignificant amounts of coke and solids were produced.

The use of low residence time cracking during co-pyrolysis offers a number of advantages: (1) the transportation costs are significantly reduced by removing contaminants close to the generation site and by reducing the bilk density of the plastics and (2) the handling of waste plastics/oils during coal co-processing would be simplified due to the low viscosities and fluidic properties of the cracked products.

5.4   CO-PYROLYSIS OF MULTIFUELS SYSTEMS NOT INVOLVING COAL

5.4.1   Lignins with Polycarbonate

Brebu and Nistor [106] examined the co-pyrolysis of bisphenol A polycarbonate with lignins of various origin and obtaining procedures (Klason annual plants, organosolv hardwood, and LignoBoost softwood lignin) with a focus on product yield and oil composition. The study presented co-pyrolysis of natural polymers with synthetic ones. Co-pyrolysis was performed at 500°C, but the volatile degradation products were passed through a layer of quartz wool maintained at 300°C. The interactions at high temperatures between lignins and PC were favorable for LignoBoost with higher oil yield (13%) and reduced amount of residue (10%–15%). The composition of co-pyrolysis depended on the starting lignins. However, phenol, bisphenol A, and their derivatives were the most important compounds in all cases. Organosolv hardwood yielded significant amount of syringol, methylsyringol, and isovanillic acid while LignoBoost lignin yielded more guaiacol and its methyl and ethyl derivatives. Residual acetic acid and sulfur compounds from the obtaining procedures were found for organosolv hardwood and LignoBoost lignin.

5.4.2 Flash Co-Pyrolysis of Biomass with Polyhydroxybutyrate

Bio-oils were obtained by Cornelissen et al. [107–109] via flash pyrolysis of biomass with polyhydroxybutyrate (PHB). The influence of a biopolymer PHB on the pyrolysis of willow was investigated using a semicontinuous pyrolysis reactor. The flash co-pyrolysis of willow/PHB blends (w/w ratio 7:1, 3:1, 2:1, and 1:1) clearly showed particular merits: a synergetic increase in pyrolysis yields, a synergistic reduction of the water content in the bio-oil, an increase in heating value, and a production of easily separable chemicals. The flash co-pyrolysis at 450°C resulted in an increased pyrolysis yield and bio-oil with a reduced water content. The occurrence of synergistic effects was observed based on a comparison between the actual pyrolysis results of the willow/PHB blends. The theoretical results were calculated from the reference pyrolysis experiments (pure willow and pure PHB) and their respective w/w ratio. A synergistic effect was observed to increase with the addition of PHB. The co-pyrolysis of 1:1 willow/PHB showed the best overall results. Also, an increased energy recuperation compared to the sum of fractional experimental values of both materials is achieved, making the co-pyrolysis of willow and PHB an energetic and economically attractive route. Finally, the flash co-pyrolysis of willow and PHB also resulted in an uninitiated phase-separation between bio-oil and crystals (crotonic acid), which are a potential source of value-added speciality chemicals.

5.4.3 Catalytic Fast Co-Pyrolysis of Wood and Alcohol

Zhang et al. [110] studied catalytic fast pyrolysis (CFP) of pine wood and alcohols (methanol, 1-propanol, 1-butanol, and 2-butanol) and their mixtures with ZSM-5 catalyst in a fluidized-bed reactor. The maximum petrochemical yield for the pine wood was 23.7%, which occurred at 600°C and 0.35/h. Both the coke and unidentified oxygenate yields decreased with increasing temperature, whereas the coke yield decreased and the unidentified oxygenate yield increased with increasing WHSV. The maximum petrochemical yield with pure methanol was 80.7%, which occurred at 400°C and 0.35/h. The yields of petrochemicals and unidentified compounds decreased with increasing temperature, and the yields of methane, CO_2, CO, and coke increased with the increasing temperature. The petrochemical yields increased rapidly at lower WHSV, while CO, CO_2, methane, and coke yields showed the opposite trend.

The study examined CFP of pinewood and methanol mixtures with different ratios at 450°C and 500°C. The results showed that the petrochemical yield increases nonlinearly with increasing H/C_{eff} ratio of the feed, and more petrochemicals can be produced from wood when methanol is added to the CFP process. $H/C_{eff} = 1.25$ is an inflection point in this process, where the petrochemical yield begins to reach an asymptote. This suggests that the amount of methanol co-fed with wood should be enough to bring the mixtures to H/C_{eff} ratio of 1.25. Adding more methanol would be a waste of methanol and increase the cost to the process. Co-CFP of pinewood and other alcohols also increase the petrochemical yield, whereas co-CFP of pine wood and methanol gave the highest increase of petrochemical yield.

The isotropic study of ^{12}C methanol and ^{13}C pinewood showed that the products were produced from the mixtures of methanol and pinewood. Benzene was made from random mixtures of methanol and pinewood. Propylene and butenes were produced more from methanol than wood. Naphthalene contained more ^{12}C and therefore were more produced from pinewood than methanol. This indicates that it is feasible to use feeds with high H/C_{eff} ratio to provide hydrogen to the hydrocarbon pool for biomass conversion.

The study concludes that the petrochemical yield produced by CFP of biomass can be enhanced by co-feeding it with the feeds that have a high H/C_{eff} ratio. In the co-feeding and co-pyrolysis process, the two feeds can be converted together with the zeolite and have a positive synergistic effect. This study provided an insight into how biomass resource can be used more efficiently to produce renewable petrochemicals.

5.4.4 PEAT AND NATURAL ALUMOSILICATES AND SYNTHETIC ZEOLITES

Kosivtsov et al. [111] showed that co-pyrolysis of peat with addition of 2% synthetic zeolites and 30% natural alumosilicates resulted in noticeable increase of the hydrocarbons in gaseous mixtures. The increase of the heat of combustion of gaseous mixture was due to the increased content of alkanes and alkenes obtained when alumosilicate materials were used. The average value of the specific heat of combustion was approximately twofold higher in comparison with the data in the absence of zeolites and alumosilicates. The bentonite clay was found to have the highest effect on peat pyrolysis.

5.4.5 GÖYNÜK OIL SHALE WITH THERMOPLASTICS

Ozdemir et al. [112] evaluated the co-pyrolysis of Turkish Göynük oil shale and PE and PS. Pyrolysis of PE resulted in 93% conversion at 550°C without catalyst. Pyrolysis oils contained about 92% aliphatic hydrocarbons (AHCs) mainly linear alkanes and alkenes (C_9–C_{34}). About 3% aromatic hydrocarbons were detected. Pyrolysis of oils shale alone resulted in 50% yield. The oil was 33% aliphatic and 14% aromatic hydrocarbons. TGA showed that oil shale weight loss started at 300°C and ended at about 550°C.

The co-pyrolysis of oils shale and PE and oil shale and PS did not result in a significant and continuous synergism at all ratios of plastic additions. However, co-pyrolysis resulted in distinctive effect of the composition of products. Co-pyrolysis of oil shale and PE gave high percentage of aliphatic structures. The ratio of alkyl benzenes to alkyl naphthalenes and alkenes/alkane as well as alkadiene/alkene changed significantly due to co-pyrolysis. On the other hand, co-pyrolysis of oil shale and PS resulted in very high pentane-soluble fraction (98%) in the product. This is a significant improvement in terms of quality of oil shale as refinery input and chemical feedstock. The 1/1 ratio co-pyrolysis of oil shale and PS gave 75% styrene, 12% ethyl benzene, and 3% 1-phenyl propene.

5.4.6   BRAZILIAN HEAVY OIL WITH POLYPROPYLENE WASTE/ PETROLEUM RESIDUE WITH SUGARCANE BAGASSE

Assumpcao et al. [113] reported that co-pyrolysis of PP with Brazilian heavy oil in the temperature range of 400°C–500°C and for various PP in an inert atmosphere resulted in about 80% of oil, half of which was diesel oil. The authors suggested a detailed investigation of this mixture for future co-pyrolysis study. Darmstadt et al. [114] studied co-pyrolysis of petroleum residue and sugarcane bagasse. They examined properties of char and activated char products.

5.4.7   ESTONIAN SHALE WITH LOW-DENSITY POLYETHYLENE

Tikma et al. [115] examined the co-pyrolysis of Estonian kukersite oil shale, its semi-coke, and *Dictyonema* shale with LDPE. The study examined the oil yield and composition from this co-pyrolysis. The amounts of gas, oil, and solid residues formed in the co-pyrolysis process consist of partial contributions from the initial objects. The composition of the co-process oil differed from that of LDPE alone. *Dictyonema* shale had a most marked effect on the LDPE destruction rate. It narrowed the HC number range of *n*-alkanes and *n*-alkenes and increased the unsaturation degree of the aliphatic HC fraction. The chemical group composition of oil depended considerably on the quantity of *Dictyonema* shale. For its quantity up to 5% of the mixture, it increased both the total oil quantity and the content of AHC fraction. At higher percentages, *Dictyonema* shale decreased the oil yield due to significant gas formation. This is due to degradation of hydrocarbons to shorter and more volatile ones. High content of AHCs and low content of aromatic ones in the oil obtained from PE change the composition of co-processing oil resulting in low content of aromatic hydrocarbons especially polycyclic ones.

Rotiwala and Parikh [116] examined co-pyrolysis of three plastics, HDPE, PP, and PS with decoiled cake of *Jatropha* (JC) at 400°C and 450°C in a batch reactor in an inert atmospheric pressure. At high temperature (450°C), the yield of liquid fractions by the pyrolysis of plastics alone was found to increase by 11%, 12.5%, and 11% for HDPE, PP, and PS, respectively. Furthermore, the gaseous fraction increased by 1.3%–2.6% and the residue generation reduced by 12.3%–15.1%. In comparison with the data for only plastics pyrolysis, the yield of liquid fraction improved by 2.0%– 4.9% for their co-pyrolysis with JC. GC–MS analysis indicated that co-processing resulted in the reduction of paraffin and olefins in the liquid fractions. The reduction was found to be in the order PS > PP > HDPE. Furthermore, the proportions of the oxygenates in the liquid products increased in the order of PP > HDPE > PS. Physical characteristics such as oxygenates, water contents, acid values, and viscosity increased during the co-pyrolysis of plastics and JC in comparison to the liquid fractions obtained from the pyrolysis of pure plastics. Also, co-pyrolysis offered a reduction in calorific values.

5.4.8 Bitumen and Oxygenates Containing Materials

Toosi [117] examined co-pyrolysis of bitumen with oxygenates containing materials at low temperatures (below 400°C). It has been known that the quality of oil produced from bitumen is superior at low temperature (below 400°C). However, at these temperatures, the rate of oil production was too low for economical production. The study examined the co-pyrolysis of bitumen with various oxygenate materials coming from coal and biomass to improve reaction rates at lower temperatures. The results of this study are described in the Masters thesis by Toosi [117].

5.4.9 Biomass with Polylactic Acid

Cornelissen et al. [109] examined co-pyrolysis of biomass with polylactic acid (PLA) in order to reduce water content of resulting bio-oil. The flash co-pyrolysis of willow/PLA blends (10:1, 3:1, 1:1, 1:2) in a semicontinuous reactor showed synergistic interaction. A higher bio-oil yield and lower water content as a function of willow/PLA ratios were obtained. Among the tested blends, the 1:2 willow/PLA blend showed the most pronounced synergy, a reduction in the production of pyrolytic water of almost 28%, accompanied by an increase in more than 37% in the production of water-free bio-oil. Additionally, PLA showed to have a positive influence on the energetic value of the bio-oil produced and the resulting energy recuperation.

5.4.10 Biomass and Waste Plastic

Xu et al. [118] examined the thermal pyrolysis of different biomass (sawdust, straw) and plastic (PP, polyvinyl chloride) and synergistic effects of co-pyrolysis of biomass and plastics in TGA. Also, in fixed-bed reactor, the influence of plastic content on co-pyrolysis of biomass and plastic was evaluated. The produced bio-oil was analyzed using elemental analysis and GC–MS. The results showed the presence of significant synergy during co-pyrolysis of biomass and plastics. The effect was most pronounced in the co-pyrolysis process of sawdust and PP. For this mixture with 80% PP, the bio-oil yield was the highest and its value was considerably higher than those for separate pyrolysis. The results of elemental analysis and GC–MS analysis showed that the bio-oil produced from co-pyrolysis had a higher hydrogen content and its calorific value was equal to that of the crude oil equivalent.

Berrueco et al. [119] examined co-pyrolysis of sawdust and PE in a fluidized-bed reactor at five temperatures: 640°C, 685°C, 730°C, 780°C, and 850°C. The thermal degradation of pure PE produced no solid residue. The addition of biomass, however, generated char and changed product distribution and increased heating value of the product gas. The gas yield increased with temperature for both pure components as well as for mixtures. The main gases produced from co-pyrolysis at low temperatures were carbon monoxide, ethylene, carbon dioxide, propylene, butadiene, methane, and pentadiene. While at high temperatures, the gas composition changed drastically, the main gas being carbon monoxide (more than 33 wt%) and others being methane, ethylene, benzene, and hydrogen. The co-pyrolysis of biomass and HDPE produced a gas of greater heating value than the ones obtained with biomass alone, with a lower production of tars. PE helped conversion of tars into hydrogen and carbon monoxide.

The higher yield of CO in co-pyrolysis at different temperatures corresponds to higher sawdust conversion. The emission of CO at lower temperatures indicate an early degradation compared to the pyrolysis of pure sawdust, which coincides with a delay of the cracking of waxes of high molecular weight and a lower production of light compounds in the liquid fraction from PE pyrolysis. The increased production of gas and decreased generation of C_5–C_{19} fractions can be due to enhanced gas-phase reactions due to char. In the co-pyrolysis, a lower amount of oxygen compounds is formed, mostly aromatic alcohols, and there was an increase in the production of water. The production of liquid other than water was lower than expected. All of these results indicate a synergistic effect in the decomposition of these two materials during co-pyrolysis, which may be aided by the char produced from sawdust during co-pyrolysis.

Oyedun et al. [120] attempted a modeling approach for co-pyrolysis of biomass and plastic waste. The thermal behavior of plastics (PS) during pyrolysis is different from that of biomass because its decomposition happens at a high temperature range (>450°C) with sudden release of volatiles compared to biomass (bamboo waste), which has a wide range of thermal decomposition. The study analyzed two different modeling approaches to evaluate possible synergistic effect in terms of energy usage between PS and bamboo. The results of the energy usage showed a reduction in energy for the mixed blends for the two approaches. The second approach, which allowed interaction between two feedstock had a more reduction in overall energy up to 6.2% when the ration of PS feed was 25%.

Pinthong [121] examined co-pyrolysis of rice husk (RH), PE, and PP mixtures using TGA. The activation energies and pre-exponential factor for HDPE, LDPE, and PP waste plastics were in the ranges of 279–455 kJ/mol and 4.95×10^{19}–1.48×10^{31}/min, respectively. The similar data obtained from co-pyrolysis experiments were lower and their values were 221–317 kJ/mol and 2.11×10^{15}–7.18×10^{21}/min, respectively.

The study examined waste plastics mixed with RH at different concentrations of 30/70, 50/50, 70/30, and 100/0 and had synergies at 485°C of about 23%, 10.97%, 9.18%, and 7.15%, respectively. The waste plastic liquor (WPL)/RH mixture of 30/70 showed the highest synergy conversion of 1.46%. Main product from pyrolysis at 485°C for all mixtures was wax. The WPL/RH mixture of 30/70, which provided the highest synergy in conversion was chosen to be pyrolyzed with H-ZSM-5 catalyst. Liquid produced from this catalytic pyrolysis contained almost no wax and had a higher heating value than the wax product. The liquid product contained 54.54% gasoline, 24.24% kerosene, 15.15% diesel, and 16.06% by volume residue.

Abnisa et al. [122] examined co-pyrolysis of PS and palm shell to produce a high-grade pyrolysis oil. The study compared the results of pyrolysis of palm shell alone with that of 1:1 (by weight) mixture of palm shell and PS. Pyrolysis was carried out in a fixed-bed reactor at temperature of 500°C, nitrogen flow rate of 2 L/m, and reaction time of 60 min. The results showed that the final oil yield of palm shell pyrolysis was 46.13 wt%. By mixing palm shell with PS, the yield of oil increased to about 61.63%. In these experiments, the high heating value was low, 11.94 MJ/kg, for oil from pyrolysis of palm shell. By contrast, the high heating value was a high,

38.01 MJ/kg, for oil from pyrolysis of mixtures. The method also consumes waste that has to be otherwise discarded for a price.

5.4.11 BIOMASS AND SEWAGE SLUDGE

Wang et al. [123] examined the co-pyrolysis of sewage sludge with two different biomasses: oat straw and bark in a TGA. The reactivity of sewage sludge in the mixtures and stand-alone pyrolysis was measured using GC–MS system. The study mainly focused on the gas production during this co-pyrolysis. The paper only presented preliminary data.

Sanchez et al. [124] examined a mixture of sewage sludge with cattle manure in a laboratory and pilot plant reactors. The pyrolysis produced largely condensable and noncondensable gas and solids. The co-pyrolysis of the mixture was successful in producing gases that can be used for downstream combustion purposes. The gas contained some sulfur and chlorine impurities (although a larger portion was with the solids), which can be removed using sodium bicarbonate as neutralizing medium to remove HCl, HF, and SO_2 effectively.

5.4.12 BIOMASS AND CATTLE MANURE

Chueluecha and Duangchan [125] examined co-pyrolysis of corncob and cattle manure with corncob/cattle manure composition of 3:1, 1:1, 1:3 by weight in the temperature range of 350°C–450°C. The results showed that the decompositions of both materials and their mixture started at 200°C and individual corncob gave the highest bio-oil yield at 400°C, while the individual cattle manure gave the highest bio-oil yield at 450°C. The corncob/cattle manure ratio of 1:3 pyrolyzed at the temperature of 400°C gave the best positive synergistic effect for bio-oil yield and negative synergies for char and gas yields. Bio-oil increased with an increase in corncob content. Also, the individual cattle manure bio-oil has a pH value higher than that of individual corncob bio-oil and the pH value of corncob/cattle manure mixture bio-oil increases with an increase in the cattle manure concentration. The results indicated that co-pyrolysis of corncob/cattle manure mixture can improve the pH value of bio-oil with a maximum of 3.71 occurred at corncob/cattle manure ratio of 1:3.

5.4.13 WASTE TIRES WITH OIL PALM EMPTY FRUIT BUNCHES

Alias et al. [126] examined the co-pyrolysis if shredded waste tires (SWTs) and shredded oil palm empty fruit bunches (SOPEFBs) in a fixed-bed reactor at 500°C with ratio 1:1. SWT and impregnated SOPEFB with 10 wt% cobalt catalyst solution were also co-pyrolyzed in the same reactor. The study examined the effect of impregnated catalyst solution on the liquid yield of SOPEFB fibers. The heating rate for the reactor was 30°C/min and the reactor was maintained at the reaction temperature for about 2 h. Both co-pyrolysis produced solid, liquid, and gas as their products. The addition of cobalt catalyst gave about 33.3% increment in the liquid yield (from 42.8% to 57.06%) compared to the yield obtained in noncatalytic co-pyrolysis. Both co-pyrolysis experiments (thermal and catalytic) gave insignificant increase in the residue production compared to individual pyrolysis. The use of an impregnated catalyst was useful in improving the bio-oil yield.

5.4.14 POLYPROPYLENE AND HYDROLYTIC LIGNIN

Sharypov et al. [127] examined the co-pyrolysis of atactic PP and hydrolytic lignin for various compositions at 400°C in an autoclave under inert conditions. The experiments were performed to optimize the light (b.p. < 180°C) hydrocarbon streams in the product. The results indicated that the addition of lignin at 20%–40% in the mixture provided a substantial increase in the yield of light hydrocarbons. The yield of light hydrocarbons from PP increases almost threefold at the level (on mass basis) of about 55% at 30 wt% of lignin in the mixture. The analysis of the liquid products by ATR spectroscopy and GC–MS showed that these products are largely formed as a result of thermal decomposition of PP. The specific contribution from lignin in the co-pyrolysis with PP is expressed as an increase in the content of C_9 hydrocarbons and beta-olefins in the light fraction.

5.4.15 PE/WOOD WITH PE/PAPER

Grieco et al. [128] examined the effects of interaction between PE and other materials by co-pyrolyzing the PE/wood and PE/paper. The co-pyrolysis experiments of mixtures containing pellets composed of pure PE, wood, and paper with 25%, 50%, and 75% PE were carried out. The yields of tar, gas, and char were measured and compared with the amount predicted as linear combination of the yields obtained from pure materials. Generally, at high heating rates, both the mixtures did not show important effects of interactions between the materials. At low heating rates (0.1°C/s), the results were quite different for PE/wood and PE/paper. In the first case, the gas yield increased by only 20% with a small reduction in the tar amount. For the mixture containing PE and paper, an increase of char yield (until 60%) was observed; tar production was lower than expected and no significant variation in the gas yield. The higher char yields during PE/paper combination is believed to be due to adsorption of tar into solid structure and transformation of adsorbed tar into char. Some tests were also performed to check the ability of paper and wood to adsorb n-hexadecane. This compound was chosen to represent fragments of the PE polymeric chains. It was found that paper char is able to significantly convert n-hexadecane into carbon. The reason for this may be the presence of large amounts of inert materials, such as metal oxides, that can promote decomposition reactions of AHCs with formation of char.

5.4.16 PLASTICS, BIOMASS, AND TIRE WASTES

Bernardo et al. [129] examined char obtained from co-pyrolysis of plastics, biomass, and tire wastes. The chars were extracted with several organic solvents in order to assess their efficiency in reducing the organic load of the chars and therefore their toxicity. The results obtained in this study indicated that the more efficient extraction solvent to be used in the organic decontamination of chars obtained in the co-pyrolysis of plastics, used tires, and biomass waste was hexane. The compounds removed from the char during the decontamination process were mainly AHCs and aromatic hydrocarbons, chemicals that may be upgraded to be used as a fuel and/ or as raw material for the organic chemical industry. However, the char obtained after the organic decontamination had high amounts of Zn, which can be a problem concerning their safe reutilization and/or final disposal. The introduction of tires in the waste mixtures (the source of Zn) must be controlled in order to obtain less toxic chars. Other decontamination treatments of Zn must be evaluated and combined with organic solvent extraction.

5.5   HYDROTHERMAL CARBONIZATION (WET PYROLYSIS)

Hydrothermal carbonization (HTC) is a thermochemical conversion process to convert biomass into a solid, coal-like product in the presence of liquid water. This process is often called a wet or hydropyrolysis process, and it results in the production of "hydro-char," which has high carbon content and low oxygen content compared to original biomass. This is characterized as a multifuel process because water in this process is a reactant and acts as a second fuel (like oil) in the overall reaction process.

The chief advantage of HTC process over conventional pyrolysis process is that it can convert wet feedstock into carbonaceous material without having to remove water with an energy-intensive and expensive drying process. In fact, the process converts benign water into an active reactant. The potential feedstock that can be used for this process are wet animal manures, human waste, sewage sludges, municipal solid waste, aquaculture and algal residues, and many other wet energy crops or their mixtures. The process is of course most beneficial when biomass is accompanied by a large amount of water.

The solids produced from this process have been given many names such as char, bio-coal or bio-char, or more accurately "hydro-char" to differentiate it from the char or coal produced by the conventional dry pyrolysis. Significant reviews of "hydro-char" have been published recently [130–143] particularly on its production processes. The interest in "hydro-char" has increased very rapidly due to its connection to understand natural coal formation [133–135,139], its use in creating new innovative materials [136], and its application in soil quality improvement [137–143].

Unlike dry pyrolysis of multifuels, HTC can be an exothermic process, which lowers both oxygen and hydrogen content of the original feedstock by mainly dehydration and decarboxylation. The overall reaction identifying the heating value of the process can be expressed as [18–28]

$$C_6H_{12}O_5 \rightarrow C_{5.25}H_4O_{0.5} + 0.75CO_2 + 3H_2O \qquad (5.1)$$

The initial phase of this overall reaction, that is, hydrolysis of cellulose is an endothermic reaction [144]. As shown in Figure 5.1, this process is not as harsh as dry pyrolysis in the reduction of H/C and O/C ratios and it produces "coal-like" material, which can be similar to bituminous or subbituminous coals. Typical effects of residence time and temperature on selected feedstock such as cellulose, peat bog, and wood are illustrated by Libra et al. [130]. Generally, the process occurs in the temperature range of 180°C–220°C at saturated pressure and for the reaction conditions

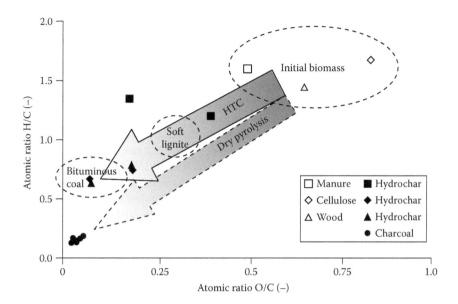

FIGURE 5.1 Comparison of H/C and O/C ratio variations for the dry and wet pyrolysis processes (HTC) for various feedstock. (From Libra, J. et al., *Biofuels*, 2, 89, 2011. With permission.)

that last for several hours. The process is accompanied by numerous reaction mechanisms such as hydrolysis, dehydration, decarboxylation, condensation polymerization, and aromatization, which are further discussed in the following sections. These are not consecutive but parallel reaction paths and the detailed nature of these mechanisms and their relative significance during the course of reaction primarily depend on the nature of the feedstock.

During HTC process, biomass/waste components are hydrolyzed to produce a large amount monomers and oligomers [130]. Simultaneously, water-soluble extractables are also produced. These monomers, oligomers, and extractables then further undergo dehydration, decarboxylation, and condensation reactions. Many intermediates such as hydroxy methyl furfural (HMF) are very reactive and some of them have high chemical values. These intermediates further undergo polymerization to produce humic acids, bitumen, and insoluble solids, some of which precipitate as HTC coal or hydro-char. Some components of biomass (e.g., crystalline cellulose) or oligomer cellulose do not hydrolyze under these reaction conditions. More details on the reaction mechanisms during HTC process are given below and in the published literature [130,132,137–142,145–147].

5.5.1 REACTION MECHANISMS

As mentioned before, HTC is accompanied by a series of chemical reactions such as hydrolysis, dehydration, decarboxylation, polymerization, and aromatization. These reaction mechanisms are briefly described below [130,132,136,148].

Hydrolysis reactions during HTC process mainly breaks down ether and ester bonds resulting in a wide range of products that include saccharides of cellulose and phenolic fragments of lignin. Along with other degradation mechanisms mentioned below, the intermediate products are also further hydrolyzed such as HMF converted to levulinic acid and formic acid. Hemicellulose is hydrolyzed around 180°C and cellulose is hydrolyzed above approximately 200°C. The detailed mechanism of cellulose hydrolysis is given by Peterson et al. [149]. While initial hydrolysis reactions are favored by the alkaline conditions, further degradation of glucose is accelerated by the acidic conditions. In a pH range of 3–7, the rate of reaction is largely independent of H+ and OH− concentrations. The hydrolysis of lignin occurs around 200°C and produces highly active low-molecular-weight substances. Some of these substances go through condensation reactions and precipitate from the solutions. In general, hydrolysis reactions are fast and transport limited. The structures produced from hemicellulose and lignin interacts with each other resulting in high solubility of aromatic structures. At high temperatures, condensation reactions are likely to occur.

Chemical dehydration of biomass generally results in the elimination of hydroxyl group and production of water. For example, the dehydration of cellulose follows [150]

$$4(C_6H_{10}O_5)_n \leftrightarrow 2(C_{12}H_{10}O_5)_n + 10H_2O \qquad (5.2)$$

The rate of decarboxylation versus dehydration is generally measured by the factor $F = $ mole of CO_2/mole of H_2O, which varies from 0.2 for cellulose and 1 for lignite. Condensation of fragments can also regenerate water during HTC process.

HTC process results in a partial elimination of carboxyl groups, which produces CO_2 and CO above 150°C [130,132,135,151]. Generally, CO_2 is produced from carboxyl groups and CO is produced from carbonyl group. One likely source for CO_2 is formic acid, which is formed in a significant amount by degradation of cellulose. CO_2 can also be produced by condensation reactions, cleavage of intramolecular bonds, and destruction of oxidized molecules at high temperatures.

In an HTC process, intermediates that are created by dehydration and decarboxylation reactions are highly active and can polymerize to produce larger molecules. Condensation reactions are also accounted for the productions of CO_2. The rate of carbonization is increasingly determined by stearic influences with a higher condensation degree of aromatics [130,132,133,135]. Thus, condensation polymerization is the main reason for the formation of bio-coal in HTC process. The condensation polymerization is most likely governed by the step growth polymerization [130,132,133,135].

Cellulosic structures are capable of forming aromatic structures under hydrothermal conditions [130,132,133,136,151]. Aromatic structures show high stability under hydrothermal reaction conditions and may be considered as a basic building block of HTC coal. Alkaline conditions favor aromatization. Cross-linking condensation of aromatic rings also make up major constituents of HTC coal. A large number of aromatic bonds reduce the effects of HTC process on the carbon content. High temperature and residence time favor aromatization. Cellulose aromatizes most in the temperature range of 200°C–300°C.

Besides the reaction mechanisms mentioned earlier, certain transformation reactions for crystalline structures in cellulose or certain oligomers are also possible. Their contribution at temperatures below 200°C appears to be small [132–134,139]. Demethylation has been used to explain conversion of phenolic structure to catechol-like structure in HTC coal. The production of a small amount of methane substantiates this hypothesis. At temperatures higher than 200°C, pyrolytic reactions may also compete with the mechanisms mentioned before, although significant amount of tar and CO (major products of pyrolysis) have not been found. Finally, FT-type reactions may also occur during HTC process.

In general, wet pyrolysis is more effective on cellulose than lignin. The literature data [130,132,133] show that during HTC of lignin, the decline in H/C ratio is not as severe as that of cellulose or even wood. The decline in H/C ratio of wood is in between that of lignin and cellulose. Wet pyrolysis will be equally effective on mixed biomass/waste as long as cellulose content of the mixed fuel is high.

5.5.2 Effects of Operating Conditions

A number of operating conditions such as (1) water and solids concentrations, (2) feed slurry pH value, (3) reaction temperature and pressure, (4) reaction residence time, and (5) nature of feed stock affect the product distributions. These effects are well described in the literature [130,132–143,145–158]. Here, we briefly summarize these literature results.

HTC process and formation of hydro-char requires the presence of water. It is known that the biomass above the water surface does not carbonize; although only small amount of water is necessary. The process of carbonization is accelerated by water because of the active role played by water as solvent, reactant, and catalyst during various steps of biomass degradation and subsequent condensation and aromatization processes. Water helps thermally driven pyrolysis. The initial step of hydrolysis is very important and the role of water for this step increases with an increase in temperature. Water facilitates the condensation polymerization of active intermediate species and also dissolves numerous compounds formed during HTC process. The amount of water can also affect the transport of fragments from the influence of reactive centers. Generally, very low concentration of biomass in water may result in very low production of precipitated carbonized solids since most biomass may be dissolved. On the other hand, excessive biomass may result in some unreacted organic materials. Generally, the rise in feed solid concentration increases the monomer concentration in the liquid phase. The key is to optimize the effect of residence time–solid concentration interplay on the extent of polymerization reaction. For each feedstock, there will be an optimum solids concentration to achieve highest yield of carbonized solids.

It is known that during HTC, pH drops because of the formations of acetic, formic, lactic, and levulinic acids [130,132,133,136]. It is also known that natural coalification requires neutral to weak acidic environment [130,132,133,136]. The effects of nature and quantity of acids and bases on the product characteristics are well described in many of the studies mentioned earlier. In general, high pH values result in the product with high H/C ratio, hydrolysis reactions are favored by the acidic conditions, and a weak acidic condition improves the overall rate of HTC [130,133–143,145–158].

Depending upon the nature of the feedstock, exothermal effect during hydrothermal operation can occur at as low as 100°C temperature (e.g., for peat). Temperature is the most important parameter in HTC process. It is known that hydrolysis of glucose with subsequent dehydration may take several seconds at 270°C but it will take up to several hours at 150°C. The rate of polymerization is also temperature dependent. The temperature has a definite influence on the nature of biomass that can be hydrolyzed. Hemicellulose hydrolyzes at 180°C and lignin at 200°C but cellulose hydrolysis requires 220°C [130,132,133,136,151]. Pyrolytic reactions also become more important at higher temperatures.

The temperature also affects all the physical and chemical properties that are important in the HTC process. The viscosity of water is also deceased by twofold when the temperature is increased from room temperature to around 350°C. Lower viscosity helps penetration of water in porous biomass media. While pressure is a more expensive process variable, it does not affect HTC process as significantly as temperature. While both dehydration and decarboxylation reactions are suppressed with an increase in pressure, this, however, does not significantly affect overall HTC. An increase in pressure facilitates (1) the removal of extractable, (2) solubilization of compressed gases and physical compacting of bio-coal, and (3) hydrogen ion transfer and condensation polymerization between solids allowing the use of higher biomass/water ratio.

Since overall HTC process is a slow process, higher residence time is important (from hours to days). In fact, HTC coal yield increases with an increase in residence time. While initial extraction and hydrolysis are fast reactions, overall HTC process is slow. Generally, it is believed that HTC reaction mechanism is rate limiting. Both diffusion-controlled transport mechanisms during biomass degradation and condensation polymerization govern the overall rate of reaction.

Finally, feedstock characteristics such as chemical composition, volatile and noncombustible fractions, moisture content, particle size, and energy content significantly affect conversion efficiency and char characteristics. These effects are well described in the several reported studies [130,132–143,145–158].

5.5.3   PRODUCT CHARACTERISTICS AND USAGES

HTC process largely produces hydro-char (solid) and liquid with dissolved nutrients. The fate of heavy metals and organic chemicals (present in the original feedstock or created by the reactions) is generally not known and must be traced during the process.

Wet pyrolysis cannot destroy heavy metals. Since they have a toxic risk potential, their fate need to be followed. If they accumulate in the solid char, which is subsequently used as soil nutrient, they can affect the food chain. Generally, except for zinc, heavy metal concentrations in char do not exceed the allowable limits [130,132,133,136,150–158].

Just like for heavy metals, a systematic knowledge of the fate of organic compounds during HTC process is not well understood. Unlike heavy metals, new organic compounds can be formed during condensation, polymerization, and aromatization reactions. The fate of compounds such as polychlorinated biphenyls (PCB) and hexa chlorobenzene (HCB) needs to be particularly followed along with all other organic chemicals during HTC process.

Both for animal manures and sewage sludge, hydro-char retains a significant level of calcium, potassium, and phosphorous. pH affects the mobility and sorption capability of the nutrients, particularly for the case of phosphorous. In the HTC process, dissolution of water-soluble minerals can be significant [130,132,133,136,150–158]; however, the nutrient content will also depend on the technique for dewatering the solid conversion product. The ratio between evaporation and dewatering governs the amount of plant nutrients that will be adsorbed or retained at the hydro-char interface. Nutrient retention should be an important parameter in the detailed process design.

Generally, an increase in temperature decreases the hydro-char yield and increases the yield of liquids and gases such as CO_2, CO, and H_2. An increase in temperature also decreases H/C and O/C ratios in the hydro-char. The maximum allowable yields (which may be obtained at very large residence time) for various feedstock are illustrated by Libra et al. [130] and others [132,133,136]. A lower biomass solid concentration (i.e., high water concentration) generally gives a lower hydro-char yield. In general, 60%–84% of the biomass carbon remains in the hydro-char. The char composition is mainly affected by the nature of the feedstock, temperature, and the reaction time. For all feedstock, an increase in temperature increases the carbon content and decreases the oxygen content of the hydro-char.

Hydro-char is often used for soil application as a fertilizer or a carbon sequester. Some discussion in this regard is given by Libra et al. [130] and others [150–158]. There are evidences that char remains in the soils for a very long period. The movement of char or its components to other ecosystems may have significant effect on its stability; the rate of mineralization in deeper layers may be insignificant compared with that of top-soils. Hydro-char promotes the fungal growth and soil aggregation [130,150–158]. It will very likely reduce the tensile strength, increase the hydraulic conductivity, and enhance the soil water-holding capacity. Hydro-chars are more acidic than bio-chars coming from dry pyrolysis. Hydro-char do undergo aging process, which can change the functional groups and thereby its effectiveness as nutrient. The hydro-char created by HTC process can also find applications as (1) activated carbon adsorbents, (2) generation of nanostructured materials, (3) catalysts support or as catalysts, (4) CO_2 sorption materials, and (5) energy production and storage materials. These applications of hydro-char are described in more details in an excellent review by Libra et al. [130].

HTC process is accompanied by a large number of intermediate products due to complex reaction mechanism. The solids coming out of HTC process represents an agglomerate of chemical substances. An elemental analysis of the hydro-char (or HTC coal) shows that it may approach lignite or even subbituminous coal depending on the reaction severity. An exception is resin whose H/C ratio remains unaffected. As mentioned earlier in Sections 5.5.1 and 5.5.2, HTC coal from lignin tends to have a lower hydrogen content and coal from cellulose tends to achieve higher carbon content [130,132,133,136]. HTC coal is soluble in benzol–alcohol mixtures, alkaline solutions, and ammonia [130]. The skeletal of HTC coal is very similar to natural coal, although HTC coal exhibits a higher amount of functional groups compared to natural bituminous coal. The removal of hydroxyl and carboxyl groups during HTC process makes the HTC coal with a lower hydrophobicity than the original materials [130,132,133,136]. While the inorganics largely remain in HTC coal, their relation with process conditions is not well known. While HTC coal has a small surface area, this area can be increased significantly (by two orders of magnitude) by removing extractables or by thermal treatment [130]. Observations of nanostructure of HTC coal reveal its potential technical applications as functionalized carbonaceous materials. As shown by Libra et al. [130] and others [150–158], with the use of proper starting materials and appropriate catalyst/template assisted treatment methods, various types of nanospheres, nanocables, nanofibers, micro-cables, submicro-tubes, and porous structures can be created from HTC coal.

5.5.4 PROCESS CONSIDERATIONS

The HTC process generates more water than carbon dioxide. Water acts as solvent and reactant and therefore carries a significant amount of inorganics and organics, many of which can be valuable chemicals. The solids in water can cause problems upon precipitation due to condensation or polymerization reactions. The wastewater of HTC process can be processed with aerobic or anaerobic treatment to lower its total organic content. Some inorganics in water may be good nutrients for soil. The gases coming out of HTC process mainly contain carbon dioxide with minor CO, CH_4, and H_2 as well as traces of C_mH_n. The dissolution of carbon dioxide in the liquid phase may affect the critical temperature condition in the process. An increase in temperature generally decreases CO and increases H_2 and CH_4. A high heating value of HTC coal product may require removal of oxygen from feed material by decarboxylation process. High carbon efficiency, however, requires low decarboxylation. Thus, the process should be optimized based on the end use of the final product. Monitoring of carbon dioxide may be a way to monitor the progress of the HTC process because decarboxylation is one of the major changes in the feedstock during HTC process [130,132,133,136].

In HTC process, the ratio of biomass to water should be kept as high as possible to enhance polymerization. Less water will also give less energy loss and less pumping costs for total throughput. The feedstock can be submerged in the water by mechanical compacting device to take advantage of the best reaction conditions. The residence time should be as large as possible to get complete reaction to occur and minimize loss or organics in wastewater. A recirculation of water is one way to achieve this objective. The reaction conditions should take advantages of the possible effects of organic acids; they may give faster polymerization and higher ash content of the produced HTC coal. Higher temperature accelerates the process and gives higher carbon content in the HTC product. High pressure required with high temperature may be expensive. Pressure–temperature relationship should be optimized based on the intended use of the end product. Since hydrolysis is fast and a diffusion-controlled reaction, small particle size of the feedstock may be beneficial. This, however, increases the energy demand and investment cost [130,132,133,136].

REFERENCES

1. Lee, S. and Shah, Y. *Biofuels and Bioenergy: Technologies and Processes*, CRC Press, Taylor & Francis Group, New York (September 2012).
2. Shah, Y. *Water for Energy and Fuel Production*, CRC Press, Taylor & Francis Group, New York (May 2014).
3. Tchapda, A. and Pisupati, S. A review of thermal co-conversion of coal and biomass/waste. *Energies*, 7(3), 1096–1148 (2014).
4. Yang, H., Yan, R., Chen, H., Lee, D.H., and Zheng, C. Characteristics of hemicellulose, cellulose and lignin pyrolysis. *Fuel*, 86, 1781–1788 (2007).
5. Raveendran, K., Ganesh, A., and Khilar, K.C. Pyrolysis characteristics of biomass and biomass components. *Fuel*, 75, 987–998 (1996).
6. Couhert, C., Commandre, J.-M., and Salvador, S. Is it possible to predict gas yields of any biomass after rapid pyrolysis at high temperature from its composition in cellulose, hemicellulose and lignin? *Fuel*, 88, 408–417 (2009).
7. Shafizadeh, F. Thermal conversion of cellulosic materials for fuel and chemicals. In *Wood and Agricultural Residues: Research on Use for Feed, Fuels, and Chemicals*, Soltes, E.J. (Ed.). Academic Press, Inc., New York (1983).
8. Zanzi, R., Sjöström, K., and Björnbom, E. Rapid pyrolysis of agricultural residues at high temperature. *Biomass Bioenergy*, 23, 357–366 (2002).
9. Kaloustian, J., Pauli, A., and Pastor, J. Decomposition of bio-polymers of some Mediterranean plants during heating. *J. Therm. Anal. Calorim.*, 61, 13–21 (2000).
10. Ahmaruzzaman, M. and Sharma, D.K. Coprocessing of petroleum vacuum residue with plastics, coal, and biomass and its synergistic effects. *Energ. Fuels*, 21, 891–897 (2007).
11. Lapuerta, M., Hernández, J.J., Pazo, A., and López, J. Gasification and co-gasification of biomass wastes: Effect of the biomass origin and the gasifier operating conditions. *Fuel Process. Technol.*, 89, 828–837 (2008).
12. Haykiri-Acma, H. and Yaman, S. Synergy in devolatilization characteristics of lignite and hazelnut shell during co-pyrolysis. *Fuel*, 86, 373–380 (2007).
13. Suelves, I., Lázaro, M.J., and Moliner, R. Synergetic effects in the co-pyrolysis of samca coal and a model aliphatic compound studied by analytical pyrolysis. *J. Anal. Appl. Pyrolysis*, 65, 197–206 (2002).
14. Zhang, L., Xu, S., Zhao, W., and Liu, S. Co-pyrolysis of biomass and coal in a free fall reactor. *Fuel*, 86, 353–359 (2007).
15. Haykiri-Acma, H. and Yaman, S. Combinations of synergistic interactions and additive behavior during the co-oxidation of chars from lignite and biomass. *Fuel Process. Technol.*, 89, 176–182 (2008).
16. Sonobe, T., Worasuwannarak, N., and Pipatmanomai, S. Synergies in co-pyrolysis of Thai lignite and corncob. *Fuel Process. Technol.*, 89, 1371–1378 (2008).
17. Edreis, E.M.A., Luo, G., Li, A., Xu, C., and Yao, H. Synergistic effects and kinetics thermal behaviour of petroleum coke/biomass blends during H_2O co-gasification. *Energy Convers. Manage.*, 79, 355–366 (2014).
18. Krerkkaiwan, S., Fushimi, C., Tsutsumi, A., and Kuchonthara, P. Synergetic effect during co-pyrolysis/gasification of biomass and sub-bituminous coal. *Fuel Process. Technol.*, 115, 11–18 (2013).
19. Park, D.K., Kim, S.D., Lee, S.H., and Lee, J.G. Co-pyrolysis characteristics of sawdust and coal blend in TGA and a fixed bed reactor. *Bioresour. Technol.*, 101, 6151–6156 (2010).
20. Wei, L.G., Zhang, L., and Xu, S.P. Effects of feedstock on co-pyrolysis of biomass and coal in a free-fall reactor. *J. Fuel Chem. Technol.*, 39, 728–734 (2011).

21. Blesa, M.J., Miranda, J.L., Moliner, R., Izquierdo, M.T., and Palacios, J.M. Low-temperature co-pyrolysis of a low-rank coal and biomass to prepare smokeless fuel briquettes. *J. Anal. Appl. Pyrolysis*, 70, 665–677 (2003).
22. Sjostrom, K., Chen, G., Yu, Q., Brage, C., and Rosen, C. Promoted reactivity of char in co-gasification of biomass and coal: Synergies in the thermochemical process. *Fuel*, 78, 1189–1194 (1999).
23. Cordero, T., Rodriguez-Mirasol, J., Pastrana, J., and Rodriguez, J.J. Improved solid fuels from co-pyrolysis of a high-sulphur content coal and different lignocellulosic wastes. *Fuel*, 83, 1585–1590 (2004).
24. Haykiri-Acma, H. and Yaman, S. Interaction between biomass and different rank coals during co-pyrolysis. *Renew. Energ.*, 35, 288–292 (2010).
25. Yuan, S., Dai, Z.-H., Zhou, Z.-J., Chen, X.-L., Yu, G.-S., and Wang, F.-C. Rapid co-pyrolysis of rice straw and a bituminous coal in a high-frequency furnace and gasification of the residual char. *Bioresour. Technol.*, 109, 188–197 (2012).
26. Xu, C., Hu, S., Xiang, J., Zhang, L., Sun, L., Shuai, C., Chen, Q., He, L., and Edreis, E.M.A. Interaction and kinetic analysis for coal and biomass co-gasification by TG–FTIR. *Bioresour. Technol.*, 154, 313–321 (2014); *Energies*, 7, 1136 (2014).
27. Pedersen, L.S., Nielsen, H.P., Kiil, S., Hansen, L.A., Dam-Johansen, K., Kildsig, F., Christensen, J., and Jespersen, P. Full-scale co-firing of straw and coal. *Fuel*, 75, 1584–1590 (1996).
28. Coughlin, R.W. and Davoudzadeh, F. Coliquefaction of lignin and bituminous coal. *Fuel*, 65, 95–106 (1986).
29. Howaniec, N. and Smoliński, A. Steam co-gasification of coal and biomass—Synergy in reactivity of fuel blends chars. *Int. J. Hydrogen Energ.*, 38, 16152–16160 (2013).
30. Feng, Z., Zhao, J., Rockwell, J., Bailey, D., and Huffman, G. Direct liquefaction of waste plastics and coliquefaction of coal-plastic mixtures. *Fuel Process. Technol.*, 49, 17–30 (1996).
31. Mastral, A.M., Murillo, R., Callén, M.S., and Garcia, T. Evidence of coal and tire interactions in coal–tire coprocessing for short residence times. *Fuel Process. Technol.*, 69, 127–140 (2001).
32. Mastral, A.M., Mayoral, M.C., Murillo, R., Callen, M., Garcia, T., Tejero, M.P., and Torres, N. Evaluation of synergy in tire rubber-coal coprocessing. *Ind. Eng. Chem. Res.*, 37, 3545–3550 (1998).
33. Taghiei, M.M., Feng, Z., Huggins, F.E., and Huffman, G.P. Coliquefaction of waste plastics with coal. *Energ. Fuels*, 8, 1228–1232 (1994).
34. Kanno, T., Kimura, M., Ikenaga, N., and Suzuki, T. Coliquefaction of coal with polyethylene using Fe(CO)5-S as catalyst. *Energ. Fuels*, 14, 612–617 (2000).
35. Jones, J.M., Kubacki, M., Kubica, K., Ross, A.B., and Williams, A. Devolatilisation characteristics of coal and biomass blends. *J. Anal. Appl. Pyrolysis*, 74, 502–511 (2005).
36. Biagini, E., Lippi, F., Petarca, L., and Tognotti, L. Devolatilization rate of biomasses and coal–biomass blends: An experimental investigation. *Fuel*, 81, 1041–1050 (2002).
37. Meesri, C. and Moghtaderi, B. Lack of synergetic effects in the pyrolytic characteristics of woody biomass/coal blends under low and high heating rate regimes. *Biomass Bioenergy*, 23, 55–66 (2002).
38. Moghtaderi, B., Meesri, C., and Wall, T.F. Pyrolytic characteristics of blended coal and woody biomass. *Fuel*, 83, 745–750 (2004).
39. Vamvuka, D., Pasadakis, N., Kastanaki, E., Grammelis, P., and Kakaras, E. Kinetic modeling of coal/agricultural by-product blends. *Energ. Fuels*, 17, 549–558 (2003).
40. Vuthaluru, H.B. Thermal behaviour of coal/biomass blends during co-pyrolysis. *Fuel Process. Technol.*, 85, 141–155 (2004).

41. Lu, K.-M., Lee, W.-J., Chen, W.-H., and Lin, T.-C. Thermogravimetric analysis and kinetics of co-pyrolysis of raw/torrefied wood and coal blends. *Appl. Energ.*, 105, 57–65 (2013).

42. Kirtania, K. and Bhattacharya, S. Pyrolysis kinetics and reactivity of algae–coal blends. *Biomass Bioenergy*, 55, 291–298 (2013).

43. Li, S., Chen, X., Wang, L., Liu, A., and Yu, G. Co-pyrolysis behaviors of saw dust and Shenfu coal in drop tube furnace and fixed bed reactor. *Bioresour. Technol.*, 148, 24–29 (2013).

44. Sadhukhan, A.K., Gupta, P., Goyal, T., and Saha, R.K. Modelling of pyrolysis of coal–biomass blends using thermogravimetric analysis. *Bioresour. Technol.*, 99, 8022–8026 (2008).

45. Pan, Y.G., Velo, E., and Puigjaner, L. Pyrolysis of blends of biomass with poor coals. *Fuel*, 75, 412–418 (1996).

46. Aboyade, A.O., Görgens, J.F., Carrier, M., Meyer, E.L., and Knoetze, J.H. Thermogravimetric study of the pyrolysis characteristics and kinetics of coal blends with corn and sugarcane residues. *Fuel Process. Technol.*, 106, 310–320 (2013).

47. Li, S., Chen, X., Liu, A., Wang, L., and Yu, G. Study on co-pyrolysis characteristics of rice straw and Shenfu bituminous coal blends in a fixed bed reactor. *Bioresour. Technol.*, 155, 252–257 (2014).

48. Gil, M.V., Casal, D., Pevida, C., Pis, J.J., and Rubiera, F. Thermal behaviour and kinetics of coal/biomass blends during co-combustion. *Bioresour. Technol.*, 101, 5601–5608 (2010).

49. Pan, Y.G., Velo, E., Roca, X., Manya, J.J., and Puigjaner, L. Fluidized-bed co-gasification of residual biomass/poor coal blends for fuel gas production. *Fuel*, 79, 1317–1326 (2000).

50. Kumabe, K., Hanaoka, T., Fujimoto, S., Minowa, T., and Sakanishi, K. Co-gasification of woody biomass and coal with air and steam. *Fuel*, 86, 684–689 (2007).

51. McIlveen-Wright, D.R., Pinto, F., Armesto, L., Caballero, M.A., Aznar, M.P., Cabanillas, A., Huang, Y., Franco, C., Gulyurtlu, I., and McMullan, J.T. A comparison of circulating fluidised bed combustion and gasification power plant technologies for processing mixtures of coal, biomass and plastic waste. *Fuel Process. Technol.*, 87, 793–801 (2006).

52. Franco, C., Pinto, F., Andre, R., Tavares, C., Dias, M., Gulyurtlu, I., and Cabria, I. Experience using INETI experimental facilities in co-gasification of coal and wastes. In *Proceedings of the Workshop Co-processing of Different Waste Materials with Coal for Energy*, Lisbon, Portugal, pp. 238–247 (November 23, 2001).

53. Collot, A.G., Zhuo, Y., Dugwell, D.R., and Kandiyoti, R. Co-pyrolysis and co-gasification of coal and biomass in bench-scale fixed-bed and fluidised bed reactors. *Fuel*, 78, 667–679 (1999).

54. Storm, C., Rudiger, H., Spliethoff, H., and Hein, K.R.G. Co-pyrolysis of coal/biomass and coal/sewage sludge mixtures. *J. Eng. Gas Turbines Power*, 121, 55–63 (1999).

55. Vamvuka, D., Kakaras, E., Kastanaki, E., and Grammelis, P. Pyrolysis characteristics and kinetics of biomass residuals mixtures with lignite. *Fuel*, 82, 1949–1960 (2003).

56. Vuthaluru, H.B. Investigations into the pyrolytic behaviour of coal/biomass blends using thermogravimetric analysis. *Bioresour. Technol.*, 92, 187–195 (2004).

57. Idris, S.S., Rahman, N.A., Ismail, K., Alias, A.B., Rashid, Z.A., and Aris, M.J. Investigation on thermochemical behaviour of low rank Malaysian coal, oil palm biomass and their blends during pyrolysis via thermogravimetric analysis (TGA). *Bioresour. Technol.*, 101, 4584–4592 (2010).

58. Hayashi, J., Mizuta, H., Kusakabe, K., and Morooka, S. Flash copyrolysis of coal and polyolefin. *Energ. Fuels*, 8, 1353–1359 (1994).

59. Onay, O., Bayram, E., and Kockar, O.M. Copyrolysis of Seyitömer lignite and safflower seed: Influence of the blending ratio and pyrolysis temperature on product yields and oil characterization. *Energ. Fuels*, 21, 3049–3056 (2007).

60. Gao, C., Vejahati, F., Katalambula, H., and Gupta, R. Co-gasification of biomass with coal and oil sand coke in a drop tube furnace. *Energ. Fuels*, 24, 232–240 (2009).

61. Ulloa, C.A., Gordon, A.L., and García, X.A. Thermogravimetric study of interactions in the pyrolysis of blends of coal with radiata pine sawdust. *Fuel Process. Technol.*, 90, 583–590 (2009).

62. Kubacki, M.L. Co-pyrolysis and co-combustion of coal and biomass, PhD thesis. University of Leeds, Leeds, U.K. (2007).

63. Chen, G., Yu, Q., Brage, C., and Sjostrom, K. Co-gasification of birch wood and Daw Mill coal in pressurized fluidized bed gasifier: The investigation into the synergies in the process. In *Proceedings of the First World Conference on Biomass for Energy and Industry*, Sevilla, Spain, pp. 1545–1548 (June 5–9, 2000).

64. Squir, K.R., Solomon, P.R., Carangelo, R.M., and DiTaranto, M.B. Tar evolution from coal and model polymers: 2. The effects of aromatic ring sizes and donatable hydrogens. *Fuel*, 65, 833–843 (1986).

65. Miura, K., Mae, K., Asaoka, S., Yoshimura, T., and Hashimoto, K. A new coal flash pyrolysis method utilizing effective radical transfer from solvent to coal. *Energ. Fuels*, 5, 340–346 (1991).

66. Miura, K., Mae, K., Sakurada, K., and Hashimoto, K. Flash pyrolysis of coal swollen by tetralin vapor. *Energ. Fuels*, 7, 434–435 (1993).

67. Miura, K., Mae, K., Yoshimura, T., Masuda, K., and Hashimoto, K. Mechanism of radical transfer during the flash pyrolysis of solvent-swollen coal. *Energ. Fuels*, 5, 803–808 (1991).

68. Lee, C., Scaroni, A., and Jenkins, R. Effect of pressure on the devolatilization and swelling behavior of a softening coal during rapid heating. *Fuel*, 70, 957–965 (1991).

69. Sun, C., Xiong, Y., Liu, Q., and Zhang, M. Thermogravimetric study of the pyrolysis of two Chinese coals under pressure. *Fuel*, 76, 639–644 (1997).

70. Seebauer, V., Petek, J., and Staudinger, G. Effects of particle size, heating rate and pressure on measurement of pyrolysis kinetics by thermogravimetric analysis. *Fuel*, 76, 1277–1282 (1997).

71. Gibbins, J.R., Gonenc, Z.S., and Kandiyoti, R. Pyrolysis and hydropyrolysis of coal: Comparison of product distributions from a wire-mesh and a hot-rod reactor. *Fuel*, 70, 621–626 (1991).

72. Pindoria, R.V., Megaritis, A., Herod, A.A., and Kandiyoti, R. A two-stage fixed-bed reactor for direct hydrotreatment of volatiles from the hydropyrolysis of biomass: Effect of catalyst temperature, pressure and catalyst ageing time on product characteristics. *Fuel*, 77, 1715–1726 (1998).

73. Pindoria, R.V., Lim, J.-Y., Hawkes, J.E., Lazaro, M.-J., Herod, A.A., and Kandiyoti, R. Structural characterization of biomass pyrolysis tars/oils from eucalyptus wood waste: Effect of H_2 pressure and sample configuration. *Fuel*, 76, 1013–1023 (1997).

74. Yu, J., Lucas, J.A., and Wall, T.F. Formation of the structure of chars during devolatilization of pulverized coal and its thermoproperties: A review. *Prog. Energ. Combust. Sci.*, 33, 135–170 (2007); *Energies*, 7, 1140 (2014).

75. Suuberg, E.M. Rapid pyrolysis and hydropyrolysis of coal, PhD thesis. Massachusetts Institute of Technology, Cambridge, MA (1978); Lee, C.W., Jenkins, R.G., and Schobert, H.H. Structure and reactivity of char from elevated pressure pyrolysis of Illinois No. 6 bituminous coal. *Energ. Fuels*, 6, 40–47 (1992).

76. Wall, T.F., Liu, G., Wu, H., Roberts, D.G., Benfell, K.E., Gupta, S., Lucas, J.A., and Harris, D.J. The effects of pressure on coal reactions during pulverised coal combustion and gasification. *Prog. Energy Combust. Sci.*, 28, 405–433 (2002).

77. De Jong, W.B., Andries, J., and Hein, K.R.G. Coal/biomass co-gasification in a pressurised fluidised bed reactor. *Renew. Energ.*, 16, 1110–1113 (1999).

78. McLendon, T.R., Lui, A.P., Pineault, R.L., Beer, S.K., and Richardson, S.W. High-pressure co-gasification of coal and biomass in a fluidized bed. *Biomass Bioenergy*, 26, 377–388 (2004).

79. Lee, C., Jenkins, R., and Schobert, H. Structure and reactivity of char from elevated pressure pyrolysis of Illinois number 6 bituminous coal. *Energ. Fuels*, 6, 40–47 (1992).

80. Sha, X.-Z., Chen, Y.-G., Cao, J., Yang, Y.-M., and Ren, D.-Q. Effects of operating pressure on coal gasification. *Fuel*, 69, 656–659 (1990).

81. Mühlen, H.J., van Heek, K.H., and Jüntgen, H. Influence of pretreatment temperature and pressure on the char reactivity during hydrogasification. *Fuel*, 65, 591–593 (1986).

82. Cetin, E., Moghtaderi, B., Gupta, R., and Wall, T.F. Influence of pyrolysis conditions on the structure and gasification reactivity of biomass chars. *Fuel*, 83, 2139–2150 (2004).

83. Cetin, E., Gupta, R., and Moghtaderi, B. Effect of pyrolysis pressure and heating rate on radiata pine char structure and apparent gasification reactivity. *Fuel*, 84, 1328–1334 (2005).

84. Aznar, M.P., Caballero, M.A., Sancho, J.A., and Frances, E. Plastic waste elimination by co-gasification with coal and biomass in fluidized bed with air in pilot plant. *Fuel Process. Technol.*, 87, 409–420 (2006).

85. Andre, R.N., Pinto, F., Franco, C., Dias, M., Gulyurtlu, I., Matos, M.A.A., and Cabrita, I. Fluidised bed co-gasification of coal and olive oil industry wastes. *Fuel*, 84, 1635–1644 (2005); *Energies*, 7, 1134 (2014).

86. Cascarosa, E., Gasco, L., García, G., Gea, G., and Arauzo, J. Meat and bone meal and coal co-gasification: Environmental advantages. *Resour. Conserv. Recycl.*, 59, 32–37 (2012).

87. Pinto, F., Franco, C., Andre, R.N., Tavares, C., Dias, M., Gulyurtlu, I., and Cabrita, I. Effect of experimental conditions on co-gasification of coal, biomass and plastics wastes with air/steam mixtures in a fluidized bed system. *Fuel*, 82, 1967–1976 (2003).

88. Priyadarsan, S., Annamalai, K., Sweeten, J.M., Holtzapple, M.T., and Mukhtar, S. Co-gasification of blended coal with feedlot and chicken litter biomass. *Proc. Combust. Inst.*, 30, 2973–2980 (2005).

89. Miller, R.S. and Bellan, J. A generalized biomass pyrolysis model based on superimposed cellulose, hemicelluloses and lignin kinetics. *Combust. Sci. Technol.*, 126, 97–137 (1997).

90. Kastanaki, E. and Vamvuka, D. A comparative reactivity and kinetic study on the combustion of coal–biomass char blends. *Fuel*, 85, 1186–1193 (2006).

91. Morgan, T.J. and Kandiyoti, R. Pyrolysis of coals and biomass: Analysis of thermal breakdown and its products. *Chem. Rev.*, 114, 1547–1607 (2014).

92. Shan, G. A comprehensive model for coal devolatilisation, PhD thesis. The University of Newcastle, Newcastle, New South Wales, Australia (2000).

93. Kriz, V. and Brozova, Z. Co-pyrolysis of coal/waste polymer mixtures. *Acta Geodyn. Geomater.*, 4(2) (146), 39–42 (2007).

94. Aboyade, A., Osibote, A., Rabiu, A., Carrier, M., and Gorgens, J. Co-pyrolysis of coal and agricultural waste. In *The International Conference on Chemistry and Chemical Engineering, IPCBEE*, Vol. 38, Jeju Island, South Korea. IACSIT Press, Singapore, pp. 88–92 (2012).

95. Nyendu, G., Agblevor, F., and Battaglia, F. Investigation of coal–biomass co-pyrolysis and co-gasification. A report from Department of Biological Engineering, Utah State University, Logan, UT (2008).

96. Sonobe, T., Pipatmanomai, S., and Worasuwannarak, N. Pyrolysis characteristics of Thai lignite and biomass blends: An in-depth experimental investigation. In *The Second Joint International Conference on Sustainable Energy and Environment* (SEE 2006), Bangkok, Thailand, p. 6 (November 21–23, 2006).

97. Onay, O. The effect of pyrolysis temperature on co-pyrolysis of lignite and pistachio seed in a fixed-bed reactor. A report from Anadolu University, Porsuk Meslek Yuksekokulu, Eskisehir, Turkey, p. 4 (2011).

98. Niu, Y., Tan, H., Wang, X., and Xu, T. Synergistic effect on co-pyrolysis of capsicum stalks and coal. *African J. Biotechnol.*, 10 (2), 174–179 (January 2011).

99. Sakurovs, R. Interactions between coking coals and plastics during co-pyrolysis. *Fuel*, 82, 1911–1916 (2003).

100. Onay, O. Copyrolysis of coal and heavy carbonaceous residue. U.S. Patent No. 3,565,766 A (September 1971).

101. Martin, O. Enhancing the properties of coal briquette using spear grass (*Imperata cylindrica*) and elephant grass (*Pennisetum purpureum*). A report from Department of Pure and Industrial Chemistry, Nnamdi Azikiwe University, Awka, Turkey (February 2014).

102. Weiland, N. Co-pyrolysis of low rank coals and biomass at transport gasifier conditions. In *A Presentation at AIChE Annual Meeting*, Pittsburgh, PA (October 28–November 2, 2012).

103. Kubacki, M. Co-pyrolysis and co-combustion of coal and biomass, PhD thesis, University of Leeds, Leeds, U.K. (August 8, 2013).

104. Johnson, L., Rostam-Abadi, M., Mirza, I., Stephenson, M., and Kruse, C. Co-pyrolysis of coal and oil shale I: Thermodynamics and kinetics of hydrogen sulfide capture by oil shale. A report of Illinois State Geological Survey. *ACS Preprint*, 71, 274–285 (1986).

105. Mulgaonkar, M., Kuo, C., and Tarrer, A. Plastics pyrolysis and coal coprocessing with waste plastics. *ACS Preprints*, 638; also a report from Department of Chemical Engineering, Auburn University, Auburn, AL (1989).

106. Brebu, M. and Nistor, M. Co-pyrolysis of various lignins with polycarbonate. *Cell. Chem. Technol.*, 48(1–2), 69–74 (2014).

107. Cornelissen, T., Jans, M., Stals, M. et al. Flash co-pyrolysis of biomass: The influence of biopolymers. *J. Anal. Appl. Pyrol.*, 85(1–2), 87–97 (May 2009).

108. Cornelissen, T., Jans, M., Yperman, J., Reggers, G., Schreurs, S., and Carleer, R. Flash co-pyrolysis of biomass with polyhydroxybutyrate: Part 1. Influence on bio-oil yield, water content, heating value and the production of chemicals. A report from University of Hasselt, Diepenbeek, Belgium (2009).

109. Cornelissen, T., Yperman, J., Reggers, G. Schreurs, S., and Carleer, R. Flash-co-pyrolysis of biomass with polylactic acid, part 1: Influence on bio-oil yield and heating value. *Fuel*, 87(7), 1031–1041 (June 2008).

110. Zhang, H., Carlson, T., Xiao, R., and Huber, G. Catalytic fast pyrolysis of wood and alcohol mixtures in a fluidized bed reactor. *Green Chem.* (part of Royal Soc. of Chemistry), 14, 98–110 (2012).

111. Kosivtsov, Y., Sulman, E., Lugovoy, Y., and Chalov, K. Catalytic co-pyrolysis of peat and polymeric wastes. A report from Department of Biotechnology and Chemistry, Tver Technical University, Tver, Russia (2007).

112. Ozdemir, M., Akar, A., Aydogan, A., Kalafatoglu, E., and Ekinci, E. Copyrolysis of Goynuk oil shale and thermoplastics. In *International Conference on Oil Shale—Recent Trends in Oil Shale*, Amman, Jordan (November 7–9, 2006).

113. Assumpcao, L., Carbonell, M., and Marques, M. Co-pyrolysis of polypropylene waste with Brazilian heavy oil. *J. Environ. Sci., Health A: Tox. Hazard Subst. Environ. Eng.*, 46(5), 461–464 (2011).

114. Darmstadt, H., Garcia-Perez, M., Chaala, A., Cao, N., and Roy, C. Co-pyrolysis under vacuum of sugar cane bagasse and petroleum residue-properties of the char and activated char products. *Carbon*, 39(6), 815–825 (May 2001).

115. Tikma, L., Luik, H., and Pryadka, N. Co-pyrolysis of Estonian Shales with low density polyethylene. *Oil Shale*, 21(1), 75–85 (2004).

116. Rotiwala, Y. and Parikh, P. Study on thermal co-pyrolysis of Jatropha deoiled cake and polyolefins. *Waste Manage. Res.*, 29(12), 1251–1261 (December 2011).

117. Toosi, E. Co-pyrolysis of bitumen and oxygenate containing materials at low temperature, PhD thesis. Department of Chemical and Material Engineering, University of Alberta, Edmonton, Alberta, Canada (June 2013).

118. Xu, Y., Chen, Y., Hua, D., Wu, Y., Yang, M., Chen, Z., and Tang, N. Co-pyrolysis of biomass and waste plastic for biofuel in fixed bed reactor. *Chem. Ind. Eng. Progr.*, 32(3), 563–569 (2013).

119. Berrueco, C., Ceamanos, J., Esperanza, E., and Mastral, J. Experimental study of co-pyrolysis of polyethylene/sawdust mixtures. *Thermal Sci.*, 8(2), 65–80 (2004).

120. Oyedun, A., Gebreegziabher, T., and Hui, C. Co-pyrolysis of biomass and plastic waste: A modeling approach. *Chem. Eng. Trans.*, 35, 883–888 (2013).

121. Pinthong, P. Co-pyrolysis of rice husk, polyethylene and polypropylene mixtures: A kinetic study, MS thesis. Department of Chemical Engineering, University of Malaya, Kuala Lumpur, Malaysia (2009).

122. Abnisa, F., Wan Daud, W., and Sahu, J. Pyrolysis of mixtures of palm shell and polystyrene: An optional method to produce a high grade pyrolysis oil. *Environ. Prog. Sustain. Energ.*, 33(3), 1026–1033 (October 2014).

123. Wang, L., Skreiberg, O., Hustad, J., and Gronli, M. Co-pyrolysis behavior of biomass fuels with sewage sludge. In *19th European Biomass Conference and Exhibition*, Berlin, Germany, VP2.5.12, pp. 1651–1657 (June 6–10, 2011).

124. Sanchez, M., Martinez, O., Gomez, X., and Moran, A. Pyrolysis of mixtures of sewage sludge and manure: A comparison of the results obtained in the laboratory (semi-pilot) and in a pilot plant. *Waste Manage.*, 27(10), 1328–1334 (2007).

125. Chueluecha, N. and Duangchan, A. Co-pyrolysis of biomass and cattle manure to produce upgraded bio-oil. In *International Conference on Chemical and Environmental Science and Engineering* (ICEEBS2012), Pattaya, Thailand (July 28–29, 2012).

126. Alias, R., Hamid, K., and Ismail, K. Co-pyrolysis and catalytic co-pyrolysis of waste tyres with oil palm empty fruit bunches. *J. Appl. Sci.*, 11, 2448–2451 (2011).

127. Sharypov, V., Beregovtsova, N., Kuznetsov, B., Baryshnikov, S., Marin, N., and Weber, J. Light hydrocarbon liquids production by co-pyrolysis of polypropylene and hydrolytic lignin. *Chem. Sustain. Dev.*, 11, 427–434 (2003).

128. Grieco, E., Baldi, G., and Fissore, D. Pyrolysis of mixtures of PE/wood and PE/paper: Effects of the interactions on the productions of tar, gas and char. In *XXXV Meeting of the Italian Section of the Combustion Institute*, p. 1, Torino, Italy (2012). doi:10.4405/35proci2012.Viii8.

129. Bernardo, M., Lapa, N., Goncalves, M., Mendes, B., and Pinto, F. Characterization of char produced in the co-pyrolysis of different waste: Decontamination and leaching studies. A report by Universidade Nova de Lisboa, Lisbon, Portugal (2011).

130. Libra, J., Ro, K., Kammann, C., Funke, A., Berge, N., Neubauer, Y., Titirici, M., Fuhner, C., Bens, O., and Emmerich, K. Hydrothermal carbonization of biomass residuals: A comparative review of the chemistry, processes and applications of wet and dry pyrolysis. *Biofuels*, 2(1), 89–124 (2011).

131. Kumar, S., Lognathan, V., Gupta, R., and Barnett, M. An assessment of U (VI) removal from groundwater using biochar produced from hydrothermal carbonization. *J. Environ. Manage.*, 92, 2504–2512 (2011).

132. Funke, A. and Ziegler, F. Hydrothermal carbonization of biomass: A literature survey focusing on its technical application and prospects. In *17th European Biomass Conference and Exhibition*, Hamburg, Germany, pp. 1037–1050 (2009).

133. Funke, A. and Ziegler, F. Hydrothermal carbonization of biomass: A summary and discussion of chemical mechanisms for process engineering. *Biofuels Bioprod. Biorefin.*, 4(2), 160–177 (2010).

134. Lee, C.W., Jenkins, R.G., and Schobert, H.H. Structure and reactivity of char from elevated pressure pyrolysis of Illinois No. 6 bituminous coal. *Energ. Fuels*, 6, 40–47 (1992).

135. Titirici, M., White, R., Falco, C., and Sevilla, M. Black perspectives for a green future: Hydrothermal carbons for environment protection and energy storage. *Energ. Environ. Sci.*, 5, 6796–6822 (2012).

136. Hu, B., Wang, K., Wu, L., Yu, S., Antonietti, M., and Titiricia, M. Engineering carbon materials from the hydrothermal carbonization process of biomass. *Adv. Mater.*, 22, 813–828 (2010).

137. Erlach, B. and Tsatsaronis, G. Upgrading of biomass by hydrothermal carbonisation: Analysis of an industrial-scale plant design. In *ECOS 2010—23rd International Conference on Efficiency, Cost, Optimization, Simulation and Environmental Impact of Energy Systems.*, EPFL (Lausanne), Switzerland (June 14–17, 2010).

138. Titirici, M.M., Thomas, A., Yu, S.H., Muller, J.O., and Antonietti, M. A direct synthesis of mesoporous carbons with bicontinuous pore morphology from crude plant material by hydrothermal carbonization. *Chem. Mater.*, 19(17), 4205–4212 (2007).

139. Sevilla, M. and Fuertes, A.B. The production of carbon materials by hydrothermal carbonization of cellulose. *Carbon*, 47(9), 2281–2289 (2009).

140. Sevilla, M. and Fuertes, A.B. Chemical and structural properties of carbonaceous products obtained by hydrothermal carbonization of saccharides. *Chem. Eur. J.*, 15, 4195–4203 (2009).

141. White, R.J., Budarin, V., Luque, R., Clark, J.H., and Macquarrie, D.J. Tuneable porous carbonaceous materials from renewable resources. *Chem. Soc. Rev.*, 38, 3401–3418 (2009).

142. Antal, M.J. and Gronli, M. The art, science, and technology of charcoal production. *Ind. Eng. Chem. Res.*, 42(8), 1619–1640 (2003).

143. Regmi, P., Moscosq, J., Kumar, S., Cao, X., Mao, J., and Schafran, G. Removal of copper and cadmium from aqueous solution using switchgrass biochar produced via hydrothermal carbonization process. *J. Environ. Manage.*, 93, 1–9 (2012).

144. Ball, R., McIntosh, A., and Brindley, J. The role of char forming process in the thermal decomposition of cellulose. *Phys. Chem. Chem. Phys.*, 1, 5035–5043 (1999).

145. Yu, S., Cui, X., Li, L., Li, K., Yu, B., Antonietti, M., and Colfen, H. From starch to metal/carbon hybrid nanostructures: Hydrothermal metal-catalyzed carbonization. *Adv. Mater.*, 18, 1636–1640 (2004).

146. Titirici, M. and Antonietti, M. Chemistry and materials options of sustainable carbon materials made by hydrothermal carbonization. *Chem. Soc. Rev.*, 39, 103–116 (2010).

147. Liu, Z., Zhang, F.-S., and Wu, J. Characterization and application of chars produced from pinewood pyrolysis and hydrothermal treatment. *Fuel*, 89, 510–514 (2010).

148. Savage, P. Organic chemical reactions in supercritical water. *Chem. Rev.*, 99, 603–621 (1999).

149. Peterson, A., Vogel, F., Lachance, R., Frolling, M., Antal, M., and Tester, J. Thermochemical biofuel production in hydrothermal media: A review of sub and supercritical technologies. *Energy Environ. Sci.*, 1, 32–65 (2008).

150. Russell, J.A., Miller, R.K., and Molton, P.M. Formation of aromatic compounds from condensation reactions of cellulose degradation products. *Biomass*, 3(1), 43–57 (1983).

151. Gullon, P., Conde, E., Moure, A., Dominguez, H., and Parajo, J. Selected process alternatives for biomass refining: A review. *Open Agric. J.*, 4, 135–144 (2010).

152. Mao, J.-D., Fang, X., Lan, Y., Schimmelmann, A., Mastalerz, M., Xu, L., and Schmidt-Rohr, K. Chemical and nanometer-scale structures of kerogen and their changes during thermal maturation investigated by advanced solid-state NMR spectroscopy. *Geochim. Cosmochim. Acta*, 74, 2110–2127 (2010).

153. Mao, J.-D., Schimmelmann, A., Mastalerz, M., Hatcher, P.G., and Li, Y. Structural features of a bituminous coal and their changes during low-temperature oxidation and loss of volatiles investigated by advanced solid-state NMR spectroscopy. *Energ. Fuels*, 24, 2536–2544 (2010).

154. Cao, X., Ro, K.S., Chappell, M., Li, Y., and Mao, J.-D. Chemical structures of swine-manure chars produced under different carbonization conditions investigated by advanced solid-state 13C NMR spectroscopy. *Energ. Fuels*, 25, 388–397 (2011).

155. Mao, J.-D., Ajakaiye, A., Lan, Y., Olk, D.C., Ceballos, M., Zhang, T., Fan, M.Z., and Forsberg, C.W. Chemical structures of manure from conventional and phytase transgenic pigs investigated by advanced solid-state NMR spectroscopy. *J. Agric. Food Chem.*, 56, 2131–2138 (2008).

156. Ramke, H.G., Blohse, D., Lehmann, H.J., and Fettig, J. In *Hydrothermal Carbonization of Organic Waste, Twelfth International Waste Management and Landfill Symposium*, Sardinia, Italy (2009).

157. Spokas, K.A. and Reicosky, D.C. Impacts of sixteen different biochars on soil greenhouse gas production. *Ann. Environ. Sci.*, 3, 179–193 (2009).

158. Mursito, A.T., Hirajima, T., and Sasaki, K. Upgrading and dewatering of raw tropical peat by hydrothermal treatment. *Fuel*, 89(3), 635–641 (2010).

6 Methane Production by Anaerobic Co-Digestion of Biomass and Waste

6.1 INTRODUCTION

In Chapters 2 through 4, we examined the merits of co-combustion, co-gasification, co-liquefaction, and co-pyrolysis as the techniques to improve the performances of single component thermochemical transformation processes. The chapters showed that for each thermochemical process, while many mixture combinations are possible, the most multifuel systems examined in recent years are some combination of coal, heavy oil, oil shale, biomass, waste, and water. Water acts as a reactant or catalyst in many high-temperature and high-pressure operations. For co-liquefaction, oil (both conventional and unconventional) also plays an important role as a part of multifuel systems.

In this chapter, we examine another example of fuel generation process that is enhanced by the use of multiple feedstock. Anaerobic digestion (AD) of waste material is one of the most energy efficient biochemical process. AD is the multistep biological process during which organic material is converted to biogas and digestate in the absence of oxygen. This process of biodegradation of materials requires anaerobic microorganisms. As shown in my previous books [1,2], the creation of methane by AD occurs in four stages, hydrolysis/liquefaction, acidogenesis, acetogenesis, and methanogenesis. These four phases occur in two parts. Although commonly not practiced, the process can generate hydrogen depending on the nature of the microorganisms. More details on the mechanism of the process are given in an excellent review by Weiland [3] among others [4–29]. The process can be applied to animal waste, food and organic waste, various wastewater treatment plants, crop and crop residues (CRs), energy crops, industrial sludge, municipal solid waste (MSW), and municipal sewage sludge among others. The importance of this process will continue to grow because of the following:

1. All wastes are growing industry because they largely depend on the human and animal populations and the quality of life.
2. Currently, the United States generates about 250 million tons of MSW/year. The MSW created by the world will soon reach about 1 billion tons/year. This number will continue to rise as population increases and quality of life improves.

3. Currently, most MSW and some other types of wastes go to landfills. Unfortunately, our landfill capacity has been decreasing. Converting this waste into fuel or energy appears to be more sustainable strategy for the long term.

4. Aqueous waste coming from industry and public works have to be treated before discharging to landfill, water streams, or other disposable sources. They cannot be discarded without appropriate pretreatment.

5. The creations of new landfills are expensive and offer significant public resistances.

The process of AD converts waste into methane, a fuel. The efficiency and stability of this process is very important for its cost-effectiveness. The process of co-digestion examined here improves the efficiency of the AD to the extent that during the last decade, co-digestion has become a norm (and not exception) in the AD process industry. This chapter examines our present state of knowledge on this subject.

As pointed out by Groenewold [30], co-digestion is a method to improve biogas yield by adding co-substrates (e.g., maize [Mz] or grass) to livestock waste products (manure). From an economical perspective, enriching a digester with energy crops might be feasible. However, using nonwaste products, explicitly grown for biogas production, also increases the potential for anthropogenic greenhouse gas (GHG) emissions in the atmosphere. In general, the GHG balance of "green electricity" from co-digesters has been evaluated using CH_4, N_2O, and CO_2 as essential parameters.

The process of AD was developed for waste disposal in the early 1970s. Due to the combination of organic material reduction and methane production, it was viewed as a very attractive process. In the agricultural sector, AD was used to reduce manure storage and resulting emission and produce electricity. It has been calculated that a farm based on 100 cows can reduce approximately 118,000–138,500 kg CO_2 eq/year depending on the manure management system applied. This range of value is approximately equivalent to the CO_2 emission of an average passenger car traveling 1 million km [4–29]. Digesting manure waste streams is an effective method to reduce GHG emissions and produce energy. While co-digestion improves energy efficiency, the leakage of CH_4 during the process contributes to the GHG. This fugitive leakage can be as high as $5.1\% \pm 1.8\%$ of total gas produced and should be reduced to the extent possible.

6.2   WHAT IS CO-DIGESTION?

Co-digestion is the simultaneous digestion of a homogeneous mixture of two or more substances. The process adds another substrate to the existing digesters to utilize the availability of free capacities. The most common situation is when the major amount of main basic substrate (e.g., sewage sludge) is mixed with and digested together with minor amounts of a single or a variety of additional substrates. The word co-digestion is used for digestion of all types of mixed substrates independent of its composition [4,7,10,17,24].

Just like single component combustion, gasification, liquefaction, and pyrolysis, AD was carried out for a single substrate until more than decade ago. For example, manure was digested to produce energy, sewage sludge was anaerobically stabilized, and industrial wastewater was pretreated before final treatment in a wastewater treatment plant. While these were convenient approaches, neither did each AD unit fully utilize its capacity, nor did it produce the methane production to the fullest level. The concept of co-digestion allows sharing of the equipment for different types of waste thus reducing any required transportation costs for individual waste and makes an optimum and most economical use of every waste treatment plant. Co-digestion also requires careful choice of multiple substrates that can give the most methane production through synergistic reaction [4,7,10,17,24].

Co-digestion can result in an important increase of the biomethane potential (BMP) when the substrates mixture is prepared with proper percentages of the different organic substrates to be digested [4,7,10,17,24]. The beneficial effect of the co-digestion is mainly due to the optimization of the nutrient balance in the substrates mixture when co-digesting nitrogen-rich substrates with carbon-rich substrates. The higher specific methane yield of 686 N m^3/tons volatile solid (VS) is achieved by co-digesting organic fraction of municipal solid waste (OFMSW) and vegetable oil (83/17 on dry weight). Also, OFMSW with animal fat (83/17 on dry weight) and cow manure (CM) with fruit and vegetable waste (FVW) (60/40 on weight) can give very high methane yield of 490 and 450 N m^3/tons VS, respectively [4–29]. Co-digestion can be synergistic, antagonistic, or neutral depending on the selection of the co-substrates.

6.3 ADVANTAGES AND DISADVANTAGES

The co-digestion process offers several advantages and disadvantages [4–29].

6.3.1 ADVANTAGES

1. It improves nutrient balance and digestion. The digestion of a variety of substrates can improve the nutrient ratio of total organic carbon/nitrogen/phosphorous, which should be ideally 300:5:1. It also maintains a reasonable mix of minerals such as Na, K, Mg, Mn, and Fe as well as balanced composition of trace metals. These types of nutrient balances help maintain stable and reliable digestion performance. It also provides a larger quantity and good quality fertilizer as the digestate.
2. In general, co-digestion enhances the methane (or biogas) production rate. The increased biogas productivity (m^3 biogas/m^3 digester volume/day) compensates for the high investment and running costs of small- and medium-sized digesters.
3. Wastes with poor fluid dynamics, aggregating wastes, particulate or bulking materials, and floating wastes can be much easily digested after improving its rheological properties through homogenization with dilute substrate such as sewage sludge or liquid manure. Good rheological properties also help the performance of the digester.

4. Co-digestion allows the mechanism to avoid overloading and underloading of a single waste that may be seasonal dependent. Through the use of multiple substrates, digestion process can be maintained at a constant rate throughout the year.

5. In agricultural digesters, the process of co-digestion can considerably improve the overall economics (payback time) of the plant. Gate fee creates the win–win situation. The provider pays significantly lower price at a farm-scale AD plant than at an incineration or composting facility (usually by a factor of 3 or 4). The farmer takes the credit of the increased biogas production along with the income from gate fee.

6. If there is a sufficient farmland available, the digestate from co-digestion can be directly recycled as a fertilizer at reasonable cost.

7. In a limiting situation of industrial wastes, energy crops may become an interesting alternative especially when plants are grown on fallow or set-aside land that attracts subsidies. Traditionally, all C_4 plants have very good growth yields. Corn has become very good co-substrate in Germany. In order to make sure that it is grown in set-aside land and is taken out of nutrition chain, it is treated after harvest with manure before it is ensiled. In Austria, Sudan grass, another C_4 plant, is grown as energy crop for co-digestion. It grows well in dry soils of the southeast Austria [4–29].

8. Co-digestion reduces fossil fuel consumption and minimize carbon footprint.

9. Co-digestion increases the plant's value to the community by recycling challenging liquid "wastes." It can be a true water resource recovery facility.

10. Co-digestion reduces operating and energy costs.

6.3.2 Disadvantages

1. Co-digestion increases the digester effluent chemical oxygen demand (COD) that requires more expensive treatment before it can be discarded.

2. Some substrates used in co-digestion require additional pretreatment. This adds to the overall cost.

3. More substrates in the digester require additional mixing. This increases the power cost for the mixing.

4. Co-digestion also increases wastewater treatment requirement. It increases Waste water treatment plant (WWTP) direct loads.

5. Some co-substrates create additional hygienization requirement. Potentially infectious materials must be hygienized by law. The hygienization can be (a) sterilization (20 min at 2 bars, 121°C), (b) pasteurization (between 70°C and 90°C for 15–60 min), and (c) sanitation (heat treatment at lower temperatures over an extended time period) [31].

6. The land use for digestate can also be a restriction in some areas.

7. Co-digestion economics very strongly depend on crop or type of co-substrate. Flexibility in choosing the right co-substrate may be limited.

8. Substrates must be trucked in, which adds to the cost.

9. Changes carbon-to-nitrogen (C/N) ratio in influent when high C, low N substrate is directed to digesters.
10. Co-digestion of manure with organic wastes can increase nitrogen by 57% compared to baseline operation, with smaller increases in ammonia and phosphorus.

The additional nutrients to the farm can exacerbate existing nutrient management systems, particularly for large operations. There are also concerns that contaminants (such as materials containing prions) or pathogens arriving with the organic waste can inhibit AD process, limit the use of AD products, or create a possibility for negative public health impacts.

6.4   MAJOR APPLICATIONS, USERS, AND RELATED ISSUES

Co-digestion is particularly applicable for agricultural biogas production from manure (which has a relatively low gas yield), which is economically not viable. Addition of co-substrates with high methane potential not only increases biogas yield (which generates more income) but also increases income through tipping fees. Another major application for co-digestion is the digester in waste treatment plant [123–127], which is usually oversized. Addition of co-substrate uses its unfilled capacity and produces more gas and consequently more electricity at only marginal additional cost. The extra electricity produced can be used for the energy needs of wastewater treatment plant at a competitive cost [4–29].

In most cases, co-digestion is applied in wet single step process such as continuous stirred tank reactor. The substrate is used in slurry of 8%–15% concentration. This does not significantly add to the cost of mixing. Wet systems are particularly useful when digestate slurry (without removal of water) can be directly used on field and green lands for soil conditioner or fertilizers.

During last decade, the percentage of co-digestion plants in waste treatment process is steadily increasing. In fact, a number of existing municipal sewage sludge digesters are already using co-substrates. New sewage plants or extension of existing plants use different types of organic and food wastes (FWs) such as biowaste, food leftovers, fat wastes, and floatation sludges as co-substrate to improve biogas production. Co-digestion plants also provide controlled organic waste disposal mechanism for fat trap contents and food leftovers for the communities.

Besides improving biogas yield, the co-digestion plants also improve the energy balance over single substrate plant. On average, for sewage sludge plant, the self-sufficiency of electricity is below 50%. With the use of a co-substrate, this number can be improved as high as 80%, and in some cases, the co-digestion plant becomes a net energy producer [4–29].

In agriculture industries, co-digestion has become a standard norm. Many small and medium-sized farm-scale digesters use mixed co-substrates with manure. According to Braun and Wellinger [24], most of about 2000 agricultural plants in Germany in 2002 used co-substrates. In other parts of Europe, such as Austria (110), Switzerland (71), Italy (>100), Denmark (>30), Portugal (>25), Sweden, France, Spain, and England, all use co-digestion process.

There are companies like CH_4 Biogas, an American Corporation, which builds co-digestion plants using manure and organic waste in the United States. The company uses Bigadan technology of a Danish corporation. According to Braun and Wellinger [24], in Denmark and to a smaller extent in Sweden and Italy, several farmer cooperatives or companies successfully operate large-scale farm digesters that use co-substrates. These companies recycle digestate to farm land and use biogas production for the generation of electricity or in some cases upgrade as a fuel.

6.5   FEED PRETREATMENT

Co-digestion process fundamentally produces two products: biogas and usable digestate. The digestate is an important and valuable organic fertilizer, but it is only marketable if it is environmentally and hygienically safe and free of visible contaminants such as plastics, stones, and metals.

The best way to obtain usable digestate is to have clean and unpolluted co-substrates with high gas potential. If the substrate contains undesirable materials, they have to be taken out preferentially before it is used in co-digestion. These materials can cause pipe blockages, scum formation or bottom layers in the digesters, or damage pumps and mixing devices. The removal of these impurities may require sophisticated and expensive equipment and additional operating costs. The types of collection vessel (bins, plastic bags, etc.), the region of collection, people's habits, the season, etc., affect the contents of undesirables in co-substrate.

The selection of a proper pretreatment process should be waste and digestion process specific and adjusted to the product quality required. For the use of organic household waste, two major digestion techniques, dry and wet digestion processes, are applied. Dry digestion technique uses solids concentration >20% and requires less pretreatment for contaminant removal. Dry digestion processes retain higher concentration of particulate matters. For most of the more problematic waste, dry digestion offers the easier solution since the requirement for the separation of the materials is small as long as the dry matter content remains above 25% solids [4–33].

The wet separation process, which uses less than 10% solids require more sophisticated approaches for the contaminant removal. It uses wet pretreatment processes to separate light solids from heavy solids. Generally, wet separation processes achieve higher level of impurity removal compared to dry separation. However, the processes are more laborious and expensive. Only few organic wastes like biowastes from separate collection, garden and yard wastes, expired food, and leather industry wastes require extended preconditioning for wet co-digestion.

In general, the pretreatment steps include size reduction of the substrate, removal of indigestible components, and hygienization. The extended pretreatment requires chopping; sieving; removal of metals; removals of glass, sand, stones, etc.; and homogenization. As mentioned in Section 6.2, hygienization includes sterilization, pasteurization, and sanitation. For most biowastes, thermophilic operation is sufficient for hygienization. Other suggestions for hygienization of biowaste are (1) using thermophilic digester operation at 55°C for at least 24 h with minimum residence time of at least 20 days; (2) in the case of mesophilic digestion, pretreating substrates at 70°C for 60 min or posttreatment of the digestate at 70°C for 60 min; and (3) composting the digestate [4–33].

Esposito et al. [8] studied co-digestion of organic waste and concluded that several pretreatment methods can be applied to further increase the biogas production of a co-digestion process, such as mechanical comminution, solid–liquid separation, bacterial hydrolysis, and alkaline addition at high-temperature, ensilage, alkaline, ultrasonic, and thermal pretreatments. However, other pretreatment methods, such as wet oxidation and wet explosion, can result in a decrease of the methane production efficiency.

You et al. [32] showed that anaerobic co-digestion of corn stover with swine manure (SM) can be considerably improved with NaOH pretreatment. This pretreatment shortened digestion time and improved biogas yield. Different NaOH concentrations (2%, 4%, and 6%) at various temperatures (20°C, 35°C, and 55°C) and 3 h of pretreatment time were tested for corn stover pretreatment. A C/N ratio of 25:1 in the substrate of corn stover and SM was employed. The results showed that lignin removal rates of 54.57%–79.49% were achieved through the NaOH pretreatment. The highest biogas production rate was observed for corn stover pretreated at 6% NaOH at 35°C for 3 h. This was 34.59% higher than that from the untreated corn stover. The increase of methane yield was from 276 to 350 mL/g VS. Digestion time of pretreated corn stover was shortened from 18 to 12–13 days. NaOH pretreatment thus increased biogas production, reduced digestion time, and separated lignin. The study recommended pretreatment of corn stover with 6% NaOH at 35°C for 3 h for all its future use in co-digestion.

In general, digestate contains nondigestible and residual particulates, liquid organic and inorganic waste constituents, and bacterial biomass. In large agricultural digesters, the digested slurry is used as fertilizer for the land. The amount of nitrogen and heavy metals that can be introduced in the land and the nature of agricultural land for digestate use are often restricted [4–33].

6.6 PROCESS AND ECONOMIC CONSIDERATIONS

Depending on the quality and nature of the waste to be used for co-digestion and the size of the operation, some additional equipment may be required for (1) delivery of the waste, (2) homogenization and mixing of co-substrates, (3) prevention of excessive foaming and scum layer formation, and (4) removal of sediments from the digester [34–60].

Generally, co-substrates are delivered by trucks and stored in a pressurized space to avoid odor emissions. The trucks need to be cleaned and sanitized for its reuse. The co-substrate is passed through a premixing and homogenization/pasteurization tank. The co-substrate requiring the extensive pretreatment is generally restricted for large centralized plants. In any given plant, all the information in processing of co-substrate pre, during, and post digester operations are always recorded.

Co-digestion significantly improves the economics of AD processes through additional biogas production and tipping fees paid by the generator of the organic wastes to the digester owner. Various commercial and laboratory-scale studies have shown that depending upon the type, concentration, and flow rate of organic wastes used, biogas production can be enhanced by as much as 25%–400% [34–60]. An economic analysis of an AD facility installed on a 700-cow dairy in northwest Washington state showed that co-digestion with 16% organic wastes more than doubled biogas production and almost quadrupled annual digester revenues compared to a manure-only baseline, with 72% of all receipts directly attributable to the addition of organic waste [34–60].

6.7 PROCESS MONITORING, CONTROL, SCALE-UP, MODELING, AND OPTIMIZATION

6.7.1 PROCESS MONITORING AND CONTROL

The monitoring of the process depends on the scale of the operation; large scale will require more attention than the small scale. The most important control parameters are the overall daily substrate flow rate (tons/day or m^3/day) and the quantity of daily production of biogas (m^3/day). The efficiency of the digester is measured by the amount of biogas produced per unit substrate flow. For the proper control, the determination of the CH_4 concentration is highly desirable. In the case of co-digestion, additional type and amount of separated impurities need to be recorded and controlled. In large-scale operations, amount of volatile fatty acids (VFAs) and ammonia concentration should be monitored and controlled [61–104].

The addition of co-substrate causes scum layers and bottom sediments. These need to be controlled for the stable operation. For each type and amount of waste, the time and temperature conditions for the sterilization should be monitored. Most importantly, the government regulations for the quality assurance of digestate for each type of waste should be followed [61–104].

Lin et al. [61] examined anaerobic co-digestion of FVW and FW at different mixture ratios for 178 days at the organic loading of 3 kg VS/(m^3 day). The dynamics of Archaeal community and the correlations between environmental variables and methanogenic community structure were analyzed by polymerase chain reactions–denaturing gradient gel electrophoresis (PCR–DGGE) and redundancy analysis (RDA), respectively. PCR–DGGE results demonstrated that the mixture ratio of FVW to FW altered the community composition of Archaea. As FVW/FW ratio increased, *Methanoculleus*, *Methanosaeta*, and *Methanosarcina* became the predominant methanogens in the community. RDA results indicated that the shift of methanogenic community was significantly correlated with the composition of acidogenic products and methane production yield. The different mixture ratios of substrates led to different compositions of intermediate metabolites, which may affect the methanogenic community. These results suggested that the analysis of microbial community could be used to diagnose and control anaerobic process.

6.7.2 SCALE-UP

Sell [62] at Iowa State University examined a scale-up procedure for substrate co-digestion in anaerobic digesters through the use of substrate characterization, bio-methane potentials (BMPs), anaerobic toxicity assays (ATAs), and sub-pilot-scale digesters. The digestion system was cattle manure with co-substrates of numerous different types of organic wastes. Based on his study, he wrote a series of three papers. The objective of the first paper was to analyze multiple substrates using various laboratory techniques so that optimum mixture ratios could be formed. The BMPs and ATAs were used to select and in some cases rule out substrates based on their contribution to methane production. Mixtures were created using constraints arising from the full-scale system. This included the use of all available manure, keeping total solids (TSs) below 15% to facilitate pumping, maintaining pH between 6.5 and 8.2 for microbial ecology, providing high COD concentrations to maximize methane production, and limiting ammonia levels to avoid toxicity [23]. The BMP and ATA results from each mixture were analyzed and compared. The mixture with the best performance was selected for subsequent testing in 100 L sub-pilot-scale anaerobic digesters.

The objective of the second paper was to analyze the performance of three 100 L sub-pilot-scale anaerobic digesters. These plug flow digesters operated at a 21-day hydraulic retention time (HRT) and were fed the mixture selected in the first study in a semicontinuous manner twice weekly (six loadings per HRT). Methane production was measured using submerged tipping buckets. Methane production from the sub-pilot-scale reactor was compared to that predicted by the BMP tests. After two HRTs, the BMP maximum and minimum were observed to be valid boundaries for the sub-pilot-scale anaerobic digester methane production, with some of the variability ascribed to seasonal substrate changes.

The objective of the third and final paper was to use a series of BMPs and an ATA to predict the methane production in three 100 L sub-pilot-scale anaerobic digesters that were subjected to a potential toxicant, glycerin. A group of ATAs were performed with glycerin inclusion rates of 0.5%, 1.0%, 2.0%, 4.0%, 8.0%, 15%, 25%, and 35% by volume. A set of BMPs was performed where a baseline mixture was combined with glycerin such that glycerin was 0.0%, 0.5%, 1.0%, 2.0%, 4.0%, 8.0%, 15%, 25%, and 35% of the combined mixture by volume. In addition, BMPs of 100% glycerin and 50% glycerin/50% deionized water by volume were also performed. The three 100 L sub-pilot-scale anaerobic digesters were operated at a 21-day HRT and were each fed in a semicontinuous manner twice weekly (six loadings per HRT). Each digester was fed a combination of the mixture selected in the first study with a different amount of glycerin (1%, 2%, 4% by volume).

The ATAs showed that glycerin was toxic to methane production at all inclusion levels. The BMPs indicated no significant difference between methane production of the 0.0%, 0.5%, 1.0%, 2.0%, and 4.0% mixture combinations; however, at 8.0%, methane production tripled. In contrast, the sub-pilot-scale reactors showed signs of toxicity for 4.0% glycerin inclusion and little to no effect on methane production for 1.0% and 2.0% glycerin inclusion. Thus, neither the ATA nor the BMP proved to be an adequate predictor for the sub-pilot-scale reactors. The most likely cause was the lack of mixing within the sub-pilot-scale digester required to keep all materials in close contact with each other to ensure adequate microbial activity and methane formation.

The aforementioned study indicates that the proper scale up of digester requires good understanding of mixing phenomena with the digester. In order to scale up laboratory-scale performance in large-scale digesters, good contacting among substrates, nutrients, and microorganism is essential.

6.7.3 Computer Simulation and Model

Zaher et al. [82,99–101] examined a computer software model to assist plant design and co-digestion operation. They pointed out that feeding the digester with a combination of waste streams introduces complexities in waste characterization that requires a model to simulate optimal parameters for co-digestion. The general integrated solid waste co-digestion (GISCOD) model was developed and tested for this purpose.

Model development overcame several challenges to achieve reliable, precise simulations. Accurate characterization of macronutrients, COD, and charge for waste streams was necessary input to the International Water Association Anaerobic Digestion Model No. 1 (ADM1) [84,85]. Particulate components of carbohydrates, proteins, and lipids vary dynamically in combined solid waste streams, making it difficult to define the waste streams for accurate input. Such waste heterogeneity could be resolved by applying a general transformer model to interface the ADM1 to practical characteristics of each waste stream. In addition, the study showed that hydrolysis rates for manure only varied considerably from hydrolysis rates for FW–manure co-digestion. Thus, for co-digestion applications, it is important to consider separate hydrolysis rates for each particulate component from each waste stream. Also, hydrolysis rates of solid wastes differed from that of decaying biomass, which is mainly limited by a disintegration step for cell lysis. The separate characterization and phasing of the co-digested waste hydrolysis allowed the optimization of biogas production and defined the corresponding operation settings of the digester.

As currently designed, GISCOD can support the operational decisions necessary for digesting trucked-in wastes with wastewater sludge or, generally, optimize the feedstock and operation of biogas plants. The model, however, needs to be tested and improved with the additional data from commercial co-digestion plants. The co-digestion models reported in literature are mainly based on the ADM1, with various modifications to upgrade it from the monosubstrate to the multisubstrate system.

Esposito et al. [8] examined co-digestion of organic waste. They concluded that biomethane potential data when experimentally determined are only valid for the particular co-substrates and the operating conditions (temperature, organic loading rate [OLR], HRT, moisture content [MC], S/I ratio, etc.) applied during the experiments. However, they can be used to calibrate mathematical models capable of simulating co-digestion process, and then such models can be used to predict the biomethane potential achievable with different co-substrates and under different operating conditions. For such purpose, only mathematical models specifically aimed at simulating the co-digestion of different co-substrates are suitable, as monosubstrate models are not capable of taking into account the peculiarities of different substrates in terms of physical and biochemical characteristics and their synergistic effects (e.g., in terms of pH, alkaline, and nutrient balance).

The ultimate aim of modeling and simulation effort is to develop a user-friendly GISCOD software package that can be made available to digestion engineering firms, wastewater treatment plants, and farm digesters.

6.7.4 PROCESS OPTIMIZATION

Alvarez et al. [163] developed a methodology for optimizing feed composition of anaerobic co-digestion of agro-industrial wastes using linear programming method. Inoculum nutrient demand and inoculum acclimation can influence the substrate biodegradation and methane potential. In spite of using granular inoculum, inoculum of different source caused variations in the substrate biodegradation. An optimization protocol for maximizing methane production by anaerobic co-digestion of several wastes was carried out. A linear programming method was utilized to set up different blends aimed at maximizing the total substrate biodegradation potential (L CH$_4$/kg substrate) or the biokinetic potential (L CH$_4$/kg substrate day). In order to validate the process, three agro-industrial wastes were considered, pig manure, tuna fish waste, and biodiesel waste, and the results obtained were validated by experimental studies in discontinuous assays.

The highest biodegradation potential (321 L CH$_4$/kg COD) was reached with a mixture composed of 84% pig manure, 5% fish waste, and 11% biodiesel waste, while the highest methane production rate (16.4 L CH$_4$/kg COD day) was obtained by a mixture containing 88% pig manure, 4% fish waste, and 8% biodiesel waste. Linear programming was proved to be a powerful, useful, and easy-to-use tool to estimate methane production in co-digestion units where different substrates can be fed. Experiments in continuous operation were recommended to acquire broader information about the biodegradation and methane production rates of the blends determined by linear programming.

The review by Khalid et al. [9] for co-digestion of solid organic waste indicated that AD is one of the most effective biological processes to treat a wide variety of solid organic waste products and sludge. They, however, noted that different factors such as substrate and co-substrate composition and quality, environmental factors (temperature, pH, OLR), and microbial dynamics contribute to the efficiency of the AD process and must be optimized to achieve maximum benefit from this technology in terms of both energy production and organic waste management. The use of advanced molecular techniques can further help in enhancing the efficiency of this system by identifying the microbial community structure and function and their ecological relationships in the bioreactor.

6.8   EXAMPLES OF FEEDSTOCK EFFECTS

Worldwide, the anaerobic stabilization of sewage sludge is probably the most important AD process. While agricultural digesters are more in quantity, the volume treated is far smaller than that for sewage sludge. AD of manure is largely done for energy production. A digester in the farm has to be financed by the energy produced, whereas sewage sludge is a waste product of wastewater treatment that has to be stabilized. In this case, the digester is paid by the polluter or local government agency. The limiting availability of the industrial waste makes it clear that co-digestion of industrial waste has only limited potential. The best gain for co-digestion is the digestion of manure with energy crops.

Today, the inclusion of waste product for co-digestion is economically essential. There are several organic substances that are anaerobically degraded without major pretreatment such as leachates, slops, sludges, oils, fats, or whey. Some waste can form inhibiting metabolics like NH_3 during AD, which requires higher dilutions with substrates like manure and sewage sludge. The feedstock that requires pretreatments are source-separated municipal biowastes, food leftovers, expired food, market wastes, and harvest residue. There are also limited stocks of organics such as straw, lignin, rich yard waste, and category 1 slaughterhouse animal by-products (ABPs) that are not very suitable for AD due to high cost of pretreatment, inhibiting components, poor biodegradability, hygienic risks, or expensive transport. Approximate biogas or methane yields for various crops and organic wastes and from co-digestion of organic wastes are summarized in Tables 6.1 and 6.2. The literature on co-digestion of various mixtures of substrates is very large and cannot be covered here in its entirety. References 1–234 give illustrations of some typical published literature on co-digestion.

One substrate ideal for co-digestion is animal manure. One of the advantages of using manure for digestion is that the bacteria required for the digestion process are already present in the manure. A disadvantage of using manure alone is that it has low-energy content; therefore, a low amount of gas production is achieved per unit volume of manure. Cattle manure only contains around 3.6 GJ/tons of dry matter, and pig manure contains around 5.3 GJ/tons of dry matter. Since dry matter is only 10% of the total volume, the energy content per unit volume of manure is very low (in the order 0.3–0.5 GJ/me manure). This is equivalent to about 18–30 m³ of biogas/m³ manure. Assuming the residence time in the digester to be 30 days, this results in the production of 0.6–1 m³ biogas/m³ digester/day. Such low amount makes this process highly uneconomical. Co-digestion with energy crops or substrates with high organic content can make them useful biogas generators.

In the following discussion, examples of feedstock effects on the co-digestion process are separated into three parts: sludge/wastewater treatments, treatments of different types of manures, and other co-digestion systems.

6.8.1 CO-DIGESTION WITH DIFFERENT TYPES OF SLUDGE AND WASTEWATER

6.8.1.1 Organic Waste with Municipal Sludge

Zupancic et al. [124,138,223] studied co-digestion of organic waste of domestic refuse (kitchen refuse and scraps of waste food mixed with water for feeding to pigs) with municipal sludge. The results were very successful and showed that the organic waste was virtually completely degraded, there was no increase in volatile solid sludge (VSS) during the process, and there were no adverse impacts on the environment. The degradation efficiency increased from 71% to 81%, and an 80% increase in biogas quantity was observed. Biogas production rate increased from 0.32 to 0.67 m³/m³/day. Solid by-products increased from 0.39 to a peak of 0.89 m³/kg VSS.

TABLE 6.1
Mean Biogas Potential of Different Crops and Organic Wastes

Substrate	Mean Biogas Yield (N m^3/ton VS^a Added)
Sugar beet	750
Fodder beet	775
Maize	605
Corn cob mix	670
Wheat	660
Triticale	605
Sorghum	550
Grass	565
Red clover	575
Sunflower	480
Wheat grain	725
Rye grain	670
Cow manure (9% TS)	255
Pig manure (7% TS)	385
Chicken Manure (15% TS)	290
Corn (whole crop)	546
Harvest residue (straw, stems)	375
Sewage sludge	300
Expired animal feed	575
Food industry waste	500
Stillage from breweries	600
Green wastes (markets)	550
Biowastes (source separated)	450
Floatation sludge/animal fat	775
Waste fat	1000
Municipal solid waste	326
Fruit and vegetable wastes	380
Municipal solid waste	480
Fruit and vegetable waste and abattoir wastewater	770
Swine manure	306
Municipal solid waste	181
Food waste leachate	267
Rice straw	317
Maize silage and straw	283
Jatropha oil seed cake	383
Palm oil mill waste	553
Household waste	317

(Continued)

TABLE 6.1 (*Continued*)
Mean Biogas Potential of Different Crops and Organic Wastes

Substrate	Mean Biogas Yield (N m³/ton VS[a] Added)
Lignin-rich organic waste	181
Swine manure and winery wastewater	316
Food waste	359

Sources: Braun, R. et al., Biogas from Energy crop digestion, IEA Bioenergy, Task 37—Energy from Biogas and Landfill Gas, Denmark, 2009; Braun, R., Anaerobic digestion—A multi faceted process for energy, environmental management and rural development, in: Ranalli, P. (ed.), *Improvement of Crop Plants for Industrial End Users*, Springer, 2007; Murphy, J. et al., Biogas from crop digestion, IEA Bioenergy, Task 37—Energy from Biogas, Denmark, September 2011; Braun, R., Potential of co-digestion, IEA Task 37 Report, 2002, http://www.novaenergie. ch/iea-bioenergytask37/Dokumente/final.PDF, accessed on November 7, 2007; Braun, R. and Wellinger, A., Potential of co-digestion, IEA Bioenergy, Task 37—Energy from Biogas and Landfill Gas, 2004; Braun, R. et al., *Biotechnol. Lett.*, 3, 159, 1981; Braun, R. et al., *Appl. Biochem. Biotechnol.*, 109, 139, 2003.

[a] VS, volatile solids.

TABLE 6.2
Relative Methane Yield from the Co-Digestion of Organic Waste

Substrate	Co-Substrate	Methane Yield (L/kg VS[a])
Cattle excreta	Olive mill waste	179
Cattle manure	Agricultural waste and energy crops	620
Fruit and vegetable waste	Abattoir wastewater	611
Municipal solid wastes	Fly ash	222
	Fat, oil, and grease	350
	Waste from sewage treatment plants	
Pig manure	Fish and biodiesel waste	620
Potato waste	Sugar beet waste	680
Primary sludge	Fruit and vegetable waste	600
Sewage sludge	Municipal solid waste	532
Slaughterhouse waste	Municipal solid waste	500

Sources: Braun, R. et al., Biogas from Energy crop digestion, IEA Bioenergy, Task 37—Energy from Biogas and Landfill Gas, Denmark, 2009; Braun, R., Anaerobic digestion—A multi faceted process for energy, environmental management and rural development, in: Ranalli, P. (ed.), *Improvement of Crop Plants for Industrial End Users*, Springer, 2007; Murphy, J. et al., Biogas from crop digestion, IEA Bioenergy, Task 37—Energy from Biogas, Denmark, September 2011; Braun, R., Potential of co-digestion, IEA Task 37 Report, 2002, http://www.novaenergie.ch/iea-bioenergytask37/Dokumente/final.PDF, accessed on November 7, 2007; Braun, R. and Wellinger, A., Potential of co-digestion, IEA Bioenergy, Task 37—Energy from Biogas and Landfill Gas, 2004; Braun, R. et al., *Biotechnol. Lett.*, 3, 159, 1981; Braun, R. et al., *Appl. Biochem. Biotechnol.*, 109, 139, 2003.

[a] VS, volatile solids.

Davidsson et al. [118] successfully performed the co-digestion of sludge from grease traps and sewage sludge. While co-digestion could not reach stable methane production in the continuous digestion tests, addition of grease trap sludge when digesting sewage sludge increased the methane potential and methane yield (amount of methane per added amount of VS). In the pilot-scale tests, the increase in methane yield was 9%–27% for grease sludge (GS) amounts corresponding to 10%–30% of the total VS added.

Agdag et al. [116] examined the feasibility of anaerobic co-digestion of industrial sludge with MSW in three simulated landfilling bioreactors during a 150-day period. They noted that co-digestion stabilized the waste and the treatment of leachate release. The addition of industrial sludge to MSW gave biogas with 72% methane content while improving the leachate quality.

Gomez et al. [81] examined co-digestion of primary sludge and fruit and vegetable fraction of MSW under mesophilic conditions. The addition of fruit and vegetable fraction of MSW in primary sludge increased the biogas production. The specific gas production and biogas yield, however, did not change by co-digestion. The application of a sudden increase in the organic load of the co-digestion system led to higher gas production accompanied by downgrading of the performance of the digester.

Neves et al. [122] examined co-digestion of sewage sludge and coffee wastes from the production of instant coffee substitutes. Methane yields in the range 0.24–0.28 m^3/kg VS were obtained except for barley-rich waste where methane yield was 0.02 m^3/kg VS. Four out of five wastes gave high reduction in TS (50%–73%) and VS (75%–80%).

Fernández et al. [37] evaluated potential of mesophilic AD for the treatment of fats from different origins through co-digestion with the OFMSW. No change in performance was observed when animal fat was changed to vegetable fat with a completely different long-chain fatty acid (LCFA) profile. This indicated that no important metabolic changes are implied in the degradation of different LCFAs with an acclimatized sludge.

Fezzani et al. [45,86,87] investigated co-digestion of olive mill wastewater (OMW) with olive mill solid waste under thermophilic condition (at 55°C). They concluded that OMW can be successfully degraded in co-digestion with olive mill waste solids without previous dilution and addition of chemical nitrogen substance. The best methane productivity was 46 L CH_4/(L OMW fed) day. The process also produced the best net energy. However, the COD removal efficiency in thermophilic condition was lower than that obtained using mesophilic conditions.

6.8.1.2 Kitchen Waste and Sewage Sludge

Sharom et al. [230] examined different compositions of kitchen waste (KW) and sewage sludge for co-digestion at 35°C and pH of 7. Five different compositions were examined. The cumulative biogas production increased with co-digestion; the highest production occurred for 75% KW/25% activated sludge that produced 59.7 mL. The biogas production rate was also obtained at this composition. Pure activated sludge gave the least production rate of biogas.

6.8.1.3   Activated Sludge and OFMSW

Bolzonella et al. [180] studied the performances of full-scale anaerobic digesters co-digesting waste activated sludge from biological nutrients removal in waste-water treatment plants, together with different types of organic wastes (solid and liquid). Results showed that the biogas production can be increased from 4,000 to some 18,000 m^3/month when treating some 3–5 tons/day of organic MSW together with waste-activated sludge. On the other hand, the specific biogas production was improved from 0.3 to 0.5 m^3/kg VS fed the reactor, when treating liquid effluents from cheese factories. The addition of the co-substrates gave minimal increases in the OLR, while the HRT remained constant. Further, the potential of the struvite crystallization process for treating anaerobic supernatant rich in nitrogen and phosphorus was studied; an 80% removal of phosphorus was observed in all the tested conditions.

In a separate study, Bolzonella et al. [140], also examined two full-scale applications of the anaerobic co-digestion of activated sludge with OFMSW. The studies were carried out at Viareggio and Treviso wastewater plants in Italy. In the first plant, 3 tons/day of source-sorted OFMSW was co-digested with activated sludge. This process increased the OLR from 1.0 to 1.2 kg TVS/(m^3 day) and a 50% increase in biogas production. At Treviso WWTP, which has been working for 2 years, some 10 tons/day of separately collected OFMSW were treated using a low-energy consumption sorting line, which allowed the removal of 99% and 90% of metals and plastics, respectively. In these conditions, the biogas yield increased from 3,500 up to 17,500 m^3/month. The payback time for the new co-digestion process was estimated to be 2 years. Delia and Agdag [231] examined anaerobic co-digestion of industrial sludge with MSWs in anaerobic simulated landfilling reactors.

6.8.1.4   Sewage Sludge and Orange Peel Waste

Serrano et al. [130] examined co-digestion of sewage sludge and orange peel waste in a proportion of 70:30 (wet weight), respectively. Mesophilic AD of sewage sludge gave low methane yield, poor biodegradability, and nutrient imbalance. The co-digestion improved the viability of the process. The stability was maintained within correct parameters throughout the process, while methane yield coefficient and biodegradability were 165 L/kg VS (0°C, 1 atm.) and 76% (VS), respectively. The OLR increased from 0.4 to 1.6 kg VS/(m^3 day). Nevertheless, the OLR and methane production rate decreased at the highest loads, suggesting the occurrence of an inhibition phenomenon.

6.8.1.5   Co-Digestion of Municipal Organic Waste and
Pretreatment to Enhance Biogas Production
from Wastewater Treatment Plant Sludge

Li [175] examined co-digestion of organic waste with wastewater treatment plant sludge in four phases. The final aim of the study was to evaluate optimum co-digestion conditions by adding selected co-substrates and by incorporating optimum pretreatment strategies for the enhancement of biogas production from anaerobic co-digestion using wastewater treatment plant sludge as the primary substrate.

In the first phase, the feasibility of using municipal organic wastes (synthetic KW and fat, oil, and grease [FOG]) as co-substrates in anaerobic co-digestion was investigated. KW and FOG positively affected biogas production from anaerobic co-digestion, with ideal estimated substrate/inoculum (S/I) ratio ranges of 0.80–1.26 and 0.25–0.75, respectively. Combined linear and nonlinear regression models were employed to represent the entire digestion process and demonstrated that FOG could be suggested as the preferred co-substrate.

The effects of ultrasonic and thermochemical pretreatments on the biogas production of anaerobic co-digestion with KW or FOG were investigated in the second phase. Nonlinear regressions fitted to the data indicated that thermochemical pretreatment could increase methane production yields from both FOG and KW co-digestion. Thermochemical pretreatments of pH = 10, 55°C, provided the best conditions to increase methane production from FOG co-digestions.

In the third phase, using the results obtained previously, anaerobic co-digestions with FOG were tested in bench-scale semicontinuous flow digesters at Ravensview Water Pollution Control Plant, Kingston, ON. The effects of HRT, OLR, and digestion temperature (37°C and 55°C) on biogas production were evaluated. The best biogas production rate of 17.4 ± 0.86 L/day and methane content of about 67.9% were obtained with thermophilic (55°C) co-digestion at HRT of 24 days and OLR of about 2.43 g TVS/(L day).

In the fourth phase, with the suitable co-substrate, optimum pretreatment method and operational parameters identified from the previous phases and anaerobic co-digestions with FOG were investigated in a two-stage thermophilic semicontinuous flow co-digestion system modified to incorporate thermochemical pretreatment of pH = 10 at 55°C. Overall, the modified two-stage co-digestion system yielded a 25.14 ± 2.14 L/day (with $70.2\% \pm 1.4\%$ CH_4) biogas production, which was higher than that obtained in the two-stage system without pretreatment.

6.8.1.6 Food Waste with Sewage Sludge

Iacovidou et al. [181] examined FW co-digestion with sewage sludge in the United Kingdom. They identified the following constraints of FW co-digestion with sewage sludge.

Co-digestion of FW with sewage sludge can be limited due to the high variability of FW. A stable digestion performance depends on the composition of FW added to sewage sludge, which if changed can cause instability in the anaerobic population and consequently in the digestion process. This is because microorganisms are acclimatized in a specific mixture, and changes in this mixture may result in changes in the process reactions. In addition, CH_4 yields may also vary due to seasonal variations in FW [191,192].

The concentration of light metal ions and biodegradation intermediates plays an important role in the smooth process performance, as they can be the most potent causes of toxicity in AD. A compound can be described as toxic or inhibitory when it causes an adverse change in the microbial population or halts bacterial growth [193]. By increasing the fraction of FW, the risk of increasing the concentration of light metal ions and/or biodegradation intermediates at levels that can be toxic to the anaerobic population becomes greater.

Light metal ions, also known as cations, are present in many types of FW and, although essential for the growth of anaerobic microorganisms, at high concentrations can be a cause of toxicity in the digestion process due to the effect of osmosis [193,194]. Co-digestion of sewage sludge with FW rich in vegetables is likely to show an increase in the potassium (K) content that may inhibit the digestion process [195]. Sodium (Na) inhibition is also likely to occur since FW is a source of Na [194,195].

VFAs, LCFAs, and ammonia (NH_3) are the main biodegradation intermediates of AD. The accumulation of these intermediates beyond certain levels in the anaerobic digester can be a cause of toxicity in the co-digestion process [102,196].

The addition of FW to sewage sludge results in an initial increase in VFAs concentration due to the rapid acidification of soluble organic compounds found in FW [166,197,198]. VFAs are subsequently decreased as a result of their uptake by the anaerobic microorganisms [194,197–200]. However, when the production rate surpasses the uptake rate, the excessive levels of VFAs produced are accumulated in the digester. This accumulation can cause acidity in the digester which if not restored can lead the pH to drop to such a level that stops the digestion process from occurring [40,166,197,201,202]. Levels at which VFAs become toxic have not been documented; however, it was reported that VFAs can be present in the digester at concentrations up to 6000 mg/L without being toxic, provided that the pH is maintained in the optimal range [203].

A high lipidic content in FW may also affect the co-digestion process with sewage sludge, due to the excessive production of LCFAs. LCFAs have been shown to be toxic to the anaerobic population, and the higher their concentration in the digester, the more toxic their effect is [193,194,204]. The levels at which LCFAs can be toxic vary widely depending on the predominant form of acid present in the digester [193,194].

The degradation of protein-rich mixtures of FW and sewage sludge is the primary cause of NH_3 production and accumulation [205,206]. NH_3 and ammonium ion (NH_4^+) concentrations both exist in anaerobic digesters. NH_4^+ can inhibit the activity of methanogens and, hence, CH_4 production. However, NH_3 was reported to be more inhibitory than NH_4^+ because of its capability to penetrate through cell membranes [166,193,207,208]. However, there is an uncertainty as to the range at which NH_3 concentrations can be inhibitive [193]. This uncertainty is mainly because of the differences in operational conditions, such as alkalinity, temperature, substrate composition, and acclimation period [40,166,193]. Mixtures with a higher proportion of FW than sewage sludge have a limited possibility for NH_3 inhibition, mainly because of the higher carbon availability [166].

Operational constraints associated with FW handling are extremely important. Impurities found in FW such as plastics, metals, glass, and other packaging parts are likely to cause tremendous technical problems in the wastewater treatment line and co-digestion performance [202,209–211]. Plastics in the form of plastic bags can be wrapped around the stirring equipment in storage and reactor tanks, wear out the pumps, and form a top layer in the reactors. Furthermore, plastic contamination in the form of phthalates can change the quality of the digestate and make it unacceptable for application to agricultural land [202,212]. Metals are toxic for the bacterial biomass, whereas lignocellulosic materials such as wood and paper are digested slowly and are troublesome [211]. Mechanical problems such as clogging in the conveyor line may also arise from metal contamination [213]. These can lead to increases in operational costs and also to the loss of a relatively important fraction of FW during the pretreatment process. This loss can be associated with a reduction in the benefits for the water industry, as less biogas will be produced per ton of waste delivered [202]. Therefore, it is of critical importance that FW is free of impurities and exogenous material prior to co-digestion with sewage sludge.

Overloading the digesters must also be avoided in order for the process to operate successfully. Overloading episodes can cause a decrease in CH_4 yields, and even a failure of the whole process, while blocking of the pipes and foaming incidents in the digester are also possible [159,166,214,215]. Increased foaming, blocking of pipes, and insufficient mixing of the substrates resulting from digester overloading can ultimately lead to a loss in biogas production.

In the United Kingdom, the current operational and regulatory framework makes anaerobic co-digestion of FW with sewage sludge a rather complicated matter. Sewage sludge and FW are both covered by different regulatory regimes. When FW is co-digested with sewage sludge, regulation becomes more complex and unclear, with the process standing between two sets of regulations [216]. Waste management license requirements, the quality of the co-digestate, and renewable energy generation credits are some of the regulatory constraints and uncertainties that currently prevent the adaptation of co-digestion of sewage sludge with FW by the water industry.

The processing of FW that contains or has been contaminated by meat and any other animal materials falls into the animal by-products regulations (ABPR). The ABPR defines three categories of ABPs, with category three being the least harmful. This category includes FW originating from households, restaurants, and catering facilities [217,218]. As such, the water industry must have an ABP permit to be able to process FW. To meet the requirements set by the ABPR, FW pretreatment is required first to ensure the removal of packaging material and other impurities that may be present in FW and second to pasteurize the FW before being added to the digester [216]. The aim of the ABPR is to sanitize FW and prevent pathogen transfer. The use of co-digestate may also be covered by the ABPR although this is not yet clearly defined [219]. This also applies to other European countries that make the application of the ABPR a rather complicated matter.

The co-digestate produced from the digestion of mechanical-segregated FW with sewage sludge is considered a waste digestate in the revised Waste Framework Directive, and as thus, it cannot be applied directly to land [217]. This would leave the water industry to deal with a huge amount of co-digestate if not otherwise managed.

Economic complications related to the planning and operation of the co-digestion process can be a significant barrier to its adaptation. The building of new digesters, the upgrade of existing ones, and the installation of facilities for the delivery, pretreatment, and storage of FW [195,202,220], required for co-digestion, cannot currently be easily provided due to economic barriers and policy restrictions. These regulations impose a level of complexity that exceeds the potential of co-digestion, and initiatives to implement this process have been severely hindered. This is not surprising as individual approaches in waste management are challenging and bound to become unsustainable.

6.8.1.7 Algal Sludge and Wastepaper

Yen and Brune [128] examined co-digestion of algal sludge and wastepaper. The unbalanced nutrients of algal sludge (low C/N ratio) were regarded as an important limitation factor to AD process. Adding high carbon content of wastepaper in algal sludge feedstock to have a balanced C/N ratio was undertaken in this study. The results showed adding 50% (based on VS) of wastepaper in algal sludge feedstock increased the methane production rate to about 1170 mL/L day, as compared to about 573 mL/L day of algal sludge digestion alone, both operated at 4 g VS/L day, 35°C and HRT of 10 days. The maximum methane production rate of about 1607 mL/L day was observed at a combined 5 g VS/L day loading rate with 60% (VS based) of paper adding in algal sludge feedstock. Results suggested an optimum C/N ratio for co-digestion of algal sludge, and wastepaper in the range of 20–25/L.

6.8.1.8 Fat, Oil, Grease, and Wastewater Treatment Facility

Long et al. [63] examined co-digestion of high organic strength FOG from restaurant grease abatement devices with wastewater treatment facility. Addition of FOG substantially increased biogas production. Co-digestion of FOG with municipal biosolids at a rate of 10%–30% FOG by volume of total digester feed caused a 30%–80% increase in digester gas production in two full-scale wastewater biosolid anaerobic digesters. However, AD of high lipid wastes has been reported to cause inhibition of acetoclastic and methanogenic bacteria, substrate, and product transport limitation, sludge floatation, digester foaming, blockage of pipes and pumps and clogging of gas collection, and handling systems. Long et al. [63] also reviewed the scientific literature on biogas production, inhibition, and optimal reactor configurations and highlighted future research needed to improve gas production and overall efficiency of anaerobic co-digestion of FOG with biosolids from municipal wastewater treatment.

6.8.1.9 Wastewater Using Decanter Cake

Kaosol and Sohgrathok [190] examined the co-digestion of wastewater using decanter cake. The wastewater from agro-industry cannot produce biogas by biological treatment because of its low COD level and low organic content. The study examined the effect of three parameters, type of wastewater, mixing, and mesophilic temperature on the co-digestion process. The study measured the biogas production of wastewater alone along with various mixtures of decanter cake and wastewater. The co-digestion of decanter cake with rubber block wastewater of the R4 (wastewater 200 mL with decanter cake 8 g) produced the highest biogas yield of 3.809 L CH_4/g COD removal with maximum methane gas of 66%. The study also showed that the mixing and mesophilic temperature did not have significant effect on the biogas potential production. The co-digestion of decanter cake with rubber block wastewater provided the highest biogas yield potential production at the ambient temperature. The decanter cake can be a potential source for biogas production.

6.8.1.10   Grease Interceptor Waste in Wastewater Treatment Plant

Aziz et al. [131] showed that grease interceptor waste (GIW) could make an ideal co-digestion feedstock. GIW is comprised of the FOG, food solids, and water collected from food service establishment grease interceptors. Two ongoing research projects were attempted to address different issues related to GIW co-digestion at municipal WWTFs. The first project was an experimental evaluation of the limits of biogas production from GIW co-digestion with municipal sludge. The second project is developing a life-cycle decision support tool to explore the environmental and economic implications of GIW co-digestion. Two lab-scale anaerobic digesters were operated under mesophilic conditions. One reactor served as a control digesting only thickened waste activated sludge (TWAS), while the other reactor received a combined feed of TWAS and GIW at varying volume fractions. At 20% GIW by volume, the second reactor showed an increase of biogas production of 317% from the control. However, at 40% GIW by volume, the digester showed a drop of methane production and signs of process failure.

These preliminary results suggested great promise for enhanced biogas production with GIW as a co-digestion feed. The anaerobic co-digestion decision support tool provided an economic and life-cycle assessment framework to explore the implications of enhanced biogas production during the anaerobic co-digestion of GIW with municipal sludge. Results indicated that while the co-digestion of GIW is environmentally favorable due to the enhanced offset of natural gas and fossil-fuel-derived energy, the economics are sensitive to regional effects and available infrastructure. Despite this fact, co-digestion substantially reduces the start-up cost for AD at facilities that presently use alternative solids handling.

6.8.1.11   Food Waste and Wastewater

Cheerawit et al. [158] investigated the potential of biogas production from the co-digestion of domestic wastewater and FW. Batch experiments were carried out under various substrate ratios of domestic wastewater and FW at 10:90, 25:75, 50:50, and 70:30 at room temperature. The results showed that co-digestion of domestic wastewater with FW was very promising for the production of methane gas. The BMP and COD removal efficiency were 61.72 mL CH_4/g COD and 75.77%, respectively. Moreover, the addition of FW to the AD of domestic wastewater showed an increasing trend of the biogas production. The laboratory batch study revealed that the use of FWs as co-substrate in the AD of domestic wastewater also has other advantages: that is, the improvement of the balance of the C/N ratio and efficient process stability.

6.8.1.12   Olive Mill Wastewater and Swine Manure

Azaizeh and Jadoun [171] examined co-digestion of OMW and SM using up-flow anaerobic sludge blanket (UASB) reactor. Swine wastewater (SWW) and OMW are two problematic wastes that have become major causes of health and environmental concerns. The main objective of this work was to evaluate the efficiency of the co-digestion strategy for treatment of SWW and OMW mixtures. Mesophilic batch reactors fed with mixtures of SWW and OMW showed that the two adapted sludges,

Gadot and Prigat, exhibited the best COD removal capacity and biogas production; therefore, both were selected to seed UASB continuous reactors. During 170 days of operation, both sludges, Gadot and Prigat, showed high biodegradation potential. The highest COD removal of 85%–95% and biogas production of 0.55 L/g COD were obtained in a mixture consisting of 33% OMW and 67% SWW. Under these conditions, an organic load of 28,000 mg/L COD was reduced to 1,500–3,500 mg/L. These results strongly suggest that co-digestion technology using UASB reactors is a highly reliable and promising technology for wastewater treatment and biogas production.

6.8.1.13 Co-Digestion of Swine Wastewater with Switchgrass and Wheat Straw

Liu [183] examined anaerobic co-digestion of swine wastewater (SWW) with switchgrass (SG) and wheat straw (WS) for methane production. The effects of different TSs concentration (2%, 3%, 4%) on the methane yield from the co-digestion with SG and WS were investigated in batch mode at mesophilic temperature (35°C). The culture from a completely mixed and semicontinuously fed anaerobic reactor treating SWW and corn stover was used as the inoculum for the batch tests. The reactor had a working volume of 14 L and was operated at 35°C with a HRT of 25 days and an OLR of 0.924 kg VS/(m^3 day). Batch reactors were operated in triplicates, each with a working volume of 500 mL. Reactors were kept in thermostatic water bath maintained at mesophilic temperature (35°C ± 1°C) with an agitation speed of 270 rpm. The volume of methane produced in the experiment was measured by gas meters. COD, pH, total Kjeldahl nitrogen (TKN), total organic carbon, TSs, and VSs analyses of reactor contents were performed at the beginning and the end of the experiment.

The results indicated that with the addition of SG, the methane production substantially increased. The methane yields at 2%, 3%, and 4% TS were 0.137, 0.117, and 0.104 m^3/kg VS added, respectively. The addition of WS in the batch reactors resulted in higher accumulated methane production, and the methane yield was 0.133 m^3/kg VS added at 2% and 3% TS concentrations. However, when the TS increased to 4%, methane production decreased because VFA accumulation increased rapidly and pH dropped to below 5.5. The first-order kinetic model was evaluated for the methane production. It was found that the model fitted the experimental data well. The study concluded that batch anaerobic co-digestion of swine waste with SG and WS at low TS concentration is a commercially viable process.

6.8.1.14 Single-Source Oily Waste and High-Strength Pet Food Wastewater

Acharya and Kurian [176] focused on the revival of a formerly failed digester of 1800 m^3 volume. The most obvious cause of failure was identified to be due to capping caused by foam and scum, as a result of attempting to treat oil-rich, high-strength wastewater. The revival was affected by implementing co-digestion in a three-step remedial procedure. Though, in the typical sense, co-digestion involved separate waste streams, here, a single waste stream was manipulated to apply the concept of co-digestion. Through the virtual co-digestion, the digester succeeded to treat the daily plant effluent flow of 50 m^3 with COD > 45 g/L and around 10 tons of sludge/day that had around 20% oil and fat, exhibiting a COD removal >90%.

The digester operated at $35°C \pm 1°C$ and HRT of 30 days with loading rates of 3.4 kg COD/(m^3 day) and 1.3 kg O&G/(m^3 day). The biogas generated from this digester was sufficient to operate a 40 hp boiler at 100 psi.

6.8.1.15 Co-Digestion of Chicken Processing Wastewater and Crude Glycerol from Biodiesel

Foucault [182] examined co-digestion of chicken processing wastewater and crude glycerol from biodiesel. The main objective of the study was to examine the AD of wastewater from a chicken processing facility and of crude glycerol from local biodiesel operations. The AD of these substrates was conducted in bench-scale reactors operated in the batch mode at 35°C. The secondary objective was to evaluate two sources of glycerol as co-substrates for AD to determine if different processing methods for the glycerol had an effect on CH_4 production. The biogas yields were higher for co-digestion than for digestion of wastewater alone, with average yields at 1 atm and 0°C of 0.555 and 0.540 L/(g VS added), respectively. Another set of results showed that the glycerol from an on-farm biodiesel operation had a CH_4 yield of 0.702 L/(g VS added) and the glycerol from an industrial/commercial biodiesel operation had a CH_4 yield of 0.375 L/(g VS added). Therefore, the farm glycerol likely had more carbon content than industrial glycerol. It was believed that the farm glycerol had more impurities, such as free fatty acids, biodiesel, and methanol. The co-digestion, of chicken processing water and crude glycerol, thus increased production of methane rich biogas.

6.8.1.16 Excess Brewery Yeast Co-Digestion in a Full-Scale Expanded Granular Sludge Bed Reactor

Zupancic et al. [138,223] examined the anaerobic co-digestion of brewery yeast and wastewater. The study showed that such additional loading of the anaerobic process did not destroy or damage the operation of the full-scale system. Full-scale co-digestion at concentration of vol. $0.7\% \pm 0.05\%$ showed no negative impacts. With the additional brewery yeast (0.7%), a 38.5% increase of biogas production was detected, which resulted in an increase of the biomethane/natural gas substitute ratio in the brewery from 10% to 16%.

6.8.2 CO-DIGESTION WITH DIFFERENT TYPES OF MANURE

6.8.2.1 Cow Manure and High Concentrated Food Processing Waste

Yamashiro et al. [105] examined anaerobic co-digestion of dairy CM and high concentrated food processing waste, under thermophilic (55°C) and mesophilic (35°C) conditions. Two types of feedstock were studied: 100% DM and 7:3 mixture (wet weight basis) of dairy manure (DM) and food processing waste (FPW). The contents of the FPW as feedstock were 3:3:3:1 mixture of cheese whey, animal blood, used cooking oil, and residue of fried potato. Four continuous digestion experiments were carried out in 10 L digester. The results showed that co-digestion under thermophilic conditions increased methane production per unit digester volume. However, under mesophilic condition, co-digestion was inhibited. TKN recovered after digestion ranged from 73.1% to 91.9%, while recoveries of ammonium nitrogen (NH_4–N) exceeded 100%. The high recovery of NH_4–N was attributed to mineralization of influent organic N. The mixtures of DM and FPW showed greater recoveries of NH_4–N then DM only, reflecting its greater organic N degradability. The ratios of extractable to total calcium, phosphorus, and magnesium were slightly reduced after digestion. These results indicated that co-digestion of DM and FPW under thermophilic temperature enhances methane production and offers additional benefit of organic fertilizer creation.

Jepsen [6] examined co-digestion of animal manure with organic household waste. He found that co-digestion significantly increased biogas production. Gas yield from manure was only 15–20 m^3/tons. On average, 60% gas production increased from addition of waste. Along with household waste, industrial waste and sludge can also be used to improve biogas production. He also noted that among organic waste to improve gas yield, an order follows:

1. Concentrated fat, fish silage, etc.—200–1000 m^3/tons
2. Fish waste, fat, flotation sludge, slaughterhouse waste, dairy waste, and organic household waste—50–200 m^3/tons
3. FVW, industrial wastewater, and sewage sludge—10–50 m^3/tons

Jepsen [6] concluded that co-digestion with animal manure results in a stable process.

Zhang et al. [129] evaluated anaerobic digestibility and biogas and methane yields of the FW in order to examine its suitability as co-substrate in co-digestion. The tests were performed at 50°C in a batch fermenter. The daily average MC and the ratio of VSs to TSs were 70% and 83%, respectively. The FW contained well-balanced nutrients. The methane yield was 348 mL/g VS after 10 days and 435 mL/g VS after 28 days. The average methane content of biogas was 73%. The average VS destruction at the end of 28 days was 81%. All these data indicate that FW is a good co-substrate for AD.

6.8.2.2 Wastepaper and Cow Dung and Water Hyacinth

Yusuf and Ify [228] carried out the co-digestion of cow dung and water hyacinth in a batch reactor for 60 days with the addition of various portions of wastepaper. The biogas production was measured keeping the amount of cow dung and water hyacinth fixed and variable amounts of wastepaper. Maximum biogas volume of 1.1 L was observed at a wastepaper amount of 17.5 g, which correspond to 10% of TSs of biomass in 250 mL solution.

6.8.2.3 MSW and Cow Manure

Samani et al. [232] examined co-digestion of OFMSW and dairy CM and found that while OFMSW produced 62 m^3 CH_4/tons while digesting alone and dairy CM produced 37 m^3/tons while digesting alone, the co-digestion produced 172 m^3 CH_4/tons of dry waste. Thus, co-digestion gave higher methane yields.

Hartmann et al. [202,212] studied co-digestion of OFMSW and manure under thermophilic conditions at 55°C. Various concentrations of OFMSW and manure were examined at HRT of 14–18 days and OLR of 3.3–4.0 g VS/(L day) over a period of 6 weeks. The experiments were started with OFMSW–manure ratio of 1:1, and this ratio was gradually increased with time over 8 weeks. Use of recirculated process liquid to adjust organic loading had a stabilizing effect. When the pH raised to 8, the reactor showed stable performance with high biogas yield and low VFA. Biogas yield was 180–220 m^3 biogas/tons of OFMSW both in co-digestion configuration and in the treatment of 100% OFMSW with process liquid recirculation. VS reduction of 69%–74% was achieved when treating 100% OFMSW. None of the processes showed signs of inhibition at the free ammonia concentration of 0.45–0.62 g N/L.

6.8.2.4   Dairy Manure and Food Waste

El-Mashad et al. [226] and El-Mashad and Zhang [233] examined biogas production of cattle manure using sunlight and biogas production of different mixtures of unscreened dairy manure and FW, respectively. In the latter study, the effect of manure screening on the biogas yield of dairy manure was also evaluated. This study showed that two mixtures, (1) unscreened manure (68%) and FW (32%) and (2) unscreened manure (52%) and FW (48%), produced methane yields of 282 and 311 m^3/kg VS, respectively, after 30 days of digestion. After 20 days, approximately 90% and 95% of final biogas was obtained. The average methane content was 62% and 59% for the first and second mixtures, respectively. The predicted results from the model showed that adding the FW into manure digester at levels up to 60% of the initial VSs significantly increased the methane yield for 20 days of digestion.

Crolla et al. [184,225] studied the benefits of addition of co-substrates such as energy crops, industrial wastes, or food industry wastes to manure. They noted that the addition of co-substrate to the manure can improve C/N balance that results in a stable and sustainable digestion process. The optimum C/N ratio appeared to be somewhere between 20:1 and 30:1. Co-digestion also improved the flow qualities of the co-digested substrates. The optimum HRT for dairy manure was 12–25 days and for cattle manure was 15–35 days. The optimum OLR was around 3.5–5.5 kg VS/(m^3 day). The optimum pH for manure was also between 6.8 and 7.2. The optimum temperature for mesophilic operation was 35°C–40°C and for thermophilic operation was 55°C–63°C. Co-digestion with manure also improved biogas production rate and generated additional tipping fees for the use of additional wastes. The study also presented the data on optimum biogas and methane yields associated with co-digestion of manure with corn silage, SG, canola seed cake, whey, waste grease, FW, and corn silage. Some of the mean biogas and CH$_4$ yield data for these systems are shown in Table 6.3. Crolla and Kinsley [225] concluded that co-digestion of manure and co-substrates resulted in

1. Reduced CH$_4$ gas emission from storage reservoirs holding digestate
2. Reduced N$_2$O gas emissions from the land application of digestate
3. Reduced odors in both storage reservoirs and during land application
4. Reduced pathogens and weed seeds in the digestate
5. Improved fertilizer value of the digestate by transforming nutrients into more readily available inorganic forms

Although nutrients in the digestate were readily available for plant uptake, it can be lost in the absence of plants. Cover crops were used to hold the nutrients.

TABLE 6.3

Biogas and Methane Yields from Digestion of Dairy Manure with Co-Substrate

Feedstock Composition	Yields of Biogas/Methane (m³/kg VS$_{Initial}$)
Dairy manure	0.378/0.235
30/70 Corn stillage/dairy manure	0.465/0.305
30/70 Corn stillage/dairy manure (repeat)	0.527/0.301
40/60 Corn stillage/dairy manure	0.630/0.402
30/70 Waste grease/dairy manure	0.511/0.358
40/60 Waste grease/dairy manure	0.569/0.398
30/70 Whey/dairy manure	0.433/0.303
30/70 Switch grass/dairy manure	0.479/0.285

Source: Crolla, A. et al., Anaerobic co-digestion of corn thin stillage and dairy manure, in: *Canadian Biogas Conference and Exhibition*, London Convention Center, London, Ontario, Canada, March 4–6, 2013.

A company called ANTARES in Wyoming County, New York, also examined co-digestion of manure with substrates like biomass crops, agricultural residues, and FOG in local situation. These results can be obtained by directly contacting ANTARES.

6.8.2.5 Pig Slurry and Organic Waste from Food Industry

The main inhibitor in the AD of pig slurry is the release of free ammonia. Campos et al. [189] examined the co-digestion of pig slurry with organic waste from food industry such as wastes from fruit and olive oil refineries (pear waste and oil bleaching earth). Batch experiments at both mesophilic (35°C) and thermophilic (55°C) conditions were performed. Due to large inhibition by ammonia in thermophilic conditions, the data for mesophilic conditions were better than those for thermophilic conditions. In both temperature conditions, however, methane production was improved by the addition of a co-substrate. Higher methane production was obtained from the co-digestion of slurry and oil bleaching earth (95% and 5%, respectively). The methane yield was 344 mL CH$_4$/(g VS$_{initial}$), which was 2.4 times the methane yield for slurry (144 mL CH$_4$/[g VS$_{initial}$]).

6.8.2.6 Goat Manure with Three Crop Residues

Zhang et al. [188] examined co-digestion of goat manure (GM) with WS, corn stalks (CSs), and rice straw (RS). GM is an excellent material for AD because of its high nitrogen content and fermentation stability. The experiments were performed under mesophilic conditions. With a TS concentration of 8% and different mixing ratios, results showed that the combination of GM with CS and RS significantly improved biogas production at all C/N ratios. GM/CS (30:70), GM/CS (70:30), GM/RS (30:70), and GM/RS (50:50) produced the highest biogas yields from different co-substrates after 55 days of fermentation. Biogas yields of GM/WS 30:70 (C/N 35.61), GM/CS 70:30 (C/N 21.19), and GM/RS 50:50 (C/N 26.23) were 1.62, 2.11, and 1.83 times higher than that of CR's, respectively. These values were determined to be optimum C/N ratios for co-digestion. However, compared to GM/CS and GM/RS treatments, biogas generated from GM/WS was only slightly higher than the single-digestion GM or WS. This result was caused by the high total carbon content (35.83%) and lignin content (24.34%) in WS, which inhibited biodegradation.

6.8.2.7 Water Hyacinth with Poultry Litter versus Water Hyacinth with Cow Dung

Patil et al. [174] compared the performance of two co-digestion systems: water hyacinth with poultry litter and water hyacinth with cow dung in mesophilic conditions with temperature ranging from 30°C to 37°C in a batch digester with retention period of 60 days. The TS concentration was 8% in each sample. The results showed that co-digestion of water hyacinth with poultry litter produced more biogas then co-digestion of water hyacinth with cow dung. The overall results showed that blending water hyacinth with poultry waste had significant improvement on the biogas yield.

6.8.2.8 Slaughterhouse Waste with Various Crops, MSW, and Manure

Siripong and Dulyakasem [106] examined the co-digestion of different agro-industrial wastes. The potential of methane production and the effects of a second feed were determined in batch AD experiments. It was shown that co-digestion of SB/VC (slaughterhouse waste and various crops), SB/VC/MSW, and SB/M (manure) provided high methane potentials. The highest methane yields obtained were 592, 522, and 521 mL/g VS, respectively, in these samples. Moreover, the second feeding could increase the methane yield of some of the substrate mixtures, due to building up of an active microbial consortium. In contrast, decreasing yields or inhibition was detected in some other substrate mixtures.

The study also examined long-term effects during co-digestion of slaughterhouse waste in four continuously stirred tank reactors. In a continuous process, the start-up stage is really important, the OLR should be low, and then it should be slightly increased gradually to avoid overload in the system and for the adaptation of microorganisms to the substrate. VFAs, alkalinity, and ammonium–N concentrations were used as control parameters for the operation of the continuous systems. The methane content of the produced biogas during the digestion and co-digestion of slaughterhouse waste was obtained between 60% and 85% (lower in the beginning and higher toward the end), and the highest methane content of 76% was found from co-digestion of SB/M toward the end of the operation. Toward the end of the investigation period, average methane yields of 300, 510, 587, and 426 mL/g VS were obtained in the digestion of SB and co-digestion of SB/M, SB/VC and SB/VC/MSW, respectively. The highest average methane potential of 587 mL/g VS was found in co-digestion of SB/VC, and it is comparable to the result of 592 mL/g VS obtained from the batch digestion of the same mixture.

6.8.2.9 Co-Digestion of Dairy Manure with Chicken Litter and Other Wastes

Canas [143] studied co-digestion of dairy manure with chicken litters and other wastes. The following conclusions were drawn by the author:

1. Chicken litter can be added into a digester treating dairy manure to increase the OLR leading to a higher methane production rate. Chicken litter can be safely added up to a 33% as VS in the feedstock increasing methane production by 49.3%. Other researches [109,153] found a similar maximum chicken manure percentage in feedstock for continuous and batch reactors.
2. No synergistic effects were detected when co-digesting chicken litter with dairy manure. However, chicken litter required water to be digested.
3. The selection of the initial OLR is related to inoculum acclimation and waste composition and depends on a wide number of factors.
4. For dairy manure, two retention times seemed to be enough to reach stable conditions. However, previously, three or four retention times were suggested to reach steady-state conditions.
5. Perhaps, a combined effect of high ammonia concentration and overloading resulted in reactor's collapse. More research needed to be done evaluating the influence of high ammonia concentration under different OLRs.
6. Because of the large total alkalinity in the system, pH and VFAs were not good indicators of instability. Instead, gas production should be followed closely in order to detect any symptom of imbalance.
7. Total and free ammonia tolerance could be improved just by simply combining dairy manure with chicken litter. However, microbial adaptation for free ammonia occurred when increasing free ammonia concentrations in reactors.
8. By establishing the retention time at 20 days, it is possible to recover up to 90% of methane from substrates. In addition, this large retention time allowed the microbial population to better develop free ammonia adaptation.
9. Co-digestion seemed to have no influence in pathogen indicator (*Escherichia coli*) removal. Removal values reached typical values ranging from 68.4% to 97.2%.
10. The microbial population can adapt to lower temperatures down to 19°C, but at longer retention times, this became economically unattractive for continuous reactors. At 20 days of retention time, methane production decreases by 10% when temperature decreases from 35°C to 25°C.
11. Filtered solids from dairy can be co-digested up to a maximum percentage in the feedstock of 70% VSs to increase methane production by 114.2% as a consequence of an increase in organic loading, but the efficiency (methane yield) decreased by 59.14%. Antagonistic effects were also found. Grease trap waste (GTW) can be co-digested improving methane yield (efficiency) and VS removal of dairy manure alone by 111.5% and 76.4%, respectively.
12. Co-digestion of sawdust with dairy manure was unsuccessful.
13. By storing the substrates at a 4°C, both samples and feedstock were preserved properly.

6.8.2.10   Swine Manure with Energy Crop Residues

Cuetos et al. [110–112] examined co-digestion of SM with energy crop residues (ECRs) that contain Mz, rapeseed (Rs), and sunflower (Sf) residues. The behavior of reactors and methane productions in both batch and continuous flow reactors were examined. Three different proportions of ECRs were tested in batch experiments for co-digestion with SM: 25%, 50%, and 75% VS. On the basis of results obtained from the batch study, 50% ECR content was selected as the mixture for the second stage of the study.

The experiments at this stage were performed under mesophilic conditions in semi-continuous reactors with HRT of 30 days, and the reactors were kept under these operational conditions over four HRTs. The results showed that the addition of ECR to the co-digestion system resulted in a major increase in the biogas production. The highest biogas yield was obtained when co-digesting Rs (3.5 L/day), although no improvement was observed in specific gas production from the addition of the co-substrate.

6.8.2.11   Food Waste and Human Excreta

Dahunsi and Oranusi [178] examined co-digestion of FW and human excreta in Nigeria where there exists no centralized sewage system and both of these wastes end up in septic tank of each home. An investigation was launched into the design and construction of an anaerobic digester system from locally available raw materials using local technology and the production of biogas from FWs and human excreta generated within a university campus. The experiment lasted for 60 days using a 40 L laboratory-scale anaerobic digester. The volume of gas generated from the mixture was 84,750 cm^3 and comprised of 58% CH_4, 24% CO_2, and 19% H_2S and other impurities.

The physicochemistry of the feedstock in the digester revealed an initial drop in pH to more acidic range and a steady increase of 4.52–6.10. The temperature remained relatively constant at mesophilic range: 22.0°C–30.5°C throughout the study. The C/N ratio of the feedstock before digestion was within 139:1. Population distributions of the microflora showed aerobic and anaerobic bacteria to include *Klebsiella* spp., *Bacillus* spp., *E. coli*, *Clostridium* spp., and a methanogen of the genus *Methanococcus*. The study concluded that in most developing nations of Sub-Saharan Africa where biomass is abundant and where biogas technology is in its infant stage, the anaerobic co-digestion can be a solution.

6.8.2.12   Poultry Manure and Straw

Babee et al. [229] examined the effects of organic loading and temperature on the anaerobic slurry co-digestion of poultry manure and straw. In order to obtain basic design criteria for AD of a mixture of poultry manure and WS, the effects of different temperatures and OLRs on the biogas yield and methane contents were evaluated. Since poultry manure is a poor substrate, in terms of the availability of the nutrients, external supplementation of carbon had to be regularly introduced, in order to achieve a stable and efficient process.

The complete-mix, pilot-scale digester with working volume of 70 L was used. The digestion operated at 25°C, 30°C, and 35°C with OLRs of 1.0, 2.0, 2.5, 3.0, 3.5, and 4.0 kg VS/(m³ day) and a HRT of 15 days. At a temperature of 35°C, the methane yield was increased by 43% compared to the one at 25°C. Anaerobic co-digestion appeared feasible with a loading rate of 3.0 kg VS/(m³ day) at 35°C. At this state, the specific methane yield was calculated about 0.12 m³/kg VS with a methane content of 53%–70.2% in the biogas. The VS removal was 72%. As a result of VFA accumulation and decrease in pH, when the loading rate was less than 1 or greater than 4 kg VS/(m³ day), the process was inhibited or overloaded, respectively. Both the lower and higher loading rates resulted in a decline in the methane yield.

6.8.2.13   Food Waste with Dairy Manure

Lisboa et al. [187] examined co-digestion of FW with dairy manure. The study showed the importance of performance ATAs before possible co-digestion food products were introduced into AD environments. This study did not give any results on biochemical methane potential assay for this co-digestion system.

6.8.2.14   Cattle Manure and Sewage Sludge

Garcia and Perez [186] examined the influence of composition and temperature on co-digestion of cattle manure and sewage sludge. *Both organic wastes were* from wastewater treatment stations. Co-digestion of sewage sludge and cattle manure has the advantage of sharing processing facilities, unifying management methodologies, reducing operating costs, and dampening investment and temporal variations in composition and production of each waste separately.

The aim of the study was to select suitable operating conditions (both composition and temperature) of anaerobic co-digestion process of cattle manure and sewage sludge to optimize the process for the biogas generation. The batch tests were developed at mesophilic and thermophilic conditions to determine the anaerobic biodegradability of three different mixtures of cattle manure and sewage sludge, both in static or stirring conditions.

The results of the study indicated that the anaerobic biodegradability of raw sludge and cattle manure mixtures was more efficient at thermophilic conditions since a greater elimination of organic matter with a greater methane yield was obtained. The most efficient process corresponded to the mixture with 25% (v/v) of cattle manure and 75% (v/v) of raw sludge with values of 62% and 75.7% of COD and total organic carbon (TOC) removals, respectively, and methane yields of 2200 mL CH_4/g COD_r and 306 mL CH_4/g VS_r, during the total processing time of 12 days. Also, it was verified that higher amount of cattle manure in the mixture meant a higher alkalinity and a greater percentage of methane in biogas. The optimal composition of the mixture selected for thermophilic conditions allowed to reach values three times higher than those obtained in mesophilic conditions for all parameters examined for the generation of biogas.

6.8.2.15   MSW and Agricultural Waste with Dairy Cow Manure

Macias-Corral et al. [142] examined co-digestion of MSW and agricultural waste with dairy CM. Anaerobic co-digestion of dairy CM, the OFMSW, and cotton gin waste (CGW) was investigated with a two-phase pilot-scale AD system. The OFMSW and CM were digested as single wastes and as combined wastes. The single waste digestion of CM resulted in 62 m^3 CH$_4$/tons of CM on dry weight basis. The single waste digestion of OFMSW produced 37 m^3 CH$_4$/tons of dry waste. Co-digestion of OFMSW and CM resulted in 172 m^3 CH$_4$/tons of dry waste. Co-digestion of CGW and CM produced 87 m^3 CH$_4$/tons of dry waste. Comparing the single waste digestions with co-digestion of combined wastes, it was shown that co-digestion resulted in higher methane gas yields. In addition, co-digestion of OFMSW and CM promoted synergistic effects resulting in higher mass conversion and lower weight and volume of digested residual.

6.8.3   OTHER CO-SUBSTRATES

6.8.3.1   Effect of Inoculum Source on Dry Thermophilic Anaerobic Digestion of OFMSW

Forster-Carneiro et al. [227] evaluated the effect of inoculum source on anaerobic thermophilic digestion. They used six different inoculum sources: corn silage, restaurant waste digested mixed with rice hulls, cattle excrement, swine excrement, digested sludge, and swine excrement mixed with digested sludge (1:1). The experiments were carried out at 55°C, and other conditions were 25% inoculum and 30% TSs. Results indicated that digested sludge was the best inoculum source for anaerobic thermophilic digestion of the treatment of OFMSW at dry conditions (30% TS). After 60 days of operation, the COD removal and VS removals by the digester were 44% and 43%, respectively. In stabilization stage, digested sludge showed higher volumetric biogas generated at 78.9 mL/day with methane yield of 0.53 L CH$_4$/g VS. For this stage, cattle excrement and swine excrement with digested sludge were good inoculums.

6.8.3.2   Co-Digestion of Municipal, Farm, and Industrial Organic Waste

Alatriste-Mondragon et al. [185] reviewed 4 years of literature on co-digestion of municipal wastewater treatment plants with co-substrates like wood wastes, industrial organic wastes, and farm wastes. The review was focused on low solids concentration (<10%) systems for batch assays and bench-scale systems. The literature on digestibility of co-digestates, data for performance and monitoring of co-digestion, inhibition of digestion by co-digestates, the design of the process (single or two stages), and operation temperature (mesophilic or thermophilic) were reviewed by these authors.

6.8.3.3   Fats of Animal and Vegetable Origin and Simulated OFMSW

Fernández et al. [37] studied co-digestion of fats of different animals and vegetable origins with OFMSW. Co-digestion process was conducted at the pilot scale in semi-continuous regime under mesophilic, 37°C temperature condition and for HRT of 17 days. Dry pet food was used as OFMSW and fat used consisted of waste from food industry (animal fat) with prescribed LCFA profile. The fat concentration was raised up to 28% of the OFMSW, and then it was switched to the vegetable fat. Total fat removal throughout the experiment was 88%, whereas biogas and methane yields were very similar to those simulated for OFMSW. The co-digestion with fat increased the amount of biogas produced according to applied organic loading. Authors recommended the use of co-digestion for this system.

6.8.3.4   Co-Digestion of High-Strength/Toxic Organic Liquid

Ramsamy et al. [132] examined co-digestion of high-strength/toxic organic liquid (i.e., leachate from a hazardous landfill site [Shongweni]) with textile size effluent (Frametex size effluent). The results of ATA proved the amenability of the size effluent and landfill leachate. These results suggested that co-digestion of these wastes was possible. The results showed that co-digestion was possible at all the sample dilutions tested, that is, 4%–40% by volume. Authors conclude that the study needed some additional work.

6.8.3.5   Co-Digestion of STP-FOGW with SC-OFMSW

Martin-Gonzalez et al. [139] examined mesophilic co-digestion of fat, oil, and grease waste from sewage treating plant (STP-FOGW) with source-collected OFMSW (SC-OFMSW) at a feed ratio of 15% (VS) carried out in a 5 L lab-scale reactor that resulted in an improvement both in terms of biogas production (72% higher) and methane yield (46% higher) in comparison with anaerobic treatment of SC-OFMSW. During the co-digestion process, a stable reactor performance was observed, and there was no inhibition either in LCFA accumulation or in VFA excess. VS and TS reduction percentages were stable and around 65% and 57%, respectively, and methane content in biogas was 63%. These results suggested that anaerobic co-digestion is a feasible and efficient way of managing STP-FOGW. Moreover, it is an environmentally friendly treatment in comparison with the landfill option and allows a methane potential that is presently being wasted to be recovered.

6.8.3.6   Biologically Pretreated Nile Perch Fish Solid Waste with Vegetable Fraction of Market Solid Waste

Kassuwi et al. [177] examined anaerobic co-digestion of various organic wastes with fish wastes. Anaerobic co-digestion of various organic wastes has been shown to improve biogas yield of fish wastes. The study presented the effect of pretreating Nile perch fish solid waste (FSW) using CBR-11 bacterial culture (CBR-11-FSW) and commercial lipase enzyme (Lipo-FSW), followed by batch anaerobic co-digestion with vegetable fractions of market solid waste in various proportions, using potato waste (PW) and cabbage waste (CW) as co-substrates either singly or combined.

Results indicated that CBR-11 pretreated FSW co-digested with PW or CW in 1:1 ratio (substrate/inoculum) had positive effect on methane yield, while Lipo pretreated FSW had negative effect on methane yield. Using CBR-11-FSW–PW, the highest yield was 1.58 times more than the untreated FSW. Whereas using Lipo-FSW–CW, the highest yield was 1.65 times lower than the one for untreated FSW. Furthermore, the optimal mixture of CBR-11 pretreated FSW and PW and CW co-substrates resulted in higher methane yield of 1322 CH_4 mL/g VS using CBR-11-FSW (10)–PW (45)–CW (45) ratio. The ratio enhanced methane yield to 135% compared to control. The results demonstrated that optimal mixture of CBR-11 pretreated FSW with both PW and CW as co-substrates enhanced methane yield.

6.8.3.7   Energy Crops as Co-Substrate

The design of the fermenters can differ slightly, depending on the technical solutions applied. While the addition of energy crops improves the performance of AD process, it comes with the additional cost. The cultivation of energy crops requires heavy machinery, diesel fuel, and synthetic fertilizer besides labor, all of which adds substantial cost. Commonly, energy crops are fed together with manure or other liquid substrates (co-digestion), in order to keep homogeneous fermentation conditions. Similar to "wet digestion," the TS content of these systems has to remain below 10% in order to enable proper reactor stirring. Recirculation of digestate is required in such digesting systems in order to maintain homogeneous and well-buffered digester conditions. However, some designs of "dry fermentation" systems allow TSs contents much higher than 10% TS. Without addition of liquid, the TSs content can increase above 30%. Typically, two-step, stirred tank, serial reactor designs are applied in most digestion plants. The second digester is often combined with a membrane-type gas holder. One-step digesters are rarely used.

The AD of energy crops requires, in most cases, prolonged hydraulic residence times from several weeks to months. Both mesophilic and thermophilic fermentation temperatures are commonly applied in AD of energy crops. Complete biomass degradation with high gas yields and minimized residual gas potential of the digestate is a must in terms of proper economy, as well as ecological soundness of the digestion process. VSs degradation efficiencies of 80%–90% should be realized in order to achieve sufficient substrate use, thereby leading to negligible emissions (CH_4, NH_3) from the digestate.

Energy crops like Mz, Sf, grass, and beets are increasingly added to agricultural digesters either as co-substrates or as the main substrate. The cultivation of energy crops on fallow or set-aside land can reduce agricultural surpluses and provide new income for agriculture. The most popular crop today is Mz. From 1 ton of Mz (dry matter), approximately 400–600 m^3 biogas can be produced. Approximately 8,000–12,000 m^3 biogas (50% CH_4 content) produces about 13,200–19,800 kW h electricity. Energy crop digestion is critically dependent on obtainable price of electricity per kW h. The capital payback time of evaluated farm-scale biogas plants lies between 9 and 13 years, which is high, but reasonable. Provided low crop production costs at high yields per land area and a high biogas yield during fermentation can be achieved, energy crops digestion can become economically viable without subsidies.

6.9   TYPICAL LARGE-SCALE CO-DIGESTION PLANTS

Braun and Wellinger [24] indicated that large-scale co-digestion plants usually have more favorable economics. In other words, the economy of scale also applies to co-digestion process. Braun and Wellinger examined large-scale centralized plants in Denmark with digester volumes of about 4650–6000 m^3 and payback times between 3 and 10 years. They concluded that careful design, layout, and operation and gate fees are essential for economic success. The plant economics also depend on restrictions on usable wastes and options for reuse of digestate. Regulations on hygienization and cost of equipment to satisfy these regulations along with process

FIGURE 6.1 Large-scale centralized co-digestion plant Grindsted, Denmark. (From Braun, R. and Wellinger, A., Potential of co-digestion, IEA Bioenergy, Task 37—Energy from Biogas and Landfill Gas, 2004.)

variables such as mixing requirements, level of heat treatment, and degree of con-taminant removals can also affect the economics of large-scale co-digestion process. Pictures of two large-scale centralized co-digestion plants in Denmark are shown in Figures 6.1 and 6.2.

In North America, significant experience with co-digestion has been obtained at East Bay Municipal Utility District (EBMUD), Riverside, Inland Empire Utility Agency (IEUA), Watsonville, Millbrae, and CMSA in California, West Lafayette in Indiana, Pendleton in Oregon, and Lethbridge and Edmonton in Canada. Two promi-nent California water agencies, EBMUD and IEUA, have implemented and tested co-digestion of FW in existing wastewater treatment biodigesters at pilot scale.

A company called BioConversion Solutions (BCS) of Boston, MA [234], through its patented Advanced Fluidized Co-digestion and Co-generation (AFC^2) technology improves commercial anaerobic co-digestion plant profitability through

1. Enhanced biodegradability of feedstock organic solids that results in up to 80%–90% conversion of VSs even for difficult-to-digest feedstock
2. Higher biogas production yield from feedstock conversion that increases plant output thereby allowing for the acceptance of more feedstock and higher tip fees
3. Near-complete recovery of nutrients (nitrogen, phosphorus, potassium, etc.) contained in the feedstock. (With BCS biodegradability enhancement system, these nutrients can be recovered in solid [struvite], concentrated

FIGURE 6.2   Large-scale centralized co-digestion plant Lemvig, Denmark. (Photo courtesy of Teodorita Al Seadi, Denmark; Braun, R. and Wellinger, A., Potential of co-digestion, IEA Bioenergy, Task 37—Energy from Biogas and Landfill Gas, 2004.)

   liquid, or a combined form depending on the prevailing wholesale prices
   and fertilizer customer preference.)

   4. Up to 80% reduction of residual sludge that otherwise incurs a disposal fee
   5. The processing of mixed feedstock "recipe" including easy-to- and more-
   difficult-to-digest feedstock
   6. The creation of feedstock by cultivating and harvesting high energy crops
   on-site or nearby the owner/operator
   7. The discharge of final water output from the plant into local sewer system or
   waterways without violating nutrient management regulations. (The output
   water may also be reusable for plant or other purposes, such as irrigation
   or industrial use, resulting in additional potential revenue or reduced water
   purchase expenditures.)

The company has developed a number of commercial plants in the United States, which achieve one or more of these seven objectives [234].

   Murphy et al. [10] described an agricultural plant for co-digestion of crop with manure. The plant (see Figure 6.3) was built in 2003, and it is located on a pig breeding farm in Austria, where 20 m³/day of manure is used as co-substrate and helps to achieve homogenization of the solid crop feedstock. The crop consists of Mz silage and crushed dry crops, together with minor amounts of residue from vegetable processing. Approximately 11,000 tons/year of crops are processed together with 7,300 tons/year of manure and leachate from the silage clamps.

   Two parallel digesters are fed hourly through an automatic dosing unit. The reactors are operated at 39°C with a 77-day residence time; this corresponds to a volumetric loading rate of 4.4 kg VS/(m³ day). Reactor mixing is performed by the mechanical stirrers. Dilution of the substrate mixture to a digestable solid (DS)

FIGURE 6.3   General view of a 1 MWe. Crop co-digestion plant using two parallel digesters (left background) and a covered final digestate storage tank (center foreground). Gas storage is integrated in digester 2 (background) and in the final storage tank (foreground); further storage capacity is provided in a dry gas storage tank (background right). (From Murphy, J. et al., Biogas from crop digestion, IEA Bioenergy, Task 37—Energy from Biogas, Denmark, September 2011.)

content below 10% is required for sufficient mixing. The plant produces 4,020,000 m³ of biogas annually. Hydrogen sulfide is removed by addition of air into the head space of digesters. The biogas is collected in an integrated gas holder inside the second digester as well as in an external gas holder. Power and heat are produced in two combined heat and power (CHP) units with a total capacity of 1 MWe and 1.034 MWth. The electricity is fed to the national power grid, and the heat is used in a local district heating network.

The digestate is collected in a gas-tight final storage tank before it is used as fertilizer in a neighborhood of the farm. Additionally, the biogas collected from the final digestate storage tank is used in the two CHP units. As an annual mean value, it is possible to achieve 98% of the theoretical capacity of the CHP. Of the energy content in the original substrate, 37% is converted into electricity. Electricity demand on-site is 7% of produced electricity. Only 7.8% of the energy content of the substrate is used as heat. Heat loss equates to 50.9% of the substrate energy. Methane loss in the CHP facility is 1.8%.

Ek et al. [159] outlined 15 years of slaughterhouse waste co-digestion experience at Tekniska Verken plant in Linkoping AB, Sweden. Experiences from research and development and plant operations led to several process improving technological/biological solutions. The improvements had positive effects on energy saving, better odor control, higher gas quality, increased OLRs, and higher biogas production with maintained process stability. The study also described how much of the process stability in AD of slaughterhouse waste depends on the plant operation that allows microbiological consortia to adapt to the substrate. The study also showed that the long retention time of the plant, accomplished by a low dilution of the substrate, is a vital component of the process stability when treating high protein substrate like slaughterhouse waste.

6.10 FUTURE PROSPECTS FOR CO-DIGESTION

By all accounts, the future for co-digestion is very bright. More and more industrial scale plants using co-digestion concept will be built throughout the world. As mentioned in Section 6.1, co-digestion is not an exception but norm in AD industry. The commercialization will become easier as more knowledge on different types of substrates blend is developed. Companies such as BSC will help build turnkey plant operations. Co-digestion will not only be a method for fuel and power generation, but it will also play an important role on waste and wastewater management. Government regulations and the selection of the best substrate mixed will have to be managed at the local levels. Along with co-combustion, co-digestion is the most advanced multifuel energy systems.

REFERENCES

1. Lee, S. and Shah, Y. *Biofuels and Bioenergy: Technologies and Processes*, CRC Press, Taylor & Francis Group, New York (2012).
2. Shah, Y. *Water for Energy and Fuel Production*, CRC Press, Taylor & Francis Group, New (2014).
3. Weiland, P. Biogas production: Current state and perspectives. *Appl. Microbiol. Technol.*, 85, 849–860 (2010).
4. Braun, R., Weiland, P., and Wellinger, A. Biogas from energy crop digestion. IEA Bioenergy, Task 37—Energy from Biogas and Landfill Gas, Denmark (2009).
5. Nges, I.A., Escobar, F., Fu, X., and Björnsson, L. Benefits of supplementing an industrial waste anaerobic digester with energy crops for increased biogas production. *Waste Manage.*, 32(1), 53 (2012).
6. Jepsen, S. *Co-digestion of Animal Manure and Organic Household Waste—The Danish Experience*, Ministry of Environment and Energy, Danish EPA, Copenhagen, Denmark (2011).
7. Braun, R. Anaerobic digestion—A multi faceted process for energy, environmental management and rural development. In *Improvement of Crop Plants for Industrial End Users*, Ranalli, P. (Ed.). Springer, Dordrecht, Netherlands, pp. 335– 416 (2007).
8. Esposito, G., Frunzo, L., Giordano, A., Liotta, F., Panico, A., and Pirozzi, F. Anaerobic co-digestion of organic wastes. *Reviews in Environmental Science and Biotechnology*, Springer-Verlag, New York (2012). doi: 10.1007/s11157-012-9277-8.
9. Khalid, A., Arshad, M., Anjun, M., Mahmood, T., and Dawson, L. The anaerobic digestion of solid organic waste. *Waste Manage.*, 31, 1737–1744 (2011).
10. Murphy, J., Braun, R., Weiland, P., and Wellinger, A. Biogas from crop digestion. IEA Bioenergy, Task 37—Energy from Biogas, Denmark (September 2011).
11. Arsova, L. Anaerobic digestion of food waste—Current status, problems and an alternative product, MSc thesis. Department of Earth and Environmental Engineering, Columbia University, New York (2010).
12. Klemeš, J., Smith, R., and Kim, J.-K. (eds.). *Handbook of Water and Energy Management in Food Processing*, Woodhead Publishing Sawston, Cambridge, Subsidiary of Elsevier, New York (2008). ISBN: 1845691954.

13. Luque, R., Campelo, J., and Clark, J. (eds.). *Handbook of Biofuels Production—Processes and Technologies*, Woodhead Publishing Sawston, Cambridge, Subsidiary of Elsevier, New York (2011). ISBN-13 1845696795, ISBN-10 1845696794.
14. Monnet, F. An introduction to anaerobic digestion of organic wastes. Remade Scotland, Scotland (2003).
15. Persson, M., Jonsson, O., and Wellinger, A. Biogas upgrading to vehicle fuel standards and grid injection. Task 37—Energy from biogas and landfill gas. IEA Bioenergy, Petten, the Netherlands (2006).
16. Polprasert, C. *Organic Waste Recycling—Technology and Management*, 3rd edn., IWA Publishing, London, U.K. (2007).
17. Braun, R. Potential of co-digestion. IEA Task 37 Report. http://www.novaenergie.ch/iea-bioenergytask37/Dokumente/final.PDF, accessed on November 7, 2007, Petten, the Netherlands (2002).
18. Mata-Alvarez, J., Mace, S., and Llabres, P. Anaerobic digestion of organic solid wastes. An overview of research achievements and perspectives. *Bioresour. Technol.*, 74, 3–16 (2000).
19. Gunaseelan, V.N. Anaerobic digestion of biomass for methane production: A review. *Biomass Bioenerg.*, 13, 83–114 (1997).
20. Wu, W. Anaerobic co-digestion of biomass for methane production: Recent research achievements, personal communication (2013).
21. Morton, C. Co-digestion charge: Is waste water's new best friend? A publication by Water World, UK's Anaerobic Digestions and Biogas Association (2013).
22. Ryan, K. Anaerobic co-digestion of farm-based manure and food waste—Are there environmental or economic benefits vs. landfilling? Press release, Mary Ann Liebert, Inc., New Rochelle, NY (2012).
23. Speece, R.E. *Anaerobic Biotechnology*, Archae Press, Nashville, TN (1996).
24. Braun, R. and Wellinger, A. Potential of co-digestion. IEA Bioenergy, Task 37—Energy from Biogas and Landfill Gas, IEA Bioenergy, Petten, the Netherlands (2004).
25. Kangle, K., Kore, S., Kore V., and Kulkarni, G. Recent trends in anaerobic co-digestion: A review. *Univ. J. Environ. Res. Technol.*, 2(4), 210–219 (2012).
26. Co-digestion of bio-waste. A Report from California Sustainability Alliance (2012).
27. Froom, M. Co-digestion: A sustainable solution for sewage sludge. *Renewable Energy World Magazine* (November 14, 2011).
28. Totzke, D. Co-digestion: A developing concept and market. A Report from The Resource, A publication of Applied Technologies, Brookfield, WI (2011).
29. Wallis, M.J., Ambrose, M.R., and Chan, C.C. Climate change: Charting a water course in an uncertain future. *J. Am. Water Works Assoc.*, 100, 70–79 (2008).
30. Groenewold, H. Anaerobic co-digestion in the Netherlands—A system analysis on greenhouse gas emissions from Dutch co-digesters, MS thesis. University of Groningen, Groningen, the Netherlands (2013).
31. Burton, C.H. and Turner, C. Anaerobic treatment options for animal manures, Chapter 7 pp. 269–320. In *Manure Management—Treatment Strategies for Sustainable Agriculture*, 2nd edn., Edwards, C., Burton, C. and Turner, C. (eds.), Silsoe Research Institute, Bedford, U.K. (2003).
32. You, Z., Wei, T., and Cheng, J. Improving anaerobic co-digestion of corn stover using sodium hydroxide pretreatment. *Energy Fuels*, 28(1), 549–554 (2014).
33. Wang, Y., Zhang, Y., Meng, L., Wang, J., and Zhang, W. Hydrogen–methane production from swine manure: Effect of pretreatment and VFAs accumulation on gas yield. *Biomass Bioenerg.*, 33(9), 1131–1138 (2009).
34. Redman, G. *A Detailed Economic Assessment of Anaerobic Digestion Technology and Its Sustainability to UK Farming and Waste Systems*, 2nd edn., The Andersons Center, A Project Funded by DECC and Managed by the NNFCC, Melton, Mowbray, Leicestershire, U.K. (March 2010).

35. Bayr, S., Rantanen, M., Kaparaju, P., and Rintala, J. Mesophilic and thermophilic anaerobic co-digestion of rendering plant and slaughterhouse wastes. *Bioresour. Technol.*, 104, 28–36 (2012).

36. Benabdallah El Hajd, T., Astals, S., Galí, A., Mace, S., and Mata-Álvarez, J. Ammonia influence in anaerobic digestion of OFMSW. *Water Sci. Technol.*, 59(6), 1153–1158 (2009).

37. Fernández, A., Sánchez, A., and Font, X. Anaerobic co-digestion of a simulated organic fraction of municipal solid wastes and fats of animal and vegetable origin. *Biochem. Eng. J.*, 26(1), 22–28 (2005).

38. FOE. *Briefing: Anaerobic Digestion*, Friends of the Earth Limited, UK's Environmental Campaigning Organization (September 2007).

39. Garcia-Pena, E.I., Parameswaran, P., Kang, D.W., Canul-Chan, M., and Krajmalnik-Brown, R. Anaerobic digestion and co-digestion processes of vegetable and fruit residues: Process and microbial ecology. *Bioresour. Technol.*, 102(20), 9447–9455 (2011).

40. Gerardi, M.H. *The Microbiology of Anaerobic Digesters*, John Wiley, Hoboken, NJ (2003).

41. Hansen, T.L., Schmidt, J.E., Angelidaki, I., Marca, E., Jansen, J.C., Mosbaek, H., and Christensen, T.H. Method for determination of methane potentials of solid organic waste. *Waste Manage.*, 24(4), 393–400 (2004).

42. Hilkiah Igoni, A., Ayotamuno, M.J., Eze, C.L., Ogaji, S.O.T., and Probert, S.D. Designs of anaerobic digesters for producing biogas from municipal solid-waste. *Appl. Energ.*, 85(6), 430–438 (2008).

43. Clemens, J., Trimborn, M., Weiland, P., and Amon, B. Mitigation of greenhouse gas emissions by anaerobic digestion of cattle slurry. *Agric. Ecosyst. Environ.*, 112, 171–177 (2006).

44. Desai, M., Patel, V., and Madamwar, D. Effect of temperature and retention time on bio-methanation of cheese whey–poultry waste–cattle dung. *Environ. Pollut.*, 83, 311–315 (1994).

45. Fezzani, B. and Cheikh, R. Thermophilic anaerobic co-digestion of olive mill wastewater with olive mill solid wastes in a tubular digester. *Chem. Eng. J.*, 132(1–3), 195–203 (2007).

46. Deng, W.-Y., Yan, J.-H., Li, X.-D., Wang, F., Zhu, X.-W., Lu, S.-Y., and Cen, K.-F. Emission characteristics of volatile compounds during sludges drying process. *J. Hazard. Mater.*, 162, 186–192 (2009).

47. He, P.J., Shao, L.M., Guo, H.D., Li, G.J., and Lee, D.J. Nitrogen removal from recycled landfill leachate by ex situ nitrification and in situ denitrification. *Waste Manage.* (Amsterdam, the Netherlands), 26, 838–845 (2006).

48. Kang, M.S., Srivastava, P., Tyson, T., Fulton, J.P., Owsley, W.F., and Yoo, K.H. A comprehensive GIS-based poultry litter management system for nutrient management planning and litter transportation. *Comput. Electron. Agric.*, 64, 212–224 (2008).

49. Komilis Dimitris, P., and Ham Robert, K. Carbon dioxide and ammonia emissions during composting of mixed paper, yard waste and food waste. *Waste Manage.* (New York), 26, 62–70 (2006).

50. Paillat, J.-M., Robin, P., Hassouna, M., and Leterme, P. Predicting ammonia and carbon dioxide emissions from carbon and nitrogen biodegradability during animal waste composting. *Atmos. Environ.*, 39, 6833–6842 (2005).

51. Criddle, C.S. The kinetics of co-metabolism. *Biotechnol. Bioeng.*, 41(11), 1048–1056 (1993).

52. Mudunge, R. Comparison of an anaerobic baffled reactor and a completely mixed reactor, Masters thesis. School of Chemical Engineering, University of Natal, Durban, South Africa (2001).

53. Peres, C.S., Sanchez, C.R., Matumoto, C., and Schmidell, W. Anaerobic biodegradability of the organic components of municipal solid wastes. *Water Sci. Technol.*, 25(7), 285–293 (1992).

54. Purcell, B.E. and Stentiford, E.I. Co-digestion—Enhancing recovery of organic wastes. http://www.orbit-online.net/journal/archiv/01-01/0101_07_text_html, accessed on May 24, 2001 (2000).

55. Albhin, A. and Vinnerås, B. Biosecurity and arable use of manure and biowaste—Treatment alternatives. *Livestock Sci.*, 112, 232–239 (2007).

56. Braun, R., Huber, P., and Meyrath, J. Ammonia toxicity in liquid piggery manure digestion. *Biotechnol. Lett.*, 3, 159–164 (1981).

57. Bryant, M.P. Commentary on the Hungate technique for culture of anaerobic bacteria. *Am. J. Clin. Nutr.*, 25, 1324–1328 (1972).

58. McCarty, P.L. and McKinney, R.E. Salt toxicity in anaerobic digestion. *J. Water Pollut. Contr. Federat.*, 33, 399–415 (1961).

59. Mladenovska, Z., Dabrowski, S., and Ahring, B.K. Anaerobic digestion of manure and mixture of manure with lipids: Biogas reactor performance and microbial community analysis. *Water Sci. Technol.*, 48, 271–278 (2003).

60. Møller, H.B., Sommer, S.G., and Ahring, B.K. Methane productivity of manure, straw and solid fractions of manure. *Biomass Bioenerg.*, 26, 485–495 (2004).

61. Lin, J., Zuo, J., Ji, R., Chen, X., Liu, F., Wang, K., and Yang, Y. Methanogenic community dynamics in anaerobic co-digestion of fruit and vegetable waste and food waste, *J. Environ. Sci.*, 24(7), 1288–1294 (2012).

62. Sell, S. A scale-up procedure for substrate co-digestion in anaerobic digesters through the use of substrate characterization, BMPs, ATAs, and sub pilot-scale digesters, MS thesis. Agricultural Engineering, Biorenewable Resources and Technology, Iowa State University, Annes, IA (2011).

63. Long, J., Aziz, T., de Los Reyes III, F., and Ducoste, J. Anaerobic co-digestion of fat, oil and grease (FOG): A review of gas production and process limitations. *Proc. Safety Environ. Prot.*, 50(3), 231–245 (May 2012).

64. Soroushian, F. Operational considerations for co-digestion. In *SARBS Seminar on Math, Operations and Maintenance for Biosolids Systems*, California Water Environment Association, CH2M HILL, Englewood, CO (September 2011).

65. Navaneethan, N. and Zitorer, D. Anaerobic co-digestion increases net biogas production by increasing microbial activity, Central State Water Environment Association (CSWEA), Technical Program, Marquette University, Milwaukee, WI (May 14–17, 2012).

66. Liu, K., Tang, Y.Q., Matsui, T., Morimura, S., Wu, X.L., and Kida, K. Thermophilic anaerobic co-digestion of garbage, screened swine and dairy cattle manure. *J. Biosci. Bioeng.*, 107(1), 54–60 (2009).

67. Lossie, U. and Pütz, P. Targeted control of biogas plants with the help of FOS/TAC. *EnerCess:* A report from Hach Lange GmbH, Duesseldorf, Germany (2012).

68. Padilla-Gasca, E., López-López, A., and Gallardo-Valdez, J. Evaluation of stability factors in the anaerobic treatment of slaughterhouse wastewater. *J. Bioremed. Biodegrad.*, 2(1), 114 (2011).

69. Pagés Díaz, J., Pereda Reyes, I., Lundin, M., and Sárvári Horváth, I. Co-digestion of different waste mixtures from agro-industrial activities: Kinetic evaluation and synergetic effects. *Bioresour. Technol.*, 102(23), 10834 (2011).

70. Palatsi, J., Viñas, M., Guivernau, M., Fernandez, B., and Flotats, X. Anaerobic digestion of slaughterhouse waste: Main process limitations and microbial community interactions. *Bioresour. Technol.*, 102(3), 2219–2227 (2011).

71. PCD. *Guideline for Utilization of Waste from Slaughter Process*, Pollution Control Department, Ministry of Natural Resources and Environment, Bangkok, Thailand (2012).

72. Salminen, E., Einola, J., and Rintala, J. Characterisation and anaerobic batch degradation of materials accumulating in anaerobic digesters treating poultry slaughterhouse waste. *Environ. Technol.*, 22(5), 577–585 (2001).

73. Salminen, E.A. and Rintala, J.A. Semi-continuous anaerobic digestion of solid poultry slaughterhouse waste: Effect of hydraulic retention time and loading. *Water Res.*, 36(13), 3175–3182 (2002).

74. Sluiter, A., Hames, B., Hyman, D., Payne, C., Ruiz, R., Scarlata, C., Sluiter, J., Templeton, D., and Wolfe, J. *Biomass and Total Dissolved Solids in Liquid Process Samples*, Laboratory Analytical Procedure (LAP), National Renewable Energy Laboratory, Golden, CO (2008).

75. Sluiter, A., Hames, B., Ruiz, R., Scarlata, C., Sluiter, J., and Templeton, D. *Determination of Ash in Biomass*, Laboratory Analytical Procedure (LAP), National Renewable Energy Laboratory, Golden, CO (2005).

76. Straka, F., Jenicek, P., Zabranska, J., Dohanyos, M., and Kuncarova, M. Anaerobic fermentation of biomass and wastes with respect to sulfur and nitrogen contents in treated materials. In *Proceedings of the 11th International Waste Management and Landfill Symposium*, CISA, Environmental Sanitary Engineering Centre, Cagliari, Italy (2007).

77. Verma, S. Anaerobic digestion of biodegradable organics in municipal solid wastes, MSc thesis. Department of Earth & Environmental Engineering, Columbia University, New York (2002).

78. Wang, L., Zhou, Q., and Li, F. Avoiding propionic acid accumulation in the anaerobic process for biohydrogen production. *Biomass Bioenerg.*, 30(2), 177–182 (2006).

79. Gelegenis, J., Georgakakis, D., Angelidaki, I., and Mavris, V. Optimization of bio gas production by co-digesting whey with diluted poultry manure. *Renew. Energ.*, 32, 2147–2160 (2007).

80. Ghaly, A. A comparative study of anaerobic digestion of acid cheese whey and dairy manure in a two-stage reactor. *Bioresour. Technol.*, 58, 61–72 (1996).

81. Gomez, G., Cuetos, M.J., Cara, J., Moran, A., and Garcia, A.I. Anaerobic co-digestion of primary sludge and the fruit and vegetable fraction of the municipal solid wastes: Conditions for mixing and evaluation of the organic loading rate. *Renew. Energ.*, 31, 2017–2024 (2006).

82. Zaher, U., Li, R., Pandey, P., Ewing, T., Frear, C., and Chen, S. Development of co-digestion software models to assist plant design and co-digestion operation. Climate Friendly Farming, Co-digestion modeling, Chapter 5, CSANR Research Report, pp. 1–15 (2010).

83. Alvarenga, P., Palma, P., Goncalves, A.P., Fernandes, R.M., Cunha-Queda, A.C., Duarte, E., and Vallini, G. Evaluation of chemical and ecotoxicological characteristics of biodegradable organic residues for application to agricultural land. *Environ. Int.*, 33, 505–513 (2007).

84. Batstone, D.J., Keller, J., Angelidaki, I., Kalyuzhnyi, S.V., Pavlostathis, S.G., Rozzi, A., Sanders, W.T.M., Siegrist, H., and Vavilin, V.A. Anaerobic Digestion Model No. 1 (ADM1), IWA Publishing, London, U.K. (2002).

85. Bou-Najm, M. and El-Fadel, M. Computer-based interface for an integrated solid waste management optimization model. *Environ. Model. Software*, 19, 1151–1164 (2004).

86. Fezzani, B. and Ben Cheikh, R. Implementation of IWA anaerobic digestion model No. 1 (ADM1) for simulating the thermophilic anaerobic co-digestion of olive mill wastewater with olive mill solid waste in a semi-continuous tubular digester. *Chem. Eng. J.* (Amsterdam, the Netherlands), 141, 75–88 (2008).

87. Fezzani, B. and Ben Cheikh, R. Optimisation of the mesophilic anaerobic co-digestion of olive mill wastewater with olive mill solid waste in a batch digester. *Desalination*, 228, 159–167 (2008).

88. Garcia-de-Cortazar, A.L. and Monzon, I.T. Moduelo 2: A new version of an integrated simulation model for municipal solid waste landfills. *Environ. Model. Software*, 22, 59–72 (2007).

89. Kleerebezem, R. and Van Loosdrecht, M.C.M. Waste characterization for implementation in ADM1. *Water Sci. Technol.*, 54, 167–174 (2006).

90. Lubken, M., Wichern, M., Schlattmann, M., Gronauer, A., and Horn, H. Modelling the energy balance of an anaerobic digester fed with cattle manure and renewable energy crops. *Water Res.*, 41, 4085–4096 (2007).

91. Manirakiza, P., Covaci, A., and Schepens, P. Comparative study on total lipid determination using Soxhlet, Roese-Gottlieb, Bligh & Dyer, and modified Bligh & Dyer extraction methods. *J. Food Compos. Anal.*, 14, 93–100 (2001).

92. Nelder, J. and Mead, R. A simplex method for function minimization. *Comput. J.*, 7, 308–313 (1965).

93. Rosen, C., Vrecko, D., Gernaey, K.V., Pons, M.N., and Jeppsson, U. Implementing ADM1 for plant-wide benchmark simulations in Matlab/Simulink. *Water Sci. Technol.*, 54, 11–19 (2006).

94. Shanmugam, P. and Horan, N.J. Simple and rapid methods to evaluate methane potential and biomass yield for a range of mixed solid wastes. *Bioresour. Technol.*, 100, 471–474 (2009).

95. Singh, K.P., Malik, A., and Sinha, S. Water quality assessment and apportionment of pollution sources of Gomti River (India) using multivariate statistical techniques—A case study. *Anal. Chim. Acta*, 538, 355–374 (2005).

96. Kumar, V., Mari, M., Schuhmacher, M., and Domingo, J.L. Partitioning total variance in risk assessment: Application to a municipal solid waste incinerator. *Environ. Model. Software*, 24, 247–261 (2009).

97. Vanrolleghem, P.A., Rosen, C., Zaher, U., Copp, J., Benedetti, L., Ayesa, E., and Jappsson, U. Continuity-based interfacing of models for wastewater systems described by Petersen matrices. *Water Sci. Technol.*, 52, 493–500 (2005).

98. Volcke Eveline, I.P., van Loosdrecht Mark, C.M., and Vanrolleghem Peter, A. Continuity-based model interfacing for plant-wide simulation: A general approach. *Water Res.*, 40, 2817–2828 (2006).

99. Zaher, U., Buffiere, P., Steyer, J.P., and Chen, S. A procedure to estimate proximate analysis of mixed organic wastes. *Water Environ. Res.*, 81, 407–415 (2009).

100. Zaher, U. and Chen, S. Interfacing the IWA Anaerobic Digestion Model No.1 (ADM1) with manure and solid waste characteristics. In *WEFTEC'06, Conference Proceedings, 79th Annual Technical Exhibition & Conference*, Dallas, TX, October 21–25, 2006, pp. 3162–3175 (2006).

101. Zaher, U., Grau, P., Benedetti, L., Ayesa, E., and Vanrolleghem, P.A. Transformers for interfacing anaerobic digestion models to pre- and post-treatment processes in a plant-wide modelling context. *Environ. Model. Software*, 22, 40–58 (2007).

102. Angelidaki, I., Alves, M., Bolzonella, D., Borzacconi, L., Campos, J.L., Guwy, A.J., Kalyuzhnyi, S., Jenicek, P., and van Lier, J.B. Defining the biomethane potential (BMP) of solid organic wastes and energy crops: A proposed protocol for batch assays. *Water Sci. Technol.*, 59(5), 927–934 (2009).

103. Husain, A. Mathematical models of the kinetics of anaerobic digestion—A selected review. *Biomass Bioenerg.*, 14, 561–571 (1998).

104. Astals, S., Ariso, M., Galí, A., and Mata-Alvarez, J. Co-digestion of pig manure and glycerine: Experimental and modelling study. *J. Environ. Manage.*, 92, 1091–1096 (2011).

105. Yamashiro, T., Lateef, S., Ying, C., Beneragama, N., Lukic, M., Mashiro, I., Ihara, I., Nishida, T., and Umetsu, K. Anaerobic co-digestion of dairy cow manure and high concentrated food processing waste. *J. Mater. Cycle Waste Manage.*, 15, 539–547 (2013).

106. Siripong, C. and Dulyakasem, S. Continuous co-digestion of agro-industrial residues, MS thesis. School of Engineering, University of Boras, Boras, Sweden (June 2012).

107. Abouelenien, F., Kitamura, Y., Nishio, N., and Nakashimada, Y. Dry anaerobic ammonia-methane production from chicken manure. *Appl. Microbiol. Biotechnol.*, 82(4), 757–764 (2009).
108. Alvarez, R. and Lidén, G. Semi-continuous co-digestion of solid slaughterhouse waste, manure, and fruit and vegetable waste. *Renew. Energ.*, 33(4), 726–734 (2008).
109. Callaghan, F.J., Wase, D.A.J., Thayanithy, K., and Forster, C.F. Continuous co-digestion of cattle slurry with fruit and vegetable wastes and chicken manure. *Biomass Bioenerg.*, 22(1), 71–77 (2002).
110. Cuetos, M.J., Fernández, C., Gómez, X., and Morán, A. Anaerobic co-digestion of swine manure with energy crop residues. *Biotechnol. Bioprocess Eng.*, 16(5), 1044–1052 (2011).
111. Cuetos, M.J., Gómez, X., Otero, M., and Morán, A. Anaerobic digestion and co-digestion of slaughterhouse waste (SHW): Influence of heat and pressure pre-treatment in biogas yield. *Waste Manage.*, 30(10), 1780–1789 (2010).
112. Cuetos, M.J., Gómez, X., Otero, M., and Morán, A. Anaerobic digestion of solid slaughterhouse waste (SHW) at laboratory scale: Influence of co-digestion with the organic fraction of municipal solid waste (OFMSW). *Biochem. Eng. J.*, 40(1), 99–106 (2008).
113. Kacprzak, A., Krzystek, L., and Ledakowicz, S. Co-digestion of agricultural and industrial wastes. *Chem. Papers*, 64(2), 127–131 (2009).
114. Panichnumsin, P., Nopharatana, A., Ahring, B., and Chaiprasert, P. Production of methane by co-digestion of cassava pulp with various concentrations of pig manure. *Biomass Bioenerg.*, 34(8), 1117–1124 (2010).
115. Xie, S., Wu, G., Lawlor, P.G., Frost, J.P., and Zhan, X. Methane production from anaerobic co-digestion of the separated solid fraction of pig manure with dried grass silage. *Bioresour. Technol.*, 104, 289–297 (2012).
116. Agdag, O.N. and Sponza, D.T. Co-digestion of mixed industrial sludge with municipal solid wastes in anaerobic simulated landfilling bioreactors. *J. Hazard. Mater.*, 140, 75–85 (2007).
117. Alvarez, R. and Liden, G. Semi-continuous co-digestion of solid slaughterhouse waste, manure, and fruit and vegetable waste. *Renew. Energ.*, 33(4), 726–734 (April 2008). Available online June 15, 2007.
118. Davidsson, A., Lovstedt, C., la Cour Jansen, J., Gruvberger, C., and Aspegren, H. Co-digestion of grease trap sludge and sewage sludge. *Waste Manage*, 28(6), 986–992 (2008). Available online May 22, 2007.
119. Kaparaju, P. and Rintala, J. Anaerobic co-digestion of potato tuber and its industrial byproducts with pig manure. *Resour. Conserv. Recycl.*, 43, 175–188 (2005).
120. Lehtomäki, A., Huttunen, S., and Rintala, J.A. Laboratory investigations on co-digestion of energy crops and crop residues with cow manure for methane production: Effect of crop to manure ratio. *Resour. Conserv. Recycl.*, 51, 591–609 (2007).
121. Murto, M., Björnsson, L., and Mattiasson B. Impact of food industrial waste on anaerobic codigestion of sewage sludge and pig manure. *J. Environ. Manage.*, 70, 101–107 (2004).
122. Neves, L., Oliveira, R., and Alves, M.M. Anaerobic co-digestion of coffee waste and sewage sludge. *Waste Manage.*, 26, 176–181 (2006).
123. Romano, R.T. and Zhang, R. Co-digestion of onion juice and wastewater sludge using an anaerobic mixed biofilm reactor. *Bioresour. Technol.*, 99(3), 631– 637 (2008). Available online June 1, 2007.
124. Zupancic, G.D., Uranjek-Zevart, N., and Ros, M. Full-scale anaerobic co-digestion of organic waste and municipal sludge. *Biomass Bioenerg.*, 32(2), 162–167 (February 2008). Available online August 20, 2007.

125. Smith, K. *Opportunities and Constraints of Co-Digestion of Sewage Sludge with Other Organic Waste Streams in Existing Waste Water Treatments Plants*, Imperial College, London, U.K. (2013).

126. Gabel, D. Co-digestion case studies enhancing energy recovery from sludge. In *MWRD PWO Seminar*, CH2M HILL, Englewood, CO (May 23, 2012).

127. Hartmann, H. and Ahring, B.K. Anaerobic digestion of the organic fraction of municipal solid waste: Influence of co-digestion with manure. *Water Res.*, 39, 1543–1552 (2005).

128. Yen, H.-W. and Brune David, E. Anaerobic co-digestion of algal sludge and waste paper to produce methane. *Bioresour. Technol.*, 98, 130–134 (2007).

129. Zhang, P., Zeng, G., Zhang, G., Li, Y., Zhang, B., and Fan, M. Anaerobic co-digestion of biosolids and organic fraction of municipal solid waste by sequencing batch process. *Fuel Process. Technol.*, 89, 485–489 (2008).

130. Serrano, A., Lopez, J., Chica, A., Martin, M., Karouach, F., Mesfiour, A., and Bari, H. Mesophilic anaerobic co-digestion of sewage sludge and orange peel waste. *Environ. Technol.*, 35(7), 898–906 (2014).

131. Aziz, T., Wang, L., Long, H., Sawyer, H., and Ducoste, J. *Sustainable Energy from Grease Interceptor Waste Co-Digestion*, An Internal Publication by North Carolina State University, Dept. of Environmental Engineering, Raleigh, NC (2013).

132. Ramsamy, D., Rakgotho, T., Naidoo, V., and Buckley, C. Anaerobic co-digestion of high strength/toxic organic liquid effluents in a continuously stirred reactor: Start-up. In Paper presented at the *Biennial Conference of the Water Institute of Southern Africa (WISA)*, Durban, South Africa. http://www.wisa.co.za/ (May 19–23, 2002) (accessed 2002).

133. Chiu-Yue, L., Feng-Yi, B., and Jen, C. Anaerobic co-digestion of septage and landfill leachate. *Bioresour. Technol.*, 68, 275–282 (1999).

134. Purcell, B.E. and Stentiford, E.I. Co-treatment: Fuelling recovery of organic wastes. *Waste Manage.*, 20, 32–33 (August 2000).

135. Reinhart, D.R. and Pohland, F.G. The assimilation of organic hazardous wastes by municipal solid waste landfills. *J. Indust. Microbiol.*, 8, 193–200 (1991).

136. Riggle, D. Anaerobic digestion for municipal solid waste and industrial wastewater. *Biocycle*, 37(11), 77 (1996).

137. Sacks, J. Anaerobic digestion of high-strength or toxic organic effluents, Masters thesis. School of Chemical Engineering, University of Natal, Glenwood, Durban, South Africa (1997).

138. Zupancic, G., Ros, M., Klemencic, M., Oset, M., and Logar, R. *Excess Brewery Yeast Co-Digestion in a Full Scale EGSB Reactor*, Department of Microbiology and Microbial Biotechnology, University of Ljubljana, Ljubljana, Slovenia (2013).

139. Martin-Gonzalez, L., Colyurato, L., Font, X., and Vicent, T. Anaerobic co-digestion of organic fraction of municipal solid waste with FOG waste from a sewage treatment plant: Recovering a wasted methane potential and enhancing the biogas yield. *Waste Manage.*, 39, 615 (2010).

140. Bolzonella, D., Battistoni, P., Susini, C., and Cecchi, F. Anaerobic codigestion of waste activated sludge and OFMSW: The experiences of Viareggio and Treviso plants (Italy). *Water Sci. Technol.*, 53(8), 203–211 (2006).

141. Luostarinen, S., Luste, S., and Sillanpää, M. Increased biogas production at wastewater treatment plants through co-digestion of sewage sludge with grease trap sludge from a meat processing plant. *Bioresour. Technol.*, 100, 79–85 (2009).

142. Macias-Corral, M., Samani, Z., Hanson, A., Smith, G., Funk, P., Yu, H., and Longworth, J. Anaerobic digestion of municipal solid waste and agricultural waste and the effect of co-digestion with dairy cow manure. *Bioresour. Technol.*, 99, 8288–8293 (2008).

143. Canas, E. Technical feasibility of anaerobic co-digestion of dairy manure with chicken litter and other wastes, MS thesis. University of Tennessee, Knoxville, TN (2010).

144. Alvarez, R., Villca, S., and Lidén, G. Biogas production from llama and cow manure at high altitude. *Biomass Bioenerg.*, 30, 66–75 (2006).

145. Alvarez, R. and Lidén, G. Anaerobic co-digestion of aquatic flora and quinoa with manures from Bolivian Altiplano. *Waste Manage.*, 28, 1933–1940 (2008).

146. Amon, T., Amon, B., Kryvoruchko, V., Zollitsch, W., Mayer, K., and Gruber, L. Biogas production from maize and dairy cattle manure—Influence of biomass composition on the methane yield. *Agric. Ecosyst. Environ.*, 118, 173–182 (2007).

147. Callaghan, F.J., Luecke, K., Wase, D.A.J., and Thayanithy, K. Co-digestion of cattle slurry and waste milk under shock loading conditions. *J. Chem. Technol. Biotechnol.*, 68, 405–410 (1997).

148. Callaghan, F.J., Wase, D.A.J., Thayanithy, K., and Forster, C.F. Co-digestion of waste organic solids: Batch studies. *Bioresour. Technol.*, 67, 117–122 (1999).

149. Callaghan, F.J., Wase, D.A.J., Thayanithy, K., and Forster, C.F. Continuous co-digestion of cattle slurry with fruit and vegetable wastes and chicken manure. *Biomass Bioenerg.*, 27, 71–77 (2002).

150. Davidsson, A., Lövstedt, C., la Cour Jansen, J., Gruvberger, C., and Aspegren, H. Codigestion of grease trap sludge and sewage sludge. *Waste Manage.*, 28, 986–992 (2008).

151. Demirel, B., Yenigun, O., and Onay, T.T. Anaerobic treatment of dairy wastewaters: A review. *Process Biochem.*, 40, 2583–2595 (2005).

152. Misi, S.N. and Forster, C.F. Batch co-digestion of multi-component agro-wastes. *Bioresour. Technol.*, 80, 19–28 (2001).

153. Misi, S.N. and Forster, C.F. Batch co-digestion of two-component mixtures of agrowastes. *IChemE, Part B*, 79, 365–371 (2001).

154. Mshandete, A., Kivaisi, A., Rubindamayugi, M., and Mattiasson, B. Anaerobic batch codigestion of sisal pulp and fish wastes. *Bioresour. Technol.*, 95, 19–24 (2004).

155. Kaparaju, P., Luostarinen, S., Kalmari, E., Kalmari, J., and Rintala, J. Co-digestion of energy crops and industrial confectionery by-products with cow manure: Batch-scale and farm-scale evaluation. *Water Sci. Technol.*, 45(10), 275–280 (2002).

156. FAO. *Livestock Sector Brief: Thailand*, Food and Agriculture Organization of the United Nations, Rome, Italy (2005).

157. Kuang, Y. Enhancing anaerobic degradation of lipids in wastewater by addition of co-substrates, PhD thesis. Murdoch University, Perth, Western Australia, Australia (2002).

158. Cheerawit, R., Thunwadee, T., Duangporn, K., Tanawat, R., and Wichuda, K. Biogas production from co-digestion of wastewater and food waste. *Health Environ. J.*, 3(2), 1–9 (2012).

159. Ek, A., Hallin, S., Vallin, L., Schnurer, A., and Karlsson, M. Slaughterhouse waste co-digestion—Experiences from 15 years of full scale operation. In *Biogas International Conference*, Linkoping, Sweden (May 8–13, 2011).

160. Zhang, L., Woo Lee, Y., and Jahng, D. Anaerobic co-digestion of food waste and piggery wastewater: Focusing on the role of trace elements. *Bioresour. Technol.*, 102, 5048–5059 (2011).

161. Zhu, Z., Hsueh, M.K., and He, Q. Enhancing biomethanation of municipal waste sludge with grease trap waste as a co-substrate. *Renew. Energ.*, 36, 1802–1807 (2011).

162. Ersahin, M.V, Gomec, C.V., Dereli, R.K., Arikan, O., and Ozturk, I. Biomethane production as an alternative: Bioenergy source from codigesters treating municipal sludge and organic fraction of municipal solid wastes. *J. Biomed. Biotechnol.*, 8, 1–8 (2011).

163. Alvarez, J., Otero, L., and Lema, J. A methodology for optimizing feed composition for anaerobic co-digestion of agro-industrial wastes. *Bioresour. Technol.*, 101, 1153–1158 (2010).

164. Bouallagui, H., Lahdheb, H., Ben Romdan, E., Rachdi, B., and Hamdi, M. Improvement of fruit and vegetable waste anaerobic digestion performance and stability with co-substrates addition. *J. Environ. Manage.*, 90, 1844–1849 (2009).

165. Buendía, I.M., Fernández, F., Villasenor, J., and Rodriguez, L. Feasibility of anaerobic co-digestion as a treatment option of meat industry wastes. *Bioresour. Technol.*, 100, 1903–1909 (2009).

166. Heo, N., Park, S., Lee, J., Kang, H., and Park, D. Single stage anaerobic co-digestion for mixture wastes of simulated Korean food waste and waste activated sludge. *Appl. Biochem. Biotechnol.*, 107, 567–579 (2003).

167. Neves, L., Oliveira, R., and Alves, M.M. Co-digestion of cow manure, food waste and intermittent input of fat. *Bioresour. Technol.*, 100, 1957–1962 (2009).

168. Ponsá, S., Gea, T., and Sánchez, A. Anaerobic co-digestion of the organic fraction of municipal solid waste with several pure organic co-substrates. *Biosyst. Eng.*, 108, 352–360 (2011).

169. Romano, R. and Zhang, R. Co-digestion of anion juice and wastewater sludge using an anaerobic mixed biofilm reactor. *Bioresour. Technol.*, 99, 631–637 (2008).

170. Wu, X., Yao, W., Zhu, J., and Miller, C. Biogas and CH_4 productivity by co-digesting swine manure with three crop residues as an external carbon source. *Bioresour. Technol.*, 101, 4042–4047 (2010).

171. Azaizeh, H. and Jadoun, J. Co-digestion of olive mill wastewater and swine manure using up-flow anaerobic sludge blanket reactor for biogas production. *J. Water Resour. Prot.*, 2, 314–321 (2010).

172. Braun, R., Brachtl, E., and Grasmug, M. Codigestion of proteinaceous industrial waste. *Appl. Biochem. Biotechnol.*, 109, 139–153 (2003).

173. Anaerobic co-digestion dairies in Washington State. A Report from Washington State University Extension Fact Sheet—FS040E, Pullman, WA (2013).

174. Patil, J., AntonyRaj, M., Gavimath, C., and Hooli, V. A comprehensive study on anaerobic co-digestion of water hyacinth with poultry litter and cow dung. *Int. J. Chem. Sci. Appl.*, 2(2), 148–155 (June 2011).

175. Li, C. Using anaerobic co-digestion with addition of municipal organic wastes and pre-treatment to enhance biogas production from wastewater treatment plant sludge, PhD thesis. Department of Civil Engineering, Queen's University, Kingston, Ontario, Canada (September 2012).

176. Acharya, C. and Kurian, R. *Anaerobic Co-Digestion of a Single Source Oily Waste and High Strength Pet Food Wastewater: A Study of Failure and Revival of a Full Scale Digester, WEFTEC Proceedings of Water Environment Federation,* Vol. 2006, pp. 5066–5073 (2006).

177. Kassuwi, S., Mshandete, A., and Kivaisi, A. Anaerobic co-digestion of biological pre-treated Nile perch fish solid waste with vegetable fraction of market solid waste. *ARPN J. Agric. Biol. Sci.*, 7(12), 1016–1031 (December 2012).

178. Dahunsi, S. and Oranusi, U. Co-digestion of food waste and human excreta for biogas production. *Brit. Biotechnol. J.*, 3(4), 485–499 (2013).

179. Babaee, A., Shayegan, J., and Roshani, A. Anaerobic slurry co-digestion of poultry manure and straw: Effect of organic loading and temperature. *J. Environ. Health Sci. Eng.*, 11, 15 (2013). doi:10.1186/2052–336x-11-15.

180. Bolzonella, D., Pavan, P., Battistoni, P., and Cecchi, F. Anaerobic co-digestion of sludge with other organic wastes and phosphorous reclamation in wastewater treatment plants for biological nutrients removal. *Water Sci. Technol.*, 53(12), 177–186 (2006).

181. Iacovidou, E., Ohandja, D., and Voulvoulis, N. Food waste co-digestion with sewage sludge-realizing its potential in UK. *J. Environ. Manage.*, 112, 267–274 (2012).

182. Foucault, L. Anaerobic co-digestion of chicken processing wastewater and crude glycerol from biodiesel, MS thesis. Texas A&M University, College Station, TX (August 2011).

183. Liu, Z. Anaerobic co-digestion of swine wastewater with switch grass and wheat straw for methane production, MS thesis. Biological and Agricultural Engineering, North Carolina State university, Raleigh, NC (2013).

184. Crolla, A., Kinsley, C., Kennedy, K., and Sauve, T. Anaerobic co-digestion of corn thin stillage and dairy manure. In *Canadian Biogas Conference and Exhibition*, London Convention Center, London, Ontario, Canada (March 4–6, 2013).

185. Alatriste-Mondragon, F., Samar, P., Cox, H., Ahring, B., and Iranpour, R. Anaerobic codigestion of municipal, farm, and industrial organic wastes: A survey of recent literature. *Water Environ. Res.*, 78, 607–636 (2006).

186. Garcia, K. and Perez, M. Anaerobic co-digestion of cattle manure and sewage sludge: Influence of composition and temperature. *Int. J. Environ. Prot.*, 3(6), 8–15 (June 2013).

187. Lisboa, M., lansing, S., and Jackson, C. On-farm co-digestion of food waste with dairy manure. A Report from Dept. of Environmental Sci. and Technology, University of Maryland, College Park, MD (2013).

188. Zhang, T., Liu, L., Song, Z., Ren, G., Feng, Y., Han, X., and Yang, G. Biogas production by co-digestion of goat manure with three crop residues. *Mater. Meth.*, 1–7 (June 25, 2013). *PLoS ONE*, 8(6), e66845, doi:10 1371/journal pone.0066845.

189. Campos, E., Palatsi, J., and Flotats, X. Co-digestion of pig slurry and organic wastes from food industry. In *Conference on Waste*, Barcelona, Spain, pp. 192–195 (June 1999).

190. Kaosol, T. and Sohgrathok, N. Enhancement of biogas production potential for anaerobic co-digestion of wastewater using decanter cake. *Am. J. Agric. Biol. Sci.*, 8(1), 67–74 (2013).

191. Buffiere, P., Loisel, D., Bernet, N., and Delgenès, J.P. Towards new indicators for the prediction of solid waste anaerobic digestion properties. *Water Sci. Technol.*, 53, 233–241 (2006).

192. Saint-Joly, C., Desbois, S., and Lotti, J.-P. Determinant impact of waste collection and composition on anaerobic digestion performance: Industrial results. *Water Sci. Technol.*, 41, 291–297 (2000).

193. Chen, Y., Cheng, J.J., and Creamer, K.S. Inhibition of anaerobic digestion process: A review. *Bioresour. Technol.*, 99, 4044–4064 (2008).

194. Carucci, G., Carrasco, F., Trifoni, K., Majone, M., and Beccari, M. Anaerobic digestion of food industry wastes: Effect of codigestion on methane yield. *J. Environ. Eng.*, 131, 1037–1045 (2005).

195. Edelmann, W. Co-digestion of organic solid waste and sludge from sewage treatment. *Water Sci. Technol.*, 41, 213 (2000).

196. Angelidaki, I. and Ahring, B. Effects of free long-chain fatty acids on thermophilic anaerobic digestion. *Appl. Microbiol. Biotechnol.*, 37, 808–812 (1992).

197. Kim, H.-W., Han, S.-K., and Shin, H.-S. The optimisation of food waste addition as a co-substrate in anaerobic digestion of sewage sludge. *Waste Manage. Res.*, 21, 515–526 (2003).

198. Sosnowski, P., Klepacz-Smolka, A., Kaczorek, K., and Ledakowicz, S. Kinetic investigations of methane co-fermentation of sewage sludge and organic fraction of municipal solid wastes. *Bioresour. Technol.*, 99, 5731–5737 (2008).

199. Converti, A., Drago, F., Ghiazza, G., Borghi, M.D., and Macchiavello, A. Co-digestion of municipal sewage sludges and pre-hydrolysed woody agricultural wastes. *J. Chem. Technol. Biotechnol.*, 69, 231–239 (1997).

200. Demirekler, E. and Anderson, G.K. Effect of sewage sludge addition on the startup of the anaerobic digestion of OFMSW. *Environ. Technol.*, 19, 837–843 (1998).

201. Cho, J.K., Park, S.C., and Chang, H.N. Biochemical methane potential and solid state anaerobic digestion of Korean food wastes. *Bioresour. Technol.*, 52, 245–253 (1995).

202. Hartmann, H., Moller, H.B., and Ahring, B.K. Efficiency of the anaerobic treatment of the organic fraction of municipal solid waste: Collection and pretreatment. *Waste Manage. Res.*, 22, 35–41 (2004).

203. Parkin, G.F. and Owen, W.F. Fundamentals of anaerobic digestion of wastewater sludges. *J. Environ. Eng.*, 112, 867–920 (1986).

204. Cirne, D.G., Paloumet, X., Björnsson, L., Alves, M.M., and Mattiasson, B. Anaerobic digestion of lipid-rich waste e effects of lipid concentration. *Renew. Energ.*, 32, 965–975 (2007).
205. Gallert, C. and Winter, J. Mesophilic and thermophilic anaerobic digestion of source-sorted organic wastes: Effect of ammonia on glucose degradation and methane production. *Appl. Microbiol. Biotechnol.*, 48, 405–410 (1997).
206. Kayhanian, M. and Hardy, S. The impact of four design parameters on the performance of a high-solids anaerobic digestion of municipal solid waste for fuel gas production. *Environ. Technol.*, 15, 557–567 (1994).
207. El Hadj, T.B., Astals, S., Gali, A., Mace, S., and Mata Alvarez, J. Ammonia influence in anaerobic digestion of OFMSW. *Water Sci. Technol.*, 59, 1153–1158 (2009).
208. Rittman, B.E. and McCarty, P.L. *Environmental Biotechnology: Principles and Applications*, McGraw-Hill International, New York (January 1, 2001).
209. Appels, L., Lauwers, J., Degrève, J., Helsen, L., Lievens, B., Willems, K., Van Impe, J., and Dewil, R. Anaerobic digestion in global bio-energy production: Potential and research challenges. *Renew. Sustain. Energy Rev.*, 15, 4295–4301 (2011).
210. De Baere, L. Anaerobic digestion of solid waste: State-of-the-art. *Water Sci. Technol.*, 41, 283–290 (2000).
211. Lebrato, J., Pérez-Rodríguez, J.L., and Maqueda, C. Domestic solid waste and sewage improvement by anaerobic digestion: A stirred digester. *Resour. Conserv. Recycl.*, 13, 83–88 (1995).
212. Hartmann, H., Angelidaki, I., and Ahring, B.K. Increase of anaerobic degradation of particulate organic matter in full-scale biogas plants by mechanical maceration. *Water Sci. Technol.*, 41, 145–153 (2000).
213. CADDET. *Food Waste Disposal Using Anaerobic Digestion, Renewable Energy*, IEA Centre for Analysis and Dissemination of Demonstrated Energy Technologies, Centre for Renewable Energy, IEA/OECD, Harwell, Oxfordshire, U.K. (1998).
214. Ganidi, N., Tyrrel, S., and Cartmell, E. Anaerobic digestion foaming causes—A review. *Bioresour. Technol.*, 100, 5546–5554 (2009).
215. Purcell, B. and Stentiford, E.I. Co-digestion and enhancing recovery of organic waste. *ORBIT J.*, 1, 1–6 (2006).
216. Defra. *Anaerobic Digestion Strategy and Action Plan*, Department for Environment, Food and Rural Affairs, London, U.K. (2011).
217. European Parliament and Council, Regulation (EC) No. 1774/202 on laying down health rules concerning animal by-products not intended for human consumption. Official Journal of the European Union L273, pp. 1–163 (2002).
218. Legislation. *The Animal By-Products (Enforcement) (England) Regulations 2011*. The National Archives, 881pp. (2011).
219. Defra. *Accelerating the Uptake of Anaerobic Digestion in England: An Implementation Plan*, Department of Environment Food and Rural Affairs, London, U.K. (2010).
220. Krupp, M., Schubert, J., and Widmann, R. Feasibility study for co-digestion of sewage sludge with OFMSW on two wastewater treatment plants in Germany. *Waste Manage.*, 25, 393–399 (2005).
221. Frear, C., Liao, W., Ewing, T., and Chen, S. Evaluation of co-digestion at a commercial dairy anaerobic digester. *Journal of CLEAN-Soil, Air, Water*, 39(7), 697–704 (2011).
222. Steyer, J.P., Bernard, O., Batstone, D.J., and Angelidaki, I. Lessons learnt from 15 years of ICA in anaerobic digesters. *Water Sci. Technol. J. Int. Assoc. Water Pollut. Res.*, 53, 25–33 (2006).
223. Zupancic, G.D., Uranjek-Zevart, N., and Ros, M. Full-scale anaerobic co-digestion of organic waste and municipal sludge. *Biomass Bioenerg.*, 32, 162–167 (2008).
224. Kübler, H., Hoppenheidt, K., Hirsch, P., Kottmair, A., Nimmrichter, R., and Nordsleck, H. Full-scale co-digestion of organic waste. *Water Sci. Technol.*, 41, 195–202 (2000).

225. Crolla, A. and Kinsley, C. *Background on Anaerobic Digestion on the Farm*, Ontario Rural Wastewater Center, Info Sheet, University of Guelph, Guelph, Ontario, Canada (2013).
226. El-Mashad, H.M., Wilko, K.P., Loon, V., and Zeeman, G. A model of solar energy utilisation in the anaerobic digestion of cattle manure. *Biosyst. Eng.*, 84, 231–238 (2003).
227. Forster-Carneiro, T., Pérez, M., Romero, L.I., and Sales, D. Dry-thermophilic anaerobic digestion of organic fraction of the municipal solid waste: Focusing on the inoculum sources. *Bioresour. Technol.*, 98, 3195–3203 (2007).
228. Yusuf, M. and Ify, N. Effect of waste paper on biogas production from co-digestion of cow dung and water hyacinth in batch reactors. *J. Appl. Sci. Environ. Manage.*, 12(4), 95–98 (2008).
229. Babee, A., Shayegan, J., and Roshani, A. Anaerobic slurry co-digestion of poultry manure and straw: Effect of organic loading and temperature. *J. Environ. Health Sci. Eng.*, 11(1), 15 (July 3, 2013).
230. Sharom, Z., Malakahmad, A., and Noor, B. Anaerobic co-digestion of kitchen waste and sewage sludge for producing biogas. In *Second International Conference on Environmental Management*, Bangi, Malaysia (2004).
231. Delia, T. and Agdag, N. Co-digestion of industrial sludge with municipal solid wastes in anaerobic simulated landfilling reactors. *Process Biochem.*, 40, 1871–1879 (2007).
232. Samani Z., Macias-Corral, M., Hanson, A., Smith, G., Funk, P., Yu, H., and Longworth, J. Anaerobic digestion of municipal solid waste and agricultural waste and the effect of co-digestion with dairy cow manure. *Bioresour. Technol.*, 99(17), 8288–8293 (2008).
233. El-Mashad, H. and Zhang, R. Biogas production from co-digestion of dairy manure and food waste. *Bioresour. Technol.*, 101(11), 4021–4028 (June 2010).
234. BioConversion Solutions, LLC, Boston, MA. http://www.bioconversionsolutions.com/ (accessed June 2015).

7 Hybrid Nuclear Energy Systems

7.1 INTRODUCTION

Chapters 2 through 6 illustrated that independent biomass and waste-based power or fuel plants are not sustainable on large scale and are not competitive with fossil fuel–based plants. The best strategy for the penetrations of biomass and waste in the energy market on large scale is to use them as a part of integrated multifuel systems for various synthetic fuel and power generation processes. These chapters point out the advantages of raw fuel integrations (such as coal/ biomass, coal/biomass/waste/ water, biomass/waste, and coal/oil) for greater penetration of renewable energy and improved process performance.

While replacement of fossil fuel by renewable fuels like biomass and waste is important, the replacement of fossil energy by carbon free nuclear and renewable energy like solar, wind, geothermal, and water is equally important. All forecasts project an increase in world energy demand, especially as population, economic productivity, and quality of life grow across the world. Currently, about 85% of our energy comes from fossil fuel. Our strategy should be to replace existing environmentally harmful fossil energy by carbon-free nuclear and renewable energy and have future growth in demand disproportionately satisfied by nuclear and renewable energy. This strategy will provide more balanced energy and fuel portfolio covering all forms of energy and fuel. This subject and other fossil, nuclear, and renewable energy–related issues are well described in a series of reports by Internal Energy Agency (IEA), International Atomic Energy Agency (IAEA), Nuclear Energy Agency (NEA), and others [1–18].

The economics of replacement of fossil energy by various renewable energy sources are sketchy and unsustainable on a large scale. We have illustrated this point for renewable biomass and waste energy in Chapter 2 through 6. Similar conclusions can be drawn for other sources of renewable energy like solar and wind energy. Wind power, though growing, provides only a modest amount of electricity worldwide and cannot generate the heat required for alternative applications. Solar energy can provide high-temperature heat and electricity, but it is time and location dependent and it is still under development on large scale. Both of these renewable energy options require vast land and capital resources and favorable climate conditions and by themselves would be incompatible with the large-scale applications. Same conclusion can be applied to geothermal energy and energy derived from water. This chapter and Chapter 8 illustrate another level of system integration to promote growth of all carbon-free sources of energy [1–3,15–17].

Unlike renewable energy, nuclear energy can be harnessed on a larger scale at a sustainable level. In the United States, the United Kingdom, Ireland, etc., it unfortunately lacks social and political support. This chapter outlines an integration strategy to gain its increased role in the energy mix. The strategy involves increasing its role to replace fossil fuel by using advanced hybrid nuclear–renewable systems accompanied by cogeneration (combined heat and power use) method for various important nonelectrical applications. This chapter describes various end games of this strategy.

When synthetic fuels are used to generate power, a significant amount of process heat can get wasted. The loss of heat energy is also an issue in many process industries. Due to these reasons, in the past, combined use of power and heat (or commonly known as cogeneration) has been practiced for fossil fuel–based (coal, oil, or gas) power and utility industries. The cogeneration [35–52] has improved thermal efficiencies of many fuel-based power industries. In this method, excess heat is often used for numerous nonelectrical applications such as district heating and industrial heating. The concept of cogeneration has also been applied at smaller scale such as micro cogeneration where the excess heat is used at a very small scale as heating a building, school, hospital, etc. When the excess heat is used for both heating and cooling, cogeneration is often called tri- or polygeneration. A common example of cogeneration process is combined cycle gas turbine where both power and heat uses are optimized.

The concept of cogeneration can also be applied to power generation processes that use biofuel or multifuels [18]. In this chapter, it is applied to nuclear energy. Unlike integrations of raw fuels described in Chapter 2 through 5, cogeneration combines the use of both electricity and waste heat to obtain improved thermal efficiency. When this approach is combined with hybrid nuclear–renewable system for power generation, it provides high thermodynamic efficiency along with the use of renewable and nuclear energy at large and economical scale with most value for its time of use.

As pointed out by Idaho National Laboratory (INL) [20,27,29,30,32,33], advanced hybrid nuclear energy systems (HNESs) with cogeneration integrate energy conversion processes to optimize energy management, reliability, security, and sustainability. This method

1. Facilitates effective integration of renewable energy (solar and wind) with nuclear energy overcoming the challenges of intermittency and transmission constraints of solar and wind energy
2. Opens markets for nuclear energy beyond only a percentage of baseload power through its nonelectrical applications
3. Promotes better usage of carbon sources, including natural gas and biomass, for the production of transportation fuels while reducing greenhouse gas (GHG) environmental impact through cogeneration
4. Supports smooth integration and enhanced efficiency of conversion of available energy resources into infrastructure-compatible products
5. Opens up the market for small- and middle-scale nuclear energy productions through their appropriate nonelectrical applications

7.2   WHY NUCLEAR ENERGY?

Unlike fossil energy, nuclear energy decarbonizes the power industry. The uranium resources required for nuclear energy are available, and advanced nuclear technologies are either developed or under development. From a sustainable development perspective, nuclear energy has a major role to play in terms of reduction of CO_2 emissions, security of energy supply, and diversification of supply and price stability. As shown in Table 7.1, nuclear energy is one of the lowest CO_2 producer sources among all fuel and energy options [1–18].

Advocates of nuclear energy for sustainable development argue that it is a well-established noncarbon technology, as demonstrated by its 16% share of the world's electricity supply in over 30 countries and even higher share in specific countries—78% in France, for example [4–12]. The net nuclear power capacity in operation in developing countries alone is close to 14% of the total worldwide. More than 10,000 reactor-years of operating experience have been accumulated over the past five decades. Like fossil energy, nuclear energy for power supply is more economical at a larger scale.

If nuclear energy finds better social and political acceptance, it has huge growth potential. Its resource base—uranium and thorium—is substantial and has no competing application. Nuclear energy is ahead of other energy technologies in internalizing external costs. From safety to waste disposal to decommissioning, the costs of all of these are in most countries already included in the price of nuclear electricity. As shown in Table 7.1, it avoids GHG emissions. The complete nuclear power chain, from resource extraction to waste disposal and including reactor and facility construction, emits only 29 tons of CO_2 per GWh [1,14], about

---

TABLE 7.1

Relative Ranges of Greenhouse Gas Emissions from Different Electricity Generation Technologies

Substance	Mean (% Relative to Lignite)	(Lower/Upper) Range, Tons of CO_2 Eq/GWh
Lignite	100	790/1372
Coal	84.2	756/1310
Oil	69.5	547/935
Natural gas	47.3	362/891
Solar PV	8.1	13/731
Biomass	4.3	10/101
Nuclear	2.75	2/130
Hydroelectric	2.5	2/237
Wind	2.5	6/124

Sources: Marano, J.J. and Ciferno, J.P., Life-cycle greenhouse-gas emissions inventory for Fischer–Tropsch fuels, Energy and Environmental Solutions, LLC, for the U.S. Department of Energy, National Energy Technology Laboratory, Pittsburgh, PA, 2001; *CO₂ Emissions from Fuel Combustion*, IEA Statistics, 2011 edn., OECD/IEA, Paris, France, 2011.

---

the same as wind and hydroelectric and more than two orders of magnitude below coal, oil, and even natural gas.

7.3 STRATEGIES FOR GROWTH

The growth of nuclear energy will depend on its social and political acceptance. One method is to overcome faults of fission process by building a hybrid fusion–fission process. An economical, environmentally, and socially acceptable role of nuclear energy will require solutions to two fundamental problems of fission power: (1) elimination of nuclear waste and (2) the problem of limited naturally fissile U^{235} fuel supply. These problems can be overcome by breeding fuel from fertile material like U^{238} and Th^{232}. A fusion–fission hybrid reactor can lay the foundation for green and plentiful nuclear energy. The game-changing ideas, inventions, and innovations over the past two decades have led to the design of a workable highly compact intense fusion neutron source. In a hybrid fusion–fission nuclear energy, driven by such an intense fusion neutron source, several novel fuel cycles that would be generally inaccessible to pure fission become available. This possibility can solve the two problems with fission process outlined earlier.

Hybrid nuclear fission–fusion process [23,26] is a means of generating power by the use of a combination of nuclear fission–fusion process. The fusion process alone currently does not achieve sufficient gain (power output over power input) to be viable as a power source. By using the excess neutrons from the fusion reaction to in turn cause a high-yield (nearly 100%) fission reaction in the surrounding subcritical fissionable blanket, the net yield from the hybrid fusion–fission process can provide a targeted gain of 100–300 times the input energy. Even allowing for high inefficiencies on the input side, this can still yield sufficient heat output for economical electric power generation. Unlike a conventional fission reactor, the fusion hybrid can consume almost all of the uranium fuel without the need for enrichment or reprocessing. The low fuel consumption, lack of need for enrichment, and small waste volumes significantly reduce the fuel cycle costs. However, the fusion equipment required will increase the construction cost of the reactor.

If we were to harness combined fusion and fission hybrid process, nuclear energy would become even more useful with such a hybrid reactor. The resource base for nuclear fusion, for example, is huge. However, fusion is still at an experimental stage. It is unlikely to provide a substantial share of electricity to the grid before 2050 [23,26].

The second workable strategy is to create a hybrid nuclear–renewable energy system [19–22,24,25,27–33] in which nuclear power can be integrated with power generated from renewable energies (like solar or wind) with backup fossil or biofuel sources for power and an energy storage system. Such hybrid energy system can also be connected to power grid when that is possible. Such hybrid system will allow (a) steady power supply at both base load and during peak demand, (b) time- and location-dependent renewable energy sources to be operated on large scale, and (c) more use of renewable energy in the power industry. This strategy will work well in remote locations when connection to a major grid is not possible or too expensive.

The third and most attractive strategy is to combine the aforementioned hybrid energy system with combined use of heat and power generated from nuclear reactors such that excess heat from nuclear reactor during low power demand can be used for nonelectrical applications. This cogeneration approach (combined use of heat and power) will enhance thermal efficiencies of the different types of nuclear reactors. Cogeneration approach is ideally suited for nuclear energy because different types of nuclear reactors can generate different levels of temperature, making it appropriate for different types of nonelectrical applications [35–52]. A combination of hybrid system of nuclear–renewable energy for power generation mixed with combined use of heat and power gives the maximum potential for the growth of nuclear energy. While cogeneration improves thermal efficiency of the process, hybrid nuclear–renewable energy system improves capacity factor and time of use value. Hybrid combination backup by energy storage allows the maximum generation of power at the largest scale possible.

While nuclear reactors for power generations are most cost-effective at a larger scale, nonelectrical applications of the nuclear heat (through cogeneration) will allow the development of small modular reactors (SMRs) targeted toward specific applications. As shown in the literature [25,35,40,41,45,51,169,173], this will open up whole new market for small- and middle-scale nuclear energy production. The nonelectrical applications of HNESs will get more public support and reduce the fossil energy use for the same purpose. This will further enhance the growth of nuclear energy.

As mentioned in Sections 7.1 and 7.3, cogeneration method is especially suited for nuclear energy. The concept of cogeneration is not new. It has been used in combined cycle natural gas power plant where excess thermal energy is effectively used to improve thermal efficiency of the power plant. Nuclear reactors are most efficient when they are operated at full scale with or without hybrid integration with renewable energy sources. Such full-scale operation of the nuclear reactor can be accomplished by having its thermal energy available when power demand is low to replace the use of harmful fossil energy for a number of nonelectrical applications such as the productions of synthetic fuels, hydrogen, industrial process and district heating, and desalination. In these cases, the level of heat generated for combined heat and power applications (CHP, cogeneration) and the nature of nonelectrical application depend on the nature of the nuclear reactor technology used (see Figure 7.1; also Table 7.2). As shown in Figure 7.1 and in Table 7.2, different temperature levels of heat and different steam temperatures can be generated by different types of nuclear reactor technology. As shown in Figure 7.1, each level of temperature can find its usefulness for different nonelectrical applications [20,27,29,30,32,33].

7.4 ADVANCED HYBRID NUCLEAR–RENEWABLE ENERGY SYSTEMS WITH COGENERATION (COMBINED HEAT AND POWER SYSTEMS)

As mentioned earlier, the nuclear–renewable hybrid energy systems [19–33] are defined as integrated facilities comprised of nuclear reactors, renewable energy generation, backup fossil or biofuel generation, and energy storage, which can be

HES alternatives for fuels manufacturing

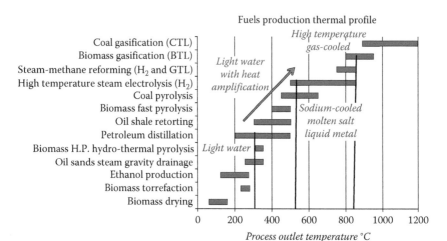

FIGURE 7.1 Types of reactor versus desired temperature. (From Antkowiak, M. et al., Summary report of the INL-JISEA workshop on nuclear hybrid energy systems, INL/EXT-12-26551, NREL/TP-6A50-55650, INL Report, prepared for US DOE, Office of Nuclear Energy, Contract No. DE-AC07-051D14517, Idaho Falls, ID, July 2012; Boardman, R., Advanced energy systems, Nuclear-fossil-renewable hybrid systems, A report to Nuclear Energy Agency, Committee for Technical and Economical Studies on Nuclear Energy Development and Fuel Cycle, INL, Idaho Falls, ID, April 4–5, 2013.)

TABLE 7.2

Temperature Capabilities of Reactor Types

Reactor Type	Typical Primary Coolant Average Temperature between Inlet and Outlet (°C)
Pressurized water reactor (PWR)	300
Boiling Water reactor (BWR)	283
Heavy-water reactor (HWR)	272
Liquid metal–cooled reactor (LMCR)	465
High-temperature gas-cooled reactor (HTGR)	725

Sources: Antkowiak, M. et al., Summary report of the INL-JISEA workshop on nuclear hybrid energy systems, INL/EXT-12-26551, NREL/TP-6A50-55650, INL Report, prepared for US DOE, Office of Nuclear Energy, Contract No. DE-AC07-051D14517, Idaho Falls, ID, July 2012; Boardman, R., Advanced energy systems, Nuclear-fossil-renewable hybrid systems, A report to Nuclear Energy Agency, Committee for Technical and Economical Studies on Nuclear Energy Development and Fuel Cycle, INL, Idaho Falls, ID, April 4–5, 2013.

attached to electrical grid if available. These types of hybrid nuclear–renewable energy systems are the best methods for the penetration of renewable sources of energy (wind and solar) in fossil energy–dominated world. This method allows the large-scale renewable energy operations and a transition to a low-carbon nuclear–renewable electricity grid. Historically, variable electricity demand was met using fossil plants with low capital costs, high operating costs, and substantial GHG emissions. However, the most easily scalable very-low-emission-generating options, nuclear and renewables (solar and wind), are capital-intensive technologies with low operating costs that should operate at full capacities to minimize costs. This method of combining nuclear and renewable energy systems requires a backup energy source such as fossil or biofuels with energy storage system to provide uniform and large-scale power without interruptions. Such a system would also satisfy the peak demand.

The renewable energy such as solar and wind produces power at a variable rate. Neither type of production matches demand. Because of the day–night and seasonal variations of sunlight, the typical capacity factor of solar devices is 18%. (The capacity factor is the actual energy output in a year divided by the potential energy output if the device were operated at full capacity for the entire period.) The capacity factor for wind is about 35%. For renewable energy sources, the mismatch between generation and demand is so large that it has been estimated that if as little as 15% of the electricity were produced by solar or wind, there would be limited economic incentive to obtain more energy from such sources, even if they are free. Nuclear energy, on the other hand, provides power at a constant rate.

Boardman [30] described nuclear energy as an energy solution "enabler," not just a heat machine. He, however, pointed out that for integrated highly coupled hybrid energy systems, it is important to [20,27,29,30,32,33]

1. Investigate challenges related to greater penetration of renewable energy and the effects that the dynamic characteristics and potential synergies among hybrid systems components may have on the overall stability of grid operations
2. Suggest and evaluate hybrid energy system options in order to address identified challenges
3. Compare dynamic setting of advanced hybrid energy systems, which not only include interactions of power generations by nuclear and renewable energy but also take into account the effect of cogeneration on the overall system dynamics

Boardman and coworkers [20,27,29,30,32,33] suggested to conduct dynamic cost analysis, including the cost of variability as well as optimum cost trade-off between heat and power, in order to assess the economic viability of the overall hybrid systems.

One method to operate both nuclear and renewable energy to their maximum capacity is to store the excess energy when the demand for power is low. Currently, fossil fuel is used as a backup storage source to renewable energy. Hydrogen can replace fossil fuels as a method to store energy; thus, it is a potential replacement as a fuel. Recent systems study by Mazza and Hammerschlag [34] has examined the production of H_2 using renewables and the subsequent conversion of that H_2 into electricity. The general conclusion is that renewables (solar cells, wind, etc.) are better suited for the direct production of electricity than for the production of H_2 for electricity purposes. The major renewable energy options are intrinsically electricity-generating devices. As shown in Chapter 8, however, the advancement in the production of solar fuels may alter this conclusion.

While the production, collection, and boosting pressure of hydrogen to pipeline pressure for hydrogen generated from renewable energy are challenging, studies by Forsberg [86] indicate that the potential of using nuclear H_2 to economically meet variable electrical demand is significant particularly at the large scale. A high-temperature steam electrolysis unit can be used to generate and store hydrogen as a fuel source to offset the time differences between energy generation and demand. This, however, requires combined use of heat and electricity from nuclear energy.

Over the last decades, several concepts of nuclear hybrid energy systems have been proposed involving the coupling of electricity and fuel production using nuclear heat [53–109]. A particular concept proposed by Forsberg and coworkers [79–83,86,109] involves the synergy between nuclear power plants, intermittent renewable energy sources like wind and solar, and hydrogen production and its storage and use in fuel cells as shown in Figure 7.2. The concept allows nuclear power plant to run at their optimal rate, which is base load, with the hydrogen system acting as an energy buffer to cope with the variations of renewable power production and electricity demand. This improves the thermal efficiency along with economic efficiency, capacity factor, and time of usage value of the overall system.

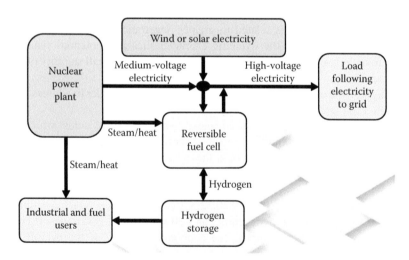

FIGURE 7.2 Example of nuclear–renewable hybrid system proposed by Forsberg. (From Ruth, M. et al., *Energy Convers. Manage.*, 78, 684, February 2014; Christensen, D., Summary of august workshop to identify potential synergies between nuclear and renewable energy opportunities, JISEA Joint Institute for Strategic Energy Analysis, NREL, December 2011, Workshop Report at: http://www.nrel.gov/docs/fy12osti/52256.pdf.)

According to Forsberg [79–83,86,109], the nuclear power plant would run continuously near its peak output. When the demand for nuclear electricity is low (because of high renewable production or when the overall demand for electricity is low), hydrogen would be produced using electrolysis (the reverse principle of fuel cell) and then stored. When there is a peak demand for electricity, instead of resorting to gas turbines, the system would produce electricity from the fuel cell. The capital cost savings made by not buying gas turbines pays for the electrolysis/fuel cell system. Forsberg [79–83,86,109] points out that the critical issues in this concept are (1) the absence of hydrogen transmission system and (2) the bulk storage of hydrogen (only underground salt deposits exist as commercial technology).

HNES research carried out by Boardman and coworkers by INL [20,27,29,30,32,33] showed that nuclear/renewable/fossil system can also be economically advantageous even without implicit or direct penalties for CO_2 emissions and before foreign fuel import cost premiums are taken into account. Nuclear/renewable/chemical plant hybrids can help stabilize the grid and allow greater penetration of both renewable and nuclear energy into the market. The benefits are achieved with the following:

1. Full utilization of capital investments.
2. Reduced requirements for electrical power storage and/or fossil fuel peak power generation units. The latter will reduce emission of GHG.
3. Coordinated production of cleaner transportation fuels particularly from renewable biomass as well as from indigenous natural gas and coal. This will reduce the use of crude oil and reduce carbon footprints on the environment.

INL analysis shows that hybrid systems can significantly increase U.S. production of biofuels and lead to sustainable energy infrastructure. INL has developed detailed process and economic models to evaluate technical and economic value of using nuclear energy beyond electric power production. The integration of nuclear energy with conventional and future fossil fuel plants to produce hydrogen, ammonia, and transportation fuels offers many advantages. These benefits can also be realized when hybrid systems are integrated with the electrical grid to synchronize clean nuclear power generation with variable grid demand and intermittent renewable power generation.

There are basic similarities in the proposed concepts of the advanced HNESs proposed by Forsberg and coworkers and researchers at INL. Both suggest improved thermodynamic efficiency of the hybrid concept by the use of excess nuclear heat during the time of low power demand for energy storage or other nonelectrical applications. As the U.S. energy system evolves, the amount of electricity from variable-generation sources is likely to increase, which could result in additional times when electricity demand is lower than available production. During these times, innovations in nuclear reactor design coupled with sophisticated control systems can allow for more complex apportionment of heat within an integrated system to carry out energy-intensive chemical, industrial, and heat-intensive processes.

A growing interest in harnessing nuclear power for nonelectrical applications through combined heat and power generation (cogeneration) method over the past two decades has been shown by many countries all over the world. The nonelectrical applications of nuclear power have been adopted for a wide range of applications [35–152] such as hydrogen production, nuclear seawater desalination, district heating, industrial process heating, oil recovery, coal conversion, and other industrial applications for the petrochemical and refinery processes, paper and textile processes, biofuel production, ship propulsion, space applications, and many other innovative applications. It is evident that the specific temperature requirements for such nonelectrical applications vary greatly from low (less than 100°C in case of heating) to very high (more than 1000°C in iron and steel industries). As shown in Figure 7.1, different types of nuclear reactors are capable of producing such a wide range of temperature. This cogeneration approach has opened up small- and middle-scale nuclear energy market.

Nuclear SMRs are expected to play a positive role in the fulfillment for the need of nonelectrical applications of nuclear power [25,35,40,41,45,51,169,173]. Recently, more than 50 concepts and designs of such innovative SMRs were developed in Argentina, Brazil, China, France, India, Japan, the Republic of Korea, Russian Federation, South Africa, and the United States. These countries believe that in order for nuclear energy to contribute effectively to future global energy supplies and sustainable development, its applications should be extended beyond electricity through various types of cogeneration systems, particularly in the transport sector where oil now supplies 95% of transport demand.

In the remaining chapters, we assess the applications of nuclear heat for the following seven cogeneration applications:

1. Use of nuclear reactor heat for oil production and refining
2. Use of nuclear reactor heat for several types of thermochemical transformations
3. Use of nuclear reactor heat for district heating
4. Use of nuclear heat for industrial process heating
5. Use of nuclear reactor heat for hydrogen production systems
6. Use of nuclear reactor heat for desalination of water
7. Use of nuclear reactor heat and power for ship, submarine, and space applications

7.5 NUCLEAR HEAT FOR OIL PRODUCTION AND REFINING

A survey of industrial energy use in France carried out by IEA [4–12] showed that about 40% of energy consumed for oil production and refining is heat at less than 400°C. The energy needs of a large oil refinery, including power, steam, hydrogen, and water, could be met by a single 600 MWe sodium-cooled fast reactor operating at 500°C–600°C. In this section, we examine applications of nuclear heat to recover and refine various types of oils. These applications replace the use of fossil fuel to provide heat for the similar applications. Thus, applications decarbonize the recovery and refining processes.

7.5.1 OIL SHALE EXTRACTION

In the oil shale system, a nuclear power plant generates load-following low-CO_2 electricity, displacing an equivalent amount of fossil-fired, CO_2-emitting power generation. When that nuclear electricity is not needed, the heat that would be used to generate it is instead diverted into heating a subsurface oil shale deposit. After several years of retorting at high temperature, the hydrocarbons in that deposit can be withdrawn as a synthetic crude oil. The credits for CO_2 reductions obtained during the nuclear electricity production can be applied to the synthetic crude to allow it to meet low emission standards. Hybrid nuclear–renewable oil shale systems potentially represent half of the market in the United States for SMRs, and thus, there are large incentives to simultaneously demonstrate the combined reactor and oil shale system. The United States is developing SMRs using light-water reactor technology [11–13].

Oil shale contains kerogen, but no liquid oil. Heating oil shale underground to ~370°C converts the kerogen to a light high-value oil, natural gas, and char. With typical proposed nonnuclear processes, a quarter of the product must be burned to heat the oil shale, resulting in a large GHG footprint per barrel of oil. A low-carbon advanced hybrid nuclear–renewable energy system with cogeneration alternative (for nuclear heat) enables expanded deployment of renewables and produces oil with one of the lowest GHG footprints of any option. For this purpose, closed steam lines would be drilled into the oil shale. The oil shale would be heated between 210°C and 250°C using steam heat. Further heating of the steam is accomplished by electric heaters to enable heating shale to 370°C [53–55]. The nuclear reactor would sell electricity to the grid at times of high prices and demand. At times of low electricity demand, (1) the reactor would provide steam for heating oil shale and (2) electricity would be bought from the grid to boost steam temperatures. The baseload nuclear plant acts as a variable source of electricity, replacing the use of fossil fuels for variable electricity production and providing the backup required when renewable inputs to the grid are low. It enables true zero-carbon renewable electricity production, eliminating the need for fossil fuel backup power, and produces liquid fossil fuels with a low carbon footprint relative to conventional sources of oil.

Forsberg [53,54] analyzed this nuclear energy cogeneration approach for underground extraction and refining of oil shale using power (electricity) or heat from a nuclear reactor. According to him, the energy requirements for retorting the oil shale are reduced by a factor of 2. Direct use of high-temperature nuclear heat avoids the conversion of heat to electricity with all the associated losses and subsequent use of electricity to produce heat. The expensive electricity is replaced by lower-cost thermal energy. He estimated that about 12 GW(t) of high-temperature heat would be required to produce a million barrel of oils per day. The required temperature would be near 700°C, which can be easily achieved using gas-cooled nuclear reactor.

7.5.2 OIL RECOVERY AND REFINING

The rapidly emerging interest in smaller-sized commercial nuclear power plants is motivated in part due to their adaptability to a broader range of energy customers, including utilities in small demand markets and industrial consumers of nonelectrical energy. The oil recovery and refining market is of special interest since it is highly energy intensive, representing roughly 7% of the total U.S. energy consumption or an average power of 200 GWt. The NuScale SMR design [153,154] is especially well suited for this market because of the small unit size of a power module (160 MWTh) and the flexible modularity of combining individual NuScale Power Module into a central plant [153,154].

A variety of processes within the oil recovery and refining industry were explored as potential application of power from a NuScale power module. Generally, field-based recovery processes such as steam-assisted gravity drain processes and in situ oil refining processes were found to be marginally suitable for coupling to a NuScale plant, due primarily to the significant disconnect between the plant lifetime and the field depletion time. Instead, the operation of traditional refineries or in-field "upgraders" appeared to be better matched to the energy output and plant lifetime of a small-scale NuScale nuclear plant. The relatively low steam temperature from a NuScale power module can be used directly for several of the processes within the refinery; other processes will require additional temperature boosting using the refinery fuel gas that is generated as a low-grade by-product of the refining process. The NuScale was found to be economically attractive for a refinery process of 250,000 barrels per day [153,154].

7.5.3 Extraction and Refining of Unconventional Oil and In Situ Light Oil Production from Soft Coal

Other possible applications of nuclear energy cogeneration technology are the use of nuclear heat for unconventional oil extraction and refining and in situ light oil production from soft coal. For unconventional oils, two categories are relevant: the extraction of heavy oil (and oil from tar and oil sands) and the extraction of the oil remaining in depleted deposits. The prospects for these applications are related to the fact that the resources of unconventional oil are larger than those of conventional oil.

Over the last three decades, methods of unconventional oil recovery have been developed, in particular in those countries in which such resources are large, such as Canada and Venezuela. Nuclear heat is applicable for oil extraction by the use of steam injection. Steam and electricity may also be needed in the course of oil processing following extraction. A nuclear reactor producing steam of the required quality and quantity is therefore needed. It has been shown [56] that high-temperature gas-cooled reactor (HTGR) technology satisfies the requirement for using nuclear energy in unconventional oil extraction. However, new methods of steam injection, such as steam-assisted gravity drainage, also make other reactor concepts applicable [53–58,70] because the required steam parameters are sufficiently low. For example, the application of steam-assisted gravity drainage in Canada using CANDU reactors is a promising concept with long-term benefits [58,70].

Independent of the reactor type, the option of supplying heat with nuclear reactors must be economically competitive with the alternative steam generation technologies, unless the amount of GHG emission influences the decision significantly.

An analysis of the latter factor conducted in References 58 and 70 for Canada shows that its role can be important. Steam can also be used for processing the oil after the extraction. In general, only dual use, where nuclear electricity can compete in the market, could make these applications economically worthwhile.

Forsberg [53,54] suggested a method for the production of light crude oil using nuclear heat without the release of large quantities of GHGs. In this method, besides addressing the recovery of light oil from old oil fields, oil shale, and tar sand, Forsberg also points out conversion of heavy oils and soft coal to char and liquid fuel by nuclear heating.

In Forsberg method, fossil deposit is heated to high temperatures using nuclear heat resulting in conversion of heavy oils and soft coal and refining of resulting oils by in situ thermal cracking. As temperature increases, the vapor generated by the distillation process will move to the recovery wells. The heavier crude will thermally crack to produce more light oils that can be recovered. The char residue will remain behind in the reservoirs. In fact, nuclear heating underground simulates distillation and thermal cracking refining process. The process has three basic advantages:

1. Large amount of heavy oil and soft coal underground resources are accessible with in situ refining process to produce usable light oils.
2. Unlike refining of heavy oil, oil shale, and soft coal on the surface, carbon dioxide emitted during underground extraction and refining process remains sequestered underground.
3. High-quality light oil generated in this process will require very little refining (with little hydrogen) to produce transportation fuels.

7.6 NUCLEAR HEAT FOR VARIOUS THERMOCHEMICAL TRANSFORMATION SYSTEMS

The heat required for many thermochemical operations involving fossil fuel, biomass, and waste can be obtained from combined nuclear heat and power (cogeneration) processes. In the following sections, few examples of the use of nuclear heat for various types of thermochemical transformations are illustrated.

7.6.1 BIOMASS PROCESSING PLANT

The advanced hybrid nuclear–renewable energy system can consist of a small-/ medium-sized nuclear reactor (SMR), a biomass processing plant, and an optional hydrogen production plant and a wind farm. The SMR operates at full power constantly and is able to switch its thermal output to supply the steam turbine for electricity generation or to supply the biomass processing plant for heat processes. The heat required for a biomass plant can vary substantially depending on whether heat is used for torrefaction process for feed preparation, pyrolysis process for the feed preparation, or the pyrolysis of entire biomass feedstock to produce bio-oil.

Such a hybrid energy system management follows variable electricity demand such that the sum of the electricity generated by both nuclear reactor and the wind farm will meet the required electricity demand. The hydrogen production plant is considered to be an electricity consumer and operates during times of wind power surplus. For the wind farm capacities of 100–2000 MW, the optimal SMR size in the hybrid energy system can vary from 200 to 300 MWe.

Forsberg [59] suggested three other alternative methods for nuclear heat–biomass liquid fuel options. Each option requires more sophisticated technology, produces more liquid fuels per unit of biomass, and requires more energy from the nuclear component of such a system [59–73].

1. The first method deals with corn to ethanol process. About two-thirds of a corn kernel is starch that becomes ethanol. The nonfermentable components that consist primarily of proteins and the by-products of fermentation process become animal feed or other useful by-products. Currently, the plant energy is provided by the natural gas. The nonsolar energy input to grow corn and convert it to ethanol is typically about 70%–80% of the energy value of ethanol [59,61,63,64,69,73]. Most of the energy is supplied by burning fossil fuels. About one-half of that energy input is in the form of low-temperature, low-pressure (150 psi) steam used within the ethanol plant [59,61,63,64,69,73].

   In a hybrid nuclear heat–ethanol plant, low-temperature steam (180°C) from nuclear reactor would first go through high-pressure turbines to produce electricity and then be sent to an ethanol plant. In the ethanol plant, steam would be condensed and warm water would be returned to the nuclear power plant. Almost all of the heat would come from steam condensation. Modern steam system would allow more than a mile of separation between the reactor and ethanol plant; thus, ethanol plants would be located beyond any security perimeters. Forsberg et al. [59,69] showed that the cost of low-temperature steam obtained from nuclear power plant would be considerably less than the cost of the natural gas needed to produce the same amount of steam. Forsberg [59,69] calculated that ethanol plants would require 19.9 GW(t) of low-temperature steam to produce ethanol with the energy equivalent of a million barrel of diesel fuel per day.

2. Forsberg et al. [59,69] also suggested the use of nuclear heat for the conversion of cellulose to ethanol. In this case, in order to maximize liquid fuel production per unit of biomass, the lignin must also be converted into liquid fuels rather than burned as boiler fuel. This can be done by using nuclear energy rather than burning lignin to provide steam for ethanol production and by commercializing methods to convert lignin to liquid fuels. Several processes to convert the lignin to hydrocarbon fuels are under development [62,63,67,68]. Most of these processes require limited quantity of hydrogen.

   Forsberg et al. [59,69] points out that if lignin is converted to liquid fuel and if steam from nuclear reactor replaces lignin that was to be burned as a fuel, the energy content of the liquid fuels per unit biomass feedstock can be potentially increased by 50% and about half of the U.S. demand for liquid transport fuel can be met. Since the nuclear reactor provides low-temperature steam at low cost, the economics are very competitive worldwide. This approach thus requires more development on cellulose to ethanol and lignin to liquid fuel technologies. Forsberg et al. [59,69] points out that the shipping expenses for cellulosic biomass can also be prohibitive in the application of this concept.

3. The last method suggested by Forsberg et al. [59,69] is the use of nuclear heat for thermochemical transformation of biomass to liquid fuel. This may be done either by torrefaction followed by gasification or by the use of nuclear heat for direct pyrolysis. The synthetic gas produced can also be converted to liquid fuel by Fischer–Tropsch synthesis, which requires high amount of heat. Here, all the carbon converted into fuel and biomass is fully converted to $((CH_2)_n)$ rather than ethanol. This option requires significant external energy input in the form of hydrogen, heat, and electricity. The economics of this approach strongly depend on nuclear heat–hydrogen production costs.

7.6.2 Synthetic Fuels

As shown in Table 7.3, liquid metal and gas-cooled reactors can generate very high temperatures, which could be used to create new synthetic fuels for energy. This will be an innovative application of nuclear energy and can considerably expand its use. This is because the transportation sector is responsible for about a quarter of the total energy use and almost 99% of this is currently supplied by organic fuel. Nuclear power can penetrate this large market through the use of electric cars and production of synthetic fuels such as methanol, ethanol, and their derivatives as well as its use in coal gasification. The productions of methanol, ethanol, and their derivatives require energy in the form of heat. The availability of a proven and economic HTGR [56] technology for this purpose will be important.

TABLE 7.3
Potentials for Various Nonelectrical Applications of Nuclear Energy

Application	Market Potential
District heating	340–7600 GWth
Desalination	Very large; high
Process heat	240–2900 GWth
Ship propulsion	40–400 million GT
Space applications	Cannot be estimated
Hydrogen production	No estimate
Coal gasification	—
Other fuel syntheses	—
Oil extraction	—
Nuclear submarines for fossil fuel transportation	—

Sources: From references 3–10 and 156–175.

7.6.3 COAL GASIFICATION

Coal gasification (i.e., conversion of solid coal into a gaseous fuel-like synthetic gas) requires very high temperatures but could be practical because the infrastructure for use of synthetic gas already exists. Coal gasification is more energy efficient, cleaner, and versatile than conventional coal combustion. The process of coal gasification is, however, quite energy intensive. HTGRs can play a role here.

Nuclear energy integrated gasification combined cycle plants are now a proven technology [60,65,69,71,72]. A 5-year program of the 100 MW(e) integrated gasification combined cycle demonstration plant at Cool Water, California, was completed in 1989; a 235 MW(e) unit at Buggenum in the Netherlands started up in 1993. There are three plants in the United States, at Wabash River in Indiana, Polk Power near Tampa in Florida, and Piñon Pine in Nevada. The largest unit is that at Puertollano in Spain, which has a capacity of 330 MW(e). In principle, these plants can also use other feedstock such as mixtures of coal/biomass or coal/waste. Nuclear energy in the form of an HTGR would be a credible, nonpolluting technology for this purpose.

The use of nuclear heat for coal gasification is significantly assessed in Germany and Poland. Poland has a plan for constructing a first nuclear reactor to be in operation in 2020s and for using the nuclear heat for synthetic fuel production from their large coal reserve [4–10]. Germany has investigated their HTR technology for coal gasification application.

7.7 NUCLEAR HEAT FOR VARIOUS HYDROGEN PRODUCTION SYSTEMS

The most promising way that nuclear energy might provide energy for transportation is through the production of hydrogen, either for direct use in fuel cell vehicles or for synfuel production [74–109]. Hydrogen research is currently on the rise. Major automobile manufacturers have set ambitious targets for putting affordable fuel cell cars on the road, and significant governmental initiatives have been launched in Europe, the United States, and Japan. Nuclear energy is part of these initiatives, but for nuclear energy to become a major future hydrogen supplier, significant efforts in both reactor technology development and installing new infrastructures will be required.

While the implications of using nuclear energy to produce electricity are reasonably well understood, the implications of using nuclear reactors to produce H_2 are less clear. Nuclear H_2 can be used to meet the existing demands for H_2. At present, the large existing markets for H_2 are (1) upgrading heavy crude oils and tar sands to liquid fuels (gasoline, diesel, and jet fuel) and (2) producing ammonia fertilizer. Many smaller markets, such as conversion of iron ore to metallic iron, also exist. Beyond these markets, many futures are possible.

The current worldwide hydrogen production is roughly 50 million tons/year. Although current use of hydrogen in energy systems is very limited, its future use could become enormous, especially if fuel cell vehicles would be deployed on a large commercial scale. For example, the Committee on Alternatives and

Strategies for Future Hydrogen Production and Use of the United States [4] estimated that fuel cell vehicles would enter into the light-duty vehicle market in 2015 in competition with conventional and hybrid electric vehicles, reaching 25% of the market around 2027 and that full replacement of gasoline light-duty vehicles with hydrogen vehicles would occur in 2050. Assuming a fleet of 300 million fuel cell–powered cars in the world, an estimated 120 million tons of hydrogen will be required annually just for supplying energy to part of the transportation sector [79–83,86,91,109].

Hydrogen production requires a primary energy source. This means that for hydrogen to become a noticeable part of the energy system, another nonhydrogen energy source is required. As the overall thermal efficiency of the energy system is likely to decrease with the use of hydrogen (because of the energy spent for hydrogen production, in particular for the case of water electrolysis), the overall amount of primary energy would be larger than for a system without hydrogen unless this conversion loss is offset by a higher efficiency in the hydrogen-fuelled power plant. (Fuel cells generally have efficiencies of at least 50%.)

Possible sources of primary energy for hydrogen production are fossil fuels, nuclear energy, and renewable energy sources. As nuclear reactors can produce both the heat and electricity required for hydrogen production, Forsberg [79–83,86,91,109] notes that the use of nuclear energy is particularly appropriate for hydrogen production. Moreover, as nuclear energy is now the most significant commercially mature non–fossil fuel energy source, there is a potential for basing large-scale hydrogen production on nuclear technologies. This would radically mitigate the emissions of GHGs from the energy sector [73,91].

While hydrogen can be produced by gasification of coal, biomass, and waste in the presence of steam or by supercritical water gasification of all carbonaceous materials, the three technological options that are most relevant for cogeneration using nuclear energy are [87–96,155]

1. Electrolysis of water using electricity from nuclear reactors in off-peak periods; also high-temperature electrolysis of steam using high-temperature heat and electricity from nuclear reactors
2. Steam or dry reforming of natural gas using high-temperature heat from nuclear reactors
3. Water dissociation by multistage thermochemical processes using high-temperature heat and electricity from nuclear reactors

7.7.1 NUCLEAR HEAT–ASSISTED ELECTROLYSIS

Electrolysis of water basically is attractive essentially when cheap electricity is available or high-purity hydrogen is required. The use of nuclear electricity in off-peak periods would be economically attractive in the light of the low marginal cost of nuclear power plants [105–108,155]. Water electrolysis at ambient pressure and temperature of 70°C–90°C is a common method for the production of high-purity hydrogen. Hydrogen can also be obtained more efficiently by significantly raising the temperature of water. The electrolysis of steam at higher temperature

(800°C–1000°C) offers several advantages including lower electricity requirement and higher efficiency resulting from lowering the activation barriers at the electrolyte surfaces. High-temperature electrolysis is a reverse reaction of the solid oxide fuel cell, where water is decomposed in the solid polymer electrolyte to hydrogen and oxygen.

7.7.2 NUCLEAR HEAT–ASSISTED REFORMING

Currently, hydrogen is produced mainly by steam reforming of natural gas/methane. This is a catalytic process involving the reaction of natural gas with steam to produce a mixture of hydrogen and CO, requiring temperatures in the range of 600°C–950°C. Both steam and dry reforming are endothermic reactions requiring external heat. Nuclear-assisted steam reforming has great potential for large-scale hydrogen production in the near term. This well-established process, however, has CO_2 as a waste product. The steam reforming system can be easily coupled to an HTGR, which can provide the necessary heat and high temperature. Considerable R&D work has been carried out in Germany for the steam reforming of methane including performing experiments in a pilot plant, EVA-I and EVA-II. Currently, work is in progress in JAERI for the HTTR14, in China for the HTR-1015, and in Russia [87–96,98–108,155].

Hydrogen can also be produced by dry reforming (CO_2 reforming) of methane. The basic CH_4 and CO_2 reaction for this process (with no addition of steam) produces CO and hydrogen. The reforming process requires high temperature (700°C–900°C) and high energy input, both of which can be provided by HTGRs. The generated CO and H_2 mixture (syngas) can be used directly as fuel for electricity generation (e.g., by fuel cells).

7.7.3 NUCLEAR HEAT–ASSISTED THERMOCHEMICAL WATER SPLITTING

Since direct thermolysis of water requires temperatures over 2500°C, the thermochemical water splitting process consists of different partial reactions, each running at a lower temperature level (800°C–1000°C). Thousands of potential thermochemical cycles have been tested to assess their viability and performance for hydrogen production, and the most promising ones for efficiency and practical applicability to nuclear heat sources have been identified as iodine–sulfur (IS), bromine–calcium (Br–Ca), and copper–chlorine (Cu–Cl) [78]. The Japan Atomic Energy Research Institute have demonstrated the feasibility of the IS process with continuous generation of hydrogen from water with recycling of the process material. An energy efficiency of 47% has been achieved in this process. In addition, a thermochemical hybrid process can be another option, which combines both thermochemical and electrolytic reactions of water splitting and has the possibility of running low-temperature reactions [87–96,98–108,155].

In comparison with the use of fossil fuels for hydrogen production, nuclear energy has the advantages of having a large resource base and of the absence of most air emissions including carbon dioxide. Moreover, the use of nuclear energy in combination with water electrolysis avoids burning natural gas, which is the dominant raw material for hydrogen production. These advantages are especially important in view of the large amount of primary energy required for an energy system with a noticeable share of hydrogen use.

In comparison with the use of renewable energy sources for hydrogen production, nuclear energy has the advantage of being a mature, available technology and has the important feature of a high energy concentration. Apart from their relative technological immaturity, renewable energy sources, although potentially large and inexhaustible, are very diffuse and available only at a low energy density. However, as shown in Chapter 8, the use of concentrated solar power to produce solar fuels (like hydrogen) is heavily researched and has significant potential. The high energy concentration in nuclear fuel enables either hydrogen production concentrated in multiproduct energy centers or distributed hydrogen production using existing electricity grids to power low-temperature electrolysis. Hydrogen and electricity are interchangeable in HNESs. Nuclear energy is, however, most suitable for large-scale hydrogen production.

Numerous countries around the world have started nuclear hydrogen programs. Along with the United States, Canada has a significant nuclear hydrogen program [96–102,155]. France has also carried out a lot of R&D on hydrogen generation from nuclear reactors [96–102]. Germany uses high-temperature heat for hydrogen production by its HTR technology. Japan and Korea have demonstrated the feasibility of nuclear hydrogen production using thermochemical iodine/sulfur cycles [75,104,155].

7.8 NUCLEAR HEAT FOR DESALINATION OF WATER

While water is essential for living, more tha a billion people—approximately 20% of the world's population—lack safe drinking water. Three billion lack access to adequate sanitation due to lack of water. Unfortunately, 94% of the world's water is saltwater and only 6% is fresh, and less than 1% of the freshwater is easily accessible (27% being in the glaciers and 72% underground). As the standard of living increases all over the globe, the demand for both energy and water is also increasing. By 2025, the number of people suffering from water stress or scarcity could increase to 3.5 billion, with 2.4 billion expected to live in water-scarce regions [110–132].

The development and use of water desalination technologies are very important. Desalinated water supply implies the production of potable water from seawater desalination plants and its injection into water transportation/storage network. Seawater desalination is a water treatment process that removes salts from seawater to obtain freshwater adequate for irrigation, drinking, and industrial use. Various forms of these technologies have been practiced for the past 50 years [110–132].

The possibility of using nuclear energy for desalination of seawater was realized as early as the 1960s. The Kazakhstan nuclear plant was the only power reactor in the world supplying heat for industrial-scale desalination at the time it was shut down in 1999. It produced 80,000 m^3/day of potable water for municipal use. The Diablo Canyon Nuclear Power Plant in the United States also produces 4500 m^3/day of freshwater from the sea for in-plant use [110–132].

7.8.1 Nuclear Desalination Technology and Its Future

Desalination of water requires energy but, as shown in Table 7.2, it can be done at relatively low temperatures. Waste heat from power plants is sufficient for this purpose. Nuclear power can play a significant role [110–132], particularly in a dual capacity, by providing water in addition to GHG-free electricity. The major commercially available desalination processes are of three kinds:

1. *Thermal processes*: heat is used to vaporize and distill freshwater from saline water; these are multistage flash distillation (MSF), multiple-effect distillation (MED), and vapor compression.
2. *Membrane processes*: suitable membranes are used for the separation of salts such as the mechanism of reverse osmosis (RO). There are also other minor processes such as freezing and solar evaporation.
3. *Hybrid processes*: process streams are exchanged in order to take advantage of the different operating temperature conditions of each plant. In particular, due to the low operating temperature requirement for MED plant, the required heat for MED plant can be supplied by an adjacent MSF plant. A number of novel technology options for various types of MSF–MED process coupling have been studied to increase energy efficiency and distillate production and minimize operational costs [110–132]. From a technical perspective, for MSF and MED, heat in the range of 100°C–130°C is required. Electricity is required for RO as the primary energy source and for MSF and MED as energy for pumping. Highly reliable backup source of energy may be required on the site of the desalination facility. The total plant capacity may range from 100 to 60,000 m^3/day depending on the needs of the water customer. In 2003, globally about 26 million m^3/day of freshwater was produced by desalination (including both brackish and seawater plants), largest being in Saudi Arabia, about 21%. The United States produced approximately 17%, 80% of which was achieved by membrane processes [110–132].

The nuclear reactor may be used solely for desalination or may be operated in a cogeneration mode. The use of nuclear heat requires a close location of the nuclear plant to the desalination plant. While the electricity required for RO plant can be obtained from a distant nuclear reactor, electricity taken directly from the plant is cheaper than that from the electricity grid and that a distant location would not allow the use of warm water from a condenser of the nuclear reactor for the RO feed [110–132].

In most nuclear heat desalination plants, steam is produced in a secondary loop for generation of electricity, and then another tertiary loop is used to heat the seawater for desalination. The saltwater is in the fourth loop. This makes the production of freshwater far removed from the radioactive isotopes of the first, primary loop. This protects freshwater from any possible nuclear radiation [110–132].

The desalination of seawater using nuclear energy (either low-temperature heat or electricity) is a demonstrated option. It is a mature technology. Over 150 reactor-years of operating experience with nuclear desalination have been accumulated worldwide [110–132]. Several demonstration programs are underway, with technical coordination support from the IAEA, to confirm its technical and economical viability under country-specific conditions. Economic studies performed at the IAEA indicate that nuclear energy can be competitive for desalination compared with fossil-fuelled energy sources. It was generally found that (a) MSF processes cost higher than RO and MED processes, (b) RO and MED processes costs are in general comparable, (c) RO is economically more favorable for less stringent drinking standards, and (d) desalination costs are higher for smaller reactors [110–132].

The desalination capacities of the world have been doubling each decade, and hence there is a tremendous potential for nuclear desalination. Efforts are now primarily directed toward reducing production cost of desalinated water through innovations and technological enhancements. Japan has a long experience of using desalination to provide process water for the operation of its thermal power plants including some nuclear power plants. Korea has developed nuclear desalination applications for the new SMART SMR reactor [110–132]. Russia was one of the first countries to demonstrate the process of desalination using nuclear heat and electricity. Currently, major activities are taking place in India, Pakistan, and China. Canada and France are also involved in nuclear desalination research [110–132]. India has constructed a 6300 m^3/day combined MSF–RO nuclear desalination demonstration plant connected to two 170 MWe PHWR units at Kalpakkam [110–132]. Other countries that are showing interest in or going forward with nuclear desalination projects, for domestic use or for exportation, include Argentina, Egypt, Indonesia, Morocco, and Tunisia [110–132].

7.9   NUCLEAR HEAT FOR DISTRICT HEATING

District heating is an attractive and growing market suitable for low-temperature nuclear heat. It is residential and commercial building heating. District heating systems use hot water or steam in the temperature range of 70°C–150°C with steam, water, and water–water heat exchangers as needed. Usually, steam is extracted from low-pressure turbines of the nuclear power plant to provide the base heating load, and steam from the high-pressure turbine is used for the peak heat demand. Development of a heat distribution system is required but, due to heat losses in the heat distribution system, the source must be nearby, usually within a few kilometers at most [133–136,139–142,144–148]. Also, the demand for heat fluctuates with the season, being very high in cold winters and low in summer, and the source must be able to accommodate this fluctuation. The use of excess nuclear heat for district heating reduces the use of fossil energy and thereby has the additional benefit of greatly reduced CO_2 emissions.

Typically, coal and gaseous fuels are used for district heating. Various other heat sources such as biomass materials, waste incineration, and waste heat from industrial processes are also used for district heating. District heating from a nuclear power plant follows the cogeneration mode, in which waste heat from power production is reused as a source of district heat.

From a technical perspective, district heating has the following requirements [110–132]:

1. Heat distribution network transporting steam or hot water generally has a temperature range of 80°C–150°C. The heat source must be within 10–15 km of the customer to avoid significant heat losses during transportation.
2. The heat generation capacities depend on the size of the customer. The capacity of 600–1200 MWth is normal for large cities, whereas it is considerably smaller for small communities.
3. Since heat is mostly supplied during winter, the annual load factor is normally not higher than 50%. The steady and reliable supply often requires a backup capacity or energy storage.

Nuclear district heating is in use in several countries and is technically a mature industry. The concern of leakage of radioactivity into the heating network has been taken care of by intermediate heat transfer loops operating at higher pressures than the steam loop from the turbines and by constant monitoring. The safe and reliable operation of several district heating networks (e.g., in Sweden, Finland, Czech Republic, Bulgaria, Hungary, Slovakia, Russia, Ukraine, and Switzerland) has proven their effectiveness. All existing reactor types (light water, heavy water, fast breeder, gas cooled, and high temperature) are potentially applicable to cogeneration [133–136,139–142,144–148]. The power capacity of heat networks is estimated to be about 600–1200 MWt in large cities and 10–50 MWt in small communities. At present, nuclear district heating appears to be most promising in countries that already have heat distribution networks [133–136,139–142,144–148].

The market potential for district heating has been estimated to be between 340 and 7600 GWt. Nuclear power provides only about 4.4 GWt. Since there are various sources for heat such as oil, coal, and natural gas, unless nuclear power is economical in the open market, it cannot make a big dent in commercializing nuclear district heating [133–136,139–142,144–148]. The scale of heat supply is also important for nuclear district heating as it is more expensive at lower power. At 500 MWt and above, the nuclear option shows good chances to be competitive even at higher discount rates [133–136,139–142,144–148]. Its future expansion will be determined by the size and growth of the demand for space and water heating, competition between heat and nonheat energy carriers for space and water heating, and competition between nuclear and nonnuclear heating. The availability of a heat distribution network is an important factor for nuclear district heating [133–136,139–142,144–148].

Finland, which has a long tradition of district heating using fossil fuel and biomass, is examining their replacement by nuclear power plants. They have started numerous test sites including one 100 km from Helsinki. France has also looked at the feasibility of connecting existing district heating networks in Paris and Lyon areas to neighboring nuclear power plants [133–136,139–142,144–148]. They have also considered the feasibility and economics of transporting heat from Nogent-sur-Seine nuclear power plant to the Paris district heating system. The NRG company in the Netherlands is also exploring cogeneration of heat and electricity from a nuclear power plant. Russia has a long experience of operating district heating using heat provided by nuclear power plants [133–136,139–142,144–148].

7.10   NUCLEAR HEAT FOR INDUSTRIAL PROCESS HEAT APPLICATIONS

There are five primary areas of industrial heat applications: food processing, paper industry, chemical industry, petroleum and coal processing, and primary metal industries [134,138,143,151,152,156]. In the United States, close to 60% of the total process heat is used in petroleum, coal processing, and chemical industries [134,138,143,151,152,156].

Industrial process heat is mainly used in the form of steam at appropriate temperature and pressure conditions. For process heat supply, there is a wide range of required heat parameters that determine the applicability of different reactor concepts. One particular concept, the HTGR, covers almost the whole temperature range and is therefore considered to be a leading candidate for the supply of nuclear process heat. The development and demonstration of an HTGR with a relatively small capacity would provide a strong impetus for process heat applications.

From a technical perspective, process heating has the following requirements:

1. The heat source must be close to the customer in order to avoid excessive heat losses during heat transmission.
2. Since process heat demand does not depend on climatic conditions, its annual load factor is around 70%–90%, a number higher than the one for district heating. The reliable and secure supply will require a backup source or energy storage device.

There have been some experiences in providing process heat for industrial purposes with nuclear energy in Canada, Germany, Norway, and Switzerland [134,138,143,151,152,156]. In Canada, CANDU reactors supplied steam for industries such as food processing and industrial alcohol production [134,138,143,151,152,156]. The application to the heavy-water production facility in Bruce, Canada, was the largest use of nuclear process heat, and it has operated very successfully for over 20 years. Six other industries that the Bruce complex provided process heat have been plastic film manufacturing, ethanol plant, apple juice concentration plant, alfalfa dehydration, cubing and pelletizing plant, a greenhouse, and an agricultural research facility [134,138,143,151,152,156].

In Germany, the Stade PWR supplied steam for a salt refinery located 1.5 km from the plant [134,138,143,151,152,156]. In Norway, the Halden Reactor supplied steam to a nearby factory for many years. In Switzerland, since 1979, the Gösgen PWR has been delivering process steam to a cardboard factory located 2 km from the plant [134,138,143,151,152,156]. As in district heat, it should be noted that about 77% of the total use is concentrated in Russia. The shares of China, Eastern Europe, the United States, and Western Europe are 14%, 6%, 4%, and 3%, respectively [134,138,143,151,152,156].

The total potential market of industrial process heat is large and of the same order of magnitude as district heating. It is estimated to be between 240 and 2900 GWt [134,138,143,151,152,156]. However, the demand in terms of size varies; some 50%

of the users need less than 10 MWt, 40% need sizes from 10 to 50 MWt, and only 10% need sizes greater than 50 MWt [134,138,143,151,152,156]. Very few need a large amount of process heat as in the Bruce example in Canada. The market is also very competitive as small fossil fuel units can provide the needed steam. The development of nuclear SMRs for this purpose has a significant potential.

7.11 USE OF NUCLEAR REACTOR HEAT AND POWER FOR SHIP, SUBMARINE, AND SPACE TRANSPORTATIONS

The transportation sector in the world consumes about 25% of the global energy including road, aviation, rail, pipeline, and navigation [152]. About 10% of these are for international and national seaborne transportation, for which nuclear energy can be directly utilized [152]. In terms of the quantity of transported goods and fleet capacity, two services have occupied the largest share of the market over the past decade: the transportation of oil and oil derivatives and the bulk transportation of goods such as grain, coal, and iron ore [3–12,152].

The first application of nuclear energy to ships was for military submarines with PWRs. The first civil application of nuclear-powered ship propulsion was for nuclear icebreakers in the former USSR, which built about nine icebreakers [3–12,152]. While market potential can be anywhere higher than 2400 ships, other considerations will determine the penetration of nuclear energy into this sector [3–12,152].

There have been some experiences of nuclear-powered civil on-surface ships in three countries: the United States (Savannah), Germany (Otto Hahn), and Japan (Mutsu) [3–12,152]. These three merchant ships were all equipped with PWRs with a containment around the reactor cooling system. At present, only Russia and Japan are conducting studies for the development of civil ship reactors [152]. The use of sail training vessels (STVs) for transportation of fossil fuels and goods in arctic conditions has been considered in the Russian Federation. The technical and economical assessment of STV carried out by Russia recommended a container transporter and a tanker as possible future vessels [3–12,152].

Qualitatively, the most suitable application for nuclear reactors for space application is that for missions requiring high-power levels (10–200 kWe) and/or long operating times (7–10 years) [3–12,152]. Based on predominant Russian experience, *future* nuclear option is more likely for medium- and long-duration space missions, where, in terms of power level, load weight, and mission duration, there is no energy carrier that is comparable with nuclear energy [3–12,152].

7.12 MARKET POTENTIAL AND FUTURE PROSPECTS FOR HNES

Based on the descriptions in Sections 7.4 to 7.11, the market potentials and future prospects for HNES appear to be bright [157–175]. While electrical and nonelectrical applications of nuclear energy systems carry a wide range, some of the important conclusions and perspectives for the future are briefly delineated below.

The analysis of the market potential for the nonelectrical applications of nuclear energy leads to the following conclusions regarding the potentials of various HNESs (with and without co-generation):

1. For the foreseeable future, power generation will remain the main application of nuclear energy, the main reasons being the advanced status of nuclear power production technologies and an increasing share of electricity in the final energy demand.

2. Currently, nuclear power has little penetration of the nonelectric energy market. However, a large demand for nonelectric nuclear energy is expected to emerge and grow rapidly, owing to
   a. Increased energy use due to population growth and industrial development
   b. The finite availability of fossil fuels and the replacement of the direct use of fossil fuels
   c. An increased sensitivity to the environmental impacts of fossil fuel combustion

3. Due to dominance of power generation need, nuclear penetration into the markets for nonelectric services will proceed with cogeneration applications wherever possible. Dedicated reactors for heat generation could eventually emerge for some applications.

4. Many nonelectrical applications require energy sources that are relatively small (100–1000 MW[th]) in comparison with the size of existing power reactors. The development of nuclear reactors of small and medium size would therefore facilitate the nonelectrical applications of nuclear energy.

5. Some nonelectrical applications require a close siting of the nuclear plant to the customer. This will require specific safety features appropriate to the location.

6. Economically, the nonelectrical applications of nuclear energy are subject to the same trends as nuclear power generation. Growing capital costs of nuclear plants have affected the cost estimations of most nonelectrical applications. Evolutionary and innovative design improvements in nuclear reactor concepts, coupled with stable nuclear fuel prices, will result in an improved competitiveness of the nonelectric nuclear applications.

7. Depending upon the regions and conditions, nuclear energy is already competitive for district heating, desalination, and certain process heat applications.

8. Using nuclear energy to produce hydrogen is likely to facilitate the indirect application of nuclear energy in transportation markets, most of which are not readily amenable to the direct use of nuclear reactors.

9. Nonelectrical applications of nuclear energy are most likely to be implemented in countries already having the appropriate nuclear infrastructure and institutional support.

10. The implementation of some nonelectrical applications (e.g., desalination) is likely to enhance the public acceptance of nuclear energy.

Table 7.3 summarizes market potentials of important nonelectrical applications of the nuclear reactor. In several areas, more R&D work is needed to make suitable estimate of market potential.

The future prospects for HNESs are bright [157–175]. It is clearly a method to enhance penetration of renewable energies like wind and solar in the electricity market. Nuclear energy also allows this penetration to occur on a larger scale making electricity generated from hybrid nuclear–renewable sources competitive with the one generated by fossil fuels. Although a combination of nuclear energy and renewable energy sources can provide a sustainable electricity demand, often as a safeguard for peak energy demand, energy storage system is needed. Hydrogen or synthetic fuels created from residual heat from nuclear–renewable hybrid energy systems during low demand for electricity can provide such a safeguard.

As mentioned above, the use of nuclear energy for the combined heat and power generations is going to be very appealing in future for numerous applications.

For industrial process heat supply, there is a wide range of required heat parameters that determine the applicability of different reactor concepts. One particular concept, the HTGR, covers practically the whole temperature range and is therefore considered to be a prime future candidate for nuclear process heat supply. The development and demonstration of such a reactor would provide a strong impetus for the process heat applications of nuclear energy.

The non-electrical applications will require smaller, modular and distributed nuclear reactors. Significant efforts are being made for such development. While smaller reactors are not as economical for power production as larger reactors, they do require lower capital costs. Smaller reactors will open up more distributed market for the nuclear heat and will allow better heat integration to the local needs.

There are enormous environmental benefits to the use of advanced hybrid nuclear energy systems with co-generation. Nuclear heating can reduce emissions from the domestic heat sector. Production of hydrogen, other synthetic fuels, unconventional oil extraction and ship, submarine or space transportation by nuclear energy can lower the carbon footprint of the transport sector. Process heat delivered to industry could reduce the need for fossil-fired steam production. In the future, as fresh water resources become scare, an increasing number of countries will have a resort to seawater desalination to produce fresh water for their needs. Unless powered by low carbon energy such as nuclear or renewable technologies, use of desalination will lead to increased carbon emissions. For all these reasons and applications, the future of advanced hybrid nuclear energy systems with co-generation is very bright.

The nonelectrical applications will require smaller, modular, and distributed nuclear reactors. Significant efforts are being made for such development.

REFERENCES

1. Marano, J.J. and Ciferno, J.P. Life-cycle greenhouse-gas emissions inventory for Fischer–Tropsch fuels. Energy and Environmental Solutions, LLC, for the U.S. Department of Energy, National Energy Technology Laboratory, Pittsburgh, PA (2001).

2. Arisolobehere, S., Beer, J., Deutch, J., Eleman, A., Friedman, J., Hergog, H., Jacoby, H. et al. *The Future of Coal: Options for a Carbon-Constrained World*, a report from MIT joint program on the Science and policy of global change, Massachusetts Institute of Technology, Cambridge, MA (2007).

3. Plasynski, S., Deel, D., and Lityanski, J. *Carbon Sequestration Atlas of the United States and Canada*, Office of Fossil Fuels, National Energy Technology Laboratory, U.S. Department of energy. Washington, DC (2007). Available from: http://www1.eere.energy.gov/hydrogenandfuelcells/mypp/pdfs/production.pdf (accessed June 2008).

4. IEA. *Combined Heat and Power: Evaluating the Benefits of Greater Global Investment*, OECD/IEA Publications, Paris, France (2008).

5. IEA. *Deploying Renewables: Principles for Effective Policies*, OECD/IEA Publications, Paris, France (2008).

6. IEA. *Empowering Variable Renewables: Options for Flexible Electricity Systems*, OECD/IEA, Paris, France (2008).

7. IEA. *Energy Policies of IEA Countries: Sweden*, OECD/IEA Publications, Paris, France (2008).

8. IEA. *World Energy Outlook 2009*, OECD/IEA Publications, Paris, France (2009).

9. IEA. *Energy Balances of Non-OECD Countries*, OECD/IEA Publications, Paris, France (2010).

10. IEA. *Energy Technology Perspectives 2010*, OECD/IEA Publications, Paris, France (2010).

11. Akimoto, H., Berbey, P., Bertel, E. et al. IAEA. *Status of Advanced Light Water Reactor Designs 2004*, IAEA-TECDOC-1391 (2004).

12. Bertel, E., Bussurin, Y., Cleveland, J. et al. IEA, NEA, and IAEA. Innovative Nuclear Reactor Development: Opportunities for International Cooperation, Three-Agency Study OECD/IEA report, Paris, France (2002).

13. McDonald, A., Riahi, K., and Rogner, H.-H. Elaboration SRES scenarios for nuclear energy. In Paper presented at the *International Conference on Energy Technologies for post Kyoto Targets in the Medium Term*, Risø National Laboratory, Roskilde, Denmark (May 19–21, 2003).

14. Vander Hoeven, M. *CO_2 Emissions from Fuel Combustion*, IEA Statistics, 2011 edn., Personnel and Finance Division OECD/IEA, Paris, France (2011).

15. *Energy Technology Perspectives 2012—Pathways to a Clean Energy System*, OECD/IEA, Paris, France (2012).

16. Kupitz, J., D'Haeseleer, W., Herring, S., Taylor, M. Paillere, H. and Cameron, R. The role of nuclear energy in a low carbon energy future. A report from Nuclear Energy Agency, Organization for Economic Co-Operation and Development, NEA No. 6887, Paris, France, (2012). Available at: http://www.oecd-nea.org/nsd/reports/2012/nea6887-role-nuclear-low-carbon.pdf (accessed June 2012).

17. Anderson, R.E., Doyle, S.E., and Pronske, K.L. Demonstration and commercialization of zero-emission power plants. In *Proceedings of the 29th International Technical Conference on Coal Utilization & Fuel Systems*, Clearwater, FL (April 18–22, 2004).

18. Farrell, A.E. and Gopal, A.R. Bioenergy research needs for heat, electricity, and liquid fuels. *MRS Bull.*, 33(4), 373–380 (2008).

19. Tonaka, N. and Echovari, L., Technology roadmap-nuclear energy joint report by OECD/NEA and OECD/IEA Paris, France (2010).

20. Ruth, M., Zinaman, O., Antkowiak, M., Boardman, R., Cherry, R., and Bazilian, M. Nuclear-renewable hybrid energy systems: Opportunities, interconnections and needs. *Energy Convers. Manage.*, 78, 684–694 (February 2014).

21. Aumeier, S., Cherry, R., Boardman, R., and Smith J. Nuclear hybrid energy systems: Imperatives, prospects and challenges. A report from Energy Systems and Technology Division, INL, Idaho Falls, ID (2010).

22. Shropshire, D. Synergizing nuclear and renewable energy systems. A report from IAEA Planning and Economic Studies Section. In A paper presented at *Technical Meeting on Technology Assessment for SMRs*, Chengdu, China (September 2–4, 2013).

23. Mahajan, S. Fusion–fission hybrids—Path to green and plentiful nuclear energy. A report from Institute of Fusion Studies, University of Texas, Austin, TX (2010).

24. Buchheit, K. Process modeling of a coal, wind and nuclear hybrid energy system. In Paper 604q, *12th Annual AIChE Meeting*, Computing and Systems Technology Division, Pittsburgh, PA (October 29–November 2, 2012).

25. Papaloannou, I., Purvins, A., Shropshire, D., and Carlsson, J. Role of hybrid energy system comprising a small medium size nuclear reactor and a biomass processing plant in a scenario with a high deployment of onshore wind farms. *J. Energy Eng.*, 140(1), 04013005 (2014).

26. Bathe, H., The fusion hybrid, Physics Today, American Institute of Physics, 3(1), 44–51 (May, 1979); 0031–9228/9050044–06.

27. Antkowiak, M., Ruth, M., Boardman, R., Bragg-sitton, S., Cherry, R., and Shunn, L. Summary report of the INL-JISEA workshop on nuclear hybrid energy systems, Salt lake City, Utah (July 1, 2012) Prepared under Task No. 6A50.1027, NREL/TP-6A50-55650, Golden, Colorado (July 2012). Report available at: http://www.nrel.gov/docs/fy12osti/55650.pdf, presentations of workshop available at: https://inlportal.inl.gov/portal/server.pt?tbb=hybrid (accessed October 2012).

28. Forsberg, C.W., Gorensek, M.B., Herring, S., and Pickard, P. Next generation nuclear plant phenomena identification and ranking tables (PIRTs). In *Process Heat and Hydrogen Co-Generation PIRTs*, Vol. 6. Forsberg, C. Gorensek, M., Herring, S., Piakard, P., Basu, S. (Eds.), U.S. Nuclear Regulatory Commission, Office of Nuclear Regulatory Research, Washington, D.C. NUREG/CR-6944, ORNL/TM-2007/147 (2007).

29. Antkowiak, M., Boardman, R., Bragg-Sitton, S., Cherry, R., Ruth, M., and Shunn, L. Summary report of the INL-JISEA workshop on nuclear hybrid energy systems, INL/EXT-12-26551, NREL/TP-6A50-55650, INL Report, prepared for US DOE, Office of Nuclear Energy, Contract No. DE-AC07-051D14517, Idaho Falls, ID (July 2012).

30. Boardman, R. Advanced energy systems, Nuclear-fossil-renewable hybrid systems. A report to Nuclear Energy Agency, Committee for Technical and Economical Studies on Nuclear Energy Development and Fuel cycle, INL, Idaho Falls, ID (April 4–5, 2013).

31. Forsberg, C. Sustainability of combining nuclear, fossil and renewable energy sources. *Progr. Nucl. Energy*, 153, 1–9 (2008).

32. Ruth, M. Report of proceedings of the nuclear and renewable energy synergies workshop, NREL Golden, CO (2011).

33. Rath, M., Antkowiak, M., and Gossett, S. Nuclear and Renewable Energy synergies workshop: Report of Proceedings Technical Report NREL/TP-6A30-52256 Golden, Colorado (December 2011). Available at: http://www.nrel.gov/docs/fy12osti/52256.pdf (accessed March 2012).

34. Mazza, P. and Hammerschlag, R. Carrying the energy future—Comparing hydrogen and electricity for transmission, storage and transportation. ILEA Report, Seattle, WA (June 2004).

35. On the role and economics of nuclear cogeneration in a low carbon energy future. A Report by NEA/NDC, 22 Section 4.5, Washington, DC (2012).

36. von Lensa, W. and Verfondern, K. 20 Years of German R&D on nuclear heat applications. In *Fourth International Conference on IGCC and XtL Technologies*, IFC2010, Dresden, Germany (May 3–6, 2010). Available at: http://www.gasificationfreiberg.org/PortalData/1/Resources/documents/paper/03-2,3-von_Lensa.pdf. (accessed on 2011).

37. Non-electricity products of nuclear energy, OECD/NEA, report prepared under guidance of committee for technical and economic studies on nuclear energy development and fuel cycle (NDC) Paris, France (2004). Available at: http://www.oecd-nea.org/ndd/reports/2004/non-electricity-products.pdf (accessed December 2004).

38. IAEA. Non-electric applications of nuclear power: Seawater desalination, hydrogen production and other industrial applications. In *Proceeding of an International Conference*, Oarai, Japan (April 16–19, 2007). IAEA-CN-152, Vienna, Austria (2009).

39. IAEA. Advanced Applications of Water Cooled Nuclear Power Plants, IAEA-TECDOC-1584, International Atomic Energy Agency, Vienna, Austria (2008).

40. Kuzentsov, V. Innovative. Small and medium sized reactors Design features, safety approaches and R&D trends. Final report of a technical meeting held at Vienna, June 7–11, 2004. International Atomic Energy Agency, (Report available May 2005). Available at: www.iaea.org/NuclearPower/SMR/ (accessed November 2004).

41. Kuznetsov, H, Lokhov, A., Cameron, R. and Cometto, M. NEA. Current status, technical feasibility and economics of small nuclear reactors, OECD/NEA, Paris, France (June 2011). Available at: www.oecd-nea.org/ndd/reports/2011/current-status-small-reactors. pdf (accessed November 2011).

42. Peterson, P. Co-generation of electricity and process heat and other applications of SMRs. Forum on small and medium reactors, Department of Nuclear Engineering, University of California, Berkley, CA (June 19, 2010).

43. Verfondem, K. Overview of nuclear cogeneration in high temperature industrial process heat applications. In *A Report at OECD-IAEA Workshop*, Paris, France (April 4–5, 2013).

44. Konishi, T and Woite, G. IAEA-Tecdoc 1056. Nuclear heat applications: Design aspects and operating experience Proceedings of four technical meetings held between December 1995–April 1998, Vienna, Austria (November 1998).

45. Carlsonn, J., Shropshire, D., van Heek A., and Fütterer, M. Economic viability of small nuclear reactors in future European cogeneration markets. *Energy Policy*, 43, 396–406 (April 2012).

46. Forsberg, C. Rethinking the nuclear energy role in a carbon-constrained world: Coupling fuels and electricity production. In *International Congress on Advances in Nuclear Power Plants, ICAPP'12*, Chicago, IL (June 24–28, 2012).

47. Madlener, A. Economics of high temperature reactors for industrial cogeneration: A utility's perspective. In *12th IAEE European Energy Conference*, Ca' Foscari University of Venice, Venice, Italy (September 9–12, 2012).

48. Veerapen, J. Co-generation and renewable: Solutions for a low carbon energy future. OECD/IEA report, Paris, France (2011).

49. Cofély Centre-Ouest. *Efficacité énergétique et environnementale agro-industrielle: Implantation d'une centrale de cogénération biomasse à Grand-Couronne (76)*, Dossier de presse, Cofély Centre-Ouest, Rennes, France (2010).

50. Dalkia Est. *Cogénération Biomasse: Source d'énergie verte au service de l'industrie et du territoire*, Dossier de presse, Dalkia Est., Essey-lès-Nancy, France (October 23, 2009).

51. Khamis, I. IAEA programme on non-electric applications of nuclear energy. In *Workshop on Technology Assessment for SMR for Near-Term Deployment* (December 5–9, 2011), IAEA, Vienna, Austria. Available at: http://www.iaea.org/NuclearPower/ Downloads/Technology/meetings/2011-Dec-5-9-WSSMR/Day-4/25_IAEA_Khamis_ NEA_SMRDec2011.pdf. (accessed February 2012).

52. Veerapen, J. Cogeneration and renewables: Solutions for a low-carbon energy future, OECD/IEA, Paris, France (2011).

53. Forsberg, C.W. High-temperature nuclear reactors for in-situ recovery of oil from oil shale. In *Proceedings of the 2006 International Congress on Advances in Nuclear Power Plants (ICAPP'06)*, Reno, NV (June 4–8, 2006). American Nuclear Society, La Grange Park, IL.

54. Forsberg, C.W. High-temperature reactors for underground liquid-fuels production with direct carbon sequestration. In *Proceedings of the International Congress on Advanced Nuclear Power Plants*, Anaheim, CA (June 8–15, 2008).

55. Smith, J., Boardman, R., Smoot, L., Omar, K., Coats, R., and Hatfield, K. Oil shale nuclear hybrid transportation fuels. In *29th Oil Shale Symposium*, Colorado School of Mines, Colorado Energy Research Institute, Golden, CO (October 19–23, 2009).

56. Smith, J. Using nuclear heat for in-situ recovery of unconventional hydrocarbons: A case for the high temperature gas reactor (HTGR). In A paper presented at *30th Oil Shale Symposium*, Colorado School of Mines, Golden, CO (October 18–22, 2010).

57. Bersak, A. and Kadak, A. Integration of nuclear energy with oil sands projects for reduced greenhouse gas emissions and natural gas consumption. MIT White Paper (2007). Available at: http://web.mit.edu/finana/Public/oilsands/MITWhitePaper.pdf. (accessed 2008).

58. Rolfe, B. CANDU plants for oil sand applications. In *IAEA Conference*, Oarai, Japan (April 16–19, 2007).

59. Forsberg, C.W., Rosenbloom, S., and Black, R. Fuel ethanol production using nuclear-plant steam. In *Proceedings of the International Conference on Non-Electrical Applications of Nuclear Power: Seawater Desalting, Hydrogen Production, and Other Industrial Applications*, Oarai, Japan (April 16–19, 2007).

60. Gary, J.H., Handwerk, G.E., and Kaiser, M.J. *Petroleum Refining: Technology and Economics*, 5th edn., CRC Press, Boca Raton, FL (2007).

61. McAloon, A., Taylor, F., Yee, W., Ibsen, K., and Wooley, R. Determining the cost of producing ethanol from corn starch and lignocellulosic feedstocks. NREL/TP-580-28893, National Renewable Energy Laboratory, Golden, CO (2000).

62. Montague, L. Lignin process design confirmation and capital cost evaluation. NREL/SR-510-31579, National Renewable Energy Laboratory, Golden, CO (2003).

63. Perlack, R.D., Wright, L.L., Turhollow, A.F., Graham, R.L., Stocks, B.J., and Erbach, D.C. Biomass as feedstock for a bioenergy and bioproducts industry: The technical feasibility of a billion-ton annual supply. DOE/GO-102995-2135, ORNL/TM-2005/66, U.S. Department of Energy, Washington, DC (2005).

64. Ragausk, A.J., Williams, C.K., Davison, B.H. et al. The path forward for biofuels and biomaterials. *Science*, 311, 484–489 (2006).

65. Self, F.E., Ekholm, E.L., and Bowers, K.E. Refining overview of petroleum, processes, and products. CD-ROM, American Institute of Chemical Engineers, New York (2007); Shinnar, R. and Citro, F. A roadmap to U.S. decarbonization. *Science* 313, 1243 (2006).

66. U.S. Department of Agriculture, National Agricultural Statistics Service. Trends in U.S. Agriculture. U.S. Department of Agriculture, National Agricultural Statistics Service, Washington D.C. (2003). Available from: www.usda.gov/nass/pubs/trends/. (accessed 2004).

67. U.S. Department of Energy, Biomass Program: Lignin-Derived Co-Products. Office of Energy Efficiency and Renewable Energy, Golden, CO (2007). Available from: http://www1.eere.energy.gov/biomass/lignin_derived.html. (accessed 2008).

68. Nifenecker, H. Hybrid nuclear systems for energy production and waste management. *Nucl. Phys. News*, 4(2), 21–23 (1994).

69. Forsberg, C. Nuclear power: Energy to produce liquid fuels and chemicals. *Chem. Eng. Progr.*, 49, 41 (July 2010).

70. Duffy, R., Kuran, S., and Miller, A., Applications of Nuclear Energy to oil sands and hydrogen production, IAEA conference on opportunities and challenges for WCR in the 21st century, Vienna, Austria (Oct. 23–29, 2009)

71. Park, K. SMART: An early deployable integral reactor for multi-purpose applications. In *INPRO Dialogue Forum on Nuclear Energy Innovations: CUC for Small & Medium-Sized Nuclear Power Reactors* (October 10–14, 2011), Vienna, Austria. Available at: http://www.iaea.org/INPRO/3rd_Dialogue_Forum/index.html. (accessed 2012).

72. Pienkoswski, L. University of Warsaw High temperature reactor for nuclear-coal synergy in Europe. Conference in Piaski, Poland (September 1–7, 2008). Available at: http://www.euroschoolonexoticbeams.be/site/files/2008_Pienkowski_lecture.pdf. (accessed 2008).

73. Hill, J., Nelson, E., Tilman, D., Polasky, S., and Tiffany, D. Environmental, economic, and energetic costs and benefits of biodiesel and ethanol biofuels. *Proc. Natl. Acad. Sci. U.S.A.*, 103(30), 11206–11210 (2006).

74. Yvon, P., Carles, P., and Le Naour, F. French research strategy to use nuclear reactors for hydrogen production. NEA report, Paris, France (2010).

75. Hino, R., Petri, M., Chang, J., Duffey, R., Hori, M., NEA report. *Third Information Exchange Meeting on Nuclear Production of Hydrogen*, Oarai, Japan (October 5–7, 2005). NEA Report No. 6122, OECD/NEA, Paris, France (2006).

76. Meneley, D., Miller, A., Xu, J. et al. NEA report. *Fourth Information Exchange Meeting on Nuclear Production of Hydrogen*, Oakbrook, IL (April 14–16, 2009). NEA Report No. 6805, OECD/NEA, Paris, France (2010).

77. Miller, A. and Duffey, R. Sustainable and economic hydrogen cogeneration from nuclear energy in competitive power markets. *Energy*, 30(14), 2690–2702 (November 2005).

78. Naterer, G., Suppich, S., Stoleberg, L. et al. Canada's program on nuclear hydrogen production and the thermochemical Cu–Cl cycle. *Int. J. Hydrogen Energy*, 35, 10905–10926 (October 2010).

79. Forsberg, C.W. Nuclear hydrogen for peak electricity production and spinning reserve. ORNL/TM-2004/194, Oak Ridge National Laboratory, Oak Ridge, TN (2005).

80. Forsberg, C.W. Meeting U.S. liquid transport fuel needs with a nuclear hydrogen biomass system. In *Proceedings of the American Institute of Chemical Engineers Annual Meeting*, Salt Lake City, UT (November 4–9, 2007).

81. Forsberg, C.W. Economics of meeting peak electricity demand using nuclear hydrogen and oxygen. In *Proceedings of the International Topical Meeting on the Safety and Technology of Nuclear Hydrogen Production, Control, and Management*, Boston, MA, June 24–28, 2007. American Nuclear Society, La Grange Park, IL (2007).

82. Forsberg, C.W. and Conklin, J.C. Hydrogen-or-fossil-combustion nuclear combined-cycle systems for base- and peak-load electricity production. ORNL-6980, Oak Ridge National Laboratory, Oak Ridge, TN (2007).

83. Forsberg, C.W. Is hydrogen the future of nuclear energy. *Nucl. Technol.* (2008).

84. Summers, W.A. Centralized hydrogen production from nuclear power: Infrastructure analysis and test-case design study. Interim Project Report, Task A.1, Nuclear Hydrogen Plant Definition, NERI Project 02-0160, WSRC-TR-2003-00484. Savannah River Technology Center, Aiken, SC (2003).

85. U.S. Department of Energy. Hydrogen, fuel cells & infrastructure technology program: Multi-year research, development and demonstration program, Office of Energy Efficiency and Renewable Energy, Washington, DC (2007).

86. Forsberg, C. Futures of hydrogen produced using nuclear energy. In A paper presented at *First International Conference on Innovative Nuclear Energy Systems for Sustainable Development of the World* (COE INES-1), Session 2B1: Innovative Energy transmutation, Tokyo Institute of Technology, Tokyo, Japan (October 3–November 4, 2004).

87. IAEA. Hydrogen as an energy carrier and its production by nuclear power. IAEA-TECDOC-1085 Vienna, Austria (May 1999).

88. Shiozawa, S., Tachibana, Y., Baba O., and Ogawa, M. Present status and perspective of HTGR in Japan. In *Proceedings of the International Seminar on Status and Prospects for Small and Medium Sized Reactors*, Cairo, Egypt (May 27–31, 2001).

89. National Academy of Engineering, Board on Energy and Environmental Systems. *The Hydrogen Economy: Opportunities, Costs, Barriers, and R&D Needs*, National Academic Press, Washington, DC (2004).

90. Hydrogen Posture Plan: An Integrated Research, Development, and Demonstration Plan U.S. Department of Energy and Department of Transportation, Washington, DC (December 2006).

91. Forsberg, C.W., Peterson, P.F. and Pickard, P., Molten salt cooled advanced high temperature reactor for production of hydrogen and electricity, Nuclear Technology, American Nuclear Society, 144 (3), 289–302 (December 2003) DOE contact No. DE-AC05-000R22725, Washington, D.C.

92. IAEA. Nuclear heat applications: Design aspects and operating experience. In *Proceedings of Four Technical Meetings Held between December 1995 and April 1998.* Vienna, Austria IAEA–TECDOC-1056 (November 1998)

93. IEA. Moving to a hydrogen economy: Dreams and realities. IEA/SLT5 Paris, France (2005).

94. IEA. Hydrogen coordination group: Policy analysis. IEA/CERT/HCG/3 Paris, France (2004).

95. OECD/NEA. *Nuclear Production of Hydrogen, First Information Exchange Meeting*, Paris, France (October 2–3, 2000). Report published in 2002.

96. OECD/NEA. *Nuclear Production of Hydrogen, Second Information Exchange Meeting*, Argonne, IL (October 2–3, 2003). Report published in 2004.

97. Rogner, H.-H. and Scott, D.S. Building sustainable energy systems: The role of nuclear-derived hydrogen. In *Nuclear Production of Hydrogen, First Information Exchange Meeting*, Paris, France (October 2–3, 2000). OECD/NEA (2002). Final report published in 2003.

98. Schultz, K.R. Use of the modular helium reactor for hydrogen production. In *World Nuclear Association Annual Symposium* London, U.K. (September 3–5, 2003).

99. National Hydrogen Energy Roadmap: Production, Delivery, Storage, Conversion, Applications, Public Education and Outreach, Department of Energy, Washington, DC (2002).

100. Nuclear Energy Research Advisory Committee & GIF. Generation IV Roadmap: Crosscutting Energy Products R&D Scope Report, GIF-008-00, Department of Energy, Washington, DC (2002).

101. Wade, D.C., Doctor, R., and Peddicord, K.L. STAR-H2: The secure transportable autonomous reactor for hydrogen production and desalination. In *Proceedings of 10th International Conference on Nuclear Engineering*, Arlington, TX (April 14–18, 2002).

102. Wald, M.L. Questions about a hydrogen economy. *Sci. Am.* (May 2004). Available at: http: //www. Scientific american. com/article/questions about-a-hydrogen/.

103. Yildiz, B., Petri, M., Conzelmann, G., and Forsberg, C. Configuration and technology implications of potential nuclear hydrogen system applications, Published by Argonm National lab, ANL 05-30 (July 31, 2005). Available at: http://www.dis.anl.gov/pubs/54478.pdf (accessed November 2005).

104. Chang, J. and Lee, W. Status of the Korean nuclear hydrogen production project, NEA London, U.K. (2010).

105. Gomez, A.., Azzaro-Pantel, C., Pibouleau, L., Domenech, S., Latge, C., Dumaz, P., and Haubensack, D. Optimization of electricity/hydrogen cogeneration from generation IV nuclear energy systems. In *Seventeenth European Symposium on Computer Aided Process Engineering*, ESCAPE17 (2007). Plesu, V. and Agachi, P.S., (Eds.)., Elsevier, New York (2007) Available at: http://www.nt.ntnu.no/users/skoge/prost/proceedings/escape17/papers/T5-36.pdf (accessed December 2007).

106. Economical and technical optimization of multi-objectives energy conversion systems (December 8, 2008). Available at: http://ethesis.inp-toulouse.fr/archive/00000786/01/gomez.pdf (accessed March 2009).

107. Larsen, R. and Wang, M., Might Canadian oil sands hydrogen promote hydrogen production technologies for transportation? Greenhouse gas emission implications of oil sands recovery and upgrading, *World Resour. Rev.* 17 (2), 220–242 (2005).

108. Rivera-Tinoco, R. Mansilla, C., Werkoff, F., and Bonallou, C. Technico-economic study of hydrogen production by high temperature electrolysis coupled with an EPR, SFR or HTR—Water steam production and coupling possibilities. In *Thirteenth International Conference Series on Emerging Nuclear Energy Systems*, ICENES Istanbul, Turkey (June 3-8, 2007). Available at: http://www.icenes2007.org/icenes_proceedings/manuscripts.pdf/Session%2010A/TECHNOECONOMIC.Pdf (accessed December 2007).

109. Forsberg, C. Economic implications of peak vs. base load electricity costs on nuclear hydrogen systems. In *AiChE Annual Meeting*, Boston, MA (2006).

110. IAEA. Economics of nuclear desalination: New developments and site specific studies final results of a coordinated research project 2002–2006, IAEA-TECDOC-1561, Vienna, Austria (2007). Available at: http://www-pub.iaea.org/MTCD/publications/PDF/te_1561_web.pdf (accessed December 2007).

111. IEA Committee on DHC and CHP. District heating and cooling: Environmental technology for the 21st century. *Policy Paper* (2002). http://www.iea-dhc.org.

112. IAEA. Examining the economics of seawater desalination using the DEEP code. IAEA-TECDOC-1186 Vienna, Austria (2002).

113. IAEA. Design concepts of nuclear desalination plants. IAEA-TECDOC-1326 Vienna, Austria (2002).

114. Techno-economic study of the feasibility of cogeneration of electricity and desalinated salt water using concentrated solar power. The CSP–DSW project, The Cyprus Institute, Nicosia, Cyprus (July 2010).

115. The Scottish Government. *News Releases* (2008). Retrieved from www.scotland.gov.uk/News/Releases/2008/07/18105154 (March 10, 2011) (accessed November 2011).

116. Tokyo Gas/Osaka Gas. *Press Release* (2010). Retrieved from www.tokyo-gas.co.jp/Press_e/20100514-01e.pdf (March 09, 2011) (accessed October 2011)

117. Trieb, F. and Muller-Steinhagen, H. Concentrating solar power for seawater desalination in the Middle East and North Africa. *Desalination*, Elsevier, the Netherlands, 220 (1–3) 165–183 (March 2008).

118. Al Taher, A., Trieb, F., Muller–Steinhagen, H., Kern, J., Scharfe, J., and Kabariti, M., Technologies for large scale sea water desalination using concentrated solar radiation, Desalination, Elsevier, New York. 235, 23–43 (2009).

119. UK DECC. Absorption cooling technology. Retrieved from http://chp.decc.gov.uk/cms/absorption-cooling-technology/ (December 15, 2010) (accessed 2011).

120. Belessiotis, V., Papanicolaou, E., and Delyannis, E. Nuclear desalination: A review on past and present. *Desal. Water Treat.*, 20, 45–50 (August 2010).

121. Non-electric applications of nuclear power: Sea water desalination, hydrogen production and other industrial applications. *Proceedings of the International Conference*, Oral, Japan (April 16–19, 2007). IAEA, Vienna, Austria (April 2009).

122. Mazumdar, D. Desalination and other non-electric applications of nuclear energy. Lectures given at the Workshop on Nuclear Reaction Data and Nuclear Reactors: Physics, Design and Safety Trieste, LNSO520005 IAEA, Vienna, Austria (February 25–March 28, 2002).

123. Misra, B. and Konishi, T. *Introduction of Nuclear Desalination—A Guidebook*. IAEA Technical Reports Series No. 400, Vienna, Austria (2000).

124. Konishi, T., Faibish, R., and Gasparini, M. Application of nuclear energy for seawater desalination—Design concepts of nuclear desalination plants. In *Proceedings of 10th International Conference on Nuclear Engineering*, Arlington, VA (April 14–18, 2002).

125. Kupitz, Y. and Konishi, T. *Encyclopedia of Desalination and Water Resources*, EOSS Publications, Oxford, U.K. (2000).

126. IAEA. Optimization of the coupling of nuclear reactors and desalination systems. IAEA-TECDOC, Vienna, Austria (2004).

127. Megahed, M.M. An overview of nuclear desalination: History and challenges. *Int. J. Nucl. Desal.*, 1(1) 2–18 (2003) ISSN 1476–914X.

128. Minato, A. and Hirai, M. Present and future activities of nuclear desalination in Japan. *Int. J. Nucl. Desal.*, 1(2) 259–270 (2004) ISSN 1476–914X.

129. Misra, B.M. and Kupitz, J. The role of nuclear desalination in meeting the potable water needs in water scarce areas in the next decades. *Desalination*, 166, 1–9 (2004).

130. Konishi, T. Global water issues and nuclear seawater desalination, World Nuclear University Summer Institute, Oxford, U.K. (2009).

131. Maciver, A. Hinge, S., Andersen, B., and Nielsen, J. New trends in desalination for Japanese nuclear power plants, based on multiple effect distillation, with vertical titanium plate falling film heat transfer configuration. *Desalination*, 182, 221–228 (2005). Available at: http://www.desline.com/articoli/6659.pdf (accessed 2005).

132. Nisan S. and Dardour, S. Economic evaluation of nuclear desalination systems. *Desalination*, 205, 231–242 (2007).

133. Safa, H. Cogeneration with district heating and cooling. In *IAEA Consultant Meeting*, Vienna, Austria (December 19–22, 2011). Available at: http://www.ecolo.org/documents/documents_in_french/cogeneration-Safa-Nogent-2011.pdf (accessed 2012).

134. Safa, H. Heat recovery from nuclear power plants. *Int. J. Electr. Power Energy Syst.*, 42(1), 553–559 (November 2012).

135. Le Pierres, N., Lou, L., Berthiaud, J., and Mazet, N. Heat transportation from the Bugey power plant. *Int. J. Energy Res.*, 33, 135–143 (2009).

136. Krivitskii, I. Nuclear district heating in Russia's regions: Problems and prospects. *Therm. Eng.*, 52(1), 25–28 (2005).

137. Kuhr, R., Bolthrunis, C., and Corbett, M. Economics of nuclear process heat applications. In *Proceedings of ICAPP '06*, New York (2006).

138. IAEA. Advances in nuclear power process heat applications. IAEA-TECDOC-1682, IAEA, Vienna, Austria (2012). Available at: http://www-pub.iaea.org/MTCD/Publications/PDF/TE_1682_web.pdf (accessed November 2012).

139. Day, A.R., Jones, P.G., and Maidment, G.G. Forecasting future cooling demand in London. *Energy Build.*, Elsevier, the Netherlands, 41 (9), 942–948 (2009).

140. EDF Energy. Consultation: Operating in 2020. National Grid (2009). Retrieved from: www.nationalgrid.com/NR/rdonlyres/3CFFA9C2-31D4-4291-B4A7-BF696051C6F5/38385/EDFEnergy.pdf (accessed August 03, 2010).

141. Ericsson, K. and Svenningsson, P. *Introduction and Development of the Swedish District Heating Systems*, Lund University, Lund, Sweden (2009).

142. European Union, Sixth Framework Programme Project Factsheet. Community Research and Development Information Service—CORDIS, European Commission (2008). Retrieved from: http://cordis.europa.eu/fetch?CALLER=FP6_PROJ&ACTION=D&DOC=22&CAT=PROJ&QUERY=012a4294 c234:6dd3:2166f93f&RCN=85711 (accessed July 29, 2010).

143. *Financial Times*. Private sector: Industry and state co-operation is a crucial factor in supply mix. *Financial Times* Special Reports: South African Power and Energy (2010). Retrieved from: www.ft.com/cms/s/0/cdfecc8a-fc17-11df-b675-00144feab49a.html#axzz17Pu1EbUr (accessed February 2011).

144. French Republic. *Journal Officiel de la République Française* (2008). Retrieved from Legifrance: http://www.legifrance.gouv.fr/affichTexte.do?cidTexte=JORFTEXT00001 9109471&dateText# (accessed July 2011).

145. IEA. *Co-Generation and District Energy: Sustainable Energy Technologies for Today and Tomorrow.* OECD/IEA Publications, Paris, France (2009).

146. Kalt, G. Kalt, G., Kranzl, L., Haas, R., Muller, A., and Biermayr, P. *Renewable Energy in the Heating and Cooling Sector in Austria (Including Regional Aspects of Styria)*, Energy Economics Group, Vienna, Austria (2009).

147. Kelly, S. and Pollitt, M. *Making Combined Heat and Power District Heating (CHP-DH) Networks in the United Kingdom Economically Viable: A Comparative Approach*, Electricity Policy Research Group, University of Cambridge, Cambridge, U.K. (2009).

148. Csik B. and Kupitz, J. Nuclear power applications: Supplying heat for home and industries. *IAEA Bulletin*, IAEA, Vienna, Austria, 39(2) (1997).

149. Bredimas, A. European industrial heat market study. In *EUROPAIRS Workshop*, Brussels, Belgium (May 26, 2011). Available at: http://www.europairs.eu/ (open workshop presentations) (accessed November 2011).

150. Bredimas, A. Preliminary study on the viability of HTR cogeneration for industrial purposes. In *EUROPAIRS Workshop*, Brussels, Belgium (May 26, 2011). Available at: http://www.europairs.eu/ (open workshop presentations) (accessed November 2011).

151. Taylor J. and Shropshire, D. Dynamic complexity study of a nuclear reactor and process heat application integration. In *Global 2009* INL/CON–09–15285 preprint, INL, Idaho also, Proceedings of Global 209. Paper 9451, September 6–11, 2009, Paris, France. (September 2009).

152. Agrawal, R., Singh, N.R., Ribeiro, F.H., and Delgass, W.N. Sustainable fuel for the transport sector. *Proc. Natl. Acad. Sci. U.S.A.*, 104(12), 4828–4833 (2007).

153. Subki, H. and Reitsma, F. Advances in small modular reactor technology developments a supplement to IAEA Advanced Reaction Information Systems (ARTS), Division of Nuclear Power, IAEA, Department of Nuclear Energy, IAEA, Vienna, Austria (September, 2014).

154. Ingersoll, D., Honghton, Z., Brown, R. and Desportes, C. Nuscale small modular reactor for cogeneration of electricity and water Desalination, 340, 84–93 (2014).

155. Shah, Y. *Water for Energy and Fuel Production*, CRC Press, Taylor & Francis Group, New York (May 2014).

156. Kononov, S. Market potential for no-electric applications of nuclear energy. Technical Report Series No. 410, International Atomic Energy Agency, Vienna, Austria (2002).

157. Non-electricity products of nuclear energy. A report by Nuclear Energy Agency of Organization for Economic Cooperation and Development, report prepared by OECD/ NEA with guidance of the committee for technical and economic studies on nuclear energy development and fuel cycle (NDC) Paris, France (2004).

158. IAEA. Market potential for non-electric applications of nuclear energy. Technical Report Series No. 410, IAEA, Vienna, Austria (2002).

159. Von Lensa, W. Michelangelo Network WP4: First results & recommendations on non-electricity applications of nuclear energy. Briefing Material (2002).

160. Rogner, H.-H. and McDonald, A. Long-term cost targets for nuclear energy. In *2003 Annual Symposium of the World Nuclear Association*, London, U.K. (2003).

161. IAEA. Guidance for the evaluation of innovative nuclear reactors and fuel cycles: Report of Phase 1A of the International Project on Innovative Nuclear Reactors and Fuel Cycles (INPRO), IAEA-TECDOC-1362 Vienna, Austria (2003).

162. IEA. *World Energy Investment Outlook,* Paris, France (2003).

163. Price, R.R., Blaise, J.R., and Vance, R.E. Uranium production and demand: Timely mining decisions will be needed. *NEA News*, 22(1) (2004).

164. OECD/NEA. Nuclear competence building. Summary Report, OECD, Paris, France (2004).

165. U.S. DOE Nuclear Energy Research Advisory Committee & GIF. A Technology Roadmap for Generation IV Nuclear Energy Systems Department of Energy, Washington, DC (2002).

166. Halbe, C. Output flexibility and technology choice in nuclear power plant investments. Master thesis, Tilburg University, Tilburg, the Netherlands (December 15, 2008). Available at: http://www.tilburguniversity.edu/research/institutes-and-researchgroups groups/tilec/energy/pdf/20081215_Halbe.pdf (accessed 2009).

167. INL. High temperature gas-cooled reactor—Projected markets and preliminary economics, INL/EXT-10-19037, Revision 1 (August 2011). Available at: http://www.osti.gov/bridge/product.biblio.jsp?osti_id=1031697 (accessed 2012).

168. NEA. *Projected Costs of Generating Electricity*, OECD/NEA, Paris, France (2010).

169. Kuznetsov, V. and Lokhov, A. NEA. *Current Status, Technical Feasibility and Economics of Small Nuclear Reactors*, OECD/NEA, Paris, France (June 2011). Available at: http://www.oecd-nea.org/ndd/reports/2011/current-status-small-reactors.pdf (accessed December 2011).

170. Rosen, M. Energy- and exergy-based comparison of coal-fired and nuclear steam power plants. *Exergy Int. J.*, 1(3), 180–192 (2001).

171. Rosen, M. Energy and environmental advantages of cogeneration with nuclear and coal electrical utilities. In *Proceedings of the Fourth IASME/WSEAS International Conference on Energy & Environment* Cambridge, U.K. (February 24–26, 2008). Available at: http://www.wseas.us/e-library/conferences/2009/cambridge/EE/EE24.pdf (accessed 2009).

172. Rosen, M. and Bulucea, C. Using exergy to understand and improve the efficiency of electrical power technologies. *Entropy*, 11, 820–835 (2009). Available at: http://www.mdpi.com/1099-4300/11/4/820 (accessed 2009).

173. Shropshire, D. Economic viability of small to medium-sized reactors deployed in future European energy markets. *Progr. Nucl. Energy*, 53(4), 299–307 (May 2011).

174. Szargut, J. Exergy analysis. *Academia*, 3(7) 31–33 (2005). Available at: http://www.english.pan.pl/images/stories/pliki/publikacje/academia/2005/07/31-33_szargut.pdf.

175. Paillere, H. Final summary record of the joint NEA/IAEA expert workshop on the technical and economic assessment of non-electric applications of nuclear energy, OECD Conference Center, Paris, France (April 4–5, 2013).

176. Christensen, D. Summary of august workshop to identify potential synergies between nuclear and renewable energy opportunities. JISEA Joint Institute for Strategic Energy Analysis, NREL, Golden, CO (December 2011). Workshop Report at: http://www.nrel.gov/docs/fy12osti/52256.pdf (accessed 2012).

8 Hybrid Renewable Energy Systems

8.1 INTRODUCTION

As mentioned in Chapter 7, unlike fossil energy systems such as coal, oil, and gas, both nuclear and renewable energy systems for electric (and for some nonelectric) applications require large amount of investment costs. While fossil energy systems have lower capital costs and an advantage of existing infrastructure, they incur higher operating costs. The electricity generations from fossil energy are competitive because fossil energy systems take advantage of economy of scale. Fossil fuels can also act as a backup storage system for the peak power demand. As shown in Chapter 7, nuclear energy systems are also large-scale systems, and the nuclear technologies are matured and well developed. Fossil energy systems suffer from harmful greenhouse gas emissions, and nuclear energy systems (fission based) suffer from the issues of nuclear waste treatment and finite life of fission process.

The world is moving toward a more balanced energy portfolio where more penetration of renewable energies like biomass, waste, solar, wind, geothermal, and water in fossil energy–dominated market is required [1–8]. While renewable energies possess several environmental and economic benefits, the commercialization of these sources of energy for both electric and nonelectric applications faces many challenges in highly competitive energy market. First, the capital costs for renewable energies are high. Second, building of large-scale systems such that price of unit energy (electrical or nonelectrical) is competitive with fossil energy is problematic. Third, fossil energy has an advantage of existing infrastructure for transport and storage of raw materials and products as well as infrastructure for product delivery to the consumers. With biomass and waste, there are serious issues of raw material storage, transportation, and preparation. With solar and wind, there are issues of time and location dependencies. Advanced geothermal systems require more infrastructures and large-scale testing, and the energy from water is not yet commercially well developed. While independent developments of these sources of energy can be risky and slow, one strategy that will work well is the development of hybrid systems that can overcome significant barriers to the use of renewable energy sources in fossil energy–dominated market. Hybrid renewable energy systems appear to be the best and most economical way for renewable energy to penetrate overall energy market. In Chapter 7, we outlined the use of nuclear energy to support hybrid renewable energy systems like solar and wind energy. In this chapter, we examine various other types of hybrid renewable energy systems. Like nuclear energy, the application of co-generation concept for solar energy is also high, and it will be illustrated in this chapter.

In Chapters 2 through 6, we showed how biomass and waste can penetrate energy and fuel market by co-fuel strategy (coal/biomass, coal/waste, biomass/waste, etc.) for important thermochemical (combustion, gasification, pyrolysis, and liquefaction) and fermentation (co-digestion) processes. These chapters clearly showed significant momentum toward co-fuel strategies, although currently it is more developed for combustion and liquefaction than gasification and pyrolysis. Co-combustion is well practiced in many parts of the world. Co-liquefaction is not too far behind because co-processing of coal and heavy oil is already practiced commercially in many two-stage commercial liquefaction processes. The use of supercritical water for multifuel liquefaction/extraction is already investigated at pilot scale and has enormous potential. Co-gasification is being rapidly developed at a larger scale. The use of supercritical water for co-gasification and reforming is already practiced at pilot scale, and this technology is gradually becoming more important. Co-pyrolysis will gain significant importance with new development of wet pyrolysis to produce a variety of clean solid fuels. Co-digestion of mixed substrate is becoming so important that it has become more of a norm than exception in anaerobic digestion industry. Overall, multifuel strategy is the best way for biomass and waste to penetrate energy market on large scale.

As shown in Chapter 7, hybrid nuclear energy also has a strong future because of its importance in both electric and nonelectric market. In electric market, nuclear energy will always be the most cost effective at large scale, and the technology for this is well developed. The use of large-scale nuclear reactors to produce hydrogen can be economical. Chapter 7 illustrated that advanced hybrid nuclear energy systems accompanied by co-generation can be very useful in (a) promoting renewable energy sources (like solar and wind energy) on large scale and (b) using nuclear heat for various nonelectric applications thus replacing harmful fossil energy usages in these applications. Since most nonelectric applications are very valuable to society and possess growing markets, the use of nuclear energy will grow through these hybrid systems. The nonelectric applications of nuclear energy will also open up a whole new market for small- and medium-sized nuclear reactor development, which has not been found to be economical in the electric market.

In this chapter, we focus on numerous other methods for building hybrid renewable energy systems that can provide sustainable and economical electric and nonelectric needs of the society. We will focus on four renewable resources, solar, wind, geothermal, and water, since we already addressed the use of biomass and waste in Chapters 2 through 6. As shown in my previous book [2], water is both fuel and a direct source of energy. The use of water as a fuel is illustrated in Chapters 2 through 6. The use of water as a means for energy storage is illustrated in Chapter 9. The use of water for hydrogen production was illustrated in Chapter 7 and is also further illustrated in this chapter. As shown by few examples in this chapter, water can also be used in a hybrid manner with all other renewable energy sources like wind, solar, and geothermal. Thus, this chapter mainly focuses on hybrid solar, wind, and geothermal renewable energy systems with the role of water in each case. This chapter illustrates with concrete examples how various types of hybrid renewable energy systems have either been implemented on smaller and midscales or have potentials to be implemented even on larger scales.

8.2 HYBRID WIND ENERGY SYSTEMS

Hybrid wind energy systems are most often used to generate electrical power in a sustainable manner [9–39]. Since wind energy is time and location dependent, its use requires either a backup system or connection to a major electric grid. Both approaches have been investigated and implemented. Wind energy can be harnessed along with solar, hydro, and nuclear energy. Figure 8.1 shows a typical hybrid wind and solar energy system setup [9].

8.2.1 Hybrid Hydroelectric and Wind Energy Power System with Hydraulic Energy Storage

A wind–hydro system generates electric energy by combining wind turbines and pumped storage. In this hybrid system, high wind energy source (compared to demand) is used to pump water in reservoirs at an elevation. This water can be used during high power demand by releasing it into a hydropower plant when needed [9–11,13,14,28,33,34,36]. This combination is particularly suited to remote islands that are not connected to larger grids [9].

Elistratov [36] examined hybrid energy system that consists of hydroelectric power system and wind energy power system with hydraulic energy storage. He concluded that this hybrid system provides reliable and high-quality power supply for consumers at both individual level and grid-connected level. He also concluded that efficiency of hydraulic energy storage based on hydroelectric power system is one of the best among the other storage systems and can be as high as 90% including operation losses of 1%–2%.

Jaramillo et al. [28] presented a theoretical study to show how wind power can be complemented by hydropower. This paper presented a conceptual framework for a hybrid power station that produces constant power output without the intermittent

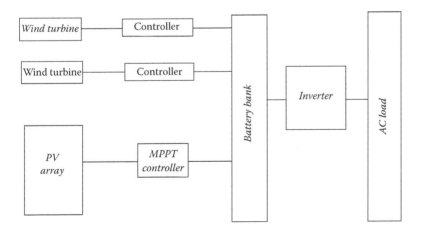

FIGURE 8.1 Block diagram of a PV/wind hybrid energy system. (From "Wind Hybrid Power Systems," https://en.wikipedia.org/wiki/Wind_hybrid_power_systems, last modified on May, 12, 2015.)

fluctuations inherent when using wind power alone. This chapter evaluated a case study for hybrid energy systems that consist of (1) hydropower plant located on the "Presidente Benito Juarez" dam in Jalapa del Marques, Oaxaca, Mexico, and (2) wind farm located near "La Venta" an area in Juchitan, Oaxaca, Mexico. The model showed that a hybrid plant combining these two energy systems can provide close to 20 MW of firm power to the electrical distribution system. On a technoeconomic basis, this combination gave the competitive production cost.

Rozali et al. [29] carried out power pinch analysis for optimization of pumped hydro storage system for hybrid renewable energy power system (e.g., hybrid wind energy system). Storage technology in hybrid power systems is urgently required to adapt with the mismatch between the renewable energy production and the time distribution of load demands. Different storage systems incur different types and amount of losses depending on the power conditioning as well as storage system efficiencies. The study focused on the design of HPS with pumped hydro storage systems with power pinch analysis. With the help of a case study, the authors showed that the application of pumped hydro storage in HPS yield lower total losses in the system. The maximum power demand target is reduced, while the targeted maximum storage capacity is increased compared to when battery storage is applied.

8.2.2 HYBRID WIND–HYDROGEN SYSTEM

In this hybrid system, the energy is stored in hydrogen, which is produced by electrolysis of water with excess wind energy. The hydrogen can be reused when the demand for power is high. The energy in the stored hydrogen can be converted to electric power through fuel cell technology or a combustion engine linked to an electric generator. Hydrogen storage and embrittlement of materials used in power systems are some of the issues in this hybrid system. Some of the known test sites for this hybrid system are described in Table 8.1 [9,15,17–20,37,39]. A schematic of this type

---

TABLE 8.1
Summary of Wind–Hydrogen Hybrid Systems

Community	Country	Wind (MW)
Ramea, Newfoundland and Labrador	Newfoundland, Canada	0.3
Prince Edward Island Wind-Hydrogen Village	PEI, Canada	
Lolland	Denmark	
Bismarck	North Dakota, United States	
Koluel Kaike	Santa Cruz, Argentina	
Ladymoor Renewable Energy Project	Scotland	
Hunterston Hydrogen Project	Scotland	
RES2H2	Greece	0.50
Unst	Scotland	0.03
Utsira	Norway	0.60

Source: "Wind Hybrid Power Systems," https://en.wikipedia.org/wiki/Wind_hybrid_power_systems, last modified on May, 12, 2015.

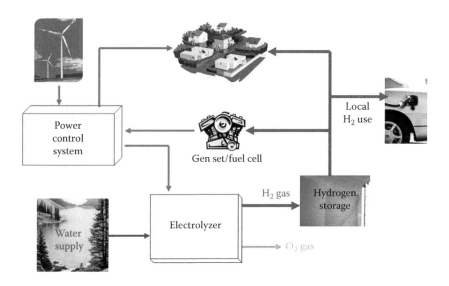

FIGURE 8.2 Hybrid wind–hydrogen energy system. (From "Wind Hybrid Power Systems," https://en.wikipedia.org/wiki/Wind_hybrid_power_systems, last modified on May, 12, 2015.)

of hybrid system is illustrated in Figure 8.2. In general, however as indicated in Chapter 7, wind energy is generally not very efficient for direct electrolysis of water.

8.2.3 HYBRID SOLAR–WIND–HYDRO ENERGY SYSTEM

Sirasani and Kamdi [34] and Bica et al. [33] theoretically analyzed a distributed hybrid solar–wind–hydro energy system. Sirasani and Kamdi [34] presented several models, which can be used for the simulation purposes of this hybrid power system. Bica et al. [33] examined the importance of renewable energies in Romania for stable power production. They suggested that with proper models, this combination can provide high-quality power with or without connections to electrical grid.

Mousa et al. [23] examined an optimum design of hybrid solar–wind power plant. They tried to take advantage of complementary characters of these energy resources. The study developed an optimal design for a hybrid solar–wind energy plant, where the variables that are optimized included the number of photovoltaic (PV) modules, the wind turbine height, the number of wind turbines, and the turbine rotor diameter. The overarching goal was to minimize costs. Simulation studies and sensitivity analysis revealed that the hybrid plant was able to exploit the complementary nature of two energy sources and deliver power reliably throughout the year.

Arjun et al. [24] developed a solar–wind hybrid system and took the detailed measurement and analysis of the data for a period of 1 month. Their results showed that the site (southern India) they examined was abundant in renewable energy, and the hybrid nature increased the overall reliability and reduced the dependence on one single source.

There are numerous other studies on hybrid solar–wind energy systems. These systems can either be connected to electric grid for stability or connected to nuclear power plants or a diesel generator for backup power. In general, an off-grid hybrid system is practical if

1. The average wind speed is at least 9 mph or 4 m/s.
2. A grid connection is not available, or it can only be made through an expensive extension. The cost of running a power line in a remote site to connect with the utility grid can be prohibitive, ranging from $15,000 to more than $50,000/mile depending on the terrain.
3. The goal is to generate clean power and have an independence from the utility.

8.2.4 SOLAR–WIND–BIOMASS RESOURCES FOR RURAL ELECTRIFICATION

Dhass and Harikrishnan [35] examined a hybrid renewable energy system combining generation of power through solar, wind, and biomass systems at Pongalur, Tamil Nadu, India. This system had an adequate solar insulation, wind velocity, and biomass fuel availability. The wind velocity was varied from 2.82 to 8.04 m/s at selected sites, and biomass gasifier feedstock rates were varied as 135 and 350 kg/h for 100 and 250 kW, respectively. The technical feasibility of PV integrated with wind and biomass systems for the required demand was analyzed. The study showed that for all load demands, the life cycle cost and life cycle unit cost for solar–biomass, wind–biomass, and PV–wind–biomass hybrid systems were always lower than those of stand-alone systems. The study concluded that PV–wind–biomass hybrid system was a technoeconomically feasible option for rural electrification. The new hybrid combination was suggested for implementation in remote places where installation of power grid was too expensive.

8.2.5 HYBRID WIND–DIESEL ENERGY SYSTEM

Wind–diesel hybrid power system combines diesel generators and wind turbines [9,13,15,16,20,26,31,32,37,38] usually alongside ancillary equipment such as energy storage, power converters, and various control components to generate electricity. They are designed to increase capacity and reduce cost and environmental impacts of electrical generation in remote communities and facilities that are not linked to power grid [9,13,15,16,20,26,31,32,37,38]. Wind–diesel hybrid systems reduce reliance on diesel fuel, which creates pollution and is costly to transport [9,13,15,16,20,26,31,32,37,38]. The successful integration of wind energy with diesel-generating sets relies on complex controls to ensure correct sharing of intermittent wind energy and controllable diesel generation to meet variable load demand. The term wind penetration implies the percentage of energy generated by wind turbines. Sites such as Mawson Station, Antarctica, as well as Coral Bay and Bremer Bay in Australia have peak wind penetration of around 90%. The variable wind output is usually controlled by (a) using variable speed wind turbines, (b) controlling demand such as heating load, or (c) storing energy in a flywheel. A list of wind–diesel (WD) systems used around the world is given in Table 8.2.

TABLE 8.2
Summary of Worldwide Wind–Diesel Hybrid Energy Systems

Community	Country	Diesel (MW)	Wind (MW)	Population	Date Commissioned	Wind Penetration (Peak) (%)
Mawson Station	Antarctica	0.48	0.60		2003	>90
Ross Island	Antarctica	3	1		2009	65
Bremer Bay	Australia	1.28	0.60	240	2005	>90
Cocos	Australia	1.28	0.08	628		
Coral Bay	Australia	2.24	0.60		2007	93
Denham	Australia	2.61	1.02	600	1998	>70
Esperance	Australia	14.0	5.85		2003	
Hopetoun	Australia	1.37	0.60	350	2004	>90
King Island	Australia	6.0	2.50	2,000	2005	100
Rottnest Island	Australia	0.64	0.60		2005	
Thursday Island, Queensland	Australia		0.45			
Ramea	Canada	2.78	0.40	600	2003	
Sal	Cape Verde	2.82	0.60		2001	14
Mindelo	Cape Verde	11.20	0.90			14
Alto Baguales	Chile	16.9	2.00	18,703	2002	20
Dachen Island [29]	China	1.30	0.15			15
San Cristobal, Galapagos Island [30]	Ecuador		2.4		2007	
Berasoli [31]	Eritrea	0.08	0.03			
Rahaita	Eritrea	0.08	0.03			
Heleb	Eritrea	0.08	0.03			
Osmussaar	Estonia		0.03		2002	
Kythnos	Greece	2.77	0.31			
Lemnos	Greece	10.40	1.14			
La Désirade	Guadeloupe	0.88	0.14			40
Sagar Island	India	0.28	0.50			
Marsabit	Kenya	0.30	0.15			46
Frøya	Norway	0.05	0.06			100
Batanes	Philippines	1.25	0.18		2004	
Flores Island	Portugal		0.60			60
Graciosa Island	Portugal	3.56	0.80			60
Cape Clear	Ireland	0.07	0.06	100		70
Chukotka	Russia	0.5	2.5			
Fuerteventura	Spain	0.15	0.23			
Saint Helena	United Kingdom		0.48		1999–2009	30
Foula	United Kingdom	0.05	0.06	31		70
Rathlin Island	United Kingdom	0.26	0.99			100

(Continued)

TABLE 8.2 (*Continued*)
Summary of Worldwide Wind–Diesel Hybrid Energy Systems

Community	Country	Diesel (MW)	Wind (MW)	Population	Date Commissioned	Wind Penetration (Peak) (%)
Toksook Bay, Alaska	United States	1.1	0.30	500	2006	
Kasigluk, Alaska	United States	1.1	0.30	500	2006	
Wales, Alaska	United States		0.40	160	2002	100
St. Paul, Alaska	United States	0.30	0.68			100
Kotzebue, Alaska	United States	11.00			1999	35
Savoonga, Alaska	United States		0.20		2008	
Tin City, Alaska	United States		0.23		2008	
Nome, Alaska	United States		0.90		2008	
Hooper Bay, Alaska	United States		0.30		2008	

Source: "Wind Hybrid Power Systems," https://en.wikipedia.org/wiki/Wind_hybrid_power_systems, last modified on May, 12, 2015.

Kornbluth et al. [32] presented an optimization study for a small hybrid WD grid in the Galapagos Islands. In 2007, 2.4 MW of wind turbines began operation on San Cristobal Island in the Galapagos, but there was a significant mismatch between wind resource and electric demand, which had to be fulfilled by diesel fuel. The study outlines opportunities to overcome traditional issues associated with hybrid grids involving renewable energy sources, that is, energy storage and spinning reserve versus reliability, without costly dedicated energy storage system. The study showed that often the implementation of energy efficiency opportunities is the most cost effective and should be investigated first. In small remote electrical grids with high fuel costs and an over abundance of off-peak renewable energy availability, load shifting can also be cost effective. For the case examined in this study, the replacement of lighting in addition to a demand response system with permanent load shifting offered the most diesel savings and lower overall operating cost for the government. The study also considered the effects of other variables (both human and energy storage and demand) on the results of the simulation.

The power generation from wind energy can be effectively managed by connecting it to the major large electricity grids. This is extensively illustrated in Chapter 9. The potentials for wind energy connected to small and isolated grids are also enormous, but power production from wind energy fed into small and isolated grids has been economically unattractive due to its inefficiency and technical problems associated with the necessary diesel generation backup system. Normally, when diesel engine operates at less than 30% of its load capacity, the diesel engine will have poor fuel economy and will be choking, leading to lower engine efficiency and faster deterioration. A company called Danvest Energy (in Denmark) has built a hybrid WD system that is capable of up to 100% wind penetration. With Danvest hybrid WD concept, remote areas and islands are offered the opportunity to utilize wind energy in combination with diesel-generated power for their power production. The Danvest WD stand-alone system is able to operate with an estimated 50%–85% fuel saving compared to power supply from diesel generation alone depending on the actual wind conditions and size of the wind turbines. The savings more than covers additional cost related to the wind turbines and the WD equipment.

Power production, which is fed to isolated grids, differs considerably from power production fed into major grids due to economical, technical, and environmental reasons. Since power production from diesel engines is very high, power fed into small isolated grids is more expensive than the conventional power plant supplying to major grids. When the wind turbines are connected to a major grid, variations and fluctuations in the wind power are absorbed by the strong grid, thus controlling frequency and voltage of power generation. With small and isolated grids, the power balance between production and consumption has to be continuously maintained to keep frequency and voltage of the small grids within predefined limits. Finally, small diesel combustion units that provide backup for small grids generally have poor combustion; more emissions of CO_2, SO_2, and NO_x gases; and more noise and poor visual impact to the surroundings.

So far, the necessity to control wind turbines by major grid has limited the contribution of wind energy into major grids to 20%–30% of the total power demand. For small isolated grids (off-grid), Danvest [31] has solved the problem of controlling the quality and quantity of wind energy, whereby the power supply from the wind energy can be increased to 100% of the demand and thereby substitute the diesel power completely during periods of high wind. The Danvest [31] WD concept has developed a unique controlling system for the WD hybrid power production, which replaces the normal controlling function of the major grids.

The system is based on standard diesel generator sets fitted with the patented WD equipment for backing up wind turbines when they are supplying less than 100% of the power demand. The WD system automatically controls the continuous operation in the following ways [31]:

1. When the wind energy is sufficient to supply all demand, diesel engine is shut down. The alternator remains connected to the bus bar for control of reactive power, voltage, frequency, and diesel engine conditions. When the wind speed is decreasing or consumption is increasing, the diesel engine is started and is connected within 1–3 s.
2. The dump load control secures a fast dynamic load balance between the fluctuating wind power and the variations in power consumptions.
3. The diesel engines that are fitted with the WD equipment are able to operate at low load for longer periods, thus maintaining normal service intervals.

More details of this novel hybrid Danvest WD concept is given in the company's website (http://www.danvest.com/).

8.2.6 Simulation of a Hybrid System Wind Turbine–Battery–Ultracapacitor

Bludszuweit et al. [30] simulated a hybrid wind energy–battery–ultracapacitor system with Simulink®. The study was aimed to improve the predictability of power generation from wind and short-term energy storage in batteries combined with ultracapacitors. The objective of the simulation was to determine the system configuration that is able to guarantee a predictable energy output to the grid during a given time interval. Within this time, the storage system should be able to compensate all types of rapid changes originated by wind speed fluctuations. Another important issue was the integration of the ultracapacitor bank. Its influence on the charge and discharge stress of the battery was investigated.

The study demonstrated that quick changes in the wind power generation can be covered by a reasonable size of battery capacity for a prediction time interval of 10 min. Very short transients in energy flow (1–10 s) are absorbed by the ultracapacitor bank, which minimizes the deterioration of the battery because important current peaks do not reach the battery. The study proposed a final system design based on measurements at a 600 kW constant speed wind energy conversion system.

8.2.7 Hybrid Wind–Water Current Power System

Mitsui Ocean Development & Engineering Company has developed the Savonius Keel & Wind Turbine Darrieus power generation system, which is a floating system that combines wind power with the underwater current power in a hybrid manner [27]. This 500 kW device shares a floating axis, and the power generation assembly sits between the wind turbine and underwater current device at a deck level, anchored by a set of rubber mounts to a floating submersible platform. The system can generate power from either or both sources. The wind turbine has a height of 154 ft and a rotor diameter of 49 ft.

The device is stable because the tidal turbine acts as ballast making the assembly self-righting, and the location of the generator at deck level creates a low center of gravity as well as offers easy access for operations and maintenance. The Savonius tidal turbine is especially suited to harvesting energy from weak currents, and its rotation can be used to start the wind turbine spinning in low wind conditions. This rather interesting device is graphically illustrated in the company's website [27], and it is installed on the cost of Japan.

8.2.8 Hybrid Wind-Compressed Air Systems

Hybrid wind energy–compressed air energy storage (CAES) power station uses excess electrical energy obtained during high wind (and low demand) to compress air and store it in underground facilities such as caverns or abandoned mines. During later periods of high electrical demand, the air is released to power turbines, generally using supplemental natural gas [9,22]. Power stations that make significant use of CAES are operational in McIntosh, Alabama, Germany, and Japan [42].

This hybrid system has some energy losses in the CAES process; also, the need for supplemental use of fossil fuels such as natural gas means that these systems do not completely make use of renewable energy and they emit CO_2 [22]. A new commercial operation for this type of hybrid system will begin in 2015 at the Iowa Stored Energy Park [9].

8.3 HYBRID GEOTHERMAL ENERGY SYSTEMS

Ever since the implementation of enhanced geothermal systems, hybrid use of geothermal energy with solar, wind, and nuclear energy has rapidly expanded. Many small- and large-scale plants have evolved using backup energy storage systems or connections to the major electrical grid [40–52]. In this section, we examine some of the innovative hybrid geothermal energy systems generating power and heating and cooling.

8.3.1 HYBRID SOLAR–GEOTHERMAL POWER PLANT

The world's first geothermal–solar hybrid power plant was dedicated in Fallon, Nevada, in 2013 [41]. This Stillwater geothermal project, which received $40 million in tax support under the recovery act, has harnessed innovative technologies to add solar energy to the facility and provide 59 MW of combined capacity to power more than 50,000 local homes.

Currently, utilities often use solar power during hours of peak consumer demand and combine it with a "baseload" coal or natural gas plant to ensure a steady power supply. The Stillwater facility combines 26 MW of PV solar-generating capacity with 33 MW of baseload geothermal power, demonstrating how a single power plant can deliver renewable peak and baseload power to American homes, business, and whole communities. The solar energy is provided by Enel Green Power (EGP) North America who installed more than 89,000 polycrystalline PV panels on a 240 acre parcel of land adjacent to Stillwater geothermal plant.

By combining the two forms of energy in one location, the power plant makes use of multiple forms of renewable energy that will increase the level of zero emission energy that is produced. It is estimated that the power plant will produce enough clean energy to avoid the emission of around 140,000 metric tons of CO_2 into the atmosphere each year. The added advantage of creating this hybrid mix is that infrastructure is already in place reducing costs and the power transmission connection has already been established [41].

8.3.2 HYBRID THERMAL SOLAR-/GEOEXCHANGE FOR POWER AND HEAT PUMP

A company called Building Tech Services [49] offers a hybrid thermal solar-/geoexchange heat pump for homes. Hybrid systems use some low-temperature solar energy to supplement the geoexchange earth energy, raising the ground temperature and lowering the amount of work the compressor must do, thereby saving kWh. The company has installed a number of these units and claims its high efficiency.

Kuyumcu et al. [51] carried out theoretical analysis of hybrid geothermal–solar system. Performance of air-cooled binary cycle organic rankine cycle (ORC) geothermal power system was inversely related to ambient temperature, where summer temperature extremes can cause performance drop of up to 70% from design. Concentrating solar thermal power generation systems act inversely almost in harmony reaching peak efficiency during most of these ambient temperature extremes. The two thermal generation systems constitute suitable candidates for hybridization, as a way of "hedging production against ambient temperature fluctuations." Kuyumcu et al. [51] analyzed this concept theoretically. A company called BM Holding is further developing this concept in Gumuskoy, where 6.6 MWe geothermal power plant will be complemented by concentrating solar power (CSP) system of adequate size. The plant will operate in the near future.

Morrison [52] carried out a study of simulation and optimization of hybrid geothermal energy systems using solar thermal collectors as a supplemental heat source. The study concluded that hybrid systems reduce ground heat exchanger size requirements. Simulation/optimization method resulted in sizing components so that ground temperatures stabilized compared to normal hybrid systems. The study also concluded that cost competitiveness of hybrid system depends on the drilling costs and solar collector costs.

8.3.3 HYBRID GEOTHERMAL–SOLAR CHIMNEY POWER PLANT

Alrobaei [48] and others [51–55] examined hybrid geothermal–solar energy technology for power generation. The hybrid geothermal–solar chimney power plant (SCPP) has generated interest because it offers an innovative way for continuous 24 h operation and improves the maneuver characteristics of grid-connected SCPP. The main target of geothermal–solar chimney power plant (GSCP) design approach was to achieve high renewable share with little or no fossil fuel backup requirements in electrical power grid. Moreover, there is an increase in the useful operating time of SCPP by replacing the daily start-up and shutdown times due to continuous operation of the power conversion system. The initially proposed operation strategy of GSCP is very simple: the plant operates during sunny periods as conventional grid connected SCPP. During the cloudy periods and nights, the plant will operate in the following modes:

1. Full geothermal mode of operation (air underneath a glass ceiling is heated by geothermal energy so it rises through a chimney).
2. Hybrid geothermal–solar mode of operation (air underneath a glass ceiling is heated by geothermal energy and solar radiation so that it rises through a chimney). This will fully exploit the potential of geothermal–solar combination to provide a major contribution to the basic supply of renewable electricity.

Alrobaei [48] also showed a modification of GSCP by using a PV system to further improve the overall efficiency. Alrobaei [48] schematically presented these two options for SCPP. The analytical details of this approach are also discussed in detail by others [51–55].

8.3.4 Hybrid Geothermal–Biomass System

A technical and economic analysis of a proposed hybrid geothermal–biomass system for district heating (DH) was carried out for Cornell University (Ithaca, NY) [42,47,55]. In these studies, low-temperature geothermal resource was considered. The researchers evaluated two design scenarios: (1) the geothermal source was applied only for DH, and (2) the geothermal source was used for the combined generation of heat (for DH) and electrical energy through an ORC plant (DH-ORC). The enhanced geothermal system (EGS) reservoir and the DH network are hydronically divided by a heat exchanger. Furthermore, a backup steam-to-water heat exchanger was inserted. In both scenarios, a boiler fed with torrefied biomass was used to produce the additional heat for the DH when needed. Heating and domestic hot water needs were fulfilled by the DH water supply pumped to building substations, from which roughly the 30% of return DH water was cascaded to greenhouse substations. The cycle was closed by pumping the return water to the central heat exchanger.

In the proposed DH, pressurized liquid water at temperatures ranging from 90°C to 140°C was used. The temperature of the DH supply can be lowered as in this case study it was constrained by the existing space-heating system. Substations of the proposed system used plate and frame water-to-water heat exchangers. The hybrid system also had water circulation pumps and roughly 20% of the pipeline network was formed by preinsulated pipes. In the greenhouses, a soil-heating system with a fluid temperature ranging from 30°C to 60°C was considered. The greenhouses network consisted in polyethylene tubes and a finned coil air heater. Furthermore, the ORC unit consisted of two low-capacity ORC power plants operating only during low-heat demand conditions. For the ORC unit, the researchers considered 25 different fluids and selected R32 fluid. The details of these cycles are schematically described in [47,55].

The research on this design is ongoing [41–43,47,55–68]. For example, EGP [69] is focusing in particular on geothermal in combination with biomass and solar energy, and the "European Technology Platform on Renewable Heating and Cooling" has a focus group specifically for renewable energy hybrid systems.

Another hybrid geothermal–biomass plant is located in Honey Lake, California, and has power output of 35 MW. It is a steam cycle power plant operated from wood waste [55,60,70]. For heating feed water, 22 kg/s of geothermal water of 118°C is used. These parameters can give the maximum geothermal output of about 8 MW. Hybrid combined cycle plant described here can save 90 million m^3 of natural gas, which would bring reduction of CO_2 by 220,000 tons/year. The real savings could be higher depending on the emissions of the plant being replaced. This plant and concept has been given significant attention in Slovakia. Hybrid combined cycle power plants are the new generation of combined cycle power plants. Boszormenyi and Boszormenyi [50] have analyzed this hybrid system and its relevance to Slovakia.

Bombarda et al. [56] proposed a hybrid geothermal–biomass plant based on ORC technology. They considered a geothermal source with temperatures ranging from 110°C to 150°C. Synthetic oil was selected as a medium to transfer heat between the biomass-fired furnace and the power unit. Furthermore, to test the proposed hybrid plant, data for the biomass source were taken from a plant located in the Molasse basin (Germany) that generates electrical and thermal energy throughout the year and according to seasonal demand, respectively. A cascade cycle was selected with octamethyltrisiloxane, regenerative, and R245fa as working fluids for the upper and lower cycle, respectively. The results highlighted a higher net power production compared to separate plants. Borsukiewicz-Gozdur [57] and others [55] assessed two hypothetical hybrid geothermal–biomass systems. For both systems, the assumed geothermal water had low temperatures (80°C–120°C) with a constant flow rate (30 kg/s). The first plant consists of a dual-fluid-hybrid (DFH) power plant formed by an upper and lower cycle [55,57].

In the upper cycle (Hirn cycle), the working fluid (water) was preheated and vaporized in a biomass boiler and subsequently directed to a steam turbine, where the fluid expanded and performed work. The fluid was then condensed in a condenser–vaporizer exchanger and subsequently was moved to the biomass boiler by the circulating pump that pressurized the fluid to the appropriate value. In the lower cycle (ORC), which was connected to the upper one by the condenser–vaporizer, after preheating using geothermal sources, the working fluid evaporated in the condenser–vaporizer and then was directed to the vapor turbine that was linked to an electrical energy generator. Subsequently, the working fluid was directed to the condenser, to the circulation group and, finally, to the organic liquid heater. The working fluids that were tested are R365mfc-water, R245fa-water, and R236fa-water [55,57].

The study also considered a second hybrid plant consisting of a single-fluid hybrid power plant. This hybrid plant was fed by combustion of biomass in the upper part, and a geothermal source is used to preheat the working fluid. For this proposed hybrid system, cyclohexane and water were tested as working fluid. The author highlighted the advantage of the combined use of geothermal and biomass against only geothermal sources and that the best option that uses the least share of biomass-derived energy was a DFH power plant with R236fa as working fluid [55,57]. The schematics of both cycles are illustrated in detail in References 55 and 57 and partially discussed in [58–67].

8.3.5 GEOTHERMAL–BIOGAS HYBRID SYSTEM

The geothermal energy and natural gas electricity production is one of the best combinations regarding CO_2 emissions (global warming effects). The electricity production with combination of geothermal and gas energy has significant positive impacts on gas emission [40,43]. In principle, it is possible to use the concepts outlined for geothermal–biomass or geothermal–solar energy except in this case biogas replaces biomass or solar energy. It is possible to combine natural gas–biogas and geothermal energy in many ways. One suggested way is shown by Najjar [185] and others [55].

These and other studies combined gas engines with ORC process. In this way, heat from exhaust gases can be combined with geothermal energy in ORC process [40,43,46,47,61–66,55,185].

8.3.6 DUAL-PRESSURE SYSTEM WITH DIRECT-STEAM COLLECTORS USING TWO DIFFERENT SOLAR COLLECTORS

In this hybrid system, the organic fluid is directly heated by the evacuated tube collector (ETC) and then by the parabolic trough collectors (PTCs) fields. The ETCs are used only for liquid preheating and phase transition with a recirculation loop-producing vapor. A vessel is used for separating the saturated vapor phase, part of which is directed to the low-pressure turbine (LPT). An appropriate amount of the saturated liquid phase in the separation vessel is compressed and directed to the high-pressure loop using PTCs. After LPT, the organic fluid is cooled in desuper-heater and condensed to start the cycle again [48,53,55,57].

In all solar power systems, the solar field represents the main cost item, account-ing for about 80% of the electricity cost, which leaves the solar source still uncom-petitive with fossil fuels for power generation. However, integration with a solar field could improve the attractiveness of the many low-enthalpy geothermal sources [55,67,68,71,72].

8.3.7 HYBRID GEOTHERMAL AND FUEL CELL SYSTEM

A hybrid energy system that includes a fuel cell and geothermal heat pump has been developed by the University of Minnesota that improves the efficiency of a ground-source heat pump by using low-grade heat from the fuel cell [73]. The sys-tem takes the waste heat from the fuel cell and combines it with the heat from the geothermal heat exchange system. The fuel cell provides electricity to the geother-mal system and the digital control unit, allowing the system to operate off the grid. The fuel cell can also provide electrical power to the building. The temperature of the fuel cell is regulated by a digital control unit. Fuel cells are more efficient than generators. However, fuel cells typically waste about 50% of the input energy in the form of heat.

In the heating mode, the waste heat from the fuel cell is pumped into the building for heating purposes. In the cooling mode, the waste heat is pumped into the ground where it is absorbed by the earth. This technology can improve the efficiency of the heating and cooling systems in the building and also has applications in remote or backup power generation systems for homes and commercial buildings. The overall benefits of this hybrid system are twofold.

1. The geothermal heat exchange recovers the heat loss by the fuel cell and increases efficiency of the geothermal heat pump.
2. By using the fuel cell, the system can be operated either on or off the grid.

This concept has also been analyzed by other investigators [74,75].

8.3.8 Hybrid Geothermal Power Plant Using CO_2

Buscheck and coworkers [76–85] and others [86] developed a hybrid geothermal power plant using CO_2. The design contrasts with conventional geothermal plants in a number of important ways. Typical geothermal power plants tap into hot water that is deep underground, pull the heat off the hot water, use that heat to generate electricity, and then return the cooler water back to the deep subsurface. Here, the water is partly replaced with CO_2 and/or another fluid. There are benefits to using CO_2, because it mines heat from the subsurface more efficiently than water. This combined approach can be at least twice as efficient as conventional geothermal approaches and expand the reach of geothermal energy in the United States to include most states west of the Mississippi River.

The research team used computer simulations to design the system. In the simulations, a system of four concentric rings of horizontal wells about 3 miles belowground, with the outer ring being a little more than 10 miles in diameter, produced as much as a half gigawatt of electrical power—an amount comparable to a medium-sized coal-fired power plant, and more than 10 times bigger than the 38 MW produced by the average geothermal plant in the United States.

The simulations also revealed that a plant of this design might sequester as much as 15 million tons of CO_2/year, which is roughly equivalent to the amount produced by three medium-sized coal-fired power plants in that time. One of the key objectives of this new technology is to find a way to help make CO_2 storage cost effective while expanding the use of geothermal energy.

During the past year, Buscheck et al. [87–89] added another gas—nitrogen—to the mix, resulting in a design that he and his colleagues believe will enable highly efficient energy storage at an unprecedented magnitude (at least hundreds of gigawatt hours) and unprecedented duration (days to months), provide operational flexibility, and lower the cost of renewable power generation. Nitrogen has several advantages. It can be separated from air at lower cost than captured CO_2. It is plentiful. It is not corrosive and will not react with the geologic formation in which it is being injected. Also, nitrogen is readily available; it can be injected selectively. Thus, much of the energy required to drive the hot fluids out of the deep subsurface to surface power plants can be shifted in time to coincide with minimum power demand or when there is a surplus of renewable power on the electricity grid.

The investigator believes that because energy is stored in the form of pressurized fluids, this concept can be further improved by selectively producing hot fluids when power demand is high, as well as reduce or stop that production when power demand is low. What makes this concept transformational is that it can deliver renewable energy to customers when it is needed, rather than when the wind happens to be blowing, or when spring thaw causes the greatest runoff.

The new geothermal plant design is so efficient at extracting heat that even smaller-scale "hot spots" throughout the western United States could generate power. The geothermal plant, however, would probably have to be connected to a large CO_2 source, such as a coal-fired power plant. Buscheck added that a pilot plant based on this design could initially be powered solely by nitrogen injection, in order to prove the economic viability of using CO_2. The study also showed that this design can work effectively with or without CO_2, broadening where this approach could be deployed. The research team is currently working on more detailed computer model simulations and economic analyses for specific geologic settings in the United States.

8.4 HYBRID SOLAR ENERGY SYSTEMS

As shown earlier in Sections 8.2.1, 8.2.3 and 8.2.4, solar energy can be combined with wind energy, geothermal energy, and nuclear energy to provide power in isolated areas (with backup energy storage) or with the support of grid connection. Unlike wind, water, and geothermal energy, solar energy can generate very high temperatures like nuclear energy and that can be put to use for nonelectrical and chemical fuel productions applications through co-generation approach. Since solar energy is time and location dependent, one of the best applications of solar energy has been to produce solar fuels that can be used for electric and transportation applications as well as for energy storage.

Chapter 7 showed that nuclear energy can be used for a variety of nonelectrical applications by the use of nuclear heat at different temperatures for different applications. Solar energy similarly can be used for a variety of nonelectric applications at different temperatures. Unlike nuclear energy, since solar energy is intermittent, its use for power generation requires a backup energy source (like the ones shown earlier), energy storage such as hydrogen or other types of fuels or connections to the main power grid. During the last fifteen years, significant efforts are being made to use co-generation method for solar energy to generate solar fuels at large or small scales along with power. Some of these efforts are briefly described in the succeeding text.

8.4.1 Solar Energy–Water Hybrid Systems

The use of solar energy for home heating has been in existence for a long time. In this method, solar panels installed on the top of the roof of the houses or buildings can absorb solar heat, and this heat is stored in water and steam circulating under the solar panels. This heat can be stored and used continuously for the environmental control in residential houses and industrial buildings.

While solar energy can be stored and used in a number of different ways, water plays an important role in harnessing solar energy. Solar hot water systems use sunlight to heat water. In low geographical latitudes, domestic hot water use at moderate temperatures can be provided by solar water heating systems [90–95]. There are at least three types of solar heaters; ETCs (most widely used), glazed flat plate collectors (used for domestic water heating), and unglazed plastic collectors (mainly used for the swimming pools in the United States [90–95]). In 2007, worldwide solar water heater system capacity was 154 GW, led by China (70 GW), Israel, and Cyprus [90–95].

Water is also heavily used in solar energy–driven heating, ventilation, and air-conditioning (HVAC) systems in residential home as well as in industrial building. Water can be a good solar energy storage device, which can be used to provide heating and cooling on as needed basis for daily and seasonal durations. Solar distillation can be used to make saline and brackish water potable. Solar energy can be used for water disinfection and water stabilization pond to treat wastewater. Solar-concentrating technologies such as parabolic dish, trough, and Scheffler reflectors can provide process heat for commercial and industrial applications. For example, 50% of process heating, air conditioning, and electrical requirement for a clothing factory in Shenandoah, Georgia, is provided by a solar energy project [90–95]. Many of these applications are similar to the ones for nuclear energy.

Finally, power generated by solar energy using PV systems needs to be stored. Off-grid PV systems have traditionally used rechargeable batteries to store excess electricity. Another approach is the use of pumped storage of hydroelectricity, which stores energy in the form of water pumped when the energy is available from a lower-elevation reservoir to a higher-elevation one. The energy is recovered when demand is high by releasing the water to run through a hydroelectric power generator [90,96–104,196]. Solar energy can also be stored by producing solar fuels like hydrogen using numerous techniques described in the succeeding text. The production of solar fuels mostly involves dissociation of water. Hydrogen can also be produced using solar reforming of fossil and biofuels using steam. Different techniques required to accumulate concentrated solar power are described in [90,96–104,196].

8.4.2 Echo Solar Hybrid System

The Echo+ solar hybrid system [90–92] is only available when purchased as part of a new home. Just like an Echo solar hybrid system, an Echo+ solar hybrid system generates electricity and thermal energy. But Echo+ solar hybrid uses the thermal energy not only for water heating, but also home heating and home cooling. Thus, this is an example of trigeneration.

Echo+ solar hybrid systems utilize a solar array made of solar electric "PV" panels and solar thermal panels. The solar electric panels generate direct current (DC) electricity directly from the sun. The electricity is routed to an inverter, which converts the solar DC electricity into household alternating current electricity and feeds it into your electrical service panel. Electricity not used in the home is fed back into the utility grid for credit.

Sunlight shines on the solar panels and heats them. Outside air is drawn through an air space or plenum between the solar panels and the roof. The movement of the air under the solar panels efficiently transfers the thermal energy from the solar panels to the air. The heated air is directed through insulated ducts into a solar appliance called energy transfer module (ETM). Here, after passing through the filter, the hot air moves across a copper tube/aluminum fin coil heat exchanger. Cold water from the home's water tank is also fed into the heat exchanger. The heat exchanger then extracts the thermal or heat energy from the air and transfers it to the water.

Often, the ETM also directs the hot air to either the inside of the home (typically, via HVAC ducts) or to the outside of the building (when home heating is not required). For cooling to occur, the outdoor temperature must be at least 5°F (3°C) colder than the indoor temperature, and the thermostat must be set to "cool." When these conditions are met, solar home cooling can occur. Echo+ solar hybrid systems draw cooler outside air under the solar panels and into the ETM, which filters the air and directs the cool air into the home's HVAC ducts.

An Echo+ solar hybrid system supplements and enhances the regular power, heating, and cooling systems in your home; it does not replace them. Echo works seamlessly with the other systems in your home, so you can have all of the benefits of solar and all of the benefits of regular utility service.

If it is sunny outside and your system is generating electricity, you can use the electricity to power your home and appliances. If you don't need the electricity, it is automatically sent back to the utility grid and your utility will typically credit you with the energy saved. When the system is not generating electricity (e.g., at night or during very cloudy periods), you can power your home with the electricity that you purchase from your local utility company.

When your Echo+ solar hybrid system is providing home heating or home cooling, it works in cooperation with your furnace or air conditioner. If temperature conditions do not allow Echo+ solar hybrid to provide solar home heating or solar home cooling, your regular furnace or air conditioner will still operate normally.

Another company Cogenra [91–93] is also very active in the development of new combined heat and power systems for hybrid solar energy. Their approach is somewhat different from the one described earlier.

8.4.3 SOLAR FUELS

While solar energy alone or in combination with other sources of renewable, nuclear, or fossil energy can be used to generate power, solar energy can also be used to generate both power and heat in hybrid manner (co-generation). This process is in concept very similar to the one described in Chapter 7 for the nuclear energy. Solar heat can generate chemical fuels, which can be used for the solar energy storage. Solar fuels, such as hydrogen, can be used for upgrading fossil fuels, burned to generate heat, further processed into electrical or mechanical work by turbines and generators or internal combustion engines, or used directly to generate electricity in fuel cells and batteries to meet energy demands. The stored solar fuel can also be used as a backup source for solar-powered electricity generation. This will allow generation of solar power during nighttimes using stored solar fuel.

The generation of solar fuel is a hybrid process where two sources of energy (solar) and fuel (water, carbonaceous fuel) work together to generate hydrogen or other chemical fuels. Sustainable, large-scale fuel production relies on the success of solar thermal power generation (CSP) in which the incident solar radiation is concentrated with reflecting mirrors to provide high-temperature process heat for driving efficient thermochemical processes in compact centralized plants. While the process has been proven technically, on a commercial scale, it is not yet proven to be economical [96–103].

8.4.3.1 Solar Fuels Production and Economics

The conversion of solar energy into solar fuels (hydrogen, syn gas and other chemical fuels) opens up numerous possibilities [105–111]. There are basically five routes that can be used alone or in combination for producing storable and transportable fuels from solar energy.

1. *Electrochemical*: Solar electricity made from photovoltaic or concentrating solar thermal systems followed by an electrolytic process [102–112].
2. *Photochemical/photobiological*: Direct use of solar photon energy for photochemical and photobiological processes [113–133].
3. *Thermal*: Direct dissociation of water by thermal forces [2,82,92,93,184]
4. *Thermochemical*: Solar heat at high temperatures followed by an endothermic thermochemical process [134–159].
5. *De-carbonization of carbonaceous materials by solar energy*: Solar gasification, reforming and cracking of carbonaceous materials [2,88,162–188].

The thermochemical route offers some thermodynamic advantages with direct economic implications. Irrespective of the type of fuel produced, higher-reaction temperatures yield higher-energy conversion efficiencies; but this is accompanied by greater losses by reradiation from the solar cavity receiver.

The economic competitiveness of solar fuel production is closely related to two factors: the cost of fossil fuels and the necessity to control the world climate by drastically reducing CO_2 emissions. The economics of large-scale solar hydrogen production indicates that the solar thermochemical production of hydrogen can be competitive compared with the electrolysis of water using solar-generated electricity. It can even become competitive with conventional fossil fuel–based processes, especially if credits for CO_2 mitigation and pollution avoidance are applied. Conceptually, the dual use of solar energy for power and fuel or heat production is similar to the one examined for nuclear energy in Chapter 7.

8.4.3.2   Solar Electrolysis

The process of solar electrolysis involves generation of solar electricity via PV or CSP followed by electrolysis of water. This process is considered to be a benchmark for other thermochemical solar processes for water splitting, which offers potential for energy efficient large-scale production of H_2. For solar electricity generated from PV cell and assuming solar thermal efficiencies at 15% or 20% and electrolyzer efficiency at 80%, the overall solar-to-hydrogen conversion efficiency will range from 12% to 16% [2,96–103,112–122]. If we assume solar thermal electricity cost of \$0.08/kWh, the projected cost of H_2 will range from 0.15\$ to 0.20\$/kWh [2,96–103,112–122], that is, from 6\$ to 8\$/kg H_2. For PV electricity, costs are expected to be twice as high. High-temperature electrolysis process can significantly reduce electricity demand if it is operated at around 800°C–1000°C via solid oxide electrolyzer cells. The high-temperature heat required for such a process can be supplied by concentrating solar power system [97].

8.4.3.3   Water Splitting by Photocatalysis and Photobiological Processes

Duonghong et al. [186] were one of the first ones to investigate the splitting of water by utilizing microsystems. In this system, the colloidal particles are made up of suitable conductor materials, for example, TiO_2. Two metallic substances, for example, ruthenium oxide and platinum, are deposited on these colloids. When the system is irradiated, hydrogen is evolved on the platinum and oxygen on ruthenium oxide. Each colloidal particle is a microphotocell. Using small TiO_2 particles, a large area of TiO_2

can be exposed to light. The system needs to be heated to last more than several hours. There are some doubts whether or not equal production of hydrogen and oxygen is achieved and whether oxygen is engaged in side reactions. It is difficult to measure the efficiency of this system. Also, hydrogen and oxygen come off from water together, and their separations add extra cost. Furthermore, the simultaneous presence of oxygen and hydrogen in water can give rise to chemical catalysis and recombination to water.

Over the last several decades, a number of catalysts have been tested to improve the efficiency of photocatalysis. These include chemically doped titanium oxide, tantalates and niobates, transition metal oxides, metal oxides, metal nitrides, metal sulfides, and oxynitrides [123,124,127–123]. While some improvements have been noted, more work is needed to make the approach economically viable. This approach is further discussed by Shah [2] and Navarro et al. [132,133].

The water splitting can also be carried out photobiologically. Biological hydrogen can be produced in an algae bioreactor. In the late 1990s, it was discovered that if the algae are deprived of sulfur, it will switch from the production of oxygen (a normal mode of photosynthesis) to the production of hydrogen. It seems that the production is now economically feasible by the energy efficiency surpassing 7%–10%.

Hydrogen can also be produced from water by hydrogenase-catalyzed reduction of protons by the electrons generated from photosynthetic oxidation of water using sunlight energy. During the last fifteen years, the use of a variety of algae to produce hydrogen from water has been extensively investigated and reviewed [125,135–143]. These reviews mention the use of sulfur deprivation with *Chlamydomonas reinhardtii* to improve hydrogen production by algae [125,135–143]. Also, certain polygenetic and molecular analyses were performed in green algae [125,135–143]. These methods, however, did not significantly improve the rate and the yield of algal photobiological hydrogen production. Solar-to-hydrogen energy conversion using algae has efficiency less than 0.1% [7]. The rate and yield of algal photobiological hydrogen production is limited by (a) proton gradient accumulation across the algal thylakoid membrane, (b) competition from carbon dioxide fixation, (c) requirement for bicarbonate binding at photosystem II (PSII) for efficient photosynthetic activity, and (d) competitive drainage of electrons by molecular oxygen. Recently, Lee [141–143] outlined two inventions: (a) designer proton channel algae and (b) designer switchable PSII algae for more efficient and robust photobiological production of hydrogen from water. These two new inventions not only eliminate the four problems mentioned earlier but also eliminate oxygen sensitivity of algal hydrogenase and H_2–O_2 gas separation and safety issue. More work in this area is needed. The details of the two new inventions are described by Lee [141–143].

8.4.3.4 H_2 from H_2O by Solar Thermolysis

The direct thermal dissociation has been examined since the 1960s [2,97,101,102]. In direct thermal decomposition, the energy needed to decompose water is supplied by heat only. This requires a minimum of at least 2200°C (even for partial decomposition) and as high as about 4700°C temperature, and this makes the process somewhat unrealistic. At this temperature, about 3% of all water molecules are dissociated as H, H_2, O, O_2, and OH. Other reaction products like H_2O_2 or HO_2 remain minor. At about 3200°C, about half of the water molecules are dissociated. It is well known that an initiation of thermal splitting of water even at low pressure requires 2000 K.

At an atmospheric pressure, 50% dissociation requires 3500 K. This temperature can be reduced to less than 3000 K at 0.01 atm pressure. As shown later, the catalysts can accelerate the dissociation at lower temperature. The lower total pressure favors the higher partial pressure of hydrogen, which makes the reactor to operate at pressures below an atmospheric pressure very difficult.

One of the problems for thermal dissociation of water is the materials that can stand temperatures in the excess of at least 2200°C–2500°C. Several materials such as tantalum boride, tantalum carbide, tungsten, and graphite are possible. However, at these temperatures, only oxides are stable. Graphite is chemically unstable in the presence of hydrogen and oxygen at these high temperatures. Tungsten and tungsten carbide get oxidized at these temperatures. The effect of hydrogen on oxide catalysts at these temperatures is not known. Ceramic materials like boron nitride can also be useful if its oxidation can be controlled. Recent studies have shown that a low amount of dissociation is possible [2,97,101,102]. The separation of oxygen and hydrogen can be carried out in a semipermeable membrane of palladium or ZrO_2–CeO_2–Y_2O_3, which removes oxygen preferentially. Lede et al. [197] used a ZrO_2 nozzle through which steam is forced into a thermal stream and decomposed and unreacted water is quenched suddenly to remove water and oxygen. The resulting gas contained only a small amount (about 1.2 mol%) of hydrogen. Another possible solution is the use of heat-resistant membrane made of Pd or ZrO_2, both of which selectively permeate hydrogen. The gas can also be separated using a magnetic field. The source of heat is also an issue. Solar or nuclear sources are the possibilities. They are although at the early stages of development and at the present time only possible on a smaller scale.

8.4.3.5 H_2 from H_2S by Solar Thermolysis

Hydrogen can also be obtained from hydrogen sulfide. H_2S can be thermally decomposed above 1500°C to produce H_2 and sulfur (S), which, after quenching, can be separated due to phase change. This thermal decomposition reaction can be carried out using solar energy with high degrees of chemical conversion [97,198–200].

8.4.3.6 Thermochemical Decomposition of Water

Thermochemical cycles have been intensely investigated over the last few decades. In this method, two-, three-, or four-step chemical reactions aided by a source of heat such as nuclear or solar can dissociate water and separate hydrogen and oxygen at temperatures around 800°C–900°C. The method has some inherent issues that are as follows:

1. The original concept [2,97,144–165] was that since the method avoided the formation of electricity by the conversion of heat to mechanical work, it would avoid Carnot cycle, as this is the fundamental difficulty in reducing the price of hydrogen production by electrolysis method. This thinking was fallacious because the methods have to have reactions carried out at different temperatures in order that the entropic properties of the partial reactions in each cycle can be used to maximum advantage.

2. With three to four cycles and need to change apparatus for each, plant capital costs for unit hydrogen production is likely to be more than the ones that

occur in the electrolysis. Furthermore, at temperatures like 800°C–900°C, the corrosion will cause the plant life to be short.

3. Generally, it is assumed [2,97,101,143–165] that the reaction would take place along the free-energy pathway, but in reality, they take path down a reaction rate pathway and not necessarily on thermodynamic pathway. Also, because of possible side reactions, the final product may not be what was intended.

In spite of these arguments, a considerable investigation on thermochemical cycles to produce hydrogen at the temperatures lower than one required for the thermal dissociation has been carried out. The moderate temperatures used in these cycles, in general, also cause less material and separation problems. More than 300 different types of chemical cycles have been proposed and tested.

Previously, thermochemical cycles were characterized as the ones which use process heat at temperatures below 950°C. These are expected to be available from high-temperature nuclear reactors. These cycles required three or more chemical reaction steps, and they are challenging because of material problems and inherent inefficiency involved with heat transfer and product separation in each step.

New advancement in the development of optical systems for large-scale solar concentrations capable of achieving mean solar concentration ratio that exceeds 5000 suns allows high-radiation fluxes capable of getting temperature above 1200°C. Such high temperatures allowed the development of efficient two-step thermochemical cycles using metal oxides redox reactions. Some of the important oxide cycles operated by solar energy are briefly described in the succeeding text. More details on other thermochemical cycles are described by Shah [2].

8.4.3.6.1 Zn–ZnO Cycle

One of the most researched metal oxide redox pairs is Zn/ZnO [2,97,158,167]. Since the products of ZnO decomposition at high temperature (viz., Zn and oxygen) readily recombine, the quenching of the product is necessary. The electrothermal process to separate Zn and oxygen at high temperatures has been experimentally demonstrated in small-scale reactors. Such high-temperature separation allows recovery of sensible and latent heats of the products to enhance the energy efficiency of the entire process. A high-temperature solar chemical reactor was developed for this process, and solar tests were carried out at Paul Scherrer Institute (PSI) solar furnace in Switzerland [2,97,158]. These tests allowed surface temperature to reach 1700°C in 2 s with very low thermal inertia of the reactor system. In 2010, solar chemical reactor concept for thermal dissociation of ZnO was demonstrated in a 100 kW pilot plant in a larger solar research facility [2,97,158,167].

In more recent studies on this cycle, a novel concept of hydrolyzer (hydrolysis step) was investigated for the hydrogen production step of $Zn + H_2O \rightarrow ZnO + H_2$. This technique showed that the water-splitting technique works at reasonable rates above 425°C. In this reaction, Zn can be melted either by the heat generated due to the reaction or by supplying molten Zn from a nearby quencher unit of the solar plant. The heat of reaction also generates steam for the reaction. The transportation of Zn to the reaction site eliminates the need for transportation and storage of hydrogen generated by the process.

8.4.3.6.2   SnO/SnO₂ Cycle

Another successful thermochemical cycle involves SnO/SnO_2 where exergy and energy efficiencies of 30% and 36%, respectively, can be obtained. The work carried out at Odeillo, France, in 1 kW solar reactor at atmospheric and reduced pressure has shown that SnO_2 reduction can be efficiently carried out at 1500°C and SnO hydrolysis can be carried out at 550°C [2,97,160,163].

8.4.3.6.3   Mixed Iron Oxide Cycle

Other metal oxides such as manganese oxide or cobalt oxide, as well as mixed oxides redox pairs, mainly based on iron, have also been considered. The mixed iron oxide cycle was demonstrated at the 10 kWth level by *HYDROSOL* (A project to deliver technology for hydrogen production via direct solar water splitting) (2002–2005) project. The reactor used in this study was a ceramic multichanneled monolith that was coated with mixed iron oxide nanomaterials, which were activated by heating to 1250°C by solar radiation. The reactor dissociated water, trapped oxygen, and allowed hydrogen to leave the reactor. *HYDROSOL 2* (2005–2009) project built a 100 kWth dual-chamber pilot reactor at the PSA, Spain.

8.4.3.6.4   Carbothermal Reduction of Metal Oxides

Carbothermal reductions of metal oxides like iron oxide, manganese oxide, and zinc oxide with carbon and natural gas to produce the metals and syngas were demonstrated in solar furnaces [97,166,168,169,173]. In EU's *SOLZINC* (A project for Zinc production via carbothermic reduction of Zno using solar energy) (2001–2005) project, the process was tested at the solar tower research facility of WIS in Israel in a 300 kWth solar reactor, at temperatures ranging from 1000°C to 1200°C. The process gave an energy conversion efficiency of around 30%. The process can be CO_2 neutral if charcoal is used for reducing ZnO.

8.4.3.6.5   High-Flux Solar-Driven Thermochemical Dissociation of CO₂ and H₂O Using Nonstoichiometric Ceria

Chueh, Haile et al. [197,201–204] noted that low-conversion efficiencies, particularly with CO_2 reduction, as well as utilization of precious materials, have limited the practical generation of solar fuels. By using a solar cavity receiver reactor, they combined the oxygen uptake and released the capacity of cerium oxide and facile catalysis at elevated temperatures to thermochemically dissociate CO_2 and H_2O, yielding CO and H_2, respectively. Stable and rapid generation of fuel was demonstrated over 500 cycles. Solar-to-fuel efficiencies of 0.7%–0.8% were achieved and shown to be largely limited by the system scale and design rather than by chemistry.

The *solar reactor* used in this study [201–204] produced hydrogen gas (H_2) and carbon monoxide (CO) from water (H_2O) and carbon dioxide (CO_2) via a two-step thermochemical cycle and it had a quartz lens that concentrated both infrared and ultraviolet radiation. This concentrated solar energy heated metal oxides within the deeper chamber. At the heart of the solar reactor was a cylindrical lining of cerium oxide or ceria, which has an incredible ability to "exhale" oxygen from its crystalline framework at high temperatures and "inhale" oxygen when cooled down. Thus, during multiple heating–cooling cycles, cerium has the ability to strip water or air of its oxygen molecules. As the CO_2 and H_2O are thermochemically dissociated, carbon monoxide (CO) and hydrogen gas (H_2) are separated out. Hydrogen gas (H_2) can be used to fuel hydrogen fuel cells and carbon monoxide (CO) combined with hydrogen gas and can be used to create synthetic gas, which can be converted to transportation fuel via Fischer–Tropsch synthesis. The reactor can also be used in a "zero CO_2 emissions" cycle: H_2O and CO_2 would be converted to methane and would fuel electricity-producing power plants that generate more CO_2 and H_2O, to keep the process going.

The process uses a solar-driven metal oxide thermochemical cycle, which switches between high-temperature and low-temperature operation. In this process, a metal oxide is stripped of its oxygen (in other words, gaining electrons) using thermal energy and is subsequently reoxidized (losing electrons) with water and/or CO_2 at a lower temperature to produce fuel. The authors indicated that the efficiency of the reactor was uncommonly high for CO_2 splitting, in part, because the process used the whole solar spectrum, and not just particular wavelengths.

8.4.3.7 Solar Gasification Technology

Solar gasification generally deals with upgrading and decarbonization of fossil fuels. Such gasification is often carried out in the presence of steam. Successful solar gasification of carbonaceous materials was first reported in the 1980s in which coal, activated carbon, coke, and coal/biomass mixtures were employed in a fixed bed–windowed reactor. Charcoal, wood, and paper were gasified with steam in a fixed bed reactor. More recently, steam gasification of oil shale and coal, biomass, waste tires and plastics, and coal in a fluidized bed reactor as well as steam gasification of petroleum coke (petcoke) and vacuum residue in fixed, fluidized, and entrained bed reactors were examined [170–184]. In the last type of reactor, dry coke particles, coal–water slurries, and vacuum residues were tested for the steam gasification.

In a conceptual solar gasification process [170–184] using steam, biomass is heated rapidly in a solar furnace to achieve flash pyrolysis at temperatures of about 900°C. Some steam is added to the pyrolyzer to increase the gas yield relative to char. The char constituting about 10%–20% of the biomass by weight is steam gasified with external heating at temperatures of 900°C–1000°C; all of the volatile hydrocarbons are then steam reformed in a solar reformer. Steam for the process is generated from heat recovered from the product gas. The composition of the synthesis gas (syngas) is adjusted to the user's needs utilizing conventional operation involving the water–gas shift reaction and CO_2 stripping. This conceptual process can be modified in a number of different ways depending on specific needs.

A number of gasification experiments were carried out using small quantities of biomass, coal, oil shale, and residual oil with external heat supplied by the sun [2,170–184]. These experiments included cellulose gasification and oil shale [175–184] gasification with carbon recovery approaching nearly 100% at temperature of 950°C and short residence times. A number of different types of solar steam gasification reactors have also been examined in the literature [2,97,175–184]. The reactor configuration examined by Z'Graggen [2,174,175] at ETH (Swiss Federal Institute of Technology) at Zurich consisted of a cylindrical cavity receiver 21 cm in length and 12 cm in inside diameter and contained 5 cm diameter aperture for solar beams. The cavity-type geometry was designed to effectively capture the incident solar radiation, and its apparent absorption is estimated to exceed 0.95. The cavity was made of Inconel 601, lined with Al_2O_3, and insulated with an Al_2O_3/ZrO_2 ceramic foam.

Piatkowski et al. [172], Piatkowski [173], and Piatkowski and Steinfeld [170] used a packed-bed solar steam gasification reactor. This reactor was specially designed for beam-down incident solar radiation as obtained through a Cassegrain optical configuration that made use of a hyperbolic reflector at the top of the solar tower to redirect the sunlight collected by a heliostat field to a receiver located at the ground level. The reactor had two cavities in series. The upper one absorbed the solar radiation and contained a small aperture to gather concentrated solar radiation. The lower cavity contained carbonaceous materials on top of a steam injector. An emitter plate separated the two cavities.

A 3D compound parabolic concentrator was incorporated in the aperture of the reactor further augmenting the incident solar flux before passing it through a quartz window in the upper cavity. The emitter plate acted as a transmitter of the radiation to the lower cavity thus avoiding direct contact between the quartz window and the reactants and products. This setup also provided uniform temperature in the lower cavity and a constant supply of radiant heat through the upper cavity, which can act as energy storage that was needed due to intermittent supply of radiant heat. The reactor can be operated with a wide variety of particle sizes, and as the reaction proceeds, both particle size and the packed-bed reactor volume decrease. The detailed dimensions and operation of this type of reactor are given by Piatkowski and Steinfeld [170]. Piatkowski et al. [172] and Perkins et al. [171] also showed an effective use of such a reactor to produce syngas from coal, biomass, and other carbonaceous feedstock. Z'Graggen [174] and Z'Graggen et al. [175] produced hydrogen from petcoke using solar gasification process. A 500 kWth scale-up of the vortex solar reactor for steam gasification based on the work done during an industrial project *SYNPET* (2003–2009) (an industrial project for developing the solar chemical technology for the steam gasification of heavy crude oil derivatives) was tested at the Plataforma Solar de Almería (PSA), Spain. The reactor has reached the hydrogen production efficiency of 60% working at 1500 K [105–111,170–184].

The solar energy is also used as the heat carrier for pressurized coal gasification process. In this process, finely powdered coal is fed by a specially designed injection system. The oxidizing and fluidizing agent is superheated steam. The heat required for the endothermic gasification reaction is introduced by means of a tubular heat exchanger assembly immersed in the fluidized bed. The technical feasibility of a solar power tower and pressurized gasifier integration has been demonstrated in small pilot plant [170–175, 182–186]. Solar energy has also been used to gasify biomass in different types of reactors [176–181].

8.4.3.8 Solar Reforming

Solar reforming of natural gas, using either steam or CO_2, has been extensively studied in solar-concentrating facilities with small-scale solar reactor [187–194]. This solar-reforming process developed within the EU's project *SOLREF* (solar reforming)

(2004–2009) has been scaled up to power levels of 300–500 kWth and tested at 850°C and 8–10 bars in a solar tower [187–197]. Solar dry reforming of methane (CH_4) with CO_2 in an aerosol solar reactor with residence times around 10 ms and temperatures of approximately 1700°C, methane and CO_2 conversions of 70% and 65%, respectively, were achieved in the absence of catalysts. The high temperatures required for solar reforming effectively limit the nature of solar energy collector. The bulk energy production required for reforming must be carried out on a large scale to compete with fossil fuels and probably requires the tower (central receiver) solar technology. Solar reforming can be carried out using different processes such as direct and indirect, each requiring a different type of reformer configuration [187–197].

8.4.3.8.1 ASTERIX: Solar Steam Reforming of Methane

An earlier joint Spanish–German project, Advanced Steam Reforming of Methane in Heat Exchange (ASTERIX) experiment, examined steam reforming of methane using solar-generated high-temperature process heat [188–190,192,194] using an indirectly heated reformer. The specific objectives of the ASTERIX experiments were to collect and store an amount of solar energy to obtain maximum conversion of methane and to produce consistently high-quality syngas. The experiment used gas-cooled solar tower (GAST) system to produce hot air (up to 0.36 kg/s at 1000°C and 9 bars) to drive separate steam reformer. This air was then fed back into the GAST cycle. The GAST technology program is described in References 2, 97, 116, and 187.

During normal operation, the heating medium, air, is taken from the GAST circuit (receiver) at a temperature of 1000°C over a suitable bench line and fed through the electric heater to the reforming reactor inlet. The process gas mixture is heated by air from 500°C to about 850°C as it passes through the catalyst bed. The endothermic reforming reaction results in the production of hydrogen and carbon monoxide with 3:1 ratio. More details of the ASTERIX experiment are given by Shah [2].

8.4.3.8.2 Weizmann Institute Tubular Reformer/Receiver

The Weizmann Institute of Science (WIS) operated a solar central receiver for development of high-temperature technology including the storage and transport of solar energy via methane reforming. WIS had designed facility for testing reformers up to about 480 kW absorbed energy. The facility was designed for either steam or carbon dioxide reforming and it can accommodate reformers that operates between 1 and 18 bars. A cavity receiver containing eight vertical reformer tubes (2 in. schedule 80) and 4.5 m long was designed. The overall dimension of the device was about 5 m high, 4.5 m wide, and 3 m deep. The reactor was designed to produce syngas at 800°C [2,187].

8.4.3.8.3 Soltox Process

In the Soltox process, a parabolic dish is used to concentrate sunlight through a quartz window into an internally insulated aluminum reactor vessel where it is absorbed on a rhodium-coated reticulated ceramic foam absorber. Concentrated organic waste and steam are mixed and made to flow through the hot (>1000°C) catalyst bed, where they react completely in fractions of a second to produce hydrogen, carbon dioxide, carbon monoxide, and halogen acids (which are easily neutralized to simple salts). The extremely good heat and mass transfer within the reactor result in a compact, highly efficient system.

When a vaporized organic waste is mixed with steam and passed through the reactor, highly specific, irreversible, endothermic reforming reactions take place on the catalyst-coated surface of the radiantly heated absorber to quantitatively destroy the waste. For example, TCE reacts with steam to produce hydrogen, carbon monoxide, and hydrogen chloride. Variable absorber thickness and adjustable gas flow rates meant that residence times within the absorber and thus reaction times and destruction efficiency can be controlled [2,97,187,191].

8.4.3.8.4   Open Loop Solar Syngas Production

The applications of open loop solar syngas production include the following:

1. Natural gas reforming for power plants. A number of European countries have imported natural gas via pipelines from north Africa and have reformed this gas to either syngas or hydrogen, increasing its calorific value by about 25% before combustion in gas turbine or fuel cell power plants.
2. Syngas production from municipal, agricultural, and organic industrial waste. In Sunbelt countries, concentrated waste streams can be gasified to syngas with solar energy at potentially acceptable costs and with essentially no emissions to the atmosphere.
3. Soltox-type processing provides an option for environmentally acceptable disposal of a number of toxic organic materials.

Open loop syngas production can also be used for the generation of syngas that is being supplied worldwide for the productions of hydrogen, methanol, ammonia, and oxyalcohols [2,97,187,191].

8.4.3.8.5   Other Solar-Reforming Processes

A number of studies have focused on the production of hydrogen by steam reforming of methane and other hydrocarbons using solar reactor [2,97,110,187–194]. Watanuki et al. [194] examined methane steam reforming using a molten salt membrane-reforming reactor. In this type of the reactor, the reforming reaction takes place in tubular reactors that consist of selective membranes, generally palladium, which separates hydrogen as it is produced. The principal advantages of a solar membrane-reforming process compared to the conventional reforming process are the following:

1. The reforming is carried out at a lower temperature, 550°C. This means a significant reduction in the energetic consumption. Low-temperature reactors also use less costing materials for the reforming reactor tubes.
2. Hydrogen is obtained with a higher purity due to highly efficient membrane separation process.
3. Methane conversions up to 90% can be reached due to high-hydrogen extraction through the membrane.
4. A big part of $CO–CO_2$ conversion is produced inside the reactor itself.

5. Emissions are reduced by about 34%–53% due to the use of concentrated solar energy to obtain the process heat.

The aforementioned described process can be easily adapted to solar gasification, but for this case, heavy hydrocarbons are used as feedstock. These are transformed into cleaner fuels for a combined cycle or in the process that can produce hydrogen. SOLREF process [187–194] and its various options for solar reforming of natural gas by steam are described by Muller [188].

8.4.3.9 Solar Cracking

This process thermally decomposes natural gas in a high-temperature solar chemical reactor, yielding two valuable products: a H_2-rich gas and a marketable high-value nanomaterial called "carbon black" (CB). Thus, both H_2 and CB are produced by renewable energy, resulting in potential emission reduction of 14 kg of CO_2 and energy saving of 277 MJ/kg H_2.

This decomposition process was successfully tested in a novel 10 kWth tubular solar reactor at a temperature range of 1430°C–1800°C at the solar furnace of CNRS PROMES (French national R&D laboratory on solar concentrating systems. It is CNRS lab closely associated with University of Perpignan) in Odeillo, France. A high dissociation of 98% was achieved with a 90% H_2 yield [97]. The process was scaled up to 50 kWth under SOLHYCARB (high temperature solar chemical reactor for Co–production of hydrogen and carbon black from natural gas cracking) EU project during 2006–2010. Solar cracking was also tested in a tubular aerosol reactor at National Renewable Energy Laboratory in Golden, Colorado, (a part of Regional Department of Energy Laboratory) where methane conversion of up to 90% was obtained. Also, methane cracking by heterogeneous carbon catalyzed reaction in a vortex reactor was carried out at ETH Zurich, Switzerland [97].

8.4.4   HYBRID SOLAR ENERGY PROJECTS SUPPORTED BY ARPA-E FOCUS

Recently, the U.S. Department of Energy supported 12 hybrid solar energy projects through Advanced research projects Agency–Energy of the Department of Energy (ARPA-E) program [205]. The Full-Spectrum Optimized Conversion and Utilization of Sunlight program is aimed at developing new hybrid solar energy converters and hybrid energy storage systems that can deliver low-cost, high-efficiency solar energy on demand. Some of these projects are briefly described in the following text:

1. Sharp Labs of America will develop a hybrid solar converter that could enable utilities to provide on demand heat and solar electricity.
2. Microlink devices based in Niles, Illinois, will develop high-efficiency solar cells that can operate at temperatures above 750°F and can extract the most energy possible from sunlight when integrated with hybrid solar converters.
3. Arizona State University will develop a solar cell that can operate at temperatures above 450°C using semiconducting materials used in today's light-emitting diode industry.

4. Massachusetts Institute of Technology will develop a hybrid solar converter that integrates thermal absorber and a solar cell into a layered stack. The design will allow focused sunlight to heat fluid piped through layers of optically transparent thermal insulation.
5. The University of Tulsa will develop a hybrid solar converter that captures nonvisible wavelengths of light to heat a fluid containing light-absorbing nanoparticles that are too small to be seen by the naked eye.

These and other project will significantly improve our ability to make use of electricity and heat from sunlight in a very effective manner [186].

8.4.5 SOLAR TWO-POWER TOWER HYBRID SYSTEM

Solar Two—a demonstration power tower located in the Mojave Desert—can generate about 10 MW of electricity [195,196]. In this central receiver system, thousands of sun-tracking mirrors called heliostats reflect sunlight onto the receiver. Molten salt at 554°F (290°C) is pumped from a cold storage tank through the receiver where it is heated to about 1050°F (565°C). The heated salt then moves on to the hot storage tank. When power is needed from the plant, the hot salt is pumped to a generator that produces steam. The steam activates a turbine/generator system that creates electricity. From the steam generator, the salt is returned to the cold storage tank, where it is stored and can be eventually reheated in the receiver. By using thermal storage, power tower plants can potentially operate for 65% of the year without the need for a backup fuel source. Without energy storage, solar technologies like this are limited to annual capacity factors near 25%. The power tower's ability to operate for extended periods of time on stored solar energy separates it from other renewable energy technologies.

8.5 FUTURE OF HYBRID RENEWABLE ENERGY SYSTEMS

The wind and solar energy are omnipresent, freely available, and environmental friendly. The wind energy systems may not be technically viable at all sites because of low wind speeds and being more unpredictable than solar energy. The combined utilization of these renewable energy sources are therefore becoming increasingly attractive and are being widely used as alternative of oil-produced energy. Economic aspects of these renewable energy technologies are sufficiently promising to include them for rising power generation capability in developing countries. A renewable hybrid energy system consists of two or more energy sources, a power conditioning equipment, a controller, and an optional energy storage system.

These hybrid energy systems are becoming popular in remote area power generation applications due to advancements in renewable energy technologies and substantial rise in prices of petroleum products.

Research and development efforts in solar, wind, and other renewable energy technologies are required to be continued for improving their performance, establishing techniques for accurately predicting their output, and reliably integrating them with other conventional backup sources like diesel and grid.

Reaching the nonelectrified rural population is currently not possible through the extension of the grid, since the connection is neither economically feasible nor encouraged by the main actors. Further, the increases in oil prices and the unbearable impacts of this energy source on the users and on the environment are slowly removing conventional energy solutions, such as fuel genset–based systems, from the rural development agendas.

The infrastructure investments in rural areas have to be approached with *cost-competitive*, *reliable*, and *efficient tools* in order to provide a sustainable access to electricity and to stimulate development. Renewable energy sources are currently one of the most, if not the only, suitable option to supply electricity in fragmented areas or at certain distances from the grid.

Hybrid systems have proven to be the best option to deliver "high-quality" community energy services to rural areas at the lowest economic cost and with maximum social and environmental benefits. Indeed, by choosing renewable energy, developing countries can stabilize their CO_2 emissions while increasing consumption through economic growth.

Unlike wind energy, solar energy has been applied for nonelectric applications such as productions of solar fuels and its use in district and industrial heating. The use of solar energy for the production of hydrogen has a very bright future and continues to advance across the world. Hydrogen as a future source of energy as well as a method for energy storage for power applications has enormous potential.

As enhanced geothermal system expands, the use of geothermal energy for both electric and nonelectric applications will expand. The new enhanced geothermal systems will allow the access of high-temperature heat that can help both electric and nonelectric applications. Future extraction of geothermal energy can also be integrated to the existing underground oil and gas pipeline infrastructure, thereby saving costs for creating new infrastructure for geothermal energy.

Although, the uses of geothermal, solar, and wind energy have very bright future, most pragmatic approach to increase their penetrations in the energy market is to use hybrid energy systems that can provide scale, stability, and competitive price for their implementations. The use of distributed hybrid renewable energy systems is also the best strategy to provide electricity in the remote and poor areas.

REFERENCES

1. Lee, S. and Shah, Y. *Biofuels and Bioenergy: Technologies and Processes*, CRC Press, Taylor & Francis Group, New York (September 2012).
2. Shah, Y. *Water for Energy and Fuel Production*, CRC Press, Taylor & Francis Group, New York (May 2014).
3. IEA. *Deploying Renewables: Principles for Effective Policies*, OECD/IEA Publications, Paris, France (2008).
4. IEA. *Empowering Variable Renewables: Options for Flexible Electricity Systems*, OECD/IEA, Paris, France (2008).
5. IEA. *Energy Policies of IEA Countries: Sweden*, OECD/IEA Publications, Paris, France (2008).
6. IEA. *World Energy Outlook 2009*, OECD/IEA Publications, Paris, France (2009).

7. IEA. *Energy Balances of Non-OECD Countries*, OECD/IEA Publications, Paris, France (2010).
8. IEA. *Energy Technology Perspectives 2010*, OECD/IEA Publications, Paris, France (2010).
9. "Wind hybrid power systems," *Wikipedia*, https://en.wikipedia.org/wiki/Wind_hybrid_power_systems (last modified on May, 12, 2015).
10. Papaefthymiou, S., Karamanou, E., Papathanassiou, S. and Papadopoulos, M. A wind-hydro-pumped storage station leading to high RES penetration in the autonomous island system of Ikaria. IEEE transactions on sustainable energy, 1 (3), 163–172 (2010).
11. Gonzalez, G., Deha Muela, R., Santos, L. and Gonzalez, A., Stochastic joint optimization of wind generation and pumped-storage units in an electricity market. IEEE transactions on power systems 23 (2), 460–468 (May 2008).
12. Hau, E. *Wind Turbines: Fundamentals, Technologies, Application, Economics*, Birkhäuser, pp. 568, 569 (2006). Springer, New York. (2013).
13. Iqbal, T. *Feasibility Study of Pumped Hydro Energy Storage for Ramea Wind-Diesel Hybrid Power System*. The Harris Center, Memorial University of Newfoundland, Newfoundland, Canada.
14. LaRoche, S. and LeBeau, T. Lower Brule Sioux tribe Wind-pump storage feasibility study project. Report by Innovation Investment LLC, Ross, CA (2007). Accessed April 17, 2011.
15. Development and commercializing of assets including wind, solar, biomass and clean energy fuels. Report by WHL energy Ltd., Perth, Western Australia, Australia. (accessed April 17, 2010).
16. Baring-Gould, I. and Corbus, D., Status of wind-diesel application in Arctic climates, NREL/CP–50042401, Presented at Arctic Energy Summit Technology Conference, Anchorage, Alaska (October 15–18, 2007) Golden CO (December 2007).
17. Prince Edward Island wind-hydrogen village Renew, ND. Department of Environment, Energy and Forestry, Prince Edward Island, Canada. (accessed October 30, 2007).
18. First Danish hydrogen energy plant is operational. Announcement by Palo Alto, CA and Holby, Denmark through renewable energy/access.com. (accessed June 8, 2007).
19. Williams M. North Dakota has first wind to hydrogen plant in nation-Renew ND. (accessed July 18, 2007).
20. Renewable Energy Commercialization in Australia—Wind projects—Advanced high-penetration wind-diesel power system (2012).
21. Madrigal, A. Bottled wind could be as constant as coal. Wired Science, March 9, 2010. (accessed July 15, 2011).
22. Gardner, J. and Haynes, T. Overview of compressed air energy storage. A report by Sustainability Research, Boise State University, Office of Research, Policy and Campus Sustainability (ER-07-001), Boise, ID. (accessed July 15, 2011).
23. Mousa, K., Alzubi, H., and Diabat, A. Design of a hybrid solar-wind power plant using optimization. A Report from Masdar Institute of Science and Technology, Abu Dhabi, UAE (2011).
24. Arjun, A., Athul, S., Mohamed, A., Neethu, R., and Anith, K. Micro-hybrid power systems—A feasibility study. *JOCET*, 7, 27–32 (2013). doi:10.7763/JOCET.2013. VI.7.
25. Shivrath, Y., Narayana, P., Thirumalasetty, S., and Laxmi Narsaiah, E. Design and integration of wind-solar hybrid energy system for drip irrigation pumping application. *Int. J. Mod. Eng. Res.*, 2(4), 2947–2950 (2012).
26. PV-wind turbine-diesel generator hybrid system for remote island electrification. A Publication by LEONICS (2012).
27. Hybrid wind-tidal turbine to be installed off Japanese Coast. Renewable Energy World editors comments (July 12, 2013).

28. Jaramillo, O., Borja, M., and Huacuz, J. Using hydropower to complement wind energy: A hybrid system to provide firm power. *Renew. Energy*, 29, 1887–1909 (2004).

29. Rozali, N., Alwi, S., Manan, Z., Klemes, J., and Hassan, M. Optimization of pumped-hydro storage system for hybrid power system using power pinch analysis. *Chem. Eng. Trans.*, 35, 85 (2013).

30. Bludszuweit, H., Fandos, J., Dominguez, J., Llombart, A., and Sanz, J. Simulation of a hybrid system wind turbine-battery-ultracapacitor. A Report from Department of Electrical Engineering, University of Zaragoza, Zaragoza, Spain (2005).

31. Danvest hybrid wind diesel concept introduction. A Publication by Danvest Group, Hellerup, Denmark (January 2013).

32. Kornbluth, K., Hinokuma, R., Johnson, E., and McCaffrey, Z. Optimizing wind energy for a small hybrid wind/diesel grid in the Galapagos islands. A Report from Davis Energy Efficiency Center, University of California, Davis, CA (2009).

33. Bica, D., Dumitru, C., Gligor, A., and Duka, A. Isolated hybrid solar-wind-hydro renewable energy systems. A Report from Electrical Engineering Department, "Petru Maior" University of Tg, Mureg, Romania (2010).

34. Sirasani, K. and Kamdi, S. Solar wind hydro hybrid energy system simulation, Personal communication (2011).

35. Dhass, A. and Harikrishnan, S. Cost effective hybrid energy system employing solar-wind-biomass resources for rural electrification. *Int. J. Renew. Energy Res.*, 3(1), 222 (2013).

36. Elistratov, V. Hybrid system of renewable energy sources with hydro accumulation. A Report from Department of Renewable Energy Sources and Hydro Power Engineering, Saint-Petersburg State Polytechnic University, Saint Petersburg, Russia (2009).

37. Hybrid renewable energy system. The Azimuth Project (2012).

38. Wind-diesel hybrid systems. A Report from Renewable Energy Research Laboratory, Wind Energy Center (2012).

39. Stillerand, C. and Michalski, J. Utilisation of excess wind power for hydrogen production in Northern Germany. In *18th World Hydrogen Energy Conference 2010 (WHEC'10)* Essen, Germany (2010).

40. Schellschmidt, R., Sanner, B., Pester, S. and Schulz, R. Geothermal energy use in Germany. In *Proceedings of the World Geothermal Congress 2010*, International Geothermal Association, Bali, Indonesia (April 25–30, 2010).

41. World's first hybrid solar-geothermal power plant in Nevada, also geothermal energy/stillwater hybrid power plant. Reports from Clean Technica website (May 18, 2012).

42. *Enhanced Geothermal Systems*, Sustainable Campus, Cornell University Publication, Ithaca, NY (2013).

43. "Enhanced geothermal systems," *Wikipedia*, https://en.wikipedia.org/wiki/Enhanced_geothermal_system (last accessed June 14, 2015).

44. Hackel, S. *Hybrid Geothermal Systems: Less is More*, Energy Center of Wisconsin, Free on-demand webcast Madison, Wisconsin (2008) http://www.ecw.org/hybrid.

45. Hackel, S. Optimization of hybrid geothermal heat pump systems. A Report from University of Wisconsin, Solar Energy Laboratory, Madison, WI, also presented at *Ninth International IEA Heat Pump Conference*, Zurich, Switzerland (2008).

46. Singh, J. and Foster, G. Advantages of using the hybrid geothermal option. A Report from Conectiv Solutions, Vineland, NJ (2000).

47. Lukawski, M., Vilaetis, K., Gkogka, L., Beckerts, K., Anderson, B., and Tester, J. A proposed hybrid geothermal-natural gas-biomass energy system for Cornell University. Technical and economic assessment of retrofitting a low temperature geothermal district heating system and heat cascading solutions. In *Proceedings of the 38th Workshop on Geothermal Reservoir Engineering*, Stanford University, Stanford, CA (February 11–13, 2013).

48. Alrobaei, H. Hybrid geothermal/solar energy technology for power generation. A Report from Higher Institute of Engineering, Houn, Libya (2006).

49. Residential HVAC. A Report from Building Tech services; also see company website (2012).

50. Boszormenyi, L. and Boszormenyi, G. Hybrid energy technologies for an efficient geothermal heat utilization. A Report from Department of Civil Engineering, University of Kosice, Kosice, SK (2003).

51. Kuyumcu, O., Solaroglu, U., Akar, S., and Serin, O. Hybrid geothermal and solar thermal power plant case study: Gumuskoy GEPP. *GHC Bull.*, 19 (January 2013).

52. Morrison, D. Simulation and optimization of hybrid geothermal energy systems using solar thermal collectors as a supplemental heat source. A Report from Dayton University, Dayton, OH (2012).

53. Nizetic, S., Ninic, N., and Klarin, B. Analysis and feasibility of implementing solar chimney power plants in the Mediterranean region. *Energy*, 8, 886 (2008). doi:10.1016/j.energy.2008.05.012.

54. Dhahri, A. and Omri, A. A review of solar chimney power generation technology. *Int. J. Eng. Adv. Technol.*, 2(3), 1 (February 2013). ISSN: 2249-8958.

55. Review of technologies for hybrid uses of geothermal and additional RES. *GeoSEE*, South East Europe Transitional Cooperation Program, Co-founded by European Union and coordinated by UM (2012).

56. Bombarda, P., Astolfi, M., Gaia, M., Macchi, E., Pietra, C., and Romano, M. Geothermal hybrid plants—Hybrid geothermal biomass plant, hybrid geothermal solar plant. Risorse geotermiche di media e bassa temperatura in Italia. Potenziale, Prospettive di mercato, Azioni. Sessione POSTER (2010). http://www.unionegeotermica.it/pdfiles/geothermexpo-2010/Poster/Bombarda-et-al-Geothermal-hybrid-plants.pdf (accessed 2011).

57. Borsukiewicz-Gozdur, A. Dual-fluid-hybrid power plant co-powered by low-temperature geothermal water. *Geothermics*, 39(2), 170–176 (2010).

58. *Geothermal Electricity and Combined Heat & Power*. EGEC (2009).

59. Hung, T. Waste heat recovery of organic Rankine cycle using dry fluids *J. of Energy Conversion and Management, Elsevier, New York.*, vol. 42(5), pp. 539–553 (2001).

60. Zolkowski, J. Waste heat recovery. *Energy Eng.*, 106(5), 63–74 (2009).

61. Hung, T., Shai, T., and Wang, S. A review of organic Rankine cycles (Rocs) for the recovery of low-grade waste heat. *Energy*, 22(7), 661–667 (1997).

62. Bell, L. Cooling, heating, generating power, and recovering waste heat with thermoelectric systems. *Science*, 321(5895), 1457–1461 (2008).

63. Mica, H. *An Introduction to the Kalian Cycle*, American Society of Mechanical Engineers, New York (1996).

64. Gripsack, S. Integration of Kalian cycle in a combined heat and power plant, a case study. *Appl. Therm. Eng.*, 29(14), 2843–2848 (2009).

65. Karthäuser, J. and Öström, T. Method for conversion of low-temperature heat to electricity and cooling, and system therefore. WO Patent 2,012,128,715 (2012).

66. Wei, D., Lu, X., Lu, Z., and Gu, J. Dynamic modeling and simulation of an Organic Rankine Cycle (ORC). *Appl. Therm. Eng.*, 28, 1216–1224 (2008).

67. Geothermal power generation. Mitsubishi Heavy Industries, https://www.mhi-global.com/discover/earth/technology/geothermal.html. (accessed April 2010).

68. NREL (National Renewable Energy Laboratory). *Concentrating Solar Power Commercial Application Study: Reducing Water Consumption of Concentrating Solar Power Electricity Generation*. U.S. Department of Energy, Washington, DC (2008).

69. Thurston, C. Enel moves hybrid renewables forward again with world's first biomass and geothermal project. A publication in renewable energy world.com. (accessed December 14, 2014).

70. Staub, V. and Fogleman, S. Utilization of geothermal energy in a hybrid wood waste geothermal power plant—Engineering, Geothermal potential of the cascade mountain range: Exploration and development, Geothermal Resource Council, Davis, CA, pp. 39–43 (1981).

71. Prayon. Dossier de presse—Sulfine, Prayon, Engis, Belgium. RWE/Tullis Russell. *RWE Press Information* (2010). www.rwe.com/web/cms/mediablob/en/512218/data/250908/1/rwe-npower-renewables/blob.pdf. Accessed March 10, 2011.

72. IEA. *Technology Roadmap: Concentrating Solar Power*, OECD/IEA Publications, Paris, France (2010).

73. Licari, J. and Ottesen, H. *Hybrid Geothermal and Fuel Cell System*, Office for Technology Commercialization, US Patent 7334406, University of Minnesota, Minneapolis, MN (2014).

74. Bendaikha, W. and Larbi, S. Hybrid fuel cell and geothermal resources for air conditioning using an absorption chiller in Algeria. *Energy Proc.*, 28, 190–197 (2012).

75. Byun, J., Jeong, D., Choi, Y., and Shin, J. Analysis of fuel cell driven ground source heat pump systems in community buildings. *Energies*, 6, 2428–2445 (2013).

76. Buscheck, T., Chen, M., Sun, Y., Hao, Y., and Elliot, T. Two stage integrated geothermal CO_2 storage reservoirs: An approach for sustainable energy production, CO_2 sequestration security and reduced environmental risk. LLNL-TR-526952, Lawrence Livermore National Laboratory, Livermore, CA (February 3, 2012).

77. Buscheck, T., Elliot, T., Celia, M., Chen, M., Sun, Y., Hao, Y., Lu, C., Wolery, T., and Aines, R. Integrated geothermal-CO_2 reservoir systems: Reducing carbon intensity through sustainable energy production and secure CO_2 storage. *Energy Proc.*, 37, 6587–6594 (2013).

78. Buscheck, T.A. Active management of integrated geothermal-CO_2-storage reservoirs in sedimentary formations: An approach to improve energy recovery and mitigate risk. Proposal in response to DE_FOA-0000336: Energy Production with Innovative Methods of Geothermal Heat Recovery, LLNL-PROP-463758 (2010).

79. Buscheck, T.A., Elliot, T.R., Celia, M.A., Chen, M., Hao, Y., Lu, C., and Sun, Y. Integrated, geothermal-CO_2 storage reservoirs: Adaptable, multi-stage, sustainable, energy-recovery strategies that reduce carbon intensity and environmental risk. In *Proceedings for the Geothermal Resources Council 36th Annual Meeting*, Reno, NV (2012).

80. Bourcier, W.L., Wolery, T.J., Wolfe, T., Haussmann, C., Buscheck, T.A., and Aines, R.D. A preliminary cost and engineering estimate for desalinating produced formation water associated with carbon dioxide capture and storage. *Int. J. Greenhouse Gas Control*, 5, 1319–1328 (2011).

81. Buscheck, T.A., Sun, Y., Chen, M., Hao, Y., Wolery, T.J., Bourcier, W.L., Court, B., Celia, M.A., Friedmann, S.J., and Aines, R.D. Active CO_2 reservoir management for carbon storage: Analysis of operational strategies to relive pressure buildup and improve injectivity. *Int. J. Greenhouse Gas Control*, 6, 230–245 (2012).

82. Buscheck, T.A., Sun, Y., Hao, Y., Wolery, T.J., Bourcier, W.L., Tompson, A.F.B., Jones, E.D., Friedmann, S.J., and Aines, R.D. Combining brine extraction, desalination, and residual-brine reinjection with CO_2 storage in saline formations: Implications for pressure management, capacity, and risk mitigation. *Energy Proc.*, 4, 4283–4290 (2011).

83. Buscheck, T.A., Elliot, T.R., Celia, M.A., Chen, M., Sun, Y., Hao, Y., Lu, C., Wolery, T.J., and Aines, R.D. Integrated geothermal-CO_2 reservoir systems: Reducing carbon intensity through sustainable energy production and secure CO_2 storage. In *Proceedings of the International Conference on Greenhouse Gas Technologies (GHGT-11)*, November 18–22, Kyoto, Japan (2012).

84. Buscheck, T.A., Chen, M., Lu, C., Sun, Y., Hao, Y., Celia, M.A., Elliot, T.R., Choi, H., and Bielicki, J.M. Analysis of operational strategies for utilizing CO_2 for geothermal

energy production. In *Proceedings of the 38th Workshop on Geothermal Reservoir Engineering*, February 11–13, Stanford University, Palo Alto, CA (2013).

85. Elliot, T.R., Buscheck, T.A., and Celia, M.A. Active CO_2 reservoir management for sustainable geothermal energy extraction and reduced leakage. *Greenhouse Gases: Sci. Technol.*, 1, 1–16 (2013). doi:1002/ghg.

86. Field, B., Bachu, S., Innovates, A., Bunch, M., Funnel, R., Holloway, S., Richardson, R. et al., Interaction of CO_2 storage with subsurface resources. IEAGHG Report 2013-08, Paris, France (April 2013).

87. Buscheck, T., Bielicki, J., Randolph, J., Chen, M., Hao, Y., Edmunds, T., Adams, B., and Sun, Y. Multi-fluid geothermal energy systems in stratigraphic reservoirs: Using brine, N_2, and CO_2 for dispatchable renewable power generation and bulk energy storage. In *Proceedings of the 39th Workshop on Geothermal Reservoir Engineering*, February 24–26, Stanford University, Stanford, CA, SGP-TR-202 (2014).

88. Buscheck, T.A., Chen, M., Hao, Y., Bielicki, J.M., Randolph, J.B., Sun, Y., and Choi, H. Multi-fluid geothermal energy production and storage in stratigraphic reservoirs. In *Proceedings for the Geothermal Resources Council 37th Annual Meeting*, September 29–October 3, Las Vegas, NV (2013).

89. Buscheck, T.A. Systems and methods for multi-fluid geothermal energy systems. U.S. Patent Application Filed (2014).

90. Echo Solar hybrid systems for existing homes-frequently ask questions. SunEdison copyright (September, 2013).

91. Kho, J., contributor Solar heat and power systems on the horizon, an announcement by Renewable Energy world. com (June 17, 2010).

92. Lewis, M. Two companies two different "hybrid solar" methods for creating power. Industrial Market Trends website (August 1, 2012).

93. Richard, M. Cogenra's "hybrid" solar system captures 80% of the sun's energy to generate electricity and hot water. *Energy/Renewable Energy* website (November 2010).

94. Margeta, J. and Glasnovic, Z. Solar energy, water storage and sustainable electric energy production. *Int. J. Energy Sci.*, 2(4), 119–126 (2012).

95. Romero, R. How to use solar power for electricity and water heating. A website publication in azcentral, The Arizona Republic. (accessed July 30, 2013).

96. Zweibel, K., Mason, J., and Fthenakis, V. A solar grand plan. *Sci. Am.*, 298(1), 64–73 (2008).

97. Meier, A. and Sattler, C. Solar fuels from concentrated sunlight, *Solar Power and Chemical Energy Systems* (*Solar PACES*), IEA Report (August 2009).

98. Taylan, O. and Berberoglu, H. *Fuel Production Using Concentrated Solar Energy*, Chapter 2, INTECH, Princeton, New Jersey. pp. 33–67 (2013).

99. Klausner, J., Petrasch, J., Mei, R., Hahn, D., Mehdizadeh, A., and Auyeung, N. Solar fuel: Pathway to a sustainable energy future. In *Florida Energy Summit*, University of Florida, Gainesville, FL (August 17, 2012).

100. "Concentrated solar power," *Wikipedia*, https://en.wikipedia.org/wiki/Concentrated_solar_power (last modified July 9, 2015).

101. Bockris, J., Gutmann, F., and Craven, W. On the splitting of water, In *Hydrogen Energy Progress IV*, Veziroglu, T., Van Vorst, W., and Kelley, J. (Eds.). International Association for Hydrogen Energy, Miami, FL, p. 1475 (1982).

102. "Water splitting," *Wikipedia*, https://en.wikipedia.org/wiki/Water_splitting (last modified June 30, 2015).

103. Steinfeld, A. and Meier, A. *Solar Fuels and Materials*, Elsevier, Amsterdam, the Netherlands, pp. 623–637 (2004).

104. Becker, M. and Bohmer, M. (Eds.). *Proceedings of the Final Presentation GAST, the Gas Cooled Solar Tower Technology Program*, Lahnstein, FRG, May 30–31, 1988, Springer-Verlag, Berlin, Germany (1989).

105. Smitkova, M., Janicek, F., and Riccardi, J. Analysis of the selected processes for hydrogen production. *WSEAS Trans. Environ. Dev.*, 11(4), 1026–1035 (2008).
106. Funk, J.E. and Reinstrom, R.M. Energy requirements in the production of hydrogen from water. *I&EC Proc. Res. Dev.*, 5(3), 336–342 (July 1966).
107. Funk, J.E. Thermodynamics of multi-step water decomposition processes. In *Proceedings of the Symposium on Non-Fossil Chemical Fuels (ACS 18), 163rd National Meeting*, Boston, MA, pp. 79–87 (April 1972).
108. Abe, I. Hydrogen productions from water. *Energy Carriers Convers. Syst.*, 1, 1–3 (2001).
109. Ris, T. and Hagen, E. Hydrogen production—Gaps and priorities. IEA Hydrogen Implementing Agreement (HIA), pp. 1–111 (2005).
110. Suarez, M., Blanco-Marigorta, A., and Peria-Quintana, J. Review on hydrogen production technologies from solar energy. In *International Conference on Renewable Energies and Power Quality (ICREPQ'11)*, April 13–15, La Palmas de Gran Canaria, Spain (2011).
111. Nowotny, J., Sorrell, C., Sheppard, L., and Bak, T. Solar-hydrogen: Environmentally safe fuel for the future. *Int. J. Hydrogen Energy*, 30(5), 521–544 (2005).
112. Naterer, G.F. Economics and synergies of electrolytic and thermochemical methods of environmentally benign hydrogen production. In *Proceedings of WHEC*, May 16–21, Essen, Germany (2010).
113. Laoun, B. Thermodynamics aspect of high pressure hydrogen production by water electrolysis. *Revue des Energies Renouvelables*, 10(3), 435–444 (2007).
114. Funk, J.E. Thermochemical and electrolytic production of hydrogen. In *Introduction to Hydrogen Energy*, Vezeroglu, T.N. (Ed.). International Association for Hydrogen Energy, Miami, Florida. pp. 19–49 (September 1975).
115. Onda, K., Kyakuno, T., Hattori, K., and Ito, K. Prediction of production power for high pressure hydrogen by high pressure water electrolysis. *J. Power Sources*, 132, 64–70 (2004).
116. McKellar, M., Harvego, E., and Gandrik, A. System evaluation and economic analysis of a HTGR powered high temperature electrolysis production plant. In *Proceedings of HTR*, Paper 093, October 18–20, Prague, Czech Republic (2010).
117. Laguna-Barcero, M., Skinner, S., and Kilner, J. Performance of solid oxide electrolysis cells based on Scandia stabilized zirconia, Personal communication. UKERC Publication, Imperial College, London, U.K. (2012).
118. Herring, J., O'Brien, J., Stoots, C., Lessing, P., and Anderson, R. High temperature solid oxide elctrolyzer system. Hydrogen, Fuel Cells, and Infrastructure Technologies, DOE Progress Report, pp. 1–5 (2003).
119. Herring, J., O'Brien, J., Stoots, C., Lessing, P., and Hartvigsen, J. High temperature electrolysis for hydrogen production. In Paper presented by *Idaho National Laboratory for Materials Innovations in an Emerging Hydrogen Economy*, Hilton Oceanfront, Cocoa Beach, FL (February 26, 2008).
120. Scaife, D. Oxide semiconductors in photoelectrochemical conversion of solar energy. *Solar Energy*, 25(1), 41–54 (1980).
121. Shyu, R., Weng, F., and Ho, C. Manufacturing of a micro probe using supersonic aided electrolysis process, DTIP of MEMS and MOEMS (April 9–11, 2008).
122. Bockris, J. and Murphy, O. One-unit photo-activated electrolyzer. U.S. Patent No. US4790916A (December 13, 1988).
123. Sato, J., Saito, N., Nishiyama, H., and Inoue, Y. Photocatalytic water decomposition by RuO_2 loaded antimonates, $M_2Sb_2O_7$ (M= Ca, Sr), $CaSb_2O_6$ and $NaSbO_3$ with d10 configuration. *J. Photochem. Photobiol. A: Chem.*, 148, 85–89 (2002).
124. Zou, Z., Ye, J., Sayama, K., and Arakawa, H. Photocatalytic hydrogen and oxygen formation under visible light irradiation with M-doped $InTaO_4$ (M = Mn, Fe, Co, Ni or Cu) photocatalysts. *J. Photochem. Photobiol. A: Chem.*, 148, 65–69 (2002).

125. Lee, J. Designer transgenic, algae for photobiological production of hydrogen from water, Chapter 20. In *Advanced Biofuels and Bioproducts*, Lee, J. (Ed.). Springer Science, New York, pp. 371–404 (2012).

126. Correa, N. Applicability of photocatalytic water splitting, 121,124, hydrophobic nano-structures for the prevention of biofouling, personal communication. Stevens Institute of Technology, Hoboken, NJ (2009).

127. Sato, J., Kobayashi, H., Saito, N., Nishiyama, H., and Inoue, Y. Photocatalytic activities for water decomposition of RuO_2 loaded $AInO_2$ (A = Li, Na) with d10 configuration. *J. Photochem. Photobiol. A: Chem.*, 158, 139–144 (2003).

128. Zou, Z. and Arakawa, H. Direct water splitting into H_2 and O_2 under visible light irradiation with a new series of mixed oxide semiconductor photocatalysts. *J. Photochem. Photobiol. A: Chem.*, 158, 145–162 (2003).

129. Harda, H., Hosoki, C., and Kudo, A. Overall water splitting by sonophotocatalytic reaction: The role of powdered photocatalyst and an attempt to decompose water using a visible light sensitive photocatalyst. *J. Photochem. Photobiol. A: Chem.*, 141, 219–224 (2001).

130. Abe, R., Hara, K., Sayama, K., Domen, K., and Arakawa, H. Steady hydrogen evolution from water on Eosin Y-fixed TiO_2 photocatalyst using a silane coupling reagent under visible light irradiation. *J. Photochem. Photobiol. A: Chem.*, 137, 63–69 (2000).

131. Fujishima, A., Rao, T., and Tryk, D. Titanium dioxide photocatalysis. *J. Photochem. Photobiol. C: Photochem. Rev.*, 1, 1–21 (2000).

132. Navarro, R., Sanchez-Sanchez, M., Alvarez-Galvan, M., del Valle, F., and Fierro, J. Hydrogen production from renewable sources: Biomass and photocatalytic opportunities. *Energy Environ. Sci.*, 2, 35–54 (2009).

133. Navarro, R., Yerga, J., and Fierro, G. Photo catalytic decomposition of water. *ChemSusChem*, 2, 471–485 (2009).

134. Prince, R. and Kheshgi, H. The photobiological production of hydrogen: Potential efficiency and effectiveness as a renewable fuel. *Crit. Rev. Microbiol.*, 31, 19–31 (2005).

135. Ghysels, B. and Franck, F. Hydrogen photo-evolution upon S deprivation stepwise: An illustration of microalgal photosynthetic and metabolic flexibility and a step stone for future biological methods of renewable H_2 production. *Photosynth. Rev.*, 106, 145–154 (2010).

136. Mells, A., Zhang, L., Forestier, M., Ghirardi, M., and Seibert, M. Sustained photobiological hydrogen gas production upon reversible inactivation of oxygen evolution in the green alga *Chlamydomonas reinhardtii*. *Plant Physiol.*, 122, 127–135 (2000).

137. Ghirardi, M., Zhang, L., Lee, J., Flynn, T., Seibert, M., Greenbaum, E., and Melis, A. Microalgae: A green source of renewable H_2. *Trend Biotechnol.*, 18, 506–511 (2000).

138. Nguyen, A., Thomas-Hall, S., Malnoe, A., Timmins, M., Mussgnug, J., Rupprecht, J., Krause, O., Hankamer, B., and Schenk, P. Transcriptome for photobiological hydrogen production induced by sulfur deprivation in the green alga *Chlamydomonas reinhardtii*. *Eukaryot. Cell*, 7(11), 1965–1979 (2008).

139. Timmins, M., Thomas-Hall, S., Darling, A., Zhang, E., Hankamer, B., Marx, U., and Schenk, P. Phylogenetic and molecular analysis of hydrogen producing green algae. *J. Exp. Bot.*, 60(6), 1691–1702 (2009).

140. Berberoglu, H. and Pilon, L. Maximizing the solar to H_2 energy conversion efficiency of outdoor photobioreactors using mixed cultures. *Int. J. Hydrogen Energy*, 35(2), 500–510 (2010).

141. Lee, J. Designer proton channel transgenic algae for photobiological hydrogen production. PCT International Patent Application Publication Number: WO2007/134340 A2 (2007).

142. Lee, J. Designer proton channel transgenic algae for photobiological hydrogen production. U.S. Patent 7,932,437 B2 (2011).
143. Lee, J. Switchable photosystem-II designer algae for photobiological hydrogen production. U.S. Patent 7,642,405 B2 (2010).
144. Inoue, M., Uehara, R., Hasegawa, N., Gokon, N., Kaneko, H., and Tamaura, Y. Solar hydrogen generation with $H_2O/ZnO/MnFe_2O_4$ system. *ISES, Solar World Congress*, pp. 1723–1729 (2001).
145. Brown, L.C., Besenbrauch, G.E., Schultz, K.R., Showalter, S.K., Marshall, A.C., Pickard, P.S., and Funk, J.F. High efficiency generation of hydrogen fuels using thermochemical cycles and nuclear power. In *Spring National Meeting of AIChE*, Nuclear engineering session THa01 139—Hydrogen production and nuclear power, New Orleans, LA (March 11–15, 2002).
146. Rosen, M.A. Developments in the production of hydrogen by thermochemical water decomposition. *Int. J. Energy Environ. Eng.*, 2(2), 1–20 (Spring 2011).
147. Steinfeld, A. and Palumbo, R. Solar thermochemical process technology. In *Encyclopedia of Physical Science and Technology*, Vol. 15, Meyers, R.A. (Ed.). Academic Press, pp. 237–256 (2001).
148. Funk, J.E., Conger, W.L., and Cariy, R.H. Evaluation of multi-step thermochemical processes for the production of hydrogen from water. In *THEME Conference Proceedings*, Miami Beach, FL, p. S11 (March 1974).
149. Chao, R.E. Thermochemical water decomposition processes. *I&EC Proc. Res. Dev.*, 13(2), 94–101 (June 1974).
150. Funk, J. Thermochemical processes for the production of hydrogen from water, Personal communication (2011).
151. Bamberger, C. and Richardson, D. Thermochemical decomposition of water based on reactions of chromium and strontium compounds, personal communication. OakRidge National Laboratory, Oak Ridge, TN (2011).
152. Weimer, A. II.1.2 Fundamentals of a solar-thermal hydrogen production process using a metal oxide based thermochemical water splitting cycle. DOE Contract No. DE-FC36-05G015044, Annual Progress Report (2006).
153. Roeb, M. and Sattler, C. Hycycles-materials and components of hydrogen production by sulfur based thermochemical cycles (FPS-Energy-212470). German Aerospace Center (DLR)—Solar Research Report (2010).
154. Law, V.J., Prindle, J.C., and Gonzales, R.B. Level 3 analysis of the copper sulfate cycle for the thermochemical splitting of water for hydrogen production. Contract No. 6F-003762, Argonne National Laboratory (August 2006).
155. Roeb, M., Sattler, C., Kluser, R., Monnerie, N., Oliveira, L., Konstandopoulos, A., Agrafiotis, C., Zaspalis, V., Nalbandian, L., Steel, A., and Stobbe, P. Solar hydrogen production by a two step cycle based on mixed iron oxides. *J. Solar Energy Eng.*, 128, 125–133 (May 2006).
156. T-Raissi, A., Huang, C., and Muradov, N. Hydrogen production via solar thermochemical water splitting. NASA/CR-2009-215441, Report of research for period March 2004 to February 2008 (2009).
157. Steinfeld, A. and Palumbo, R. Solar thermochemical process technology. *Encycl. Phys. Sci. Technol.*, 15(1), 237–256 (2001).
158. Steinfeld, A. Solar hydrogen production via a two-step water-splitting thermochemical cycle based on Zn/ZnO redox reactions. *Int. J. Hydrogen Energy*, 27(6), 611–619 (2002).
159. Funke, H.H., Diaz, H., Liang, X., Carney, C.S., Weimer, A.W., and Li, P. Hydrogen generation by hydrolysis of zinc powder aerosol. *Int. J. Hydrogen Energy*, 33(4), 1127–1134 (2008).

160. Abanades, S., Charvin, P., Lemont, F., and Flamant, G. Novel two-step SnO_2/SnO water-splitting cycle for solar thermochemical production of hydrogen. *Int. J. Hydrogen Energy*, 33(21), 6021–6030 (2008).

161. Meier, A. and Steinfeld, A. Solar thermochemical production of fuels. *Adv. Sci. Technol.*, 74(1), 303–312 (2011).

162. Steinfeld, A. Solar thermochemical production of hydrogen—A review. *Solar Energy*, 78(5), 603–615 (2005).

163. Aoki, A., Ohtake, H., Shimizu, T., Kitayama, Y., and Kodama, T. Reactive metal-oxide redox system for a two-step thermochemical conversion of coal and water to CO and H_2. *Energy*, 25(3), 201–218 (2000).

164. Kromer, M., Roth, K., Takata, R., and Chin, P. *Support for Cost Analyses on Solar-Driven High Temperature Thermochemical Water-Splitting Cycles*, TIAX, LLC2011, Lexington, MA (February 22, 2011). Report No. DE-DT0000951.

165. Abanades, S., Charvin, P., Flamant, G., and Neveu, P. Screening of water-splitting thermochemical cycles potentially attractive for hydrogen production by concentrated solar energy. *Energy*, 31, 2805–2822 (2006).

166. Osinga, T., Olalde, G., and Steinfeld, A. Solar carbothermal reduction of ZnO: Shrinking packed-bed reactor modeling and experimental validation. *Indust. Eng. Chem. Res.*, 43(25), 7981–7988 (2004).

167. Osinga, T., Frommherz, U., Steinfeld, A., and Wieckert, C. Experimental investigation of the solar carbothermic reduction of ZnO using a two-cavity solar reactor. *J. Solar Energy Eng.*, 126(1), 633–637 (2004).

168. Epstein, M., Olalde, G., Santén, S., Steinfeld, A., and Wieckert, C. Towards the industrial solar carbothermal production of zinc. *J. Solar Energy Eng.*, 130(1), 014505 (2008).

169. Wieckert, C., Frommherz, U., Kräupl, S., Guillot, E.,Olalde, G., Epstein, M., Santén, S. et al. A 300 kW solar chemical pilot plant for the carbothermic production of zinc. *J. Solar Energy Eng.*, 129(2), 190–196 (2007).

170. Piatkowski, N. and Steinfeld, A. Solar driven coal gasification in a thermally irradiated packed-bed reactor. *Energy Fuels*, 22, 2043–2052 (2008).

171. Perkins, C.M., Woodruff, B., Andrews, L., Lichty, P., Lancaster, B., Bringham, C., and Weimer, A. Synthesis gas production by rapid solar thermal gasification of corn stover. Report of Midwest Research Institute, Kansas, under Contract No. DE-AC36-99GO10337, Department of Energy (2009).

172. Piatkowski, N., Wieckert, C., and Steinfeld, A. Experimental investigation of a packed bed solar reactor for the steam gasification of carbonaceous feedstocks. *Fuel Process. Technol.*, 90, 360–366 (2009).

173. Piatkowski, N. Solar driven steam gasification of carbonaceous feedstocks, Personal communication. Solar Fuels and Materials Division, ETH (2012).

174. Z'Graggen, A. Solar gasification of carbonaceous materials—Reactor design, modeling and experimentation, Doctorate thesis. Dissertation, ETH No. 17741, ETH (2008).

175. Z'Graggen, A., Haueter, P., Trommer, D., Romero, M., De Jesus, J., and Steinfeld, A. Hydrogen production by steam-gasification of petroleum coke using concentrated solar power—II. Reactor design, testing, and modeling. *Int. J. Hydrogen Energy*, 31(6), 797–811 (2006).

176. Gordillo, E. and Belghit, A. A bubbling fluidized bed solar reactor model of biomass char high temperature steam-only gasification. *Fuel Process. Technol.*, 92(3), 314–321 (2010).

177. Gordillo, E. and Belghit, A. A downdraft high temperature steam-only solar gasifier of biomass char: A modelling study. *Biomass Bioenergy*, 35(5), 2034–2043(2011).

178. Hathaway, B.J., Davidson, J.H., and Kittelson, D.B. Solar gasification of biomass: Kinetics of pyrolysis and steam gasification in molten salt. *J. Solar Energy Eng.*, 133(2), 021011 (2011).

179. Lichty, P., Perkins, C., Woodruff, B., Bingham, C., and Weimer, A. Rapid high temperature solar thermal biomass gasification in a prototype cavity reactor. *J. Solar Energy Eng.*, 132(1), 011012 (2010).

180. Flechsenhar, M. and Sasse, C. Solar gasification of biomass using oil shale and coal as candidate materials. *Energy*, 20(8), 803–810 (1995).

181. Weimer, A., Perkins, C., Mejic, D., Lichty, P., and inventors. WO Patent WO/2008/027980, as signee. Rapid solar-thermal conversion of biomass to syngas. U.S. Patent WO2008027980 (June 3, 2008).

182. Zedtwitz, P. and Steinfeld, A. The solar thermal gasification of coal—Energy conversion efficiency and CO_2 mitigation potential. *Energy*, 28(5), 441–456 (2003).

183. Groeneveld, M.J. The co-current moving bed gasifier, PhD thesis. Twente University of Technology, Enschede, the Netherlands (1980).

184. Phillips, J. Different types of gasifiers and their integration with gas turbines. A Report from EPRI, CA (2003).

185. Najjar, Y. Efficient use of energy by utilizing gas turbine combined systems. *Appl. Therm. Eng.*, 21, 407–438 (2001).

186. Duonghong, D., Borgarello, E., and Gratzel, M. *J. Am. Chem. Soc.*, 103, 4085 (1981).

187. Spiewak, I., Tyner, C., and Langnickel, U. Applications of solar reforming technology. SANDIA Report SAND93-1959, UC-237, Sandia National Lab., Albuquerque, NM (November 1993).

188. Muller, W.D. Solar reforming of methane utilizing solar heat. In *Solar Thermal Energy for Chemical Processes, German Studies on Technology and Applications.* Solar Thermal Utilization, Vol. 3, Baker, M. (Ed.). Springer-Verlag, Berlin, Germany, pp. 1–179 (1987).

189. Klein, H.H., Karni, J., and Rubin, R. Dry methane reforming without a metal catalyst in a directly irradiated solar particle reactor. *J. Solar Energy Eng.*, 131(2), 021001 (2009).

190. Sattler, C. and Raeder, C. SOLREF—Solar steam reforming of methane rich gas for synthesis gas production. Final Activity Report, DLR, Cologne, Germany (June 2010).

191. Ogden, J.M. Review of small stationary reformers for hydrogen production. A Report for IEA, Agreement on the production and utilization of hydrogen, Task 16, hydrogen from carbon containing materials, IEA/H2/TR-02/002 (2002).

192. Padban, N. and Becher, V. Clean hydrogen rich synthesis gas. Literature and state of art review (Re: Methane Steam Reforming), Report No. CHRISGAS, WP11 D89 (October 2005).

193. Olsson, D. Comparison of reforming process between different types of biogas reforming reactors. Project Report, 2008 MVK 160 Heat and Mass Transport, Lund, Sweden (May 2008).

194. Watanuki, K., Nakajima, H., Hasegawa, N., Kaneko, H., and Tamaura, Y. Methane-steam reforming by molten salt membrane reactor using concentrated solar thermal energy. In *Proceedings of WHEC*, Vol. 16, Lyon, France, pp. 13–16 (June 2006).

195. Solar power towers. Sun lab snapshot, a brochure prepared by Energy Efficiency and Renewable Energy Cleaning house, A NREL under contact no. DOE/GO-100097-406 Golden CO (April 1998).

196. Tyner, C., Sutherland, J., and Gould, W. Solar Two: A molten salt power tower demonstration. A Report for Department of Energy under Contract Number DE-AC04-94AL8500 (1995).

197. Lede, J., Lapicque, F., and Villermaux, J. *Int. J. Hydrogen Energy*, 8, 675 (1983).

198. Kappauf, T. and Fletcher, E.A. Hydrogen and sulfur from hydrogen sulfide—VI. Solar thermolysis. *Energy*, 14(8), 443–449 (1989).

199. Zaman, J. and Chakma, A. Production of hydrogen and sulfur from hydrogen sulfide. *Fuel Process. Technol.*, 41(2), 159–198 (1995).

200. Harvey, W.S., Davidson, J.H., and Fletcher, E.A. Thermolysis of hydrogen sulfide in the temperature range 1350–1600 K. *Indust. Eng. Chem. Res.*, 37(6), 2323–2332 (1998).

201. Chueh, W., Falter, C., Abbott, M., Scipio, D., Furler P., Haile, S., and Steinfeld, A. High-flux solar-driven thermochemical dissociation of CO_2 and H_2O using nonstoichiometric Ceria. *Science*, 330, 1797 (December 24, 2010).

202. Chueh, W. and Haile, S. A thermochemical study of ceria: Exploiting an old material for new modes of energy conversion and CO_2 mitigation. *Philos. Trans. R. Soc. Lond. A*, 368, 3269 (2010).

203. Chueh, W. and Haile, S. Ceria as thermochemical reaction medium for selectively generating syngas or methane from H_2O and CO_2. *ChemSusChem*, 2, 735 (2009).

204. Haile, S. and Chueh, W. Thermo chemical synthesis of fuels for strong thermal energy, U.S. Patent Application 20,090,107,044 (2009).

205. 12 Hybrid solar energy projects get $30 million from ARPA-E focus. A Report from Clean Technica, announcement by Department of Energy, Washington, DC (2013).

9 Energy and Fuel Systems Integration by Gas, Heat, and Electricity Grids

9.1 INTRODUCTION

In Chapters 2 through 6, we examined raw fuel integration for various types of energy and fuel generation systems. These chapters showed that a combination of fuels such as coal-biomass, coal-waste, coal-heavy oil, biomass-waste, biomass-water, etc., enhances our ability to create more sustainable, environment-friendly, and efficient thermochemical processes such as combustion, gasification, liquefaction, and pyrolysis. Similarly, a combination of different types of waste can enhance the productivity of anaerobic digestion (codigestion) process. While the concept of multifuel processing is well advanced for the cocombustion, coprocessing, and codigestion processes, their successes have also pushed faster developments and implementation of other multifuel thermochemical processes. These chapters clearly demonstrated that a multifuel strategy will

1. Reduce the use of fossil fuels (or a particular type of fossil fuel), thus preserving our fossil fuel resources for a longer period
2. Reduce the emissions of CO_2 in the environment
3. Allow penetrations of renewable fuels like biomass and waste for energy generation on a larger, economical and sustainable scale
4. Alleviate the need for new and independent costly infrastructure development for energy and fuel generations from biomass and waste
5. Provide maximum flexibility for usages of all available raw materials (fossil, biomass, and waste) for generations of energy and fuels
6. Allow the maximum use of synergies among various feedstock for optimum product yield
7. Develop new energy and fuel infrastructure that is capable of handling diverse and mixed feedstock. Such infrastructure will have a longer sustainability

Chapters 7 and 8 dealt with the second level of energy and fuel systems integration. These chapters dealt with hybrid energy systems for nuclear and renewable energy sources, which can be used for both electric and nonelectric applications through combined use of heat and power (CHP) (cogeneration). In this case, systems integration can occur at the feed side (e.g., nuclear–renewable, renewable–renewable, or renewable–fossil) or at the product side (heat and power). This hybrid energy system strategy will allow deeper penetrations of nuclear and renewable sources of energy like solar, wind, geothermal, and water for power generation. The approach also makes the use of CHP for numerous nonelectrical applications. The approach allows economical use of nuclear and renewable (solar and wind) energy at smaller scales for their specific nonelectrical applications or for the generation of hydrogen. In summary, the Chapters 7 and 8 point out the importance of hybrid energy systems strategy for (1) more penetrations of carbon-free energy sources in power production, (2) reduction in the usage of fossil energy sources and thereby reduction of CO_2 emissions, and (3) better use of nuclear and renewable energy sources for nonelectric applications both at smaller and larger scales.

While multifuel and hybrid energy systems (with or without cogeneration) are very valuable for coupling two or more fuel feedstock or energy sources or coupling heat and power (cogeneration), varying supply and demands of energy (electricity) and fuel (particularly gas) are best managed by holistic systems integration of various fuel and energy generation and storage systems with multiple users [1–4]. This concept has been commonly called "grid" (or network) approach. Natural gas grid and electric power grid (or now the smart grid [SG]) are the ultimate examples of an integrated approach to manage energy and fuel supply and demand from multiple sources to multiple users from dispersed community. The heat grids have been used in local processes, and they are becoming more important as many energy systems (nuclear-, solar-, and gas-powered plants) make use of combined heat and power (CHP) generations (cogeneration) for both electrical and nonelectrical applications. Heat is also a good source of energy storage (ES).

While the concept of networking can be applied to other sources of fuels such as coal, biomass, waste, etc., these fuel systems are much more complex and diverse in their generations, transport, storage, and usage mechanisms. These complex and diverse mechanisms do not allow a coherent and easily manageable implementation of the unified "grid" concept. The transportation network for oil closely follows that for gas and will not be independently examined here. Furthermore, the versatility of gas and electricity will make them most sought after sources of energy and fuel in the future. Their usages are also quite intertwined, rapidly expanding, and closely connected to heat network. This chapter, therefore, illustrates the importance of the third level of integration (i.e., holistic integration) through the workings of gas, heat, and electrical grids. While gas can be relatively easily stored, the storage of electricity is an important issue. This chapter also evaluates various methods of ES and their importance in the management of gas, heat, and electrical grids. Both gas and electrical grids are becoming increasingly more complex and smarter, and they will be the methods of handling constantly changing supply and demand of the gas and electricity.

Networks (or grids) are an important feature of modern economies. Reduced to their most basic level, they consist of patterns of interconnections between different things. In the case of gas, heat, and electricity, networks are essential for transmitting valuable fuel, heat, or electricity from the point of supply to the point of end use. The most basic is a unidirectional link in which gas, heat, and electricity are moved from one point to another point. More advanced gas, heat, and electricity networks involve hub and spoke structures and more transmission options, more dynamic demand response, more use of information, and communication technologies, and some even are designed to permit flows in reverse so that they can be bidirectional.

An important feature of gas, heat, and electric networks is how their values change as they become denser and more complex. Networks tend to become increasingly valuable with size due to reinforcing positive feedback that occurs as more entities join the network. These characteristics facilitate the development of adjacent networks, uncovering hidden opportunities to create value as new links are established. Networks create opportunities to bring buyers and sellers together. Network growth contributes to lowering costs and creating greater flexibility in the system, which in turn fosters further growth. Capturing these economies, or what economists refer to as economies of scope and scale, serves to lower transaction costs and create more efficient markets. As the networks (or grids) become more complex, they become more flexible, but their control becomes more complex.

There can, of course, be downsides to grids. For example, grids, or portions of grids, can become less valuable if they become congested with excessive use. Parts of a grid can become stranded, or overbuilt, as resources deplete or markets shift, creating cost recovery challenges. This is where appropriate pricing, investment, optionality, and storage become critical to optimizing the scale of the grid to match supply and demand. These are some of the reasons for the creations of SGs for gas, heat, and electricity. These issues also suggest the importance of microgrids and satellite operations [4,108,137,138,144,145], which can regionally control supply and demand of electricity, heat, and gas and can be isolated from overall grid if they are malfunctioned.

The world needs both resilient and sustainable energy and fuel systems. These require efficient, manageable, and responsive fuel, heat, and electric power grids. A resilient energy system optimizes the strengths of individual energy networks, and it harmonizes their operations to the benefit of consumers. An energy system that relies heavily on any one energy network will be more vulnerable. Furthermore, networks that are optimized to be more efficient with lower losses and reduced transaction cost are more sustainable [284,285]. Gas grid is well-positioned to work with other energy sources like heat and power (and also oil) to improve overall energy system resilience. Furthermore, as gas grid becomes more intelligent with new digital and software technologies, its interactions with heat and electricity will expand.

Gas networks, which are often underground, in contrast to heat and power grids (which are above the ground), can often provide stable service during severe weather events. In this way, gas can contribute broadly to economic resiliency by providing diversification, redundancy, and backup systems different from those provided by electricity. Heat interacts with both gas and electricity. The workings of gas and electrical grids clearly complement each other [4,23]. Gas, heat, and electricity support new concepts for grids-within-a-grid and multisource microgrids (or satellite operations) to increase resilience and reaction time in the face of disruptions. In addition, distributed power systems built around gas, heat, and electrical networks can provide fast power recovery for public utilities such as hospitals, waterworks, and government agencies, which is very important in all different types of disasters. This is why gas, power, heat, and liquid fuel networks are best when optimized to support each other. This creates another layer of systems integration [4,23].

Technology innovation can contribute to gas, heat, and electricity network development. There are innovations taking place upstream, which focus on supply side of gas, heat, and electricity. This includes technologies to improve efficiency and lower costs of large-scale remote gas projects such as deep water gas and sour gas. New technologies are also developed for large-scale electricity generation and ES projects. More innovative technologies are developed to produce biogas from biomass and waste and unconventional gas like shale gas and coal bed methane [8–12]. Newer processes are also adopted to improve combustion efficiency for power generation [4]. Methane, hydrogen, and syngas can be produced from a variety of sources such as steam reforming; supercritical water gasification; high-temperature electrolysis; thermochemical water dissociation using nuclear and solar energy; anaerobic digestion of aqueous waste; and recovery of gas hydrates, all leading to more gas with different compositions and different raw materials for power generation and storage [6,7,13]. More efforts are being made to improve the thermal efficiency of various synthetic fuel operations. New nuclear reactor technologies and innovative solar cells for capturing solar energy more efficiently are constantly developed. These can affect the level of heat and its downstream electrical or nonelectrical applications. All these new developments will require constant upgrading of gas, heat, and electricity grids. The use and control of CHP systems and the connections to the transportation sector will also require adjustments in the operation of gas, heat, and electrical grids.

Technologies will also be developed to secure, integrate, and optimize gas, heat, and electrical systems from an efficiency as well as environmental and safety standpoint. The new technologies that help integrate and transform small-scale liquid natural gas (LNG) and compressed natural gas (CNG) systems into "virtual pipelines" will be important to the rapid development of new gas markets like the transportation sector [23]. This is analogous to the development of grid-integrated hybrid and electrical cars [297–318], which can create new markets for electricity, or the development of better heat grid for various nonelectrical applications.

There are technological innovations focused on expanding the range of applications for gas, heat, and electricity. The flexibility of natural gas, heat, and electricity makes them valuable complements to other generation sources like renewables and a variety of nonelectrical applications. Across each of these areas, advances in the industrial internet, including data analytics, machine-to-machine, and machine-to-human interfaces, will have powerful effects on the productivity of gas, heat, and electricity grids [4].

ES also needs to be integrated in network-based energy and fuel systems [227,248–296]. It can also provide an important contribution to the development and emergence of the smart grid concept at all voltage levels. ES can become an integrated part of CHP, solar thermal and wind energy systems to facilitate their integration in the grid. ES can facilitate nonelectrical applications of nuclear and solar energy. Thus, the developments of grids and ES systems are closely intertwined [227,248–296]. The demands for peak increase in fuel, heat, and power can also be resolved when ES is available at different levels of the gas, heat, and electrical systems: centralized ES as a reserve; decentralized storage in the form of demand management and demand response systems [263–267].

Thus, in the developments of holistic integrated fuel and energy systems, three parameters are important; (1) basic characteristics of smart gas, heat, or electrical grids, and their interactions with each other and other fuel grids, (2) methods for ES, and (3) coupling of smart gas, heat, or electrical grids with ES for the best management of supply and demand. This chapter briefly analyzes these topics.

9.2 NATURAL GAS GRID

Gas is fundamentally a network industry. Networks are necessary to move gas from the source of production to the customers. As more gas producers and customers are attached to the network, the network itself becomes more valuable. There are also the interconnections and interdependencies between gas networks and other networks, both physical and digital [1–27].

For years, the concept of gas grid has been applied to manage supply and demand of gas, which has been controlled by elaborate gas pipeline infrastructure and storage systems (often called a "gas network"). Approximately 89% of gas supply and demand is managed by gas pipeline infrastructure [4,23], which has a global network of over 1.4 million km [4,23]. About 70% of these pipelines are regional lines that support collection and distribution within their domestic markets. The balance consists of large-diameter continental scale pipelines found in places like the United States, Canada, Russia, and increasingly China [4,23]. Pressure is essential for moving gas through the pipeline system. This is achieved through compressors that are located at regular intervals along the pipeline. With compression, gas moves about 25 mi/h (40 km/h) through long-distance pipelines. Compared to the transmission of electricity, this is a slow process and does not allow very fast response to the long-distance demand.

The second mode of gas transport is the liquid natural gas (LNG) network, which constitutes 10% of global gas trade [4,23]. This sea-based network is made possible by the invention of LNG and large special-purpose transport ships that ply long distance in the oceans linking exporters and customers. Finally, 1% of gas can also be transported using existing road and rail infrastructure. This is made possible through compression technologies like compressed natural gas (CNG) and small-scale LNG. A gallon of CNG has about 25% of the energy content of a gallon of diesel fuel, and LNG has 60% of the volumetric energy density of diesel fuel. Both CNG and LNG are cost-competitive, and their market demands are rapidly expanding. This mode of transportation is often called "virtual" pipelines. This method allows an efficient use of CNG and LNG in transportation sector and in distributed power plants [4,23].

Natural gas grid, thus, has three subcomponents: pipelines, LNG sea network, and virtual pipelines. The cost of transportation through pipelines is much lower than that for sea-based LNG transportation and the transportation by virtual pipelines.

Economics plays a large role in determining which solution is selected to move gas from sources of supply to ultimate demand [4,23].

Over the past four decades, we have developed a sophisticated transportation and storage system for gas to meet our variable needs of gas for residential, commercial, industrial and transportation purposes. Within United States, we have several hundred thousand miles of underground pipeline infrastructure for gas at different levels of distribution, storage, and user systems and these are closely watched, controlled, and managed to support variable demands. While United States always had an elaborate transportation and storage infrastructure for gas, over the past 5 years, unconventional gas activity has thrust the nation into an unexpected position. The United States is now the largest natural gas producer, at 65 billion cubic feet (Bcf)/day, in the world [23]. At the same time, unconventional activity is spurring the growth of natural gas liquids (NGLs) production of over 500,000 barrels of oil equivalent (boe) per day in United States since 2008. Over the past 5 years, daily oil production of United States has also increased by 1.2 million barrels [23].

In order to accommodate the growth in U.S. natural gas, NGLs, and crude oil, the past 2 years have witnessed a rapid growth in direct capital investment toward oil and gas infrastructure assets. It is estimated that the capital spending in oil and gas midstream and downstream infrastructure has increased by 60%, from $56.3 billion in 2010 to $89.6 billion in 2013 [23]. This increase in capital spending is reshaping the U.S. oil and gas infrastructure landscape. A large portion of the projects being developed during this sustained infrastructure investment period will shift the United States toward being energy trade balanced and add key infrastructure segments that enable growing energy production in the Midcontinent region to reach demand centers on the U.S. Gulf Coast and Eastern seaboard.

This newly developing infrastructure will also handle different types of unconventional gas (i.e., shale gas, coal bed methane, etc., with composition very different from that of natural gas, biogas, propane, hydrogen) [5–14,17,20–23,27] so that in future, transportation and storage infrastructure of gas is capable of handling a variety of gas to meet increasing and variable demands without significantly affecting the pricing structure. The successful penetrations of unconventional gas as well as gas coming from renewable energy sources like biomass and waste will depend on the success of our transportation and storage infrastructure. A summary of our existing natural gas infrastructure is illustrated in Table 9.1.

Besides unconventional gas and gas coming from biomass and waste (biogas), future energy landscape will significantly change with the production and use of hydrogen in a variety of ways. Hydrogen is the most abundant and noncarbon energy source that can be used either as fuel or source of electrical energy. In order for hydrogen to penetrate energy market in a significant way, we will need to (1) use existing natural gas infrastructure to the extent possible or (2) develop its own transportation and storage infrastructure. At present, we can introduce 5%–15% by volume of hydrogen in natural gas pipeline. An independent development of hydrogen transportation and storage infrastructure is in the research stage [5–9,13–15]. Since hydrogen is also a good source for electric ES, hydrogen grid or natural gas grid containing hydrogen will be intertwined with the electrical grid.

TABLE 9.1

Summary of Natural Gas Transportation and Storage Infrastructure in the United States [4–29]

Topic	Number
Gas pipeline systems	210
Underground NG storage facilities	400
Hubs or interconnections market centers	24
Delivery points	11,000
Receiving points	5,000
Interconnection points	1,400
Compressor stations to maintain pipeline NG pressure	1,400
Interstate and intrastate transmission pipelines (miles)	305,000
LNG import facilities	8
LNG peaking facilities	100
Import/export locations for pipeline NG transport	49

Today, modern fossil fuel–based power plants (and especially natural gas combined cycles) are becoming more and more flexible. Their ramping-up speed in response to rapid changes in demand is increasing. They can provide reliable and flexible backup power for renewable energy sources like wind and solar. In the short term, however, ES is also needed to fill the gap between the ramping-down time of wind and solar and the ramping-up time of these backup plants. The challenge is to increase existing storage capacities and increase efficiencies [4,23].

Gas storage is closely linked to electricity storage. Some seasonal variations in demand for gas and electricity can be covered by natural gas storage. Gas is an important fuel for electricity production, and natural gas power plants have a very high efficiency (above 60% for the best available technology), a very high flexibility, and low CO_2 emissions (replacing an old coal-fired power plant by a natural gas–fired power plant reduces the CO_2 emissions per kWh up to 80%) [4,23]. In the future, injection of biogas and hydrogen into the natural gas grid [8–12], and the longer-term commercialization of carbon capture and sequestration technology, will further de-carbonize gas-powered electricity generation. Indeed, the expansion of natural gas–fired power plants, the increased efficiency and the reduced costs of flexible combined-cycle and simple-cycle natural gas turbines combined with the strong and fast-growing interconnection of the grid on a larger scale will continue to help efficiency and security of energy supply and demand needs. These discussions once again indicate that in future, natural gas grid, hydrogen grid, and electrical grid would be strongly intertwined. The development of CHP will make localized heat grids also an integral part of this mix.

9.2.1 GRID INTEGRATIONS

Gas network or grid must be connected to other networks [20,21,24–28]. These connections are based on the needed functionality. New interconnections create more supply diversity and customer options. This also makes grid more flexible. As grid becomes more complex by adding functions of gathering and processing, storage, LNG systems at large scale, LNG or CNG systems at small scale, distribution grids, and the complex control and information and communication software systems needed to monitor, track, and trade natural gas, it also becomes more powerful and smart—a system that is far more than just interconnected pipelines. It becomes a real example of holistic integration of energy and fuel systems.

In the near future, we are likely to see a natural gas grid that is complex and fully integrated both horizontally and vertically. The deepening connection between gas supply and electricity generation is an obvious example of vertical integration. Another example is the potential for gas to become more commonplace in supplying fuel to heavy vehicles. Both LNG and CNG are competitive in this market. The vertical integration of natural gas will also occur with other types of gas such as biogas, shale gas, coal bed methane, propane and hydrogen, whose compositions are considerably different from 98% to 99% pure (methane) natural gas. The use of existing pipelines for these gases will require some pretreatments. These gas supplies, with the help of new technology, will be cheaper and can be produced with lower environmental impact than ever before. Midstream and consumer technologies are also advancing [24–28]. There is a significant opportunity to further optimize and secure the sea- and land-based components of natural gas systems. Digital monitoring and control of pipelines and optimization of CNG/LNG fueling systems are two examples of innovative integrations of existing technology. The next generation of digital systems will employ technologies including: satellite, wireless, cloud storage, and software tools for remote monitoring and control and predictive analytics. This will integrate data collection, processing, reporting, and analytics in smart ways and create a grid similar to the SG for electricity [4,23].

Blending hydrogen into natural gas pipeline networks has also been proposed as a means of delivering pure hydrogen to markets, using separation and purification technologies downstream to extract hydrogen from the natural gas blend close to the point of end use [4–9,13–15]. Hydrogen can be inserted in natural gas pipeline up to the concentration of 5%–15% [23]. Anything higher than that would require its own dedicated transportation infrastructure. Hydrogen storage may also require specially designed systems. As a hydrogen delivery method, blending can defray the cost of building dedicated hydrogen pipelines or other costly delivery infrastructure during the early market development phase. This hydrogen delivery strategy also incurs additional costs, associated with blending and extraction, as well as modifications to existing pipeline integrity management systems, and these must be weighed against alternative means of bringing more sustainable and low-carbon energy to consumers. The insertion of hydrogen in natural gas grid will also enhance the interactions between gas and electricity grids.

Deeper horizontal and vertical integrations will also create the potential for greater competition across networks. Supply and demand structure will become more hybrid and interactive and provide a greater customer flexibility. As the complexity of gas systems increases, they also offer more options. Finally, deeper integration of networks, both vertically and horizontally, will enhance the overall resilience of energy systems, making them more impervious to disruption. In general, as grids become larger, they also become more efficient, more flexible, more complex, and more resilient [4,23].

Natural gas is a versatile fuel. Environmental advantages coupled with high efficiency and flexibility make natural gas a great choice for power generation and many other usages. Unlike coal or nuclear fuel, it can be cost-effective when used in both large- and small-scale applications. Furthermore, as gas networks become available, it can become an alternative to oil [4,23]. The technologies being deployed today are designed to bring flexibility to operations, help consumers capture value, and create security of supply. Effectively mobilizing these demand-side technologies is an important part of achieving the sustainability, resilience, and competitive benefits of natural gas. These technologies are also the paths for the development of more smart and improved gas grid [17–29].

Since transportation of gas is not as rapid as that of electricity, gas grid may need to be managed on a smaller scale to obtain the faster response to the changes in supply and demand. The gas grid can be distributed in satellite operations for better and faster control of supply and demand. This is not to say that various distributed (satellite) gas grids cannot be interconnected for overall better control [4]; the overriding control of supply and demand will be at regional level for gas grid. In a similar context, heat grids will also have to be managed locally, because heat cannot be transported over a long distance due to transmission losses. Thus, effective integrations among gas, heat, and electricity will have to be carried out at local and regional levels.

9.2.2 LEVELS OF GRID APPLICATIONS

The improvement in gas generation technologies at large, medium, and small scales will also improve the effectiveness of gas grid. The latest flexible combined cycle power plants are reaching thermal efficiencies such that almost two-thirds of the energy in the natural gas is converted into electricity; 750 MW gas-driven power plants can start up in fewer than 30 min and increase power output at 100 MW/min [4,23]. When necessary, they can be turned down to 14% of their base load capacity to adjust to changes in system demand or the introduction of intermittent renewables [18–24,27–29].

At a smaller scale (typically 10–120 MW), gas-fired CHP systems using gas turbines can create both electricity and steam. As shown in Chapters 7 and 8, CHP systems can be used for numerous nonelectrical applications such as heating and cooling large commercial buildings, hospitals, airports, or industrial sites, providing industrial process heat, etc. Utilizing CHP (also called cogeneration), these power plants can achieve thermal efficiencies in excess of 80%, often with lower emissions and losses than grid-supplied options [4,18–24,27–29].

CHP operations can be designed to disconnect from the larger grid in the event of a disruption, allowing critical facilities to be operated in "island" mode during natural disasters. CHP projects, however, require customized designs and are difficult to implement in existing buildings. Additionally, coordination between electric grid operators and CHP operators is critical on a number of issues including excess power flowing back to the grid, backup power requirements if the natural gas supply is disrupted, or safety concerns that might result from miscommunication [4,23].

For even smaller applications (0.3–10 MW), natural gas engine can have high efficiencies (up to 45% thermal efficiency) relative to other simple combustion technologies. These engines can be used in CHP configurations to achieve higher efficiency or in other distributed settings [4,18–24,27–29]. They have low emissions relative to their diesel-fired counterparts and can run on a wide range of natural gas qualities (from rich to lean) and a variety of other gases including biogas, landfill gas, coal bed methane, sewage gas, and combustible industrial waste gases. The ability of some gas engines to burn natural gas with higher liquids content (rich gas) is ideal for oil and gas field power generation applications such as drilling and enhanced oil recovery.

Thus, the advances in combined-cycle technology, CHP using gas turbines or gas engines, and others like fuel cells and small biogas systems will continue to expand the role of the gas grid. These improvements will also bring workings of gas grid, localized heat grid, and electricity grid more intertwined. Digital technologies and the industrial internet will also play a large role in the network improvement. Technology will thus play a critical role in supplying, securing, and growing the gas grid of the future [4,23].

9.2.3 ROLE OF GAS STORAGE

The way natural gas grid is designed is somewhat different from the SG for electricity described in Section 9.3. At present, storage of compressed gas underground is the primary technology used to store natural gas [4,23,30,31]. In the natural gas industry, the most rapid consumption of natural gas occurs in winter. However, it is uneconomical to design transcontinental pipelines and natural gas treatment plants to meet peak natural gas demands. Instead, the natural gas is produced and transported at a nearly constant rate throughout the year. A variety of different types of large underground storage systems in different geologies at locations near the customer are used to store the excess natural gas produced during the summer. This practice minimizes the cost of the long-distance natural-gas pipeline system and improves reliability by locating storage facilities near the customer. In the winter, these underground storage facilities provide the natural gas to meet customer demands.

While there are some differences in traditional gas grid and electricity grid, these differences become small as both grids mature and become more complex and interact with each other. As shown in Figure 9.1, INL evaluated two types of hybrid energy systems. In the first traditional hybrid energy system, the primary nuclear or fossil fuel power generation plant is connected to renewable power sources (solar and/or wind), with backup (diesel or gas) energy sources and ES system to generate sustainable power at large scale for an electrical grid. In the more advanced hybrid energy system, the traditional system is connected to synthetic fuel plant from natural gas using cogeneration system. Thus, in advanced hybrid energy system, gas, heat, and electricity grids are connected to simultaneously produced electricity and transportation fuel [32–36]. This example is a clear indication of the advantages of the connections among gas, heat, and electrical grids. Both electricity and transportation fuel can be used for a number of downstream applications. Such an advanced hybrid system allows the use of renewable energy for power generation at large-scale and sustainable scale and improves thermal efficiency of the overall process.

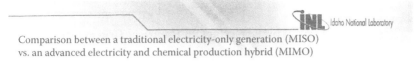

Comparison between a traditional electricity-only generation (MISO)
vs. an advanced electricity and chemical production hybrid (MIMO)

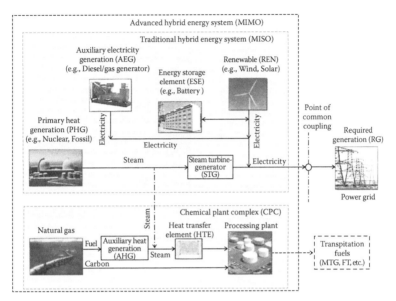

FIGURE 9.1 Interconnections among electricity, heat, and chemical fuel production. Courtesy of Ruth, M. et al. *Nuclear and Renewable Energy Synergies Workshop: Report of Proceedings*, JISEA, prepared under Task No. 6A302003, Technical Report NREL/TP 6A30–52256, Golden Colorado (December 2011); Boardman, R. Advanced energy systems-nuclear-fossil-renewable hybrid systems. A report to Nuclear Energy Agency-committee for technical and economical studies on nuclear energy development and fuel cycle, INL, Idaho Falls, ID (April 4–5, 2013).

9.3 SMART ELECTRICAL GRID

Along with the usage of gas, the consumption of electricity is rapidly expanding. ExxonMobil [37] predicts that by 2040, 40% of our energy consumption will be electrical. The supply and demand of the electricity is managed by smart electrical grid [38–43]. The national electrical grid system is continuously expanding, and it is divided into three major regional parts: eastern interconnection, western interconnection, and Texas interconnection. These larger grids provide efficiency and stability in supply and demand needs. An effort is being made to interconnect these three grids into a single U.S. grid. An *SG*, also called smart electrical/power grid, intelligent grid, future grid, etc. [44–47], is an enhancement of the twentieth century power grid. The traditional power grids are generally used to carry power from a few central generators to a large number of users or customers. In contrast, the SG uses two-way flows of electricity and information to create an automated and distributed advanced energy delivery network. Table 9.2 gives a brief comparison between the old electrical grid and the SG [41].

TABLE 9.2

Brief Comparison between the Old Electrical Grid and the SG

Old Electrical Grid	SG
Electromechanical	Digital
One-way communication	Two-way communication
Centralized generation	Distributed generation
Few sensors	Sensors throughout
Manual monitoring	Self-monitoring
Manual restoration	Self-healing
Failures and blackouts	Adaptive and islanding
Limited control	Pervasive control
Few customer choices	Many customer choices
Not optimized	Always optimized around end objectives
Less automated	Fully automated
Simple but not flexible	Complex but flexible and robust
Takes less advantages of information and communication systems	Takes full advantages of sophisticated information and communication systems

Source: Farhangi, H., *IEEE Power Energy Mag.*, 8(1), 18, 2010.

For the future electrical distribution system, grids will become more active and will accommodate bidirectional power flows and an increasing transmission of information. Some of the electricity generated by large conventional plants will be displaced by the integration of renewable energy sources. An increasing number of PV, biomass, and on-shore wind generators will feed into the medium- and low-voltage grid [32–36,253]. Conventional electricity systems must be transformed in the framework of a market model in which generation is dispatched according to market forces and the grid control center undertakes an overall supervisory role (active power balancing and ancillary services such as voltage control).

The SG is expected to control the demand side as well as the generation side, so that the overall power system can be more efficiently and rationally operated. The SG includes many technologies such as information technology, communications and control technologies [52–54,59–61,65,76,86,105–179], and electrical energy storage (EES) technologies [194–261]. The EES can be used in a number of different ways in the management of SG. This issue is discussed in details in Section 9.5.

By utilizing modern information technologies, the SG is capable of delivering power in more efficient ways and responding to wide-ranging conditions and events [48–51]. Broadly stated, the SG could respond to events that occur anywhere in the grid, such as power generation, transmission, distribution, and consumption, and adopt the corresponding strategies. For instance, once a medium-voltage transformer failure event occurs in the distribution grid, the SG may automatically change the power flow and recover the power delivery service.

Since lowering peak demand and smoothing demand profile reduces overall plant and capital cost requirements, in the peak period, the electric utility can use real-time pricing to convince some users to reduce their power demands, so that the total demand profile full of peaks can be shaped to a nicely smoothed demand profile. More specifically, the SG can be regarded as an electric system that uses information, two-way, cybersecure communication technologies [83,84,86,89], and computational intelligence in an integrated fashion across electricity generation, transmission, substations, distribution, and consumption to achieve a system that is clean, safe, secure, reliable, resilient, efficient, and sustainable.

The aforementioned description covers the entire spectrum of the energy system from the generation to the end points of consumption of the electricity [52,53]. It also presents the ultimate integration of the energy systems. While the ultimate SG is a vision, it is also a loose integration of complementary components, subsystems, functions, and services under the pervasive control of highly intelligent management-and-control systems. Given the vast landscape of the SG research, different researchers may express different visions [47,54,55,126] for the SG due to different focuses and perspectives [56,57]. The development and refinement of the SG is an ongoing task, because new functions [58–61] and more efficiency are constantly added and required.

9.3.1 Benefits and Requirements of Smart Grid

The initial concept of SG started with the idea of advanced metering infrastructure (AMI) with the aim of improving demand-side management [68,128,134,173–176, 186–190] and energy efficiency, and constructing self-healing reliable grid protection against malicious sabotage and natural disasters [85,90–95]. In some way, this concept was similar to the management of natural gas grid. However, new requirements and demands expand the initially perceived scope of SG [96–99,170]. In spite of the constantly changing final role of SG, the anticipated benefits and the basic requirements of it carry the following elements [52–54,59–61,65,68,76,86,105–179]:

1. *Improved power quality, capacity, security, and reliability*: The SG improves power reliability and quality. It also enhances the capacity and efficiency of the existing electric power networks. The smart infrastructure and management system of SG will also present opportunities to improve grid security. In short, SG manages integrated, safe, secure, efficient, and sustainable power supply and demand.
2. *Provide more choices and service expansion capabilities*: It provides more choices for the consumer and enables new markets, products, and services. More recent expansion of SG will be in the management of power requirement of plug in vehicles. The CHP will also provide new market for the SG. SG will also be required to manage new ES options.
3. *Facilitate penetration of renewable sources of power/decarbonize power industry*: By allowing more penetration of renewable energy sources in electrical industry, SG will help to decarbonize the power industry. It will

help the reduction of greenhouse gases. The expanded markets of hybrid and electric cars will also reduce emission of GHG.

4. *Facilitate and automate power distribution and provide self-control maintenance*: The SG is required to accommodate distributed power sources. It will also serve the localized needs of a shopping mall, hospital, university campus, etc., through the use of smart microgrids, which can serve as subsystems of overall SG. It will help in the reduction of oil consumption by reducing the need for inefficient generation during peak usage periods. The SG will automate maintenance and operation of power delivery system.

5. *Optimize facility utilization*: The SG will optimize the facility utilization. It will minimize the construction of backup (peak load) power plants by properly utilizing ES systems. SG will easily and effectively add or remove new or outdated sources of power generation. It will improve resilience to disruption and will allow predictive maintenance and self-healing responses to system disturbances.

These benefits and requirements will make SG development a continuous process as more functions, more demands for efficiency, and more complexities for control systems are constantly added.

9.3.2 Smart Grid Subsystems

From a technical and operational point of view, functions of SG can also be divided into three major subsystems: smart infrastructure and distribution system, smart customer-focused management system; and smart, secure, reliable, and failure protection system [179].

9.3.2.1 *Smart Infrastructure* and Distribution *System*

The operation of SG requires three types of infrastructure: energy, information, and communication [53,54,76,105–107,111,112,114–115,117,119]. The energy infrastructure [120,121] supports advanced electricity generation, delivery, and consumption [124,139–144]. It is mainly focused on the energy supply and demand with focus on renewable energy supply and two ways electrical flow. In cogeneration (CHP) systems, energy infrastructure will provide dynamic and optimum control of heat and power usage. The information infrastructure supports accurate metering, monitoring, and management of the supply and demand. Gathering of all necessary information is very important for subsequent communication and necessary control functions. Finally, communication infrastructure allows efficient communication connectivity and information transmission among systems, devices, and applications in the context of the SG. The dynamic communications among various parts of SG drive the management and control functions.

In an electrical network, power generated must be transmitted and distributed to various customers. The distribution grid is thus an important part of overall SG infrastructure. For the distribution grid, the most important problem is how to deliver power to serve the end users better. However, since many distributed generators will be integrated into the smart distributed grid, this, on one hand, increases the system flexibility for power generation and, on the other hand, makes the power flow control much more complicated. These resulting complications force to investigate and control power distribution and delivery mechanisms in a smarter way [52–54,59–61,65,68,76,86,105–179].

As an example of smart power distribution and delivery mechanism, Takuno et al. [62] proposed two in-home power distribution systems, in which the information is added to the electric power itself and electricity is distributed according to this information. The first one is a circuit switching system based on alternating current (AC) power distribution, and the other is a direct current (DC) power dispatching system via power packets. Many in-home electric devices are driven by DC power and have built-in power conversion circuits to commutate AC input voltage. The packetization of energy requires high-power switching devices. Silicon carbide junction gate field-effect transistors are able to shape electric energy packets [63].

The system proposed by Takuno et al. [62] has the potential as an intelligent power router. Using energy packet, power distribution can be easily regulated by controlling the number of sent packets. This concept can also be further modified to control smart distribution between heat and power in cogeneration mode.

The smart infrastructure system supports two-way flow of electricity and information. In the traditional power grid, the unidirectional flow of electricity transfers electricity from generation to transmission to distribution to the users. In an SG, electricity can also be put back into the grid by users. For example, users may be able to generate electricity using solar panels at homes and put it back into the grid, or electric vehicles may provide power to help balance loads by "peak shaving" (sending power back to the grid when demand is high). The same principle applies if geothermal energy is used to operate heat pump in the house. This backward flow is important in the optimization of supply and demand of the power. This type of dual flow has many other important applications. For example, it can be extremely helpful in a microgrid that has been "islanded" due to power failures [108,116,122,127,136–138,149,164,166,171,172]. The microgrid can function, although at a reduced level, with the help of the energy fed back by the customers. Smart infrastructure should be able to handle various intricacies of two ways flow of electricity. This feature will become even more important when SG is connected to plug in electric or hybrid vehicles [297–318].

9.3.2.2 *Smart* Customer-Focused *Management System*

With demand response and dynamic control of supply and demand, SG is forced to perform many functions simultaneously. Also modern grid provides advanced management and control services. The key reason why SG can revolutionize the holistic and dynamic integration of power supply and demand is the explosion of functionality based on its smart infrastructure and distribution system. With the development of new management applications and services that can leverage the technology and capability upgrades enabled by this advanced infrastructure, the grid will keep becoming "smarter." The smart management system takes advantage of the smart infrastructure and the distribution system to pursue various advanced management objectives that are customer-focused [59–61,76,86,107,150–158,177,178]. The management system should always be customer-focused with the aim to provide more flexible, efficient, and customer-friendly services.

Thus far, most of such objectives are related to energy efficiency improvement, supply and demand balance, emission control, operation cost reduction, and utility maximization [65,129–135]. In future, with vehicle grid integration of plug-in electric vehicles, the SG will have to manage power supply and demand of mobile industry along with the electric usage for stationary purposes. As the usages of gas and CHP applications for nonelectrical purposes expand, the SG management system will become more and more complex [165–169]. Smart management system allows grid to operate in an optimum manner at all time.

9.3.2.3 *Smart,* Personally Secure, Reliable, and Failure *Protection System*

The smart, secure, reliable, and failure protection system is the subsystem in SG that provides advanced grid reliability analysis, failure protection, and security and privacy protection services [52,109,110,113,118,123,159]. As a part of smart infrastructure and power distribution, both management and protection of SG are equally important. The SG must be protected against failure using failure protection mechanisms. The grid must contain mechanisms to protect against cybersecurity attacks (natural or manmade) [145–148]. Smart protection system also does not allow a failure in one part of the grid to spread to other parts [160–163]. This is a part of designing a reliable distribution network. Here the breakup of large SG into various microgrids [108,116,122,127,136–138,149,164,166,171,172] can help. Finally, the SG must preserve the privacy issues of their customers. Thus, SG must provide security and reliability both at individual customer level as well as at the system level.

9.3.3 SMART GRID TECHNOLOGY ASSESSMENT, DEVELOPMENT, STANDARDIZATION, AND OPTIMIZATION

As mentioned in Section 9.3.2, several technologies such as information, communication [74,100,103,104], control, etc., are imbedded in the development of SG. Significant research to improve these technologies has been published in the literature. This includes the critical assessment of the basic concepts and the recommendations of technologies used in SG [64–66] and their standardizations [66,67]. For example, Vasconcelos [68] outlined the potential benefits of smart meters. Brown and Suryanarayanan [69] provided an industry perspective for the smart distribution system and identified areas of research for the improvement of technologies associated with the smart distribution system. Baumeister [70] and Chen [71] assessed SG cybersecurity and privacy issues.

The use of communication networks, wireless communication, and communication architecture for power systems has also been examined in the literature [72–74,87,102,103]. Gungor and Lambert [73] provided a clear analysis of the hybrid network architecture that can provide heterogeneous electric system automation application requirements. Akyol et al. [72] analyzed how, where, and what types of wireless communications are suitable for deployment in the electric power system. Wang et al. [74] provided a survey of the communication architectures in the power systems. They also discussed the network implementation considerations and challenges in the power system settings.

In order to realize the new grid paradigm, NIST provided a conceptual model, which can be used as a reference for the various parts of the electric system where SG standardization work is taking place [67,75,77,101,125]. This conceptual model divides the SG into seven domains: customers, markets, service providers, operations, bulk generation, transmission, and distribution. Each domain encompasses one or more SG *actors (such as end users, operators, service providing organizations, managers, generators of electricity, and distributors)*, including devices, systems, or programs that make decisions and exchange information necessary for performing applications. NIST model is widely used as a standardization of various functions of SG.

The optimization of a SG depends on the ultimate objectives. Different users have different end games in mind. In general, however, the grid optimization is needed because

1. It allows maximum usage of the existing infrastructure, reduces and extends resource usage, and reduces emissions of CO_2 and other pollutants
2. It reduces overall cost and improves reliable delivery of power to the customers
3. By saving costs, it allows investments in new generation, transmission, and distribution facilities

For grid optimization [78], grid efficiency is very important. In 2006, a total of 1638 billion kWh of energy was lost on the U.S. power grid with 655 billion kWh lost in the distribution system alone. To put this in perspective, a 10% improvement in grid efficiency at the distribution level alone would have produced $5.7 billion in savings based on the 2006 national average price of electricity. It would also have saved over 42 million tons of CO_2 emissions [38,43,76].

Numerous studies [51,78,79–82,87,88] have attempted to examine different models for grid optimization. Nygard et al. [82] examined an optimization model for energy reallocation in an SG. They evaluated a self-healing problem, which takes action in near-real-time to reallocate power to minimize disruption using integer linear programming models. Simmhan et al. [88] adopted an informatics approach to demand response optimization in SGs. The informatics approach allowed them to build an intelligent and adaptive grid. Ahat et al. [81] treated the SG as a complex system, locating the problems at local as well as global levels, and solving them with coordinated methods. By studying and analyzing SG, they isolated homogeneous parts with similar behaviors or objectives and applied classical optimization algorithms at different levels with coordination. The method guaranteed the flexibility in terms of system size and allowed its applicability in different scenarios and models. The optimization of each homogeneous subsystem with specific algorithms was achieved. These were then co-ordinated to achieve global optimization.

As shown in Chapters 7 and 8, in future both nuclear and renewable energies will be applied to both electrical and nonelectrical applications through the use of excess heat generated by these power plants. As shown, this will require the management of CHP applications using smart power to heat (and vice versa) switching device. The controlled management of heat will require modifications in the working of SG for the electricity. This will be a part of future development of the SG. In future, smart electrical grid will be closely intertwined with gas and heat grids and an optimization of this holistic integration of multiple grids will also be required.

9.4 METHODS OF ENERGY STORAGE

ES is the storage of some form of energy that can be drawn upon at a later time to perform some useful electrical or nonelectrical operation. In the late 1800s, ES became a dominant factor in economic development with the widespread introduction of electricity and refined chemical fuels, such as gasoline, kerosene, and natural gas. Unlike the storage of fuels like wood (biomass), coal, oil, gas, etc., electricity must be used as it is generated. Electricity is transmitted in a closed circuit, and is difficult to store in large quantity over a sustained period of time. This meant that changes in demand could not be accommodated without either cutting supplies (e.g., blackouts) or arranging for a storage technique.

Currently, many renewable energy technologies such as solar and wind energy cannot be used for base-load power generation as their output is time and location dependent. Solar energy is not available in night, and wind energy cannot be harnessed when wind is not blowing. Batteries and other ES technologies therefore become key enablers for a steady supply of power from these renewable sources.

There are five different methods for energy and fuel storage: chemical, electrochemical, electrical, mechanical, and thermal [180–210]. The power storage is often accomplished using traditional batteries and capacitors. More research is being carried out to apply new field of nanotechnology for the development of better batteries and capacitors. Nanomaterials can significantly increase the capacity and lifetime of batteries and energy and power density of capacitors. Other storage techniques like hydrogen, flywheels, supercapacitors, etc., have also been more and more used, particularly for moderate- and large-scale storage [211–237]. These and many others are further evaluated in this section.

9.4.1 Chemical Energy Storage

Chemical fuels have become the dominant form of ES, both in electrical generation and energy and fuel transportations. Chemical fuels in common use are processed coal, gasoline, diesel fuel, natural gas, shale gas, biogas, liquefied petroleum gas (LPG), propane, butane, ethanol, biodiesel, and hydrogen. All of these chemicals are readily converted to mechanical energy and then to electrical energy using heat engines that are used for electrical power generation. They can also be directly used in vehicles and numerous residential, industrial, and commercial usages.

Liquid hydrocarbon fuels are the dominant forms of ES for use in transportation. Unfortunately, some of these produce greenhouse gases when used to power cars, trucks, trains, ships, and aircraft. Carbon-free energy carriers, such as hydrogen and some forms of ethanol or biodiesel, are being sought in response to concerns about the consequences of greenhouse gas emissions [182,184,188–192,194–197,199,200].

Chemical ES also focuses on hydrogen and synthetic natural gas (SNG), since these could have a significant impact on the storage of electrical energy in large quantities. The main purpose of such a chemical ES system is to use "excess" electricity to produce hydrogen via water electrolysis. Once hydrogen is produced, different ways are available for using it as an energy carrier, either as pure hydrogen or as SNG. Although the overall efficiency of hydrogen and SNG is low compared to storage technologies such as pumped hydroelectricity storage and Li-ion batteries, chemical ES is the only concept, which allows the storage of large amounts of energy, up to the TWh range, and for greater periods of time—even as seasonal storage. Another advantage of hydrogen and SNG is that these can be used in different sectors, such as transport, mobility, heating, and the chemical industry along with their usage in the power industry. Here we further examine the roles of biofuels, hydrogen, and SNG for chemical storage of energy [182,184,188–192,199,205].

9.4.1.1 Biofuels and Cellulosic Waste

Besides fossil fuels, various biofuels such as biodiesel, straight vegetable oil, alcohol fuels, or biomass can be used to replace fossil fuels for ES. Various chemical processes can convert biomass and waste into methane, hydrogen, and liquid biofuels, which can be suitable as replacements for existing fossil fuels [37,238]. The gasification of biomass, cellulosic waste, and other low rank feedstock can provide the necessary gaseous biofuels to produce electricity and heat to backup the intermittent renewable generation. Use of lignocellulose and cellulosic waste to generate biofuels like methanol, ethanol, butanol, and biodiesel for ES is also under significant development. During the last two decades, biofuels are preferred over fossil fuels because of their favorable environmental impact [37,238].

9.4.1.2 Hydrogen

Hydrogen is a chemical energy carrier, just like gasoline, ethanol, or natural gas. The unique characteristic of hydrogen is that it is the only carbon-free or zero-emission chemical energy carrier. Hydrogen is a widely used industrial chemical that can be produced from any primary energy source. Major methods for its productions are reforming, steam gasification of carbonaceous materials, supercritical water gasification, electrolysis and thermochemical, and photocatalytic or photobiological dissociations of water [37,238].

Water electrolysis plants on a large scale (up to 160 MW) are state-of-the-art for industrial applications. Several plants of these types have been built in Norway, Egypt, and Peru, among others, in the late 1990s. Because of high production costs and low conversion efficiency of hydrogen production by water electrolysis, hydrogen is, however, not yet in widespread use. While the efficiency of chemical production of hydrogen from natural gas is high (50%), it is accompanied by a significant production of carbon dioxide. Sometimes wind energy is used to produce hydrogen via electrolysis if the power cannot be directly fed into the grid. As shown in Chapters 7 and 8, the use of solar and nuclear energies for the production of hydrogen by numerous electrochemical, thermochemical, and photocatalytic processes is being explored all over the world.

Different approaches exist for hydrogen storage: either as a gas under high pressure, a liquid at very low temperature, adsorbed on metal hydrides, or chemically bonded in complex hydrides. However, for stationary applications, storage under high pressure is the most popular choice. Hydrogen can be stored under pressure in gas bottles or tanks, and this can be done practically for an unlimited time. Smaller amounts of hydrogen can be stored in above-ground tanks or bottles under pressures up to 900 bar. For larger amounts of hydrogen, underground piping systems or even large salt caverns under pressures up to 200 bar can be used [182,184,188–192,194–196,199].

Hydrogen can be converted to electricity using fuel cell, where an electrochemical reaction between hydrogen and oxygen produces water, heat is released, and electricity is generated. In addition to fuel cells, gas motors, gas turbines, and combined cycles of gas and steam turbines can also be considered for power generation from hydrogen. Hydrogen systems with fuel cells (less than 1 MW) [226] and gas motors (under 10 MW) [227] can be adopted for CHP generation in decentralized installations. Gas and steam turbines with up to several hundred MW could be used for the power plants. Hydrogen can also provide spinning reserve. Spinning reserve is the electrical production capacity on the electrical grid to provide power in the event of an unexpected shutdown of a power plant or grid failure.

The strongest argument for hydrogen as ES is its extremely large versatility: it can be stored (the technology is mature), used directly in cars, used to produce electricity (through engines or fuel cells), used as a primary chemical for many products, used for hydrocracking in refineries, and injected into the gas grid (up to 5%–15%, which is a major volume), etc.

9.4.1.3   Synthetic Natural Gas

SNG is another option to store electricity as chemical energy. SNG can be produced by the reaction between hydrogen and carbon dioxide in a methanation reactor. SNG can also be produced by the gasification of biomass and waste and by numerous other technologies [37,238].

As is the case for hydrogen, the SNG produced can be stored in pressure tanks, underground, or fed directly into the gas grid. Several CO_2 sources are conceivable for the methanation process, such as fossil-fuelled power stations, industrial installations, or biogas plants. To minimize losses in energy, transport of the gases CO_2 (from the CO_2 source) and H_2 (from the electrolysis plant) to the methanation plant should be avoided. The production of SNG is preferable at locations where CO_2 and excess electricity are both available. In particular, the use of CO_2 from biogas production processes is promising as it is a widely used technology [37,238].

SNG has an advantage over hydrogen due to its higher energy density. SNG can also be easily transported by natural gas network. Hydrogen concentration in the natural gas pipeline, on the other hand, cannot exceed 5%–15%. The main disadvantage of SNG is the relatively low efficiency due to the conversion losses in electrolysis, methanation, storage, transport, and the subsequent power generation. The overall power generation efficiency for SNG is lower than that for hydrogen.

9.4.2   Electrochemical Energy Storage

An early solution to the problem of storing energy for electrical purposes was the development of the battery, an electrochemical storage device. It has been of limited use in large electric power systems due to small capacity and high cost.

9.4.2.1 Batteries

A battery is a device that transforms chemical energy into electric energy. Each cell in a battery contains three basic components: an anode, a cathode, and an electrolyte, and their properties relate directly to their individual chemistries. Batteries are broadly classified as primary and secondary batteries. Primary batteries are the most common and are designed as single-use batteries, to be discarded or recycled after they run out. They have very high impedance, which translates into long life ES for low-current loads. The most frequently used primary batteries are carbon–zinc, alkaline, silver oxide, zinc air, and some lithium metal batteries [204,207,208,213,217,218,221–224,232–237,239].

Secondary batteries are designed to be recharged and can be recharged up to 1000 times depending on the usage and battery type. For most of these batteries, very deep discharges result in a shorter cycle life, whereas shorter discharges result in long cycle life. The charge time varies from 1 to 12 h, depending upon battery condition and other factors. Commonly available secondary batteries are nickel–cadmium (NiCad), lead–acid, nickel metal–hydride (Ni–MH), and lithium-ion (Li-ion) batteries. Some of the drawbacks of the secondary batteries are limited life, limited power capability, low energy-efficiency, and disposal concerns [204,207,208,213,217,218,221–224,232–237,239].

Lead–acid batteries, invented in 1859 by French physicist Gaston Planté, are the oldest type of rechargeable battery. Today, lead–acid batteries dominate in the car market. Despite having a low energy-to-weight ratio and a low energy-to-volume ratio, their ability to supply high surge currents means that the cells maintain a relatively large power-to-weight ratio. These features, along with their low cost, make them attractive for use in motor vehicles to provide the high current required by automobile starter motors, but not for propulsion [239].

In 1899, a Swedish scientist Waldmar Jungner invented a rechargeable nickel–cadmium (NiCd) battery, the first battery to use an alkaline electrolyte. In 1959, Energizer came up with a new alkaline battery that consisted of a manganese dioxide cathode and a powdered zinc anode with an alkaline electrolyte. The use of powdered zinc gave the anode a greater surface area. Toward the end of the 1980s, Stanford R. Ovshinsky invented the nickel metal–hydride (Ni–MH) battery, which (1) gave longer life spans than NiCd batteries, (2) were less toxic, and (3) less damaging to the environment [204,207,208,213,217,218,221–224,232–237,239].

Lithium is the metal with lowest density and the greatest electrochemical potential and energy-to-weight ratio. In the 1980s, a team led by an American chemist John B. Goodenough produced the lithium-ion (Li-Ion) battery, a rechargeable and more stable version of the lithium battery. In 1996, the lithium-ion polymer battery was released. These batteries (1) held their electrolyte in a solid polymer composite instead of a liquid solvent, (2) were to be encased in a flexible wrapping, allowing it to specifically fit into a particular device, and (3) had a higher energy density than normal lithium-ion batteries [204,207,208,213,217,218,221–224,232–237,239].

In 2005, Altairnano announced a nanosized titanate electrode material for lithium-ion batteries. Subsequently, Toshiba reported another fast-charging Li-ion battery, based on new nanomaterial technology that provided even faster charge times, greater capacity, and a longer life cycle. Further developments on lithium-ion battery were also reported by MIT and researchers in France in 2006. In 2011, about 95% of all lithium batteries were produced in Japan, Korea, and China [239].

In a hybrid flow battery (HFB), one of the active masses is internally stored within the electrochemical cell, whereas the other remains in the liquid electrolyte and is stored externally in a tank. Therefore, hybrid flow cells combine features of conventional secondary batteries and redox flow batteries: the capacity of the battery depends on the size of the electrochemical cell. Typical examples of an HFB are the Zn–Ce and the Zn–Br systems. In both cases, the anolyte consists of an acid solution of Zn^{2+} ions. During charging, Zn is deposited at the electrode and at discharging, Zn^{2+} goes back into solution. As membrane, a microporous polyolefin material is used; most of the electrodes are carbon–plastic composites. Various companies are working on the commercialization of the Zn–Br HFB, which was developed by Exxon in the early 1970s. In the United States, ZBB Energy and Premium Power sell trailer-transportable Zn–Br systems with unit capacities of up to 1 MW/3 MWh for utility-scale applications 5 kW/20 kWh systems for community energy storage (CES) are being developed as well.

Batteries can provide ES (the time and power is proportional to the number of modules) and other important ancillary services (e.g., voltage and frequency stabilization). Batteries can be close to wind farms or PV systems coupled with the transmission grid and the distribution grid. They can be used for stationary applications (district storage) or mobile applications (electric vehicles).

Today, there is simply no battery technology capable of meeting the demanding performance requirements of the grid: uncommonly high power, long service life time, and very low cost. Due to their excessive costs, less than 1% of the grid storage is done by batteries. The relative weakness of the Japanese grid strongly encouraged the Japanese industry to develop new battery technologies (such as sodium–sulfur) to support the grid. Also, very progressive Japanese consumer electronic industry developed successful Ni–MH and Li-ion battery technologies that may be used for the grid integration of intermittent renewable energy sources [197,204,207,208,213,217,218,221–224,232–237,239].

9.4.2.2 Fuel Cells

Although fuel cells were invented about the same time as the battery, its development has accelerated in the last two decades in an attempt to increase the conversion efficiency of chemical energy stored in hydrocarbon or hydrogen fuels into electricity. Like a battery, a fuel cell uses stored chemical energy to generate power. Unlike batteries, its ES system is separate from its function as the power generator. It produces electricity from an external fuel supply as opposed to the limited internal ES capacity of a battery [2].

Fuel cells are very useful as power sources in remote locations, such as spacecraft, remote weather stations, large parks, rural locations, and in certain military applications [4,23]. A fuel cell system running on hydrogen can be compact, lightweight, and has no major moving parts, and in ideal conditions, it can achieve up to 99.9999% reliability. It is widely used in base-load power plants, electric and hybrid vehicles, off-grid power supply, and notebook computers.

The potential low cost for the fuel cell requires: (1) economies of scale associated with the large fuel-cell facilities; (2) fewer fuel-cell design constraints than those that exist in other fuel-cell applications, for example, weight and size constraints in vehicle applications; (3) a feed of pure H_2 to boost fuel-cell performance; (4) the use of oxygen from the thermochemical H_2 production systems, rather than air, to reduce capital costs per kW; and (5) increase the efficiency of the fuel cell. The increased output of the fuel cell with oxygen is equivalent to a major reduction in the capital costs of fuel cells [226]. The projected fuel cell efficiency is about 70% with pure oxygen. While the use of oxygen increases peak operating temperatures and fuel cell efficiency, in "real world," peak operating temperatures are limited by the availability of high-temperature turbine materials.

The major technical and economic challenge in providing spinning reserve is that the additional electrical production must come online very rapidly in the case of failure of another electrical-generating plant or failure of part of the electrical grid. This is currently accomplished by having power plants at part load with their turbines spinning. Although this approach allows the rapid increase in power generation when required, it comes with high costs. Fuel cells, on the other hand, have a unique capability: in a fraction of a second, they can go from no power output to high power output. Because of this capability, one of the major existing markets for fuel cells is for computer data centers, where there is a very high cost associated with temporary power outages—even those that last a fraction of a second.

9.4.3 MECHANICAL ENERGY STORAGE

There are three types of mechanical ES systems: compressed air ES, flywheel ES, and pumped-storage hydroelectricity (PHS). Here we briefly examine each of these systems.

9.4.3.1 Compressed Air Energy Storage

Compressed air (compressed gas) ES is a technology known and used since the nineteenth century for different industrial applications including mobile ones [199,202,203,209,213,214,220]. Air is used as storage medium due to its availability. Electricity is used to compress air and store it in either an underground structure or an above-ground system of vessels or pipes. When needed, the compressed air is mixed with natural gas, and burned and expanded in a modified gas turbine. Typical underground storage options are caverns, aquifers, or abandoned mines.

At the moment, there are only two CAES plants in the world (one in the United States and one in Germany), and they are designed to take advantage of variations in the price of electricity. According to The Economist, since CAES plants are inefficient, they are commercially viable only in places where the price of power varies dramatically. In United States this concept is implemented in Iowa. British Petroleum is also considering this concept [199,202,203,209,213,214,220].

In this method of storage, when electricity supply is higher than demand, air is compressed and stored underground. When demand exceeds supply, a gas turbine is fired up. The compressed air starts up turbine rapidly. In a "normal" case, the gas turbine has to drive an air compressor that eats up a large part of the energy generated by the gas turbine (above 20%). Also the start-up of the turbine is slower as the combustion air needs to be compressed first. The use of CAES increases the efficiency and the start-up time. It is a realistic alternative to pumped hydro storage in regions that lack the mountains.

If the heat released during compression is dissipated by cooling and not stored, the air must be reheated prior to expansion in the turbine. This process is called diabatic CAES and results in low round-trip efficiencies of less than 50%. Diabatic CAES technology is well proven; the plants have a high reliability and are capable of starting without extraneous power. The advantage of CAES is its large capacity; disadvantages are low round-trip efficiency and geographic limitation of locations.

The next generation of compressed air storage with heat recovery (called adiabatic CAES) intends [216] to store the heat produced during compression. This heat produces temperatures that can exceed 650°C, and the heat exchanger can be very large. Rapid (in minutes) cycles of ambient temperature to 650°C and back to ambient temperature for 20 times a day require special attention to the material problems. A demonstration project in Germany addresses this issue, and it will be ready to start in 2015. Liquid air storage [203] is a new concept that also deserves attention. This concept is based on a new combination of modules that have been proven on industrial scale.

9.4.3.2 Flywheel Energy Storage

A flywheel is simply a device for storing energy or momentum in a rotating mass. The potter's wheel is often cited as the earliest use of a flywheel. Space crafts have long used the gyroscopic stability inherent in flywheels to control their altitude [199,228]. In flywheel ES, rotational energy is stored in an accelerated rotor, a massive rotating cylinder. The main components of a flywheel are the rotating body/cylinder (comprised of a rim attached to a shaft) in a compartment, the bearings, and the transmission device (motor/generator mounted onto the stator). The energy is maintained in the flywheel by keeping the rotating body at a constant speed. An increase in the speed results in a higher amount of energy stored. To accelerate the flywheel, electricity is supplied by a transmission device. If the flywheel's rotational speed is reduced, electricity may be extracted from the system by the same transmission device. Flywheels of the first generation, which have been available since about 1970, use a large steel rotating body on mechanical bearings.

Advanced FES systems have rotors made of high-strength carbon filaments, suspended by magnetic bearings, and spinning at speeds from 20,000 to over 50,000 rpm in a vacuum enclosure. The main features of flywheels are the excellent cycle stability and a long life, little maintenance, high power density, and the use of environmentally inert material. However, flywheels have a high level of self-discharge due to air resistance and bearing losses and suffer from low current efficiency. Today flywheels are commercially deployed for power quality in industrial and Uninterruptible power supplies (UPS) applications, mainly in a hybrid configuration. Efforts are being made to optimize flywheels for long-duration operation (up to several hours) as power storage devices for use in vehicles and power plants.

Recent failures [199,228] in an U.S. installation of fly wheels for frequency regulation raised some doubts about this technology. Another failure in a German lab added further safety issues. While their technical maturity is well advanced, fly-wheels will always stay in niche applications, as they suffer from high self-discharge (20%–100%/day) and from extremely high costs for high capacity storage. The market for flywheels will be focused more on selling ancillary services to distri-bution grids (voltage stabilization, frequency stabilization, etc). Their main market competitor is Li-ion batteries, which are capable of providing further services apart from pure power services.

9.4.3.3 Pumped-Storage Hydroelectricity

PHS has existed for a long time—the first pumped hydro storage plants were used in Italy and Switzerland in the 1890s. By 1933, reversible pump-turbines with motor generators were available. Typical discharge times range from several hours to a few days. The efficiency of PHS plants is in the range of 70%–85%. Some areas of the world have used geographic features to store large quantities of water in ele-vated reservoirs, using excess electricity at times of low demand to pump water up to the reservoirs, then letting the water fall through turbine generators to retrieve the energy when demand peaks.

Pumping water up and down to store wind turbine energy brings efficiency around 20%:80% of the electricity originally produced by the turbine is lost.

Pumped-storage plants are characterized by long construction times and high cap-ital expenditure. Main drawbacks are the dependence on topographical conditions and large land use. Pumped storage is the most widespread ES system in use on power networks. Its main applications are for energy management, frequency con-trol, and provision of reserve. Advantages of PHS are the very long lifetime and practically unlimited cycle stability of the installation.

There is over 90 GW of pumped storage in operation worldwide, which is about 3% of global generation capacity [181,182,184,219,220]. In 2000, the United States had 19.5 GW of pumped-storage capacity, accounting for 2.5% of base-load-generating capacity. In 1999, the EU had 32 GW capacity representing 5.5% of total electrical capacity in the EU. Other examples include Tiahuang ping (1800 MW) in China. This pumped hydropower plant has a reservoir capacity of 8 million m^3 with a vertical distance of 600 m. The plant runs on average at 80% efficiency and provides about 2% of China's daily electricity consumption. Dinorwig power station is a pumped-storage hydroelectric scheme (1728 MW) on the edge of the Snowdonia national park in Gwynedd, North Wales. It includes diesel generators and large batteries, which would allow the plant to restart even in the event of a complete shutdown of the grid and in turn can help restart the national grid. The plant runs on average at between 70% and 80% efficiency [179,181,182,206,210, 211,219,220].

Snowy Mountains hydroelectric scheme in Australia (4500 MW, 7 sites) is highly complex and integrated. The scheme interlocks 7 power stations and 16 major dams through 145 km of transmountain tunnels and 80 km of aqueducts. The scheme virtually reverses the flow of the Snowy River from its natural course toward the ocean and directs it inland. The scheme is the largest renewable energy generator in mainland Australia, and it also provides additional water for an irrigated agriculture industry.

9.4.4 ELECTRICAL ENERGY STORAGE

There are three types of EES systems: capacitor, supercapacitor, and superconducting magnetic ES.

9.4.4.1 Capacitor

Capacitors use physical charge separation between two electrodes to store charge. They store energy on the surfaces of metalized plastic film or metal electrodes [180–200]. When compared to batteries and supercapacitors, the energy density of capacitors is very low—less than 1% of a supercapacitor, but the power density is very high, often higher than that of a supercapacitor. This means that capacitors are able to deliver or accept high currents, but only for extremely short periods, due to their relatively low capacitance [2].

9.4.4.2 Supercapacitor

Supercapacitors (also known as electrochemical double-layer capacitors [DLC]), are a technology, which has been known for 60 years. They fill the gap between classical capacitors used in electronics and general batteries, because of their nearly unlimited cycle stability as well as extremely high power capability and their many orders of magnitude higher ES capability compared to traditional capacitors. This technology can be developed to further increase capacitance and energy density, thus enabling compact designs [199,229].

Supercapacitors are very high surface area activated carbon capacitors that use a molecule-thin layer of electrolyte as the dielectric to separate charge. The supercapacitor resembles a regular capacitor except that it offers very high capacitance in a small package. ES is by means of static charge rather than of an electrochemical process inherent to the battery. Supercapacitors rely on the separation of charge at an electrified interface that is measured in fractions of a nanometer.

The two main features of supercapacitors are (1) extremely high capacitance values, of the order of many thousand farads, and (2) possibility of very fast charges and discharges due to extraordinarily low inner resistance. These features are not available with conventional batteries. Supercapacitors are also durable, highly reliable, maintenance-free, and have a long lifetime and operation over a wide temperature range and in diverse environments (hot, cold, and moist). The lifetime reaches 1 million cycles (or 10 years of operation) without any degradation, except for the solvent used in the capacitors, which deteriorates in 5 or 6 years irrespective of the number of cycles. Supercapacitors are not suitable for the storage of energy over longer periods of time, because of their high self-discharge rate, their low energy density, and high investment costs.

The supercapacitors are environmentally friendly and easily recycled or neutralized. The efficiency is typically around 90%, and discharge times are in the range of seconds to hours. The lifetime of supercapacitors is virtually indefinite.

They can reach a specific power density, which is about 10 times higher than that of conventional batteries (only very-high-power lithium batteries can reach nearly the same specific power density), but their specific energy density is about 10 times lower. However, unlike batteries, almost all of this energy is available in a reversible process [180–200]. Because of their properties, supercapacitors are suited especially to applications with a large number of short charge/discharge cycles, where their high performance characteristics can be used. Since about 1980, they have been widely applied in consumer electronics and power electronics. A supercapacitor is also ideally suited as an UPS to bridge short voltage failures. A new application could be the electric vehicle, where they could be used as a buffer system for the acceleration process and regenerative braking.

9.4.4.3 Superconducting Magnetic Energy Storage

Superconducting magnetic energy storage (SMES) systems work according to an electrodynamic principle. The energy is stored in the magnetic field created by the flow of DC in a superconducting coil, which is kept below its superconducting critical temperature. Hundred years ago, the discovery of superconductivity required a temperature of about 4 K. Today materials are available, which can provide super-conductivity at around 100 K. The main component of this storage system is a coil made of superconducting material. Additional components include power conditioning equipment and a cryogenically cooled refrigeration system.

In an SMES system, energy is stored within a magnet that is capable of releasing megawatts of power within a fraction of a cycle to replace a sudden loss in line power. It stores energy in the magnetic field created by the flow of DC power in a coil of superconducting material that has been cryogenically cooled. The stored energy can be released back to the network by discharging the coil. The power conditioning system uses an inverter/rectifier to transform AC power to DC and vice versa with about 2% and 3% loss in each direction. SMES loses the least amount of electricity in the ES process compared to other methods of ES. The round trip efficiency of SMES systems is around 85%–95% [180–197,199,230]. The main advantage of SMES is the very quick response time: the requested power is available almost instantaneously. It can provide very high power output for a short period of time. There are no moving parts in the main portion of SMES, but the overall reliability depends crucially on the refrigeration system.

In principle, the energy can be stored indefinitely as long as the cooling system is operational, but longer storage times are limited by the energy demand of the refrigeration system. The method is particularly useful to provide industrial power quality and serve individual customers vulnerable to voltage fluctuations. SMES systems are generally installed on the exit of the power plants to stabilize output or on industrial sites where they can be used to accommodate peaks in energy consumption [180–199,230,231].

Large SMES systems with more than 10 MW power are mainly used in particle detectors for high-energy physics experiments and nuclear fusion [210]. To date, a few, rather small SMES products are commercially available; these are mainly used for power quality control in manufacturing plants such as microchip fabrication facilities [199,230,231].

9.4.5 THERMAL ENERGY STORAGE

CHP (cogeneration) power plants are usually integrated into district heating (cooling) systems and operated according to the heat demand, while the electricity produced is fed into the grid, independent of the electricity needs. Some demonstration projects have shown that the reverse operation, that is, produce electricity according to the demand of the electric grid while storing the heat has high benefits: heat storage is much cheaper than electricity storage. This indirect form of storage needs a smart energy management system over the whole district. Thermal storage can be subdivided into three different technologies: storage of sensible heat, storage of latent heat, and heat pumps including thermochemical ad and absorption storage [180–199,201,212,215,225,227,240,241].

9.4.5.1 Storage of Sensible Heat

The storage of sensible heat is one of the best-known and most widespread technologies, with the domestic hot water tank as an example. The storage medium may be a liquid such as water or thermo-oil, or a solid such as concrete or the ground. Thermal energy is stored solely through a change of temperature of the storage medium. The capacity of a storage system is defined by the specific heat capacity and the mass of the medium used.

For example, a solar pond is simply a pool of water, which collects and stores solar energy. It contains layers of salt solutions with increasing concentration (and therefore density) to a certain depth, below which the solution has a uniform high salt concentration. When sunlight is absorbed, the density gradient prevents heat in the lower layers from moving upward by convection and leaving the pond. This means that the temperature at the bottom of the pond will rise to over 90°C, while the temperature at the top of the pond is usually around 30°C. The heat trapped in the salty bottom layer can be used for many different purposes, such as the heating of buildings or industrial hot water or to drive a turbine for generating electricity. This concept was tested at University of Texas, El Paso, in 1986. Other examples of solar pond are

1. Ormat turbines; 150 kW solar pond built at En Boqeq in Israel on the Dead Sea in 1980.
2. Bhuj solar pond in state of Gujarat in India; built in 1993 and worked well until 2000.
3. Demonstration project built by RMIT University with cooperation between Geo-Eng. Australia Pty Ltd. and Pyramid Salt Pty Ltd. at Pyramid Hill in June 2001. The system is working well and a plan for commercial solar pond is being investigated [180–199,201,212,215,225,227,240,241].

In an adiabatic CAES, the heat released during compression of the air may be stored in large solid or liquid sensible heat storage systems. Various R&D projects are exploring this technology, but so far, there are no adiabatic CAES plants in operation. The solid materials can be concrete, cast iron, or even a rock bed. For liquid systems, different concepts with a combination of nitrate salts and oil have been considered. The round-trip efficiency is expected to be over 70%.

Thermal storage can be carried out in a pressurized or a nonpressurized vessel. For liquid system, a heat exchanger can be used to avoid pressurized vessel; however, this adds complexity and cost. Direct contact between pressurized air and the storage medium provides higher surface area for the heat transfer. However, the pressure vessel costs more. Generally, the cost of storage material is low. When a large temperature range of 50°C–650°C is covered, a dual media approach (salt and oil) has to be used [180–199,201,212,215,225,227,240–242].

9.4.5.2 Storage of Latent Heat

Latent heat storage is accomplished by using phase change materials (PCMs) as storage media. There are organic (paraffins) and inorganic PCMs (salt hydrates) available for such storage systems. Latent heat is the energy exchanged during a phase change such as the melting of ice. It is also called "hidden" heat, because there is no change of temperature during energy transfer. The best-known latent heat—or cold—storage method is the ice cooler, which uses ice in an insulated box or room to keep food cool during hot days. Currently, most PCMs use the solid–liquid phase change, such as molten salts as a thermal storage medium for concentrated solar power (CSP) plants. The advantage of latent heat storage is its capacity to store large amounts of energy in a small volume and with a minimal temperature change, which allows efficient heat transfer [180–199,201,212,215,225,227,240,241].

In the context of EES, it is mainly sensible/latent heat storage systems, which are important. CSP plants primarily produce heat, and this can be stored easily before conversion to electricity and thus provide dispatchable electrical energy.

State-of-the-art technology is a two-tank system for solar power plants, with one single molten salt as heat transfer fluid and storage medium. The molten salt is heated by solar radiation and then transported to the hot salt storage tank. Electricity is produced by passing hot salt through a steam generator, which powers a steam turbine. Subsequently, the cold salt (still molten) is stored in a second tank before it is pumped to the solar tower again.

The main disadvantages are the risk of liquid salt freezing at low temperatures and the risk of salt decomposition at higher temperatures. In solar rough plants, a dual-medium storage system with an intermediate oil/salt heat exchanger is preferred [225,240–242]. Typical salt mixtures such as $Na-K-NO_3$ have freezing temperatures >200°C, and storage materials and containment require a higher volume than storage systems for solar power plants. The two-tank indirect system is being deployed in "Andasol 1–3," three 50 MW parabolic trough plants in southern Spain, and is planned for Abengoa Solar's 280 MW Solana plant in Arizona [180–199,201,212,215,225,227,240,241]. Apart from sensible heat storage systems for CSP, latent heat storage is under development by a German–Spanish consortium—including DLR and Endesa—at Endesa's Litoral Power Plant in Carboneras, Spain. The storage system at this pilot facility is based on sodium nitrate, has a capacity of 700 kWh, and works at a temperature of 305°C [180–199,201,212,215,225,227,240–242].

Molten salt batteries are also used for thermal storage. These are both primary and secondary cell high-temperature electric batteries that use molten salts as an electrolyte. They can offer both higher energy and power densities through the proper selections of reactant pairs and a high conductivity molten salt electrolyte, respectively. They are used in services where both high energy density and high power density are required such as for powering electric vehicles. High operating temperatures (400°C–700°C) bring problems of thermal management and safety and place more stringent requirements on the rest of the battery components.

Latent heat thermal ES is particularly attractive due to its ability to provide high-energy storage density and its characteristics to store heat at constant temperature corresponding to the phase transition temperature of PCM. Phase change can be in the following form: solid–solid, solid–liquid, solid–gas, liquid–gas, and vice versa. In solid–solid transitions, heat is stored as the material is transformed from one crystalline to another. These transitions generally have small latent heat and small volume changes than solid–liquid transitions. Solid–solid PCMs offer the advantages of less stringent container requirements and greater design flexibility [180–199,201,212,215,225,227,240–243]. Most promising materials are organic solid solution of penta erythritol (m.p. 188°C, latent heat of fusion 323 kJ/kg), penta glycerine (m.p. 81°C, latent heat of fusion 216 kJ/kg), $Li_2 SO_4$ (m.p. 578°C, latent heat of fusion 214 kJ/kg), and KHF_2 (m.p. 196°C, latent heat of fusion 135 kJ/kg) [225,240–242]. Trombe wall with these materials could provide better performance than a plain concrete Trombe wall [225,240–242].

Although solid–gas and liquid–gas transitions have higher latent heat of phase transition, their large volume changes on phase transition can create the containment problems and rule out their potential utility in thermal-storage systems. Large changes in volume make the system complex and impractical [180–199,201,212,215,225,227,240–243]. Solid–liquid transformations have comparatively smaller latent heat than liquid–gas. However, these transformations involve only a small change (of order of 10% or less) in volume.

Solid–liquid transitions have proved to be economically attractive for use in thermal ES systems. PCMs themselves cannot be used as heat transfer medium. A separate heat transfer medium must be employed with heat exchanger in between to transfer energy from the source to the PCM and from PCM to the load. The heat exchanger to be used has to be designed specially, in view of the low thermal diffusivity of PCMs in general. The container holding PCM should be compatible with PCM and should be able to handle volume changes [225,240–242].

Any latent heat ES system requires: (1) a suitable PCM with its melting point in the desired temperature range, (2) a suitable heat exchange surface, and (3) a suitable container compatible with the PCM. A wide range of technical options are available for storing low-temperature thermal energy [225,240–242].

9.4.5.3 Heat Pumps

There are two types of heat pumps used for thermal ES: seasonal thermal storage and thermochemical heat pumps.

9.4.5.3.1   Seasonal Thermal Storage Heat Pumps

This seasonal thermal storage can be divided into two broad categories:

1. Low-temperature systems use the soil adjoining the building as a low-temperature seasonal heat storage system drawing upon the stored heat for space heating. Such systems can also be seen as an extension to the building design.
2. High-temperature seasonal heat stores are essentially an extension of the building's heating, ventilation and air conditioning (HVAC) and water heating systems. Water is normally the storage medium, stored in tanks at temperatures that can approach boiling point.

In both cases, very effective insulation of the building structure is required to minimize heat loss from the building, and hence the amount of heat that needs to be stored and used for space heating [180–199,201,212,215,225,227,240–243].

In cities, a large amount of electricity is used for driving individual electric air-conditioning systems. These units are typically switched on–off over the same period, thus causing major stress to the electric grid. Many of these units are mounted on south-facing room, which further ruins the already weak efficiency. Every year, several millions of units are mounted across the world [225,241–243].

Centralized air-conditioning systems have a much higher efficiency and if these air-conditioning systems are operated overnight, with ice-storage, cheap electricity is then used overnight and the peak demand is shifted from noon to night. This has a major positive impact on the grid, since it is one of the most efficient forms of indirect electricity storage. It is also one of the most efficient examples of the added value of systems integration, and by this method, cooling can be provided at about one-tenth of the costs [225,241–243].

9.4.5.3.2   Thermochemical Heat Pumps

Sorption (adsorption and absorption) storage systems work as thermochemical heat pumps under vacuum conditions. These types of systems have a more complex design. Heat from a high-temperature source heats up an adsorbent (e.g., silica gel or zeolite), and vapor (working fluid, e.g., water) is desorbed from this adsorbent and condensed in a condenser at low temperatures. The heat of condensation is withdrawn from the system. The dried adsorbent and the separated working fluid can be stored as long as desired. During the discharging process, the working fluid takes up low-temperature heat in an evaporator. Subsequently, the vapor of the working fluid adsorbs on the adsorbent and heat of adsorption is released at high temperatures Depending on the adsorbent/working fluid pair, the temperature level of the released heat can be up to 200°C and the energy density is up to three times higher than that of sensible heat storage with water. However, sorption storage systems are more expensive due to their complexity [225,241–247].

Thermochemical systems rely on the energy absorbed and released in breaking and reforming molecular bonds in a completely reversible chemical reaction. In this case, the heat stored depends on the amount of storage material, the endothermic heat of reaction, and the extent of conversion [225,241–247].

9.4.6 CHALLENGES FOR ENERGY STORAGE

The main challenges for ES are three fold: technological, market and government regulations and strategy. The demand for electricity is constantly increasing. This leads to the need for increasing capacity and efficiency of existing storage facilities. When new technologies for local use and decentralized or large centralized applications are developed, they need to be brought to the market.

Appropriate market signals need to be created and used to incentivize the building of new storage capacities and provision of new storage services. The regulatory issues, which can vary at different locations, need to be handled locally. Finally, a holistic integrated approach to storage capacity, new technologies, market signals, and government regulations should be developed [180,186,194].

The economic and business case for ES varies from case to case, depending on where the storage is needed: generation, transmission, distribution, or customer level. The storage location depends on its benefits for users/operators.

The existence of compensation schemes for storage also affects the storage decision. This is a key issue when some stakeholders are part of the regulated market and the others are part of the deregulated market (e.g., producers and end customers) [180,186,194].

The ES studies in both Europe and the United States demonstrated that the provision of a single service (e.g., kWh) was not sufficient to make the storage scheme cost effective; services such as frequency stabilization and voltage stabilization have a much higher commercial value. The economic viability of the future ES systems, irrespective of its location and the grid connection (transmission or distribution), requires its ownership [180,186,194].

9.5 INTEGRATION OF GRIDS (SG) AND ENERGY STORAGE

ES should not be seen as a standalone technology. Its integration with grids (gas, heat, or electricity) is very important [241,248–318]. It can complement in a number of ways to improve the grid flexibility. A whole package of integrated grids and ES system includes the following three components [248–266]:

1. Flexible centralized and decentralized generation systems
2. Large centralized and small decentralized storage
3. Back-up capacity and transmission and distribution systems to integrate all of these subsystems

For electrical grid, ES can increase grid reliability and asset utilization [186,190]. ES facilities relieve congestion and constrains within the grid. They also provide easy and more stable connection of renewable sources to the grid and make islanding possible [268]. It also allows load leveling and peak shaving [267]. While, in principle, grid integrates various sources of electricity generation and electricity distribution systems, the working of grid can be enormously facilitated by the availability of stored energy. Grid ES lets electric energy producers send excess electricity over the electricity transmission grid to temporary electricity storage sites that become energy producers when electricity demand is greater, optimizing the production by storing off-peak power for use during peak times. In principle, the peak electricity demand can also be managed by over installation of renewable energy generators or shifting usage and production on the grid from one place and time to another. These options should be considered for determining the need for ES; however, such options may not be always possible.

Photovoltaic and wind turbine users can avoid the necessity of having battery storage by connecting to the grid, which effectively becomes a giant battery. Photovoltaic operations can store electricity for the night's use, and wind power can be stored for calm times. Germany is well known as a leading country for the introduction of renewable energies. Germany has set a target to increase the share of renewable energy from less than 20% to around 60%–80% by 2030. To achieve the German target, more ES capacity is necessary. For both short-term and long-term needs, a very large amount of ES will be needed to deliver peak power. The present installed storage capacity of 40 GWh in Germany [248–266,283] by pump hydrostorage can cover only the hourly demand and a part of the daily demand. To cover the additional hourly and daily demand, electrochemical ES such as batteries can be used. For the weekly and monthly demand, CAES, H_2, and SNG storage technologies are expected. On the whole, grid ES plays a key role in the development of a low-carbon electricity system. ES can supply more flexibility and balancing to the grid, providing a backup to intermittent renewable energy. Locally, it can improve the management of distribution networks, reducing costs and improving efficiency. In this way, it can improve the security and efficiency of electricity transmission and distribution (reduce unplanned loop flows, grid congestion, and voltage and frequency variations), stabilize market prices for electricity, while also ensuring a higher security of energy supply.

ES in conjunction with power electronics that interfaces the ES and the utility grid has a great technical application and enormous impact on future SG. This results in many benefits for managing the fluctuations in energy use as well as generation. Some of these benefits are briefly summarized in the following:

1. *Grid voltage and angular stability enhancement*: ES allows maintenance of the voltages of the generation and load ends within the normal values by providing additional reactive power and injection of real power for a period up to 2 s. ES technology has the capability to keep all components in the system in synchrony with each other and mitigate system collapse by the production and consumption of active power.
2. *Power Grid Reliability and Power quality Improvement*: ES increases power grid reliability by providing uninterrupted power supply. ES improves power quality by reducing harmonic distortions, and eliminating voltage sags and surges through the provision of ride-through for momentary outages, and extended protection from longer outages, coupled with advanced power electronics. ES can suppress grid frequency excursion [270] by discharging active power for a duration of up to 30 min. This prompt spinning reserve mitigates load-generation imbalance.
3. *Energy Management Improvement/Renewable Support*: The ES can be used in the energy management applications such as load shifting, peak shaving, load leveling, and commodity storage where electricity storage

technologies are used in daily cycles for financial benefit [269]. ES can dispatch stored energy at off-peak or low cost times to manage demand on grid-sourced power. ES also facilitates the integration of RE and can improve the net power quality [271]. Hybrid ES with wind and solar PV makes ES increase the value of the PV and wind-generated electricity and provides a steady supply of electricity to the customers [272,273].

Real planning and implementation strategy should be related to the real-time control and operational functionalities of the ES in conjunction with renewable energy (RE) so as to get rapid integration process [274]. The advantages of ES outlined in Section 9.4 were substantiated in a report released in December 2013 by the department of energy [249]. The report also indicates that the investment in power system infrastructure such as transmission lines and distribution lines in certain areas at times of peak demand can be reduced with the help of ES. For residential users, using SG can be encouraged to shift their energy buying toward periods when surplus power is available. Users may accomplish this shift by either changing the time when they need electricity, or by buying and stored electricity for later use, or both. Customers can also feed electricity back to the grid using other independent sources of energy such as solar panel, heat pump, gas heating, etc., ES can facilitate all of these transactions.

Just as for electrical grid, ES needs to be integrated in all network-based energy systems such as heating and cooling network and in gas networks. The expansion of natural gas–fired power plants, a fall in gas prices, and the increased efficiency and the reduced costs of flexible combined-cycle natural gas turbines have made gas network more useful to the operation of the CHP systems and the electrical grid. In future, electrical grid will increasingly intertwine with gas and heat grids and ES systems can facilitate all interactions among these three networks (or grids).

9.5.1 Levels of Grid-Energy Storage Integration

ES should be integrated at all levels of the electricity system:

1. *Generation level*: Arbitrage, balancing, and reserve power, etc.
2. *Transmission level*: frequency control, investment deferral
3. *Distribution level*: voltage control, capacity support, etc.
4. *Customer level*: peak shaving, time of use cost management, etc.

These different locations in the power system will involve different stakeholders and will have an impact on the type of services to be provided. Each location will provide a specific share of deregulated and regulated income streams. Different ES systems will have to be considered (centralized and decentralized), and specific business models will have to be identified. Different levels of ES integration is essential in the overall vision of the SG, most importantly with respect to renewable generation [274].

Community energy storage (CES) also known as distributed energy storage system (DESS) provides voltage control, peak load management, reactive support, capacity and ancillary service market, frequency regulation, and other SG services

[277] and could be controlled in real-time utilizing feeder or substation load signals. During system outages, CES could also allow nonfaulted subsections of the grid to operate in autonomous mode with proper relaying and protection schemes [276]. The best approach to the application of CES is to use smaller systems rated 25–200 kW with discharge time of 2–4 h [275] that are connected on the low voltage side of the distribution transformer and protect the final low voltage (LV) circuits to individual end users. Another application is to use a larger 200–5000 kW system that is directly connected to various microgrids.

CES provides ultimate in voltage control and service reliability to the end users, particularly when more sophisticated electronic loads, such as plug-in hybrid electric vehicle (PHEV) charging units, are connected to SG by the customers [297–318]. Besides these changing load patterns, there is a high penetration of RE, most especially solar roof tops, which cause a growing amount of energy reverse back into the grid when RE generation exceeds customer's power demand. These functions can be easily facilitated by CES. The location of CES throughout the distribution system will store the excess energy with less line losses and redispatched back to the same customers when required. The CES could equally control the local voltage during cloud pass-over or voltage sag in solar PV. CES units can also provide peak power in case of abnormal amount of PHEV quick charges [250,262,264,276,297–318].

9.5.2 Scale of Energy Storage and Their Grid Applications

As mentioned earlier, ES technology can serve at various locations at which electricity is produced, transported, consumed, and held in reserve (backup). Depending on the location and application, storage can be large scale (GW), medium sized (MW), or microscale (kW). The nature of storage also depends on the required rate of response. For example, large ES does not require quick time response, while intermediate storage is more suitable for power application where fast response is necessary. For all small storage systems, battery technologies are preferred. ES systems are used for various power utility applications all over the world. These applications of various scales of ES are briefly summarized in Table 9.3 [250,262].

Storage capacity also depends on the size of the reservoir. This determines the time when this power is available. In the past, with one cycle per day, ES was rated mainly in GWh (energy capacity); today the same systems are used up to 10 and 20 times/day and the installed power in GW (given by the number and the size of the installed turbines) becomes more important. This dynamic behavior of existing storage will increasingly move in the direction of quick and powerful response to the needs of the grid.

9.5.2.1 Large-Scale Storage for Energy Applications (GWh)

Large-scale ES refers to the methods used to store electricity on a large scale within an electrical power grid. This storage category is characterized by capacities that normally exceed megawatt levels (in GWh level) with discharge capabilities of up to hours, days, weeks, or months. In general, these storage methods serve various functions, including energy management, peak shaving, and spinning reserve, among others. Besides thermal and chemical fuel storage, PHS and CAES are the preferred

TABLE 9.3

Applications for Various Scales of ES

Scale	Name	Capacity
Large scale		
Energy applications	Bath Country PHS Station, United States	2772 MW
	Guangdong PHS Power Plant, China	2400 MW
	Dnister PHS Power Plant, Ukraine	2268 MW
	Okutataragi PHS Power Station, Japan	1932 MW
	Ludington PHS Power plant, United States	1872 MW
	Huntorf CAES Power plant, Germany	290 MW
	McIntosh CAES Plant, Macintosh, AL, United States	110 MW
Intermediate scale		
ES power applications	20 MW flywheel plant, Stephentown, NY, United States	20 MW
	1 MW/s SMES system, Wisconsin, United States	1 MW/s
	200 kW flywheel, Dogo Island, Japan	100 kW
Small-scale energy		
Storage-battery storage applications	Golden Valley Electric Association(GVEA), Fairbanks, AK, United States	27 MW,14.6 MWh
	Sumitomo Densetsu office, Osaka, Japan	3 MW, 800 kWh
	Bus Terminal Energy Storage, Long Island, NY, United States	1.2 MW, 6.5 MWh
	AEP DES system at Chemical Station, N. Charleston, WV, United States	1.0 MW, 7.2 MWh
	Metlakatla Power and Light (MP&L), AL, United States	1.0 MW,1.4 MWh
	Pacificorp. Castle Valley, UT, United States	250 kW, 2 MWh

Source: VDE—ETG Energy Storage Task Force: Energy storage in power supply systems with a high share of renewable energy sources significance—State of the art—Need for action, Report, December 2008.

modes of storage for this category. These types of storage methods have been used for load leveling and peak shaving applications for decades. The first applications of PHS were in Italy and Switzerland around the 1890s, while the first application of CAES was in the Huntorf plant in Bremen, Germany, which has been commercially operated since 1978 [278]. Typical PHS and CAES have efficiencies of around 65%–85% and 40%–50%, respectively, and maximum outputs that exceed several thousands of megawatts (in several GW range) and a few hundreds of megawatts, respectively.

As of March 2012, pumped-storage hydroelectricity (PSH) was the largest-capacity form of grid ES available; the Electric Power Research Institute (EPRI) reports that PSH accounts for more than 99% of bulk storage capacity worldwide. Over the last 60 years of operation, PHS capacity all over the world has expanded to more than 90 GW, which is nearly 3% of the global generation capacity [278]. PHS is attractive, and essential, when networks are mainly composed of a large number of regional grids with very weak interconnections. It is likely that pumped-storage hydropower will become especially important as a balance for very large-scale photovoltaic solar energy or large-scale wind energy generations [279].

For improvement of pumped-storage hydropower, the use of underground reservoirs as lower dams has also been investigated as grid storage device. Salt mines could be used, although ongoing and unwanted dissolution of salt can be a problem. If they prove affordable, underground systems could greatly expand the number of pumped-storage sites. Saturated brine is about 20% more dense than fresh water, and it increases storage capacity. The Yanbaru seawater PHS in Japan is an example plant that operates by pumping seawater up into the reservoir [280]. A new concept in pumped storage is to utilize wind turbines to drive water pumps directly, in effect an "Energy Storing Wind Dam." This can provide a more efficient process and usefully smooth out the variability of energy captured from the wind.

CAES systems are widely accepted in the United States, but their application is limited in Europe and Japan due to economic feasibility and geographical restrictions. An example for the improvement of the CAES system is the implementation of fabricated storage tanks [281], and there are also efforts to use high-pressure underground piping to increase its efficiency to 70% [282].

For the storage of large amounts of energy, other options such as electrochemical ES would be too expensive and require too much space. An alternative is the transformation of electricity into hydrogen or synthetic methane gas for storage and distribution within the existing natural gas grid .The efficiency of full-cycle conversion of electric power to hydrogen is about 55%–75%, and to SNG about 50%–70%. In the United States and Germany, the storage capacity of the existing natural gas grid is very large. In Germany alone, it is about 200 TWh (about 400 TWh including the distribution grid) [250,262,283]. From a technical point of view, it is possible to inject up to 10% hydrogen into natural gas without any negative effects on the gas quality. Because hydrogen has one-third the energy of natural gas, it is possible to inject hydrogen containing 7 TWh of energy into the natural gas grid. At any point of the gas grid, it is possible to convert the gas back into electricity with a high-efficiency gas power plant (~60%). It is expected that in Germany, the weekly and monthly ES demand will be about 8.2 TWh in 2030, which can nearly be covered by such an injection of hydrogen into the gas grid. This solution is only possible in countries where a gas grid exists; otherwise, the hydrogen or synthetic methane must be stored in additional high-pressure vessels or caverns.

9.5.2.2 Intermediate Scale Energy Storage (MWh) or Fast Discharge Storage for Power Applications

In the intermediate level of storage, power from renewable energy sources like wind and PV solar energy is stored in batteries or other systems mentioned in Table 9.3. The Japan Wind Development Co. Ltd. has constructed a wind power generation facility equipped with a battery in Aomori, Japan (Futamata wind power plant). This facility consists of 51 MW of wind turbines (1500 kW × 34 units) and 34 MW of NaS batteries (2000 kW × 17 units). By using the NaS battery, the total power output of this facility is smoothed and peak output is controlled to be no greater than 40 MW.

In order to achieve this plan, the NaS battery system controls charging or discharging in accordance with the output of wind power generation. This facility meets the technical requirements of the local utility company to connect to the grid.

The devices for intermediate-scale ES have very fast discharge capability with high power discharge on the order of seconds or less. Flywheel, SMES, and supercapacitors are examples that fall into this category. These devices have a very high efficiency, which are in the range of 90% and above, and lifetime of 20–40 years. Due to high power discharge capability, these types of storage devices are presently not suitable for load leveling and peak shaving purposes like other high ES technologies. However, they are superior for various power quality purposes. Traditional high ES systems utilized these storage devices as a backup supply for sensitive loads, and for mitigation of voltage sags/swells, flicker, and harmonics [279,286,287].

Flywheels have been used in power systems for more than a century. Current generation flywheels have storage capacities of 100 kW, and the use of multiple flywheels in parallel developed by the Beacon Power Corporation can provide storage output at the multimegawatt level [288]. Aside from power quality applications and frequency and voltage regulation, recent developments in flywheels have shown a growing interest in their application for mitigating fluctuations of renewable energy sources [289,290].

SMES was first introduced around 1969 and has since been used for many applications, including load leveling, spinning reserve, transient and dynamic stability, power quality, and increasing transmission line capacity [241,291].

Despite a number of useful applications reported in the literature, real-life large-scale installations of SMES are still considered costly, and most of the literature describes computer and laboratory simulations. In the United States, the first SMES used for power grid applications was implemented in 1981 for power quality and grid stability on the 500 kW Pacific Intertie that interconnects California and the Northwest [292]. Similar to flywheel technology, SMES is used to enhance the performance of grids connected to wind farms, particularly in stabilizing fluctuations of the wind turbine output [293,294].

Supercapacitors are a highly viable technology that has received attention by utilities and vehicle applications. In power system applications, they have been studied for power quality purposes, such as ride-through improvement in electric motor drives during voltage sags and in custom power devices [241]. Due to its high cost, it is rarely used as single units but instead in hybrid systems where it is combined with high ES devices, such as lead–acid batteries [295].

A smart factory, smart building, smart hospital, smart store, or another intermediate-level grid with ES may be treated as a "smart microgrid." For flexibility in resisting outages caused by disasters, it is very important to deploy smart microgrid, that is, distributed smart power sources, as an element in constructing SGs. These are discreet subsystems of overall SG. ES is an essential component of a smart microgrid, which should be scalable, autonomous, and ready to cooperate with other grids. The architecture for the smart microgrid should have a single controller and should be scalable with respect to ES, that is, it should adjust smoothly to the expansion and shrinkage of ES (battery) capacities according to the application in, for example, a factory, a building, a hospital, or a store. The microgrid and ES should in general be connected to the network; even if a particular smart microgrid is not connected to a grid, for example, in the case of an isolated island, it should still have similar possibilities of intelligent adjustment, because an isolated smart microgrid can also expand or shrink.

9.5.2.3 Small-Scale Energy Storage Systems (kWh) or Battery Energy Storage Systems (BESSs)

In general, batteries can be used for both power and energy applications with discharge durations from minutes up to a few hours. Therefore, the choice of battery storage for particular purpose involves consideration of both the energy- and power-related cost. The decentralized storage of small-scale photovoltaic solar energy system requires small storage that can be handled by battery technologies. In recent years, however, the low-voltage grid is reaching its performance limit because of the increasing number of installed PV systems. The relationship between grid and PV system is two ways, wherein customer can feed back to the grid the power generated by the PV system. To encourage operators of decentralized systems, the price for self-consumed PV energy is higher. Therefore, self-consumption of power will become an important option for private households with PV facilities, especially as the price of electricity increases.

To measure the amount of energy consumed or fed into the grid, two meters are needed. One meter measures the energy generated by the PV system. The other meter works bidirectionally and measures the energy obtained from or supplied to the grid. The generated energy that is not immediately consumed is stored in the battery. Stationary batteries provide electricity at low price per kWh with low maintenance cost. BESSs refer to secondary chemical battery technologies, and they have been used in many areas for power utilities. Of the different battery technologies available, lead–acid batteries are the most popular due to their cost, technological maturity, and availability. Value regulated lead acid battery (VRLA) is currently the most popular choice for flooded lead–acid batteries, as they are maintenance-free and have a lifespan of 1000–2000 cycles at 75% depth of discharge and have an efficiency of approximately 70%–80% [295]. Compared to lead–acid batteries, NiCd, Ni–MH, and Li-ion batteries have higher energy densities, greater efficiency and require less maintenance. Ni–MH and Li ion, while expensive, are superior to NiCd batteries.

NaS batteries are primarily suitable for large-scale grid ES [296] but carries high installation and maintenance costs. They are suitable for many applications, including stabilizing intermittent renewable energy and substation applications, such as load leveling, peak shaving, and emergency power supply. Of the different types of chemical flow batteries, vanadium redox (VRB), polysulfide bromide (PSB), and zinc bromine (ZnBr) are the most common and are being actively developed and commercially installed for small-scale applications.

9.5.3 FUTURE OF GRIDS AND ENERGY STORAGE

Gas grid will continue to grow and evolve. Megapipeline and LNG projects will continue to be the anchor systems for global network growth [4,23]. Complementing these large-scale systems will be the next generation of smaller modular systems or satellites. Gas grid development will involve the interplay between these anchors and satellites around the world [4,23]. Another important trend will be the continued integration of gas with electrical grid on new projects related to CHP for nonelectrical applications, deeper penetrations of renewable energies, and more involvement of gas and electrical grids in the transportation sector. As different sources of gas (natural gas, biogas, shale and other unconventional gas, synthetic gas, and hydrogen) are further developed, they will be used more and more for power, heat, and oil substitutes. The advancement in combined cycle power plants, more use of CHP in small- and middle-scale gas turbines and gas engines, and more use of distributed power will enhance the complexity of the gas grid and gas storage systems. The role of hydrogen and its interaction with both gas and electrical grids will expand [4,23].

Distributed small- and medium-scale power is poised for rapid growth. As gas engine and smaller-scale gas turbine technology has evolved and become more efficient, distributed power options are able to compete more effectively against centralized generation. Small modular nuclear reactors and their use for more nonelectrical applications will make the use of nuclear energy at smaller and distributed scales more palatable. In emerging markets, where grids are underdeveloped or nonexistent, and regulatory and financial institutions are underdeveloped, distributed energy is an attractive choice. Five areas of significant growth potentials are [4,250,262]

1. Cogeneration applications with gas-driven, nuclear, and renewable energy power plants using CHP for nonelectrical and heating and cooling usages
2. Pipeline and processing plant compression in gas networks
3. Providing electricity in remote sites like mines and oil and gas fields, numerous off-shore operations like ships, drilling platforms, and renewable power generations as well as in the areas in developing countries with weak or unstable grids
4. Fast-track deployment of generation to meet emergency power needs
5. Next-generation fuel cell or microturbine technologies

The economics of energy transportation is greatly improved when there is an ability to leverage existing infrastructures. The existing road and rail infrastructures are very effectively used for the transportation of solid and liquid fuels. They are also used in current developments of shale gas for transportation of sand, water, and NGLs. Some of the digital and communication technologies developed for defense industry have been successfully applied for the automation of gas and electrical grids. Natural gas pipelines and LNG networks have followed a similar evolutionary path to power systems. Gas grids have been able to use some of the oil pipeline infrastructure.

The integration of different types of networks or grids is important because constraints in one type of network often provides opportunities for other networks [4,284,285]. For example, constraints in the expansion of oil pipelines can lead to expansion of road and railway transportation network. Both natural gas and electrical grids are uniquely positioned to support other energy networks. Multinetwork optimization is about the role of technology in facilitating network switching in a proactive way to find optimum solutions for the customers. The development of optimum solutions will make grids more practical, sustainable, and robust. The issues of deeper penetrations of renewable and nuclear energies and enhanced use of gas and electricity in transportation sector will cause more overlap among gas (oil), heat, and power. These topics and the need for distributed energy will provide roadmap for further integration to drive economic development, sustainability, and energy security.

Both power and gas grid operators need to appreciate potentials for these two energy sources and the need for interactions among them. Managing the increasing load variability from supply and demand sources is a growing challenge for network operators. Because electricity is created as it is used, managing its variation in supply and demand is more complex. The deeper penetration of intermittent source of renewable energy adds to this complexity. Various options are available to grid operators. The one that shows promise is the use of SG- and incentive-based mechanisms using price signals. As a result, natural gas used as a stand-by option is often the most cost-effective way to manage this variability. Such an approach, however, does require inherent flexibility of the pipeline, underground storage, and support of LNG systems. Thus, management of gas and electrical grids requires strong interactions on the issues such as assignments of responsibilities, cost, and grid improvement [4,284,285].

Hybrid and plug-in electrical vehicle's market will expand. This will bring extra burden to the management of smart electrical grid. This movement will, however, integrate more the static and mobile use of energy. Along with deeper penetrations of nuclear and renewable energy in the overall energy market, hybrid and electric cars will be the major force for decarbonization of energy usage. Smart electrical grid will play a major role in this transformation [4,77,87,297–318].

ES will continue to play an important role in the operation of the smart electrical grid. Today, and for the next decade, excess electricity will be stored in pumped hydrosystems (over 99%). The vast majority of existing large-scale ES is based on pumped hydrostorage. Pumped hydrostorage systems were built purely for electricity management. They were initially built for pumping at night (supply of electricity higher than demand) and producing electricity during daytime (supply of electricity lower than demand). These pumped hydroplants made business sense when grid was not strongly interconnected, and the main objective of the storage system was to level demand of the small regional areas. Today, grid is much better integrated and in this situation, pumped hydrostorage does not necessarily present the best option. PHS stores electricity (like a battery), and PHS loses about 15%–25% of electricity (round trip efficiency of about 75%–85%). It is worth mentioning that the start-up of a new pump hydrostorage system takes almost 10 years [249–265].

Some of these hydrodams have been extended with a smaller pumped-storage system. They are hybrid systems of renewable energy production and ES. In future, pumped hydrostorage systems will follow more and more cycles per day. This tendency will be driven by the needs of the future grid. The demand will vary more stochastically: the increased electricity production from renewables will bring additional variability to the supply. Many existing pumped hydrosystems are planned to be upgraded by adding more turbines. The overpowered storage systems will become more flexible and produce higher power outputs over a shorter time. They will produce more GW, but the GWh will stay constant. The overall storage capacity will not increase (the size of the reservoir stays the same). This will result in the higher degree of volatility in the system, which is a result of both, intermittent

RES and schedule changes in conventional generation. However, the future grid will need more storage capacity as well as more flexibility and more dynamic reaction time, since volatile generation will constitute the major part of our consumed energy [250,262].

In the medium term, excess renewable electricity may also be converted to hydrogen. This will open a multitude of new routes, such as feeding hydrogen directly into the natural gas grid up to 5%–15% (these are gigantic quantities, and the natural gas storage and distribution may be used for free). Hydrogen may also be used to power fuel cell cars. Hydrogen may also be converted to natural gas (power to gas concept), or to methanol (for cars), or converted back to electricity (through stationary fuel cells or gas engines). The hydrogen/storage route opens a very large multitude of new paths [77,87,248–265,297–318].

In the near future, development of a national SG, which connects all three regional SG in the United States, is likely to occur. This will further improve efficiency and sustainability of our electric power supply and demand system. The national SG will provide protection against failures in regional grids. The concept of national gas and electrical grids may be at some point extended to multinations or even global gas and electricity grids. The world will move more and more toward holistic energy systems integration. The integration will be both horizontal and vertical. Fuel, heat, and power networks will be more and more intertwined. Fuel and ES systems and their integrations to the grids will bring more stability and resilience to the grid operations.

REFERENCES

1. Energy storage packing some power, Technology quarterly (March 3, 2012). *The Economist*.
2. EIA. Existing capacity by energy source (January 2010).
3. Kroposki, B., Garrett, B., Macmillan, S., Rice, B., Komomua, C., O'Malley, M., and Zimmerle, D. Energy systems integration: A convergence of ideas. NREL Report, NREL/TP-6A00-55649, under contract No. DE-AC36-08GO28308 (July 2012).
4. Evans, P. and Farina, M. The age of gas & the power of networks, a report from General Electric Company, New York (2013).
5. Riis, T., Sandrock, G., Ulleberg, O., and Vie, P. Hydrogen storage-gaps and priorities. HIA HCG Storage Paper, IEA HIA Task 17, pp. 1–13 (2005).
6. Innes, W. National Research Council and National Academy of Engineering. *The Hydrogen Economy: Opportunities, Costs, Barriers and R&D Needs Section 4: Transportation, Distribution and Storage of Hydrogen*, National Academy Press, Washington, DC (2004).
7. Zhou, L. Progress and problems in hydrogen storage methods. *Renewable and Sustainable Energy Reviews*, 9, 395–408 (2005).
8. Renewable gas-vision for a sustainable gas network, a report by National grid U.S.A. service Co., Waltham, MA (2010).
9. Melaina, M., Antonia, O., and Penev, M. Blending hydrogen into natural gas pipeline networks: A review of key issues, NREL. Prepared under Task No. HT12-2010, NREL/TP-5600-51995, under DOE contract No. DE-AC36-08GO28308 (March 2013).
10. Hagen, M., Polman, E., Jensen, J., Myken, A., Jonsson, O., and Dahl, A. Adding gas from biomass to the gas grid. Report SGC 118, ISSN 1102-7371, ISRN SGC-R-118-SE, Swedish gas center (July 2001).
11. Hussey, B. Biogas injection into the natural gas grid, a report by Commission for Energy Regulation, Dublin, Ireland (September 11, 2013).

12. Goellner, J. Expanding the shale gas infrastructure, CEP, 49 (August 2012).
13. "Hydrogen storage," *Wikipedia*, https://en.wikipedia.org/wiki/Gas_networks_simulation (last modified July 9, 2015).
14. Godula-Jopek, A., Jehle, W., and Wellinitz, J. Hydrogen storage technologies, new materials, transport and infrastructure, Wiley online (November 5, 2012), doi: 10.1002/9783527649921.
15. "Gas network simulation," *Wikipedia*, https://en.wikipedia.org/wiki/Gas_networks_simulation (last modified May 13, 2015).
16. Baldwin, J. Biomethane to grid UK project review. CNG services report (March 11, 2014).
17. Hagen, M., Polman, E., Myken, A., Jensen, J., Jonsson, O., Biomil, A.B., and Dahl, A. Adding gas from biomass to the gas grid. Final report July 1998–February. 2001, under contract No. XVII/4.1030/Z/99–412 (2001).
18. Specht, D., Baumgart, F., Feigl, B., Frick, V. et al. Storing bioenergy and renewable electricity in the natural gas grid, FVEE-AEE, Topics (2009).
19. "Power to Gas," *Wikipedia*, https://en.wikipedia.org/wiki/Power_to_gas (last modified June 22, 2015).
20. Barati, F., Seifi, H., Sepasian, M., Nateghi, A., Shafie-khah, M., and Catalao, J. Multi-period integrated framework of generation, transmission, and natural gas grid expansion planning for large scale systems. *IEEE Trans. Power Syst.,* PP(99), 1–11, IEEE Power and Energy Society (October 31, 2014).
21. EU commission task force for smart grids–expert group 4-smart grid aspects related to gas. EU report EG4/SEC00601DOC (June 5, 2011).
22. Weidenaur, T., Hoekstra, S., and Wolters, M. Development options for the Dutch gas distribution grid in a changing gas market, personal communication (2009).
23. Fullenbaum, R., Fallon, J., and Flanagan, B. Oil and gas transportation and storage infrastructure: Status, trends, & economic benefits. A report by HIS global Inc., submitted to American Petroleum Institute, Washington, DC (December 2013).
24. Wolters, M. Requirements of future gas distribution networks. In *23rd World Gas Conference* (June 5–9, 2006), Amsterdam, the Netherlands (2006).
25. 2013 special reliability assessment: Accommodating an increased dependence on natural gas for electric power-phase II: A vulnerability and scenario assessment for the north American bulk power system. NERC report (May 2013).
26. Dehaeseleer, J. Gas industry views regarding smart gas grid, a report from Marcogaz-technical association of the European natural gas industry. EGATE2011, Copenhagen, Denmark (2011).
27. Weiss, J., Bishop, H., Fox-Permer, P., and Shavel, I. Partnering natural gas and renewables in ERCOT. A report by The Brattle Group. Prepared for The Texas clean energy coalition (2013).
28. North American natural gas midstream infrastructure through 2035: A secure energy future. A report by The INGAA Foundation Inc. (June 28, 2011).
29. Natural gas infrastructure-papers 1–9, prepared by the gas infrastructure subgroup of the resource and supply task group, working document of the NPC North American Resource Development study (September 15, 2011).
30. Natural gas pipeline and storage infrastructure projections through 2030, a report submitted by ICF International to the INGAA foundation Inc. F-2009-04 (October 2009).
31. EIA. The basics of underground natural gas storage. U.S. Energy Information Administration report (August 2004).
32. Antkowiak, M., Ruth, M., Boardman, R., Bragg-sitton, S., Cherry, R., and Shunn, L. Summary report of the INL-JISEA workshop on nuclear hybrid energy systems. Prepared under Task No. 6A50.1027, NREL/TP-6A50-55650 (July 2012). Report available at: http://www.nrel.gov/docs/fy12osti/55650.pdf, presentations of workshop available at: https://inlportal.inl.gov/portal/server.pt?tbb=hybrid. (Accessed December 2012)

33. Antkowiak, M., Boardman, R., Bragg-Sitton, S., Cherry, R., Ruth, M., and Shunn, L. Summary report of the INL-JISEA workshop on nuclear hybrid energy systems. INL/EXT-12-26551, NREL/TP-6A50-55650, INL Report, prepared for US DOE, Office of Nuclear Energy, Contract No. DE-AC07-051D14517, Idaho Falls, ID (July 2012).

34. Boardman, R. Advanced energy systems-nuclear-fossil-renewable hybrid systems. A report to Nuclear Energy Agency-committee for technical and economical studies on nuclear energy development and fuel cycle, INL, Idaho Falls, ID (April 4–5, 2013).

35. Ruth, M. Antkowick, M. and Gossett, S. *Nuclear and Renewable Energy Synergies Workshop: Report of Proceedings*. JISEA, prepared under Task No. 6A302003, Technical Report NREL/TP 6A30–52256, Golden Colorado (December 2011).

36. Antkowiak, M. and Gossett, S. Technical Report NREL/TP-6A30-52256 (December 2011) Available at: http://www.nrel.gov/docs/fy12osti/52256.pdf (Accessed February 2012).

37. Shah, Y. *Water for Energy and Fuel Production*, 13, CRC Press, New York (May 2014).

38. Fang, X., Misra, S., Xue, G., and Yang, D. Smart grid—The new and improved power grid—A survey, *IEEE Commun. Surveys Tutor.*, 13, 1–38 (2011).

39. Akyildiz, I., Su, W., Sankarasubramaniam, Y., and Cayirci, E. A survey on sensor networks. *IEEE Commun. Mag.*, 40(8), 102–114 (2002).

40. Akyildiz, I. and Wang, X. A survey on wireless mesh networks. *IEEE Radio Commun.*, 43(9), 23–30 (2005).

41. Farhangi, H. The path of the smart grid. *IEEE Power Energy Mag.*, 8(1), 18–28, (2010).

42. Yu, Y., Yang, J., and Chen, B. The smart grids in China—A review. *Energies*, 5, 1321–1338 (2012).

43. Shamshiri, M., Gan, C., and Tan, C. A review of recent development in smart grid and micro-grid laboratories. In *2012 IEEE International Power Engineering and Optimization Conference (PEOCO 2012)*, Melaka, Malaysia (June 6–7, 2012).

44. Gharavi, H. and Ghafurian, R. Smart grid: The electric energy system of the future. *Proc. IEEE*, 99(6), 917–921 (2011).

45. Gharavi, H. and Hu, B. Multigate communication network for smart grid. *Proc. IEEE*, 99(6), 1028–1045 (2011).

46. Forner, D., Erseghe, T., Tomasin, S., and Tenti, P. On efficient use of local sources in smart grids with power quality constraints. *IEEE SmartGridComm'10*, pp. 555–560, (2010).

47. FutuRed. Spanish electrical grid platform, strategic vision document (2009).

48. Anderson, R., Boulanger, A., Powell, W., and Scott, W. Adaptive stochastic control for the smart grid. *Proc. IEEE*, 99(6), 1098–1115 (2011).

49. Aquino-Lugo, A. and Overbye, T. Agent technologies for control application in the power grid. In *43th Hawaii International Conference on System Sciences* (January 5–8, 2010), sponsored by Shidler College of Business, University of Hawaii at Manoa, Hawaii. pp. 1–10 (2010).

50. Austin Energy. Austin Energy Smart Grid Program, http://www.austinenergy.com/About%20Us/Company%20Profile/smartGrid/indx.htm (2010) (Accessed March 20, 2014).

51. Bakker, V., Bosman, M., Molderink, A., Hurink, J., and Smit, G. Demand side load management using a three step optimization methodology. In *IEEE SmartGridComm'10*, pp. 431–436 (2010).

52. Chertkov, M., Pan, F., and Stepanov, M. Predicting failures in power grids: The case of static overloads. *IEEE Trans. Smart Grid*, 2(1), 162–172 (2011).

53. Cho, H., Yamazaki, T., and Hahn, M. Aero: Extraction of user's activities from electric power consumption data. *IEEE Trans. Consum Electron.*, 56(3), 2011–2018 (2010).

54. Coll-Mayor, D., Paget, M., and Lightner, E. Future intelligent power grids: Analysis of the vision in the European Union and the United States. *Energy Policy*, 35, 2453–2465 (2007).

55. Li, F., Qiao, W., Sun, H., Wan, H., Wang, J., Xia, H., Xu, Z., and Zhang, P. Smart transmission grid: Vision and framework. *IEEE Trans. Smart Grid*, 1(2), 168–177 (2010).

56. Moslehi, K. and Kumar, R. A reliability perspective of the smart grid. *IEEE Trans. Smart Grid*, 1(1), 57–64 (2010).

57. Ipakchi, A. and Albuyeh, F. Grid of the future. *IEEE Power Energy Mag.*, 7(2), 52–62 (2009).

58. Bressan, N., Bazzaco, L., Bui, N., Casari, P., Vangelista, L., and Zorzi, M. The deployment of a smart monitoring system using wireless sensors and actuators networks. In *IEEE SmartGridComm'10*, pp. 49–54 (2010).

59. Mohagheghi, S., Stoupis, J., Wang, Z., Li, Z., and Kazemzadeh, H. Demand response architecture: Integration into the distribution management system. In *IEEE SmartGridComm'10*, pp. 501–506 (2010).

60. Mohsenian-Rad, A. and Leon-Garcia, A. Optimal residential load control with price prediction in real-time electricity pricing environments. *IEEE Trans. Smart Grid*, 1(2), 120–133 (2010).

61. Mohsenian-Rad, A., Wong, V., Jatskevich, J., Schober, R., and Leon-Garcia, A. Autonomous demand-side management based on game-theoretic energy consumption scheduling for the future smart grid. *IEEE Trans. Smart Grid*, 1(3), 320–331 (2010).

62. Takuno, T., Koyama, M., and Hikihara, T. In-home power distribution systems by circuit switching and power packet dispatching. In *IEEE SmartGridComm'10*, pp. 427–430 (2010).

63. Takuno, T., Hikihara, T., Tsuno, T., and Hatsukawa, S. HF gate drive circuit for a normally-on SiC JFET with inherent safety. In *13th European Conference on Power Electronics and Applications (EPE2009)* (September 8–10, 2009), Lille, France, pp. 1–4 (2009).

64. Chen, S., Song, S., Li, L., and Shen, J. Survey on smart grid technology (*in Chinese*). *Power Syst. Technol.*, 33(8), 1–7 (April 2009).

65. Hassan, R. and Radman, G. Survey on smart grid. In *IEEE SoutheastCon 2010*, pp. 210–213 (March 18–21, 2010).

66. Uslar, M., Rohjansand, S., Bleiker, R., González, J., Specht, M., Suding, T., and Weidelt, T. Survey of smart grid standardization studies and recommendations—Part 2. In *IEEE PES'10*, pp. 1–6 (2010).

67. Rohjansand, S., Uslar, M., Bleiker, R., González, J., Specht, M., Suding, T., and Weidelt, T. Survey of smart grid standardization studies and recommendations. In *IEEE SmartGridComm'10*, pp. 583–587 (2010).

68. Vasconcelos, J. Survey of regulatory and technological developments concerning smart metering in the European Union electricity market, http://cadmus.eui.eu/handle/1814/9267. *EUI RSCAS PP* (2008).

69. Brown, H. and Suryanarayanan, S. A survey seeking a definition of a smart distribution system. In *North American Power Symposium'09* (October 4–6, 2009), Mississippi State University, Starkville, MS, pp. 1–7 (2009).

70. Baumeister, T. Literature review on smart grid cyber security, Technical Report, http://csdl.ics.hawaii.edu/techreports/10-11/10-11.pdf (2010).

71. Chen, T. Survey of cyber security issues in smart grids. *Cyber Security, Situation Management, and Impact Assessment II; and Visual Analytics for Homeland Defense and Security II (part of SPIE DSS 2010)*, pp. 77090D-1–77090D-11 (2010).

72. Akyol, B., Kirkham, H., Clements, S., and Hadley, M. A survey of wireless communications for the electric power system. Prepared for the U.S. Department of Energy (2010).

73. Gungor, V. and Lambert, F. A survey on communication networks for electric system automation. *Comput. Netw.*, 50(7), 877–897 (2006).

74. Wang, W., Xu, Y., and Khanna, M. A survey on the communication architectures in smart grid. *Comput. Netw.*, 55, 3604–3629 (2011).

75. Locke, G. and Gallagher, P. NIST framework and roadmap for smart grid interoperability standards, release 1.0, Office of national coordinator for smart grid interoperability, NIST special publication 1108 jointly with Department of Commerce (January, 2010), http://www.nist.gov/publicaffairs/releases/upload/smartgrid interoperability final pdf.

76. Neely, M., Tehrani, A., and Dimakis, A. Efficient algorithms for renewable energy allocation to delay tolerant consumers. In *IEEE SmartGridComm'10*, Rethymnon, Greece (August 22–27, 2004), pp. 549–554 (2010).

77. Sortomme, E., Hindi, M., MacPherson, S., and Venkata, S. Coordinated charging of plug-in hybrid electric vehicles to minimize distribution system losses. *IEEE Trans. Smart Grid*, 2(1), 198–205 (2011).

78. Caldon, R., Patria, A., and Turri, R. Optimal control of a distribution system with a virtual power plant. In *Bulk Power System Dynamics and Control Conference*, Rethymnon, Greece (August 22–27, 2004), pp. 278–284 (2004).

79. A smart grid is an optimized grid, a white paper by ABB, Car NC (2013).

80. Smart grid optimization, a white paper by ECHELON, Menwith Hill, U.K. (2013).

81. Ahat, M., Amor, S., Bui, M., Bui, A., Guerard, G., and Petermann, C. Smart grid and optimization. *Am. J. Operat. Res.*, 3, 196–206 (2013).

82. Nygard, K., Ghosen, S., Chowdhury, M., Loegering, D., and McCulloch, R. Optimization models for energy reallocation in a smart grid. In *IEEE, Information Workshop* (October 10–20, 2011), Casa de Cultura, Paraty, Brazil, p. 186 (2011).

83. Bu, S., Yu, F., and Liu, P. Stochastic unit commitment in smart grid communications. In *IEEE INFOCOM 2011 Workshop on Green Communications and Networking*, (April 10–15, 2011), Shanghai, China, pp. 307–312 (2011).

84. Bu, S., Yu, F., Liu, P., and Zhang, P. Distributed scheduling in smart grid communications with dynamic power demands and intermittent renewable energy resources. In *IEEE ICC'11 Workshop on Smart Grid Communications* (October 17–20, 2011), Brussels, Belgium (2011).

85. Rahimi, F. and Ipakchi, A. Demand response as a market resource under the smart grid paradigm. *IEEE Trans. Smart Grid*, 1(1), 82–88 (2010).

86. Sauter, T. and Lobashov, M. End-to-end communication architecture for smart grids. *IEEE Trans. Ind. Electron.*, 58(4), 1218–1228 (2011).

87. Schneider, K., Gerkensmeyer, C., Kintner-Meyer, M., and Fletcher, R. Impact assessment of plug-in hybrid vehicles on pacific northwest distribution systems. In *Power & Energy Society General Meeting* (July 20–24, 2008), Pittsburgh, PA, pp. 1–6 (2008).

88. Simmhan, Y., Giakkoupis, M., Cao, B., and Prasanna, V. On using cloud platforms in a software architecture for smart energy grid. In *IEEE Second International Conference on Cloud computing (Cloudcom)* (November 30–December 3, 2010), Indiana University, Indianapolis IN (December 2010).

89. Barmada, S., Musolino, A., Raugi, M., Rizzo, R., and Tucci, M. A wavelet based method for the analysis of impulsive noise due to switch commutations in power line communication (PLC) systems. *IEEE Trans. Smart Grid*, 2(1), 92–101 (2011).

90. Bennett, C. and Highfill, D. Networking AMI smart meters. In *IEEE Energy 2030 Conference'08* (November 17–18, 2008), Atlanta, GA, pp. 1–8 (2008).

91. Berthier, R., Sanders, W., and Khurana, H. Intrusion detection for advanced metering infrastructures: Requirements and architectural directions. In *IEEE SmartGridComm'10*, pp. 350–355 (2010).

92. Best, R., Morrow, D., Laverty, D., and Crossley, P. Synchro-phasor broadcast over Internet protocol for distributed generator synchronization. *IEEE Trans. Power Deliv.*, 25(4), 2835–2841 (2010).

93. Bobba, R., Rogers, K., Wang, Q., Khurana, H., Nahrstedt, K., and Overbye, T. Detecting false data injection attacks on DC state estimation. In *The First Workshop on Secure Control Systems'10* (April 12, 2010), Stockholm, Sweden, pp. 1–9 (2010).

94. Bonanomi, P. Phase angle measurements with synchronized clocks principle and applications. *IEEE Trans. Power Apparatus Syst.*, 100(12), 5036–5043 (1981).

95. Borghetti, A., Nucci, C., Paolone, M., Ciappi, G., and Solari, A. Synchronized phasors monitoring during the islanding maneuver of an active distribution network. *IEEE Trans. Smart Grid*, 2(1), 82–91 (2011).

96. Al-Nasseri, H. and Redfern, M. A new voltage based relay scheme to protect micro-grids dominated by embedded generation using solidstate converters. In *19th International Conference Electricity Distribution* (May 21–24, 2007), Vienna, Austria, pp. 1–4 (2007).

97. American Transmission Company. American Transmission Company Phasor Measurement Unit Project Description (2011), http://www.smartgrid.gov/sites/default/files/09-0282-atc-project-description-07-11-11.pdf (Accessed February 2012).

98. Andersen, P., Poulsen, B., Decker, M., Træholt, C., and Østergaard, J. Evaluation of a generic virtual power plant framework using service oriented architecture. In *IEEE PECon'08*, pp. 1212–1217 (2008).

99. Baldick, R., Chowdhury, B., Dobson, I. et al. Initial review of methods for cascading failure analysis in electric power transmission systems. In *IEEE Power and Energy Society General Meeting'08* (July 20–24, 2008), Pittsburgh, PA, pp. 1–8 (2008).

100. Wang, Y., Lin, W., and Zhang, T. Study on security of wireless sensor networks in smart grid. In *2010 International Conference on Power System Technology* (October 24–28, 2010), Hangzhou, Zhejiang, China, pp. 1–7 (2010).

101. Japan. Japan's roadmap to international standardization for smart grid and collaborations with other countries (2010).

102. Gungor, V., Lu, B., and Hancke, G. Opportunities and challenges of wireless sensor networks in smart grid. *IEEE Trans. Ind. Electron.*, 57(10), 3557–3564 (2010).

103. Parikh, P., Kanabar, M., and Sidhu, T. Opportunities and challenges of wireless communication technologies for smart grid applications. In *IEEE Power and Energy Society General Meeting'10* (July 25–29, 2010), Minneapolis, MN.

104. Parvania, M. and Fotuhi-Firuzabad, M. Demand response scheduling by stochastic SCUC. *IEEE Trans. Smart Grid*, 1(1), 89–98 (2010).

105. Bose, A. Smart transmission grid applications and their supporting infrastructure. *IEEE Trans. Smart Grid*, 1(1), 11–19 (2010).

106. Bou Ghosn, S., Ranganathan, P., Salem, S., Tang, J., Loegering, D., and Nygard, K. Agent-oriented designs for a self healing smart grid. In *IEEE SmartGridComm'10*, pp. 461–466 (2010).

107. Brown, R. Impact of smart grid on distribution system design. In *IEEE Power and Energy Society General Meeting* (July 20–24, 2008), Pittsburgh, PA—*Conversion and Delivery of Electrical Energy in the 21st Century*, pp. 1–4 (2008).

108. Brucoli, M. and Green, T. Fault behavior in islanded micro grids. In *19th International Conference on Electricity Distribution* (May 21–24, 2007), Vienna, Austria, pp. 1–4 (2007).

109. Cai, Y., Chow, M.-Y., Lu, W., and Li, L. Statistical feature selection from massive data in distribution fault diagnosis. *IEEE Trans. Power Syst.*, 25(2), 642–648 (2010).

110. Calderaro, V., Hadjicostis, C., Piccolo, A., and Siano, P. Failure identification in smart grids based on Petri Net modeling. *IEEE Trans. Ind. Electron.*, 58(10), 4613–4623 (2011).

111. Caron, S. and Kesidis, G. Incentive-based energy consumption scheduling algorithms for the smart grid. In *IEEE SmartGridComm'10*, pp. 391–396 (2010).

112. Chen, L., Li, N., Low, S., and Doyle, J. Two market models for demand response in power networks. In *IEEE SmartGridComm'10*, pp. 397–402 (2010).

113. Chen, X., Dinh, H., and Wang, B. Cascading failures in smart grid—Benefits of distributed generation. In *IEEE SmartGridComm'10*, pp. 73–78 (2010).

114. Clarke, E. Multi part pricing of public goods. *Public Choice*, 11(1), 17–33 (1971).

115. Cleveland, F. Cyber security issues for advanced metering infrastructure (AMI). In *IEEE Power and Energy Society General Meeting: Conversion and Delivery of Electrical Energy in the 21st Century* (July 20–24, 2008), Pittsburgh, PA, pp. 1–5 (2008).

116. Colson, C. and Nehrir, M. A review of challenges to real time power management of micro grids. In *IEEE Power & Energy Society General Meeting* (July 26–30, 2009), Calgary, Alberta, Canada, pp. 1–8 (2009).

117. Conejo, A., Morales, J., and Baringo, L. Real-time demand response model. *IEEE Trans. Smart Grid*, 1(3), 236–242 (2010).

118. D´an, G. and Sandberg, H. Stealth attacks and protection schemes for state estimators in power systems. In *IEEE SmartGridComm'10*, pp. 214–219 (2010).

119. Deep, U., Petersen, B., and Meng, J. A smart microcontroller based iridium satellite-communication architecture for a remote renewable energy source. *IEEE Trans. Power Deliv.*, 24(4), 1869–1875 (2009).

120. Department of Energy, Office of Electricity Delivery and Energy Reliability. Study of security attributes of smart grid systems—Current cyber security issues (2009), http://www.inl.gov/scada/publications/d/securing the smart grid current issues.pdf (Accessed December 2009).

121. Driesen, J. and Katiraei, F. Design for distributed energy resources. *IEEE Power Energy Mag.*, 6(3), 30–40 (2008).

122. Driesen, J., Vermeyen, P., and Belmans, R. Protection issues in microgrids with multiple distributed generation units. In *Power Conversion Conference'07* (April 2–5, 2007), Nagoya, Japan, pp. 646–653 (2007).

123. Efthymiou, C. and Kalogridis, G. Smart grid privacy via anonymization of smart metering data. In *IEEE SmartGridComm'10*, pp. 238–243 (2010).

124. Ettoumi, F., Sauvageot, H., and Adane, A. Statistical bivariate modeling of wind using first-order markov chain and weibull distribution. *Renew. Energy*, 28(11), 1787–1802 (2003).

125. European Committee for Electrotechnical Standardization (CENELEC). Smart Meters Coordination Group: Report of the second meeting held on 2009-09-28 and approval of SM-CG work program for EC submission (2009).

126. European Smart Grids Technology Platform. Vision and strategy for Europe's electricity networks of the future, (2006) http://www.smartgrids.eu/documents/vision.pdf (Accessed December 2006).

127. Fang, X., Yang, D., and. Xue, G. Online strategizing distributed renewable energy resource access in islanded microgrids. In *IEEE Globecom'11* (2011).

128. Federal Energy Regulatory Commission. Assessment of demand response and advanced metering. Staff Report. (2010), http://www.ferc.gov/legal/staff-reports/2010-dr-report.pdf (Accessed December 2010).

129. Guan, X., Xu, Z., and Jia, Q.-S. Energy-efficient buildings facilitated by microgrid. *IEEE Trans. Smart Grid*, 1(3), 243–252 (2010).

130. Hatami, S. and Pedram, M. Minimizing the electricity bill of cooperative users under a quasi-dynamic pricing model. In *IEEE SmartGrid-Comm'10*, pp. 421–426 (2010).

131. He, M., Murugesan, S., and Zhang, J. Multiple timescale dispatch and scheduling for stochastic reliability in smart grids with wind generation integration. In *IEEE INFOCOM Mini-Conference* (April 10–15, 2011), Shanghai, China, pp. 461–465 (2011).

132. Ibars, C., Navarro, M., and Giupponi, L. Distributed demand management in smart grid with a congestion game. In *IEEE SmartGridComm'10*, pp. 495–500 (2010).

133. Kim, Y.-J., Thottan, M., Kolesnikov, V., and Lee, W. A secure decentralized data-centric information infrastructure for smart grid. *IEEE Commun. Mag.*, 48(11), 58–65 (2010).

134. Kishore, S. and Snyder, L. Control mechanisms for residential electricity demand in smart grids. In *IEEE SmartGridComm'10*, pp. 443–448 (2010).

135. Kroposki, B., Margolis, R., Kuswa, G., Torres, J., Bower, W., Key, T., and Ton, D. Renewable systems interconnection: Executive summary. Technical Report NREL/TP-581-42292, U.S. Department of Energy (2008).

136. Laaksonen, H. Protection principles for future microgrids. *IEEE Trans. Power Electron.*, 25(12), 2910–2918 (2010).

137. Lasseter, R. Smart distribution: Coupled micro grids. *Proc. IEEE*, 99(6), 1074–1082 (2011).

138. Lasseter, R. and Paigi, P. Micro grid: A conceptual solution. In *PESC'04*, pp. 4285–4290 (2004).

139. Leon, R., Vittal, V., and Manimaran, G. Application of sensor network for secure electric energy infrastructure. *IEEE Trans. Power Deliv.*, 22(2), 1021–1028 (2007).

140. Li, H., Lai, L., and Qiu, R. Communication capacity requirement for reliable and secure state estimation in smart grid. In *IEEE SmartGrid-Comm'10*, pp. 191–196 (2010).

141. Li, H., Mao, R., Lai, L., and Qiu, R. Compressed meter reading for delay-sensitive and secure load report in smart grid. In *IEEE SmartGridComm'10*, pp. 114–119 (2010).

142. Li, H. and Zhang, W. Qos routing in smart grid. In *IEEE Globecom'10*, pp. 1–6 (2010).

143. Lisovich, M. and Wicker, S. Privacy concerns in upcoming residential and commercial demand-response systems. In *The TRUST 2008 Spring Conference* (April 2–3, 2008), Claremont Resort and Spa, Berkley, CA, (2008).

144. Lu, B., Habetler, T., Harley, R., and Gutiérrez, J. Applying wireless sensor networks in industrial plant energy management systems—Part I: A closed-loop scheme. *IEEE Sensors*, 5(2), 145–150 (2005).

145. Lu, Z., Lu, X., Wang, W., and Wang, C. Review and evaluation of security threats on the communication networks in the smart grid. In *Military Communications Conference'2010* (October 31–November 3), San Jose, CA, pp. 1830–1835 (2010).

146. McDaniel, P. and McLaughlin, S. Security and privacy challenges in the smart grid. *IEEE Sec. Privacy*, 7(3), 75–77 (2009).

147. McLaughlin, S., Podkuiko, D., and McDaniel, P. Energy theft in the advanced metering infrastructure. In *Fourth Workshop on Critical Information Infrastructures Security* (September 30–October 2, 2009), Bonn, Germany, pp. 176–187 (2009).

148. Metke, A. and Ekl, R. Security technology for smart grid networks. *IEEE Trans. Smart Grid*, 1(1), 99–107 (2010).

149. Mitra, J., Patra, M.S.B., and Ranade, S. Reliability stipulated microgrid architecture using particle swarm optimization. In *Ninth International Conference on Probabilistic Methods Applied to Power System* (June 11–15, 2006), Stockholm, Sweden, pp. 1–7 (2006).

150. Molderink, A., Bakker, V., Bosman, M., Hurink, J., and Smit, G. Management and control of domestic smart grid technology. *IEEE Trans. Smart Grid*, 1(2), 109–119 (2010).

151. Nikkhajoei, H. and Lasseter, R. Microgrid protection. In *IEEE Power Engineering Society General Meeting'07*, pp. 1–6 (2007).

152. Ning, J., Wang, J., Gao, W., and Liu, C. A wavelet-based data compression technique for smart grid. *IEEE Trans. Smart Grid*, 2(1), 212–218 (2011).

153. Ochoa, L. and Harrison, G. Minimizing energy losses: Optimal accommodation and smart operation of renewable distributed generation. *IEEE Trans. Power Syst.*, 26(1), 198–205 (2011).

154. Overman, T. and Sackman, R. High assurance smart grid: Smart grid control systems communications architecture. In *IEEE SmartGrid- Comm'10*, pp. 19–24 (2010).

155. Potter, C., Archambault, A., and Westric, K. Building a smarter smart grid through better renewable energy information. In *IEEE PSCE'09*, pp. 1–5 (2009).

156. Qiu, R., Chen, Z, Guo, N., Song, Y., Zhang, P., Li, H., and Lai, L. Towards a real-time cognitive radio network testbed: Architecture, hardware platform, and application to smart grid. In *Fifth IEEE Workshop on Networking Technologies for Software Defined Radio (SDR) Networks* (June 21, 2010), Boston, MA, pp. 1–6 (2010).

157. Qiu, R., Hu, Z., Chen, Z., Guo, N., Ranganathan, R., Hou, S., and Zheng, G. Cognitive radio network for the smart grid: Experimental system architecture, control algorithms, security, and microgrid test bed. *IEEE Trans. Smart Grid* 2(4), 724–740 (2011).

158. Ramachandran, B., Srivastava, S., Edrington, C., and Cartes, D. An intelligent auction scheme for smart grid market using a hybrid immune algorithm. *IEEE Trans. Ind. Electron.*, 58(10), 4603–4612 (2011).

159. Rial, A. and Danezis, G. Privacy-preserving smart metering, http://research.microsoft.com/pubs/141726/main.pdf

160. Roncero, J. Integration is key to smart grid management. In *IET-CIRED Seminar 2008: SmartGrids for Distribution*, pp. 1–4 (2008).

161. Roozbehani, M., Dahleh, M., and Mitter, S. Dynamic pricing and stabilization of supply and demand in modern electric power grids. In *IEEE SmartGridComm'10*, pp. 543–548 (2010).

162. Rusitschka, S., Eger, K., and Gerdes, C. Smart grid data cloud: A model for utilizing cloud computing in the smart grid domain. In *IEEE SmartGridComm'10*, pp. 483–488 (2010).

163. Russell, B. and Benner, C. Intelligent systems for improved reliability and failure diagnosis in distribution systems. *IEEE Trans. Smart Grid*, 1(1), 48–56 (2010).

164. Salomonsson, D., Söder, L., and Sannino, A. Protection of low-voltage dc microgrids. *IEEE Trans. Power Deliv.*, 24(3), 1045–1053 (2009).

165. Samadi, P., Mohsenian-Rad, A.-H., Schober, R., Wong, V., and Jatskevich, J. Optimal real-time pricing algorithm based on utility maximization for smart grid. In *IEEE SmartGridComm'10*, pp. 415–420 (2010).

166. Sortomme, E., Venkata, S., and Mitra, J. Microgrid protection using communication-assisted digital relays. *IEEE Trans. Power Deliv.*, 25(4), 2789–2796 (2010).

167. Souryal, M., Gentile, C., Griffith, D., Cypher, D., and Golmie, N. A methodology to evaluate wireless technologies for the smart grid. In *IEEE SmartGridComm'10*, pp. 356–361 (2010).

168. Sreesha, A., Somal, S., and Lu, I.-T. Cognitive radio based wireless sensor network architecture for smart grid utility. In *2011 IEEE Long Island Systems, Applications and Technology Conference (LISAT)* (May 6, 2011), Long Island, NY pp. 1–7 (2011).

169. Taneja, J., Culler, D., and Dutta, P. Towards cooperative grids: Sensor/actuator networks for renewables integration. In *IEEE SmartGrid-Comm'10*, pp. 531–536 (2010).

170. Tsoukalas, L. and Gao, R. From smart grids to an energy Internet: Assumptions, architectures and requirements. In *DRPT 2008*, pp. 94–98 (2008).

171. Tumilty, R., Elders, I., Burt, G., and McDonald, J. Coordinated protection, control & automation schemes for microgrids. *Int. J. Distributed Energy Res.*, 3(3), 225–241 (2007).

172. Vandoorn, T., Renders, B., Degroote, L., Meersman, B., and Vandevelde, L. Active load control in islanded microgrids based on the grid voltage. *IEEE Trans. Smart Grid*, 2(1), 139–151 (2011).

173. Wang, Z., Scaglione, A., and Thomas, R. Compressing electrical power grids. In *IEEE SmartGridComm'10*, pp. 13–18 (2010).

174. You, S., Træholt, C., and Poulsen, B. Generic virtual power plants: Management of distributed energy resources under liberalized electricity market. In *The Eighth International Conference on Advances in Power System Control, Operation and Management*, pp. 1–6 (2009).

175. Yu, Y. and Luan, W. Smart grid and its implementations (*in Chinese*). *CSEE*, 29(34), 1–8 (2009).

176. Yuan, Y., Li, Z., and Ren, K. Modeling load redistribution attacks in power systems. *IEEE Trans. Smart Grid*, 2(2), 382–390 (2011).

177. Zareipour, H., Bhattacharya, K., and Canizares, C. Distributed generation: Current status and challenges. In *NAPS'04*, pp. 1–8 (2004).

178. Zhang, P., Li, F., and Bhatt, N. Next-generation monitoring, analysis, and control for the future smart control center. *IEEE Trans. Smart Grid*, 1(2), 186–192 (2010).

179. Balijepalli, V., Pradhan, V., Khaparde, S., and Shereef, R. Review of demand response under smart grid paradigm, IEEE PES Innovative smart grid technologies-India, 978-1-4673-0315-6/11 (2011).

180. The future role and challenges of Energy Storage, a white paper of European commission directorate general for energy, pp. 1–35 (2012).

181. "Energy storage," *Wikipedia*, https://en.wikipedia.org/wiki/Energy_storage (last modified July 10, 2015).

182. Rastler, D., EPRI project manager, Electric energy storage technology options: A white paper primer on applications, costs, and benefits. (Free download) *EPRI*, Palo Alto, CA, Report No. 1020676, 2010. Accessed September 30, 2011.

183. The Boston Consulting Group: Revisiting Energy Storage, report (February 2011).

184. Wagner, L. Overview of energy storage methods, Mora Associates research report (December 2007).

185. Electrical Energy Storage, a white paper by International Electro technical Commission, Geneva, Switzerland (December, 2011).

186. Electric Power Research Institute: Electric Energy Storage Technology Options White Paper (2010).

187. Kawashima, M. Overview of Electric Power Storage, Internal paper of Tepco (2011); Kuhnhenn, E. and Ecke, J. *Power-to-Gas Stromspeicher, Gasproduktion, Biomethan Oder Flexible Last*, DVGW Energie/Wasserpraxis 7/8 (2011).

188. Bradbury, K. Energy storage technology review, an independent report (August 22, 2010).

189. Cheung, K., Cheung, S., Silva, R., Juvonen, M., Singh, R., and Woo, J. Large-scale energy storage systems, Imperial College London, London, U.K., ISE2 (2003).

190. Rastler, D. Electricity energy storage technology options, a white paper primer on applications, costs and benefits 1020676 EPRI Report, Palo Alto, CA (2010).

191. Carnegie, R., Gotham, D., Nderitu, D., and Preckel, P. Utility scale energy storage systems-benefits, applications and technologies, State utility forecasting group, pp. 1–90 (June 2013).

192. EPRI, EPRI-DOE handbook of energy storage for transmission and distribution applications. Technical Report, EPRI and U.S. Department of Energy (2003).

193. Gonzalez, A., Gallachir, B., McKeogh, E., and Lynch, K. Study of electricity storage technologies and their potential to address wind energy intermittency in Ireland. Tech. Rep., Sustainable Energy Research Group, University College Cork, Corcaigh, Ireland (2004).

194. Schoenung, S. and Hassenzahl, W. Long-vs. Short-Term energy storage technologies analysis a Life-Cycle cost study a study for the DOE energy storage systems program. Tech. Rep., SAND2003-2783, Sandia National Laboratories, Albuquerque, NM (2003).

195. Schoenung, S. and Eyer, J. Benefit/cost framework for evaluating modular energy storage study for the DOE energy storage system program. Sandia report SAND, Vol. 978 (2008).

196. Chen, H., Cong, T., Yang, W., Tan, C., Li, Y., and Ding, Y. Progress in electrical energy storage system: A critical review. *Progr. Nat. Sci.*, 19, 291312 (March 2009).

197. Furlong, E., Piemontesi, M., Prasad, P., and Sukumar, D. Advances in energy storage techniques for critical power systems, personal communication (2012).

198. *Five Minute Guide to Electricity Storage Technologies*, a publication by ARUP, www.arup.com (2010).

199. Wu, T. Energy Storage Methods, a report from Dept. of Electrical and Computer Engineering, the University of British Columbia, Vancouver, British Columbia, Canada (April 6, 2012).

200. Electricity storage, a post note from Parliamentary office of Science and Technology, No. 306 (April, 2008).

201. Hou, Y., Vidu, R., and Stroeve, P. Solar energy storage systems review, a report from California solar energy collaborative, University of California, Davis, CA (2011).

202. Pendick, D. Storing energy from the wind in compressed-air reservoirs, *New Sci.*, 195(2623), 44–47 (September 29, 2007). Accessed December 2007.

203. Bullis, K., "Liquefied air could power cars and store energy from sun and wind-a 19th century idea might lead to cleaner cars, larger-scale renewable energy", MIT Technology Review (May 20, 2013); www.technology review.com (Accessed December 2013).

204. Conway, E. World's biggest battery switched on in Alaska, *Telegraph.co.uk* (September 2, 2008).

205. "Hydrogen Infrastructure," *Wikipedia*, https://en.wikipedia.org/wiki/Hydrogen_infrastructure (last modified January 2, 2015).

206. Hydroelectric Power (PDF). United States Bureau of Reclamation. Accessed October 13, 2008.

207. Mancini, T. Advantages of Using Molten Salt, Sandia National Laboratories, Albuquerque, NM. Accessed December 2007 (http//www.scppa.org/Renewable–energy/ Solar thermal/NSTTF/Salt htm).

208. Atwater, T. and Dobley, A. *Metal/Air batteries,* Lindens Handbook of Batteries (2011).

209. Bullough, C. Advanced adiabatic compressed air energy storage for the integration of wind energy European. In *Wind Energy Conference and Exhibition*, London, U.K. (November 22–25, 2004).

210. Dötsch, C. Electrical energy storage from 100 kW—State of the art technologies, fields of use. *In Second International Renewable Energy Storage Conference*, Bonn, Germany (November 22, 2007).

211. Fujihara, T., Imano, H. and Oshima, K., Development of pump turbine for seawater pumped-storage power plant, *Hitachi Review*, 47(5):199–202 (1998).

212. Jähnig, D., Jaehnig, D., Hausner, R., Wagner, W., Isaksson, C., Thermo-chemical storage for solar space heating in a single-family house. In *10th International Conference on Thermal Energy Storage: Ecostock 2006*, Pomona, NJ (May 31–June 2, 2006).

213. Lailler, P. et al. Lead Acid Systems, INVESTIRE 2003, [per04]. M. Nakhamkin: Novel Compressed Air Energy Storage Concepts Developed by ESPC, EESAT (May 2007); http://www.tpower.co.uk/investire/ (Accessed December 2007).

214. Radgen, P. 30 Years compressed air energy storage plant Huntorf—Experiences and outlook. In *Third International Renewable Energy Storage Conference*, Berlin, Germany (November 24–25, 2008).

215. Schossig, P. Thermal energy storage, In *Third International Renewable Energy Storage Conference*, Berlin, Germany (November 24–25, 2008).

216. Tamme, R. Development of Storage Systems for SP Plants, DG TREN—DG RTD Consultative Seminar "*Concentrating Solar Power*," Brussels, Belgium (June 27, 2006).

217. Linden, D. *Handbook of Batteries*, 2nd edn., McGraw-Hill, New York (1995).

218. Fabjan, C., Garche, J., Harrer, B., JÄrissen, L., Kolbeck, C., Philippi, F., Tomazic, G., and Wagner, F. The vanadium redox-battery: An efficient storage unit for photovoltaic systems, *Electrochim. Acta*, 47(5), 825–831 (2001).

219. Martin, G. and Barnes, F. Aquifer underground pumped hydroelectric energy storage, In *Masters Abstracts International*, Vol. 46 (2007).

220. Cyphelly, I. Storage technology report ST8: Pneumatic storage. Technical Report WP ST8-PNEUMATIC STORAGE, Investire-Network (2003).

221. Sudworth, J. and Tilley, A. *The Sodium Sulfur Battery*, Springer, Berlin, Germany (1985).

222. Kamibayashi, M. and Tanaka, K. Recent sodium sulfur battery applications. In *2001 IEEE/PES Transmission and Distribution Conference and Exposition* (November 2, 2001), Atlanta, GA, Vol. 2 (2001).

223. Sudworth, J. The sodium/nickel chloride (ZEBRA) battery. *J. Power Sources*, 100,149–163 (November 2001).

224. Galloway, R. and Dustmann, C. ZEBRA battery-material cost availability and recycling. In *Proceeding of International Electric Vehicle Symposium (EVS-20)*, Long Beach, British Columbia, Canada, p. 19 (2003).

225. Sharma, A., Tyagi, V., Chen, C., and Buddhi, D. Review on thermal energy storage with phase change materials and applications. *Renew. Sustain. Energy Rev.*, 13,318–345 (2009).

226. Harkins, E., Pando, M., and Sobel, D. Electrical energy storage using fuel cell technology, Senior design project, Department of Chemical and Biomolecular Engineering, University of Pennsylvania, Philadelphia, PA (2011).

227. Ribeiro, P., Johnson, B., Crow, M., Arsoy, A., and Liu, Y. Energy storage systems for advanced power applications. *Proc. IEEE*, 89(12), 1744–1756 (2001).

228. Ruddell, A. Storage technology report ST6: Flywheel. Technical Report WP ST 6 FLYWHEEL, Investire-Network (2003).

229. Willer, B. Storage technology report ST5: Supercaps. Technical Report WP ST 3 Supercaps, Investire-Network (2003).

230. Hassenzahl, W. Superconducting magnetic energy storage. *IEEE Trans. Magnet.*, 25(2), 750–758 (1989).

231. Buckles, W. Hassenzahl, W., Div, P., Supercond, A., and Middleton, W. Superconducting magnetic energy storage. *IEEE Power Eng. Rev.*, 20(5), 16–20 (2000).

232. Hammond, R., Everingham, S., and Srinivasan, D. Batteries for stationary standby and for stationary cycling applications part 1: Standby vs. cycling—definitions and concepts. In *Power Engineering Society General Meeting* (July 13–17, 2003) Toronto, Ontario, Canada, *IEEE*, Vol. 1, p. 145 (2003).

233. McDowall, J. Batteries for stationary standby and for stationary cycling applications part 3: Operating issues. In *Power Engineering Society General Meeting, 2003, IEEE* (July 13–17, 2003) Toronto, Ontario, Canada, Vol. 1, p. 154 (2003).

234. Symons, P. Batteries for stationary standby and for stationary cycling applications part 4: Charge management. In *Power Engineering Society General Meeting, 2003, IEEE* (July 13–17, 2003) Toronto, Ontario, Canada, Vol. 1, p. 157 (2003).

235. Rodriguez, G. Operating experience with the chino 10 MW/40 MWh battery energy storage facility. In *Energy Conversion Engineering Conference, 1989. IECEC-89, Proceedings of the 24th Intersociety*, Vol. 3, pp. 1641–1645 (1989).

236. Parker, C. Lead-acid battery energy-storage systems for electricity supply networks. *J. Power Sources*, 100,18–28 (November 2001).

237. Kodama, E. and Kurashima, Y. Development of a compact sodium sulphur battery. *Power Eng. J.*, 13(3), 136–141 (1999).

238. Lee, S. and Shah, Y. *Biofuels and Bioenergy: Technologies and Processes*, CRC Press, New York (September 2012).

239. "Battery (electricity)," *Wikipedia*, https://en.wikipedia.org/wiki/Battery_(electricity) (last modified June 27, 2015).

240. "Solar pond," *Wikipedia*, https://en.wikipedia.org/wiki/Solar_pond (last modified June 19, 2015).

241. "Thermal energy storage," *Wikipedia*, https://en.wikipedia.org/wiki/Seasonal_thermal_energy_storage (last modified June 11, 2015).

242. Hesaraki, A., Holmberg, S., and Haghighat, F. Seasonal thermal energy storage with heat pumps and low temperatures in building projects—A comparative review. *Renew. Sustain. Energy Rev.*, 43, 1199–1213 (March 2015).

243. "Seasonal thermal energy storage," *Wikipedia*, https://en.wikipedia.org/wiki/Seasonal_thermal_energy_storage (last modified April 9, 2015).

244. Sadr, F. Thermochemical heat pump, Renewable energy course project, Energy Research center of Netherlands, personal communication (2013).

245. Kato, Y. Thermochemical energy storage-possibility of chemical heat pump technologies, High density thermal energy storage workshop, hosted by Advanced Research projects agency-Energy (ARPA-E), Hilton, Arlington, VA (January 31, 2011).

246. Tahat, M., Babushaq, R., O'Callaghan, P., and Probert, S. Integrated thermochemical heat pump/energy-storage. *Int. J. Energy Res.*, 19(7), 603–613 (September 1995). Article published on line March 14, 2007, Wiley & Sons, New York.

247. Bougard, J. and Jadot, R. Thermochemical energy storage and chemical heat pumps, personal communication (2013).

248. "Grid energy storage," *Wikipedia*, https://en.wikipedia.org/wiki/Grid_energy_storage (last modified July 5, 2015).

249. Energy Department Releases Grid Energy Storage Report (December 12, 2013), http://cleantechnica.com/2013/02/21/lightsail-gets-5-5-million-for-compressed-air-energy-storage/#gsc.tab=0.

250. Petinrin, J. and Shaaban, M. Implementation of energy storage in a future smart grid. *Aust. J. Basic Appl. Sci.*, 7(4), 273–279 (2013).

251. Roberts, B. The role of energy storage in development of smart grids. *Proc. IEEE*, 99(6), 1139 (June 2011).

252. Espinar, B. and Mayer, D. The role of energy storage for mini-grid stabilization, report. IEAPVPS T11-0X:2011 (2011).

253. VDE—ETG Energy Storage Task Force: Energy storage in power supply systems with a high share of renewable energy sources significance—State of the art—Need for action, report (December 2008).

254. Eyer, J. and Corey, G. Energy storage for the electricity grid: Benefits and market potential assessment guide, report, Sandia National Laboratories, Albuquerque, NM (February 2010).

255. Hoffmann, C. Design of transport- and storage capacities in energy supply systems with high shares of renewable energies, IRES, Berlin, Germany (2010).

256. International Energy Agency: Prospects for Large Scale Energy Storage in Decarbonized Grids, report (2009).

257. Farley, P. Largest solar thermal storage plant to startup. http://spectrum.ieee.org/energy/environment/largest-solar-thermal-storage-plant-to-start-up, Article 2008. Accessed July 27, 2011.

258. Patel, P. Smarter grid batteries that go with the flow, http://spectrum.ieee.org/energy/the-smarter-grid/batteries-that-go-with-the-flow. Article, May 2010. Accessed April 10, 2011.

259. Ridge Energy Storage & Grid Services LLC. http://www.ridgeenergystorage.com, Accessed August 8, 2011.

260. Shinichi INAGE: Prospective on the Decarbonized Power Grid, IEC/MSB/EES Workshop, Germany (May 31– June 1, 2011).

261. Grid energy storage, a report by U.S. Department of Energy, Washington, DC (December 2013).

262. Daud, M., Mohamed, A., and Hannan, M. A review of the integration of energy storage systems (ESS) for utility grid support. *Electric. Rev.*, ISSN: 0033-2097, R. 88 NR 10a,185–191 (2012).

263. Storage solutions allow for renewable energy on demand, a publication by Revolt and Milena Gonzalez revolt, 12, 10 (2013).

264. Energy storage: The key to a smart power grid, BBC news business, U.K. (April 23, 2014).
265. Perrin, M. et al. Investigation on Storage Technologies for Intermittent Renewable Energies: Evaluation and recommended R&D strategy, INVESTIRE. Final Technical Report of European Community project ENK5-CT-2000-20336 (2004).
266. Denholm, P., Ela, E., Kirby, B., and Milligan, M. Role of energy storage with renewable electricity generation. Technical Report NREL/TP-6A2-47187, NREL (2010).
267. Kumar, R. Assuring voltage stability in the smart grid. Innovative smart grid technologies (ISGT), IEEE PES, IEEE (2011).
268. Hamidi, V. and Smith, K. Smart grid technology review within the transmission and distribution sector. Innovative smart technologies conference Europe (ISGT Europe), IEEE PES, IEEE (2010).
269. Hamsic, N. et al. Stabilizing the grid voltage and frequency in isolated power systems using flywheel energy storage system. The Great Wall World Renewable Energy Forum, China (2006).
270. Rebours, Y. and Hirschen, D. What is spinning reserve? The University of Manchester, Manchester, U.K., pp. 1–11 (2005).
271. Noce, C., Riva, S., Sapienzal, C. and Brenna, M., Electrical energy storage in smart grid: Black start study using a real time digital simulator. In *Third IEEE International Symposium on Power Electronics for Distributed Generation Systems (PEDG), 2012, 3rd IEEE International Symposium* (June 25–28, 2012), Aalborg, Denmark (2012).
272. Choi, S. et al. Energy storage systems in distributed generation schemes. In *Power and Energy Society General Meeting-Conversion and Delivery of Electrical Energy in the 21st Century, 2008 IEEE* (July 24–28, 2008), Pittsburgh, PA (2008).
273. Styczynski, Z., Lombardi, P., Seethapathy, R., Piekutowski, M., Ohler, C., et al., Electrical energy storage and its tasks in the integration of wide scale renewable resources. In *Integration of Wide Scale Renewable Resources into the Power Delivery System, 2009 CIGRE/IEEE PES Joint Symposium, IEEE* July 29–31, 2009, Calgary, Alberta, Canada (2009).
274. Petinrin, J. and Shaaban, M., Implementation of Energy storage in a fiture smart grid. *Austr. J. Basic Appl. Sci.*, 7(4) 273–279 (2013) ISSN–1991–8178.
275. Bjelovuk, G. and Nourai, A., et al., Community energy storage (CES) and the smart grid: A game changer, AEP presentation to the electricity storage association (2009).
276. Delille, G. et al. Dynamic frequency control support: A virtual inertia provided by distributed energy storage to isolated power systems. In *Innovative Smart Grid Technologies Conference Europe (ISGT Europe), 2010 IEES PES, IEEE,* October 11–13, 2010, Gothenburg, Sweden (2010).
277. Zhou, L. and Qi, Z. Modelling and simulation of flywheel energy storage system with IPMSM for voltage sags in distributed power network. In *International Conference on Mechatronics and Automation, 2009, ICMA 2009, IEEE,* (August 9–12), 2005, Changchun, China, (2009).
278. Energy Storage Association E. Technologies. http://www.electricitystorage.org/ESA/technologies/ (Accessed July 2015).
279. Bhatia, R.S., Jain, S.P., and Singh, B. Battery energy storage system for power conditioning of renewable energy sources. In *Proceedings of International Conference on Power Electronics and Drives Systems*, pp. 501–506 (2005).
280. J Power Co E. Okinawa sea water pumped storage, Japan, http://www.jpower.co.jp/english/international/consultation/detail/se_as_japan24.pdf (Accessed July 2015).
281. Author, N. Review of electrical energy storage technologies and systems and of their potential for the U.K. *EA Technol.*, 21, 1–34 (2004).
282. Ibrahim, H., Ilinca, A., and Perron, J. Energy storage systems—Characteristics and comparisons. *Renew. Sustain. Energy Rev.*, 12(5), 1221–1250 (2008).

283. Sawin, J., Martinot, E., and Appleyard, D. Renewables continue remarkable growth, http://www.renewableenergyworld.com/rea/news/article/2010/09/renewables-continue-remarkable-growth (Accessed December 2010).

284. Sterner, M. Bioenergy and renewable power methane in integrated 100% renewable energy systems—Limiting global warming by transforming energy systems, Dissertation, University of Kassel, Kassel, Germany (July 2009).

285. Sterner, M. Jentsch, M., Gerhardt, N., Trost, T., Specht, M., Sturmer, B., Zuberbuhler, U. Power-to-Gas: Storing renewables by linking power and gas grids. In Presentation, *IEC Workshop EES*, Freiburg, Germany (May 31, 2011).

286. Omar, R. and Rahim, N.A. Implementation and control of a dynamic voltage restorer using Space Vector Pulse Width Modulation (SVPWM) for voltage sag mitigation. In *Proceedings of International Conference for Technical Postgraduates* (December 14–15, 2009), Kuala Lumpur, Malaysia, Vol. 1, pp. 1–6 (2009), published by IEEE, Curan Associates Inc. (April, 2010)

287. Virulkar, V. and Aware, M. Analysis of DSTATCOM with BESS for mitigation of flicker. In *Proceedings of International Conference on Control, Automation, Communication and Energy Conservation*, Perundurai, Tamil Nadu, pp. 1–7 (2009).

288. Moore, T. and Douglas, J. Energy storage—Big opportunities on a smaller scale. *EPRI J.*, 16–23 Spring, 2006 (2006).

289. Lazarewicz, M.L. and Ryan, T.M. Integration of flywheel based energy storage for frequency regulation in deregulated markets. In *Proceedings of IEEE Power and Energy Society General Meeting* (July 25–29, 2010), Minneapolis, MN, pp. 1–6, IEEE (2010).

290. Lu, N., Weimar, M.R., and Makarov, Y.V., Rudolph, F., Murthy, S.I, Arseneaux, S. and Loutan, C. An evaluation of the flywheel potential for providing regulation service in California. In *Proceedings of IEEE Power and Energy Society General Meeting* (July 25–29, 2010), Minneapolis, MN, pp. 1–6, IEEE (2010).

291. Hsu, C.S. and Lee, W.J. Superconducting magnetic energy storage for power system applications. *IEEE Trans. Ind. Appl.*, 29(5), 990–996 (1993).

292. Rogers, J.D., Schermer, R.I., Miller, B.L., and Hauer, J.F. 30-MJ superconducting magnetic energy storage system for electric utility transmission stabilization. *Proc. IEEE*, 71(9), 1099–1107 (1983).

293. Chen, S.-S., Wang, L., Chen, Z., Lee, W.-J., Power-flow control and transient-stability enhancement of a large-scale wind power generation system using a superconducting magnetic energy storage (SMES) unit. In *Proceedings of IEEE Power and Energy Society General Meeting—Conversion and Delivery of Electrical Energy in the 21st Century*, Pittsburgh, PA, pp. 1–6 (2008).

294. Ngamroo, I., Supriyadi, A., Dechanupapritth, S. and Mitani, R., Stabilization of tie-line power oscillations by robust SMES in interconnected power system with large wind farms. In *Proceedings of Asia and Pacific Transmission & Distribution Conference & Exposition*, Seoul, South Korea, pp. 1–4 (2009).

295. Vazquez, S., Lukic, S., Galvan, E., Franquelo, L., Carrasco, J., Energy storage systems for transport and grid applications. *IEEE Trans. Ind. Electron.*, 57(12), 3881–3895 (2010).

296. Doughty, D.H. et al. Batteries for large-scale stationary electrical energy storage. *Electrochem. Soc. Interface*, 19:49–53 (2010).

297. Eberle, U. and von Helmolt, R. Sustainable transportation based on electric vehicle concepts: Energy and Environmental Sci. 3, 689–699 (2010); DOI: 10.1039/C001674H.

298. Tomić, J. and Kempton, W. Using fleets of electric-drive vehicles for grid support. *J. Power Sources*, 168(2), 459–468 (2002).

299. Hadley, S. and Tsvetkova, A. Potential impacts of plug-in hybrid electric vehicles on regional power generation. *Electr. J.*, 22(10), 56–68 (2009).

300. Han, S., Han, S., and Sezaki, K. Development of an optimal vehicle-to grid aggregator for frequency regulation. *IEEE Trans. Smart Grid*, 1(1), 65–72 (2010).

301. Brooks, A. and Thesen, S. PG&E and Tesla Motors: Vehicle to grid demonstration and evaluation program, http://spinnovation.com/sn/Articles on V2G/PG and E and Tesla Motors—Vehicle to Grid Demonstration and Evaluation Program.pdf.

302. WINMEC, UCLA. WINSmartEV—Electric Vehicle (EV) Integration into Smart Grid with UCLA WIN Smart Grid Technology, http://www.winmec.ucla.edu/ev.asp (Accessed March 2012).

303. Saber, A. and Venayagamoorthy, G. Unit commitment with vehicle-to-grid using particle swarm optimization. In *IEEE Bucharest Power Tech Conference*, (June 28–July 2, 2009), Bucharest, Romania, pp. 1–8 (2009).

304. Saber, A. and Venayagamoorthy, G. Plug-in vehicles and renewable energy sources for wwcost and emission reductions. *IEEE Trans. Ind. Electr.*, 58(4), 1229–1238 (2011).

305. Clement, K., Haesen, E., and Driesen, J. Coordinated charging of multiple plug-in hybrid electric vehicles in residential distribution grids. In *IEEE PSCE'09*, pp. 1–7 (2009).

306. Clement-Nyns, K., Haesen, E., and Driesen, J. The impact of charging plug-in hybrid electric vehicles on a residential distribution grid. *IEEE Trans. Power Syst.*, 25(1), 371–380 (2010).

307. Donegan, P. Ethernet backhaul: Mobile operator strategies & market opportunities. *Heavy Read.*, 5(8) 1–22 (2007).

308. Hochgraf, C., Tripathi, R., and Herzberg, S. Smart grid charger for electric vehicles using existing cellular networks and sms text messages. In *IEEE SmartGridComm'10*, pp. 167–172 (2010).

309. Hutson, C., Venayagamoorthy, G., and Corzine, K. Intelligent scheduling of hybrid and electric vehicle storage capacity in a parking lot for profit maximization in grid power transactions. In *IEEE Energy 2030*, pp. 1–8 (2008).

310. Jansen, B., Binding, C., Sundström, O., and Gantenbein, D. Architecture and communication of an electric vehicle virtual power plant. In *IEEE SmartGridComm'10*, pp. 149–154 (2010).

311. Jin, T. and Mechehoul, M. Ordering electricity via Internet and its potentials for smart grid systems. *IEEE Trans. Smart Grid*, 1(3), 302–310 (2010).

312. Kempton, W. and Tomić, J. Vehicle-to-grid power fundamentals: Calculating capacity and net revenue. *J. Power Sources*, 144(1), 268–279 (2005).

313. Kempton, W. and Tomić, J. Vehicle-to-grid power implementation: From stabilizing the grid to supporting large-scale renewable energy. *J. Power Sources*, 144(1), 280–294 (2005).

314. Kempton, W., Tomić, J., Letendre, S., Brooks, A., and Lipman, T. Vehicleto-grid power: Battery, hybrid, and fuel cell vehicles as resources for distributed electric power in California. In *Prepared for California Air Resources Board and the California Environmental Protection Agency* (2001).

315. Kempton, W., Udo, V., Huber, K., Komara, K., Letendre, S., Baker, S., Brunner, D., and Pearre, N. A test of vehicle-to-grid (V2G) for energy storage and frequency regulation in the PJM system. In *Mid-Atlantic Grid Interactive Cars Consortium* (2009).

316. Lund, H. and Kempton, W. Integration of renewable energy into the transport and electricity sectors through V2G. *Energy Policy*, 36(9), 3578–3587 (2008).

317. Pan, F., Bent, R., Berscheid, A., and Izraelevitz, D. Locating PHEV exchange stations in V2G. In *IEEE SmartGridComm'10*, pp. 173–178 (2010).

318. Roe, C., Evangelos, F., Meisel, J., Meliopoulos, A., and Overbye, T. Power system level impacts of PHEVs. In *42nd Hawaii International Conference on System Sciences*, (Jan 5–8 2009), Shider College of Business, University of Hawaii, Manoa, HI, pp. 1–10 (2009).

10 Multifuel, Hybrid, and Grid-Integrated Vehicles
A Case Study

10.1 INTRODUCTION

This book outlines three levels of integration that are currently happening in the energy and fuel industry: that multifuel (co-process), hybrid (with or without co-generation), and grid integration have so many positive attributes that they will become normal part of the future energy system management. In this chapter, we outline a case study that exemplifies the social acceptance of these three levels of energy and fuel systems integration.

Over the last several decades, the vehicle industry is adapting more and more multifuel, hybrid, and grid-integrated hybrid or electric vehicles (EVs). This movement will accelerate due to the following reasons [1]:

1. Movement toward less use of fossil fuel in an automobile. The use of a mixture of renewable ethanol and gasoline is already in practice.
2. The Environmental Protection Agency (EPA) requirement of better mileage efficiency of the engine. The automobiles will be required to achieve more than 40–50 mpg in the next several decades. This may not be possible for a single fuel internal combustion engine (ICE). Hybrid cars are more likely to achieve this goal in a shorter period of time.
3. Legislations requiring less CO_2 and other harmful gas emissions. This goal is not easily attainable with a single fossil fuel with the existing ICE. Both hybrid and electric cars will achieve this goal rapidly.
4. More use of noncarbon sources such as hydrogen and electricity in the transportation industry. Hybrid and electrical cars will satisfy this requirement.
5. More use of renewable fuels like ethanol, butanol, and biofuels in the automobile. This will require some changes in the present engine design. Cars that can run on pure ethanol or 85% ethanol (e.g., gasohol) are already in the market.
6. The decline in world conventional oil production. Refining of unconventional oil will be more expensive and will emit more CO_2 in the environment.
7. Liquid natural gas, liquid propane, compressed natural gas (CNG), and propane are considered to be better fuels for larger trucks and buses. They will be used more and more compared to conventional oil or diesel oil. This is partly due to a decrease in natural gas price and an increase in its supply.

8. The market for hybrid cars is expanding worldwide. The use of multifuels and hybrid cars for larger vehicles is also rapidly expanding.
9. Since transportation industry is a large consumer of fuel and energy, significant efforts are being made to make easy connections of hybrid and plug-in electric cars to the smart grid.
10. The use of multiple fuels, multiple sources of energy, and energy storage (ES) is being more accepted in large vehicles, trains, ships, and planes worldwide.

These issues are forcing vehicle industry to manufacture future vehicles that are more (1) fuel flexible, (2) hybrid, and (3) system (grid) integrated (like hybrid or electrical cars). The movement has already made significant progress. In this chapter, we examine the present state of the art for multifuel, hybrid, and grid-integrated electric or hybrid vehicles. While the major discussion will be focused on automobile industry, the adoption of these levels of energy and fuel systems integration in other types of vehicles will also be illustrated with specific examples.

10.2 MULTIFUEL VEHICLES

Multifuel vehicles are produced in various different formats [2–16]. One type that is already a part of car industry norm is the ICE car that uses a mixture of renewable fuel like ethanol and gasoline in the same tank. In the United States, all current gasoline comes with 10% ethanol. The percentage of ethanol can vary in many flex-fuel vehicles. Besides this, numerous cars of different sizes are now equipped with dual-fuel engines where the two fuels are stored in different tanks. Some cars are also developed that can be operated with more than two fuels with the same engine. The following discussion will cover all three cases.

10.2.1 Flex-Fuel Vehicles

Flex-fuel vehicles, as dual-fuel vehicles, have an ICE capable of functioning with a mixture of fuels. Flex-fuel vehicles are therefore not the same as bi-fuel vehicles, which have separate tanks for each different fuel type they use. Flex-fuel vehicles normally run on a mixture of either ethanol or methanol and gasoline [10,14].

10.2.1.1 Advantages
1. Ethanol burns cleaner than gasoline and therefore is responsible for fewer toxic fumes, which is highly advantageous from an antipollution point of view. Ethanol does not contribute significantly to greenhouse gasses (GHGs).
2. For all compositions of the flex fuel, the vehicle can burn the fuel in its combustion chamber. Electronic sensors gauge the blend, while microprocessors adjust the fuel injection and timing.
3. Many flex-fuel vehicles make use of ethanol, which originates from corn, sugar cane, or cellulose. This is a better alternative to purchasing foreign oil or upgrading unconventional oil.

4. The flex-fuel vehicle can receive the flex-fuel tax credit, which replaced the clean-fuel burning deduction.
5. Flexible-fuel vehicles (FFVs) can use a single tank for a mixture of fuels—typically gasoline and ethanol, or methanol, or biobutanol.

10.2.1.2 Disadvantages

1. Currently, the use of ethanol from corn can be considered a disadvantage because its fuel use raises its price for food use, particularly when the crop is in short supply. Ethanol need to be produced from lignocellulosic materials that are not good candidates for food.
2. Ethanol can cause corrosion and damage to the engine, mainly because it absorbs dirt easily.
3. Ethanol is also not as economical as gasoline, in that it does not provide the same level of fuel efficiency. Also, there are currently only a few stations nationwide that supply pure ethanol.

10.2.2 E85 Fuel Vehicles

An E85 FFV or dual-fuel vehicle is an alternative fuel vehicle with an ICE designed to run on more than one fuel, usually gasoline blended with either ethanol or methanol; both fuels are stored in the same fuel tank [4,6,12,13]. Due to its lower energy content, ethanol will reduce gas mileage and it has been put forward that a driver who is burning 15% gasoline and 85% ethanol can expect to see a drop in gas mileage by 5%–15% compared to gasoline usage alone. Some of advantages and disadvantages of these vehicles are listed in Table 10.1.

10.2.3 Natural Gas Bi-Fuel Vehicles

Natural gas bi-fuel vehicles (NGVs) are bi-fuel vehicles with multifuel engines capable of running on two fuels [2,3,5,7,8,15,16]. In ICEs, one fuel is gasoline or diesel and the other is an alternate fuel such as CNG, liquefied propane gas (LPG), often designated as liquefied petroleum gas or hydrogen. CNG consists of mostly methane and is a clean burning alternative fuel. Natural gas can be formulated into CNG or liquefied natural gas (LNG) to fuel vehicles. The two fuels are stored in the

---

TABLE 10.1
Advantages and Disadvantages of E85

Advantages	Disadvantages
Reduces the use of imported petroleum because it is domestically produced	Only compatible with FFV
Lowers air pollution emissions	Lower MPG compared to gasoline vehicles
Increases resistance to engine attack	
Cost of fuel mixture very similar to that of gasoline	

TABLE 10.2

Advantages and Disadvantages of Natural Gas Vehicles

Advantages	Disadvantages
Approximately all of natural gas used in America is produced domestically.	Limited driving range compared to gasoline vehicles
Less smog-producing pollutants.	
Less greenhouse gas emissions.	
Less expensive than gasoline.	

separate tanks and the engine runs on one fuel at a time. The advantages and disadvantages of natural gas vehicles are briefly outlined in Table 10.2.

LPG (liquid petroleum gas or propane) is a clean burning fossil fuel that can be used to power ICEs. It mostly contains propane and/or butane. LPG-fueled vehicles produce fewer toxic and smog-forming air pollutants. LPG is usually less expensive than gasoline, and it mostly comes from domestic sources. LPGVs are bi-fuel vehicles with multifuel engines capable of running on two fuels. On ICEs, one fuel is gasoline or diesel and the other is either CNG, LPG, or hydrogen. The two fuels are stored in separate tanks and the engine runs on one fuel at a time. The advantages of LPGV are as follows:

1. Fewer toxic and smog-forming air pollutants.
2. Most of the LPG used in the United States comes from domestic sources.
3. Costs less than gasoline.

The disadvantages of LPGV are

1. Very few newer passenger and trucks commercially available (vehicles can be retrofitted for LPG)
2. LPG less readily available than gasoline or diesel
3. Fewer miles on a tank of fuel

General Motors and Chrysler Group are producing full-size pickups that can seamlessly switch back and forth between natural gas and gasoline. The trucks have two tanks, and drivers can choose which fuel they want to use; if they run out of one, the vehicle automatically switches to the other.

The ability to run on either fuel is an important step for winning over new customers who might be interested in the cost savings from the significantly cheaper CNG but worry about running out of fuel and not finding a compatible station.

There are only about 1000 fueling stations across the country that sell CNG, with roughly half of those open to the general public. Because of that, there were only 112,000 CNG vehicles on the U.S. roads at the end of 2010, and a large percentage are heavy-duty vehicles that return to a base each night to refuel, such as city buses, delivery trucks, or garbage trucks [2,3,5–8,10–12,19,20]. The CNG vehicles, whether pure CNG or bi-fuel, cost more than their gasoline-only counterparts. GM and Chrysler are planning to go ahead with their vehicles, due to the growing demand from businesses looking to save on operating costs.

Liquefied petroleum gas and natural gas are very different from petroleum or diesel and cannot be used in the same tanks, so it would be impossible to build a (LPG or NG) flexible fuel system. Instead, vehicles are built with two, parallel, fuel systems feeding one engine. For example, Chevys Silverado 2500 HD, which is now on the road, can effortlessly switch between petroleum and natural gas and offers a range of over 650 miles [19,20]. While the duplicated tanks cost space in some applications, the increased range, decreased cost of fuel, and flexibility where (LPG or NG) infrastructure is incomplete are significant incentives to purchase such cars. While the U.S. natural gas infrastructure is still being developed, it is increasing at a fast pace and already has 2600 CNG stations in place [5,8,16,19–21]. Due to rapid increase in supply, the natural gas price is considerably lower than that of gasoline (on mpg basis). With a growing fueling station infrastructure and lowering of natural gas price, a large-scale adoption of these bi-fuel vehicles could be seen in the near future. Some vehicles have been modified to use another fuel source if it is available, such as cars modified to run on autogas (LPG) and diesel are modified to run on waste vegetable oil that has not been processed into biodiesel [5,8,16,19–21].

10.2.4 NEW DESIGNS FOR MULTIFUEL HEAVY VEHICLES

A recent study [17] considered different options to design a multifuel engine retaining the power densities and efficiencies of the latest diesel heavy-duty truck engines while operating with various other fuels. In a first option, an igniting diesel fuel is coupled to a main fuel that may have any cetane or octane number fuel in a design where every engine cylinder accommodates a direct diesel injector, a glow plug, and the multifuel direct injector in a bowl-in-piston combustion chamber configuration. In the second option, gasoline fuel replaces the diesel fuel with similar injection mechanisms and combustion chamber configuration. In this option, engine cylinder is accompanied by a jet ignition prechamber.

Both these designs permit load control by changing the amount of fuel injected and diesel-like, gasoline-like, and mixed diesel/gasoline-like modes of operation modulating the amount of the multifuel that burn premixed or by diffusion. These new designs have the potential to deliver even better than current heavy-duty truck diesel fuel conversion efficiency for both part load and full load operating with variable quality fuels.

10.2.5 VOLVO MULTIFUEL CAR

The Volvo Multi-Fuel is a five-cylinder, 2.0 L prototype car that runs on five different fuels: (1) hythane, which is 10% hydrogen and 90% methane, (2) biomethane, (3) natural gas (CNG), (4) bioethanol E85 (85% bioethanol and 15% petrol), and (5) petrol [18]. This concept car was first introduced at the Michelin Challenge Bibendum 2006 and considered to be the only one of its kind. The car is designed for high performance and perfect for driving on any of the five different fuels. The Multi-Fuel is just as safe as all the other Volvo vehicles plus it is exceptionally clean. This type of multifuel car is the first step toward hydrogen economy.

The Volvo Car Corporation believes that the road to the future is not one but many fuels and no renewable fuel type can alone replace the fossil fuels of today [18]. Since local conditions vary, different markets need engines for different alternative fuels coupled with cleaner conventional ones. The main idea of the Multi-Fuel car is to make use of the locally produced fuels thus reducing the need to import fuel from other countries.

This Multi-Fuel car is equipped with one large and two smaller tanks containing 98 L of gaseous fuels (hythane, biomethane, and CNG) and one 29 L tank for liquid fuels (bioethanol E85 and petrol). The small gaseous fuel tanks are made of steel, while the large tank has a durable, gas-tight aluminum liner, reinforced with high-performance carbon fiber composite and an exterior layer of hardened fiber-glass composite [18].

The fuel tanks are placed under the luggage compartment floor thus preserving the full loading capacity of the Multi-Fuel car. The two fuel fillers are used to fill up all five fuel types, one for gaseous and one for liquid fuels. In order to switch between fuel types, the driver would simply have to push a button.

The Multi-Fuel car, its engine, tanks, transmission, and the fuel system, is designed to function on the five different fuels. This means that it can be started directly on gas. The Multi-Fuel is remarkably clean and meets all the emission standards of Europe and United States. It also has an alternative catalyst system that is developed to meet the tough demands on extremely low tailpipe emissions for the U.S. market [18].

The vehicle uses two catalysts, one close coupled to the engine that lowers initial start emissions while the one under the floor reduces high-speed emissions. The double catalysts and advanced engine control system lead to very low emissions. High-temperature materials in the exhaust and turbo allow extremely high exhaust gas temperatures of up to 1050°C. This enables the car to run cleaner, accelerate quicker, and operate smoother at higher speed [18].

10.3 HYBRID VEHICLES

During the last two decades, hybrid vehicles have gained significant acceptance by the customers because of its fuel efficiency, low carbon emission, and other environmental considerations [19–50]. Extra costs of the hybrid vehicles compared to the conventional vehicles are compensated by their lower fuel costs. While large hybrid vehicles have been in existence for a while, in recent years, the technology has been increasingly more adapted for the smaller car market.

10.3.1 ENVIRONMENTAL AND SAFETY BENEFITS OF HYBRID CARS

Hybrid cars provide several environmental benefits such as fuel economy and gas emission reductions, positive impact by the battery usage, and noise control. We briefly examine these three benefits.

10.3.1.1   Fuel Economy and Gas Emissions

The hybrid vehicle typically achieves greater fuel economy than conventional internal combustion engine vehicles (ICEVs), resulting in fewer emissions being generated. These savings are primarily achieved due to the following reasons:

1. Hybrid cars have a smaller engine providing average usage rather than peak power usage. The peak power needs are satisfied by both electric motors and engine. A smaller engine has lower weight and less internal losses.
2. For stop-and-go city traffic, hybrid cars are well suited due to significant battery capacity to store and reuse recaptured energy.
3. Depending on the power rating of the motor/generator, hybrid cars recapture loss in kinetic energy into electrical energy during braking process, which would have normally been wasted as heat.
4. Hybrid cars use Atkinson cycle engine instead of Otto cycle engine to provide better fuel economy.
5. Hybrid cars can shut down engine during traffic stops, or while coasting or other idle periods.
6. Hybrid cars are aerodynamically well shaped reducing air drag on the engine. This improves fuel efficiency and handling of the car. A box shaped or SUVs exert more strain on car engine resulting in poor fuel economy.
7. Hybrid cars often use special tires that are more inflated than regular tires and stiffer or by choice of carcass structure and rubber compound have lower rolling resistance while retaining acceptable grip and thus improving fuel economy irrespective of the source of power.
8. Hybrid cars power the ac, power steering, and other auxiliary pumps electrically as and when needed; this reduces mechanical losses when compared with driving them continuously with traditional engine belts.

These features make a hybrid vehicle particularly efficient for city traffic where there are frequent stops, coasting, and idling periods. In addition, noise emissions are reduced, particularly at idling and low operating speeds, in comparison to conventional engine vehicles. For continuous high-speed highway use, these features are much less useful in reducing emissions [19–21,44–51].

Hybrid vehicle emissions today are getting close to or even lower than the recommended level set by the EPA. The recommended levels they suggest for a typical passenger vehicle should be equated to 5.5 metric tons of carbon dioxide. The three most popular hybrid vehicles, Honda Civic, Honda Insight, and Toyota Prius, set the standards even higher by producing 4.1, 3.5, and 3.5 tons showing a major improvement in carbon dioxide emissions. Hybrid vehicles can reduce air emissions of smog-forming pollutants by up to 90% and cut carbon dioxide emissions in half compared to gasoline-driven ICE [19–21,31,44–51].

10.3.1.2   Environmental Impact of Hybrid Car Battery

Today, most hybrid car batteries are one of two types: (1) nickel–metal hydride or (2) lithium ion. Both types are regarded as more environmentally friendly than lead-based batteries, which constitute the bulk of conventional gasoline run car starter batteries today. Lithium ion is the least toxic of the two mentioned in this section [41]. The toxicity levels and environmental impact of nickel–metal hydride batteries—the type currently used in hybrids—are much lower than batteries like lead acid or nickel cadmium [19–21,41,44–51].

Hitachi has promoted lithium-ion battery for hybrid EVs (HEVs). In addition to its smaller size and lighter weight, lithium-ion batteries deliver performance that helps to protect the environment with features such as improved charge efficiency without memory effect. The lithium-ion batteries have the highest energy density of any rechargeable batteries and can produce a voltage more than three times that of nickel–metal hydride battery cell while simultaneously storing large quantities of electricity as well. The lithium-ion batteries also produce higher vehicle power, avoid wasteful use of electricity, and provide excellent durability. Additionally, the use of lithium-ion batteries reduces the overall weight of the vehicle and also achieves improved fuel economy of 30% better than petro-powered vehicles with a consequent reduction in CO_2 emissions [19–21,41,44].

10.3.1.3   Noise Control and Road Safety for Cyclists and Pedestrians

Hybrid cars run more quietly than ICE cars, particularly at lower speeds. This provides more noise control in slow-moving congested traffic. However, a 2009 National Highway Traffic Safety Administration report indicated that due to their low noise, HEVs were twice as likely to be involved in a pedestrian crash than ICEVs when vehicle was slowing or stopping, backing up, entering, or leaving a parking space (when the sound difference between HEVs and ICEVs is most pronounced). For crashes involving cyclists or pedestrians, there was a higher incident rate for HEVs than ICEVs when a vehicle was turning a corner. But there was no statistically significant difference in accident rates between HEVs and ICEVs when they were driving straight at normal highway speed [19–21,39,40].

Several automakers developed EV warning sounds designed to alert pedestrians to the presence of electric drive vehicles such as HEV, plug-in HEVs (PHEVs), and all-EVs traveling at low speeds. Their purpose is to make pedestrians, cyclists, the blind, and others aware of the vehicle's presence while operating in all-electric mode [54–57]. Vehicles in the market with such safety devices include the Nissan Leaf (see Figure 10.1) and Chevrolet Volt among many others [19–21,39,40,49–51].

10.3.2   Hybrid Vehicle Powertrain Configurations

There are four types of hybrid vehicle powertrain configurations: parallel hybrid, mild-parallel hybrid, series–parallel hybrid, and series hybrid. We briefly examine here the characteristics of each type of hybrid vehicle powertrain configuration. These configurations are used both in family automobiles as well as other vehicles such as bicycles, mopeds, and heavy vehicles [19–21,26–28,42,44–51].

10.3.2.1   Parallel Hybrid

In a parallel-hybrid vehicle, the single electric motor and the ICE are installed such that they can power the vehicle either individually or together. In contrast to the

FIGURE 10.1   The 2011 Nissan Leaf was the first plug-in electric car equipped with Nissan's Vehicle Sound for Pedestrians.

power-split configuration, typically only one electric motor is installed. Most commonly, the ICE, the electric motor, and gearbox are coupled by automatically controlled clutches. For electric driving, the clutch between the ICE and gearbox is open, while the clutch between electric motor and the gearbox is engaged. While in combustion mode, the engine and motor run at the same speed. The first mass production of this type of hybrid vehicle in the world market was the first-generation Honda Insight.

Parallel-hybrid powertrain configuration is also widely used in hybrid scooters and bicycles. Honda has developed an IC/electric hybrid scooter [19–21,45–51]. Yamaha has developed Gen-Ryu, which uses a 600cc engine and an additional electric motor [19–21]. Piaggio MP3 Hybrid uses a 125cc engine and an additional 2.4 kW motor [19–21,42]. Mopeds, electric bicycles, and even electric kick scooters are a simple form of a hybrid, as power is delivered both via an ICE or electric motor and the rider's muscles.

In a "parallel-hybrid bicycle," human and motor power are mechanically coupled at the pedal drivetrain or at the rear or the front wheel, for example, using a hub motor, a roller pressing onto a tire, or a connection to a wheel using a transmission element. Human and motor torques are added together. At present time, most of motorized bicycles and Mopeds are built in this manner [19–38,42].

10.3.2.2 Mild-Parallel Hybrid

These types of hybrid vehicles use compact electric motor (usually <20 kW) to provide auto-stop/start features and to provide extra power assist [31–33] during the acceleration and to generate power during the deceleration phase by the process of regenerative braking. Honda Civic Hybrid and Honda Insight second generation, among many others [19–21,22–38], use this type of power configuration.

FIGURE 10.2   Toyota Prius, series–parallel hybrid.

10.3.2.3   Power-Split or Series–Parallel Hybrid

In a power-split hybrid-electric drivetrain, there are two power drives (motors): an electric motor and an ICE. The power from these two can be shared to drive the wheels via a power splitter. The ratio can be anywhere from 0% to 100% for either combustion engine or electric motor. Often, it is split between combustion engine (say 70%) and electric motor (say 30%). The combustion engine can act as a generator charging the batteries. Toyota Prius (see Figure 10.2) and Ford Escape and Fusion, among others, use this type of power configuration. A more refined version such as the Toyota Hybrid Synergy Drive has a second electric motor/generator on the output shaft (connected to the wheels). In cooperation with the "primary" motor/generator and the mechanical power split, this provides a continuously variable transmission.

In an open road driving, the main power source is ICE, although electric motor provides assistance when required. This increases the available power for a short period, giving the effect of having a larger engine than actually installed. In most applications, the ICE is switched off when the car is running slow or is stationary, thereby reducing curbside emissions [19–21,22–38,44–51].

10.3.2.4   Series Hybrid

A "series- or serial-hybrid vehicle" has also been referred to as an extended range EV or range-extended EV (EREV/REEV); however, range extension can be accomplished with either series or parallel-hybrid layouts. Series-hybrid vehicles are driven by the electric motor with no mechanical connection to the engine. Instead, there is an engine tuned for running a generator when the battery pack energy supply is not sufficient for demands. This arrangement is not new. It is commonly used in diesel-electric locomotives and ships. Porsche used this setup in the racing cars where a motor in each of the two front wheels was used. This arrangement is also sometimes referred to as an "electric transmission." The vehicle cannot move unless the ICE is running. In 1997, the first series hybrid was sold in Japan by Toyota [33]. GM introduced the Chevy Volt series plug-in hybrid in 2010, with an all-electric range aimed at 40 miles [19–21,22–38,44–51].

Series-hybrid powertrain configuration is also extensively used in a bicycle. In a "series hybrid (SH) bicycle," the user powers a generator using the pedals. This is converted into electricity and can be fed directly to the motor giving a chainless bicycle and also to charge a battery. The motor draws power from the battery and must be able to deliver the full mechanical torque required because none is available from the pedals. Due to its simplicity in design and manufacturing, SH bicycles are commercially available [4].

Kinzel invented the first SH bicycle (U.S. Patent 3884317) in 1975 [19–21,42]. In 1994, Macdonalds conceived the Electrilite [5], which used power electronics allowing regenerative braking and pedaling while stationary. In 1996, Blatter and Fuchs built an SH bicycle [19–21,44–51] and in 1998, they mounted the system onto a Leitra tricycle (European patent EP 1165188). In 1999, Kutzke described his concept of the "active bicycle" weighing nothing and having no drag by electronic compensation. Several prototype SH tricycles and quadricycles [19–21,42] were built by Fuchs and colleagues by 2005 [19–21].

In 1900, Porsche developed a series hybrid using two motor-in-wheel-hub arrangements with a combustion generator set providing the electric power. The braking regenerative hybrid was first invented by Arthurs in 1978–1979. His home-converted Opel GT was reported to deliver as much as 75 mpg of fuel [19–21].

10.3.3 TYPES OF HYBRID VEHICLES

10.3.3.1 Motorized Bicycles
Motorized bicycles have utilized all varieties of engines, from IC two-stroke and four-stroke gasoline engines to electric, diesel, or even steam propulsion. Most motorized bicycles are based or derived from standard general-purpose bicycle frame designs and technologies.

The earliest motorized bicycles were ordinary utility bicycles fitted with an add-on motor and transmission to assist normal pedal propulsion, and it is this form that principally distinguishes the motorized bicycle from a moped or motorcycle. In the early days, pedal propulsion was increasingly replaced by constant use of a two or four-stroke gasoline engine. In countries where automobiles and/or fuels are prohibitively expensive, the motorized bicycle has enjoyed continued popularity as a primary mode of transportation [1,19–21,44–47].

The design of the motorized bicycle or motorbike varies widely according to intended use. Some motorized bicycles are powerful enough to be self-propelled, without use of the pedals. Moped has only a vestigial pedal drive fitted primarily to satisfy legal requirements and suitable only for starting the engine or for emergency use. The alternate design philosophy to the moped is the so-called "motor-assist" or "pedal-assist" bicycle [19–21,42,45,51]. These machines utilize the pedals as the dominant form of propulsion, with the motor used only to give extra assistance when needed for hills or long journeys [19–21,42,45–51].

Currently, several companies manufacture bicycles with both four-stroke and two-stroke gasoline engine designs. Among these, Golden Eagle Bike Engines currently produces a rear-engine (rack-mounted) kit using a belt to drive the rear wheel. Staton-Inc., also uses a rack mount with either a tire roller mount (friction drive) or a chain-driven, geared transmission. Other manufacturers produce kits using small two- or four-stroke gas engines mounted in the central portion of the bicycle frame and incorporating various types of belt- or chain-driven transmissions and final drives. Some of these brands include bikes from the companies like Jiangdu Flying Horse Gasoline Engine Factory Ltd. and EZ Motorbike Company, Inc., among others [45–51].

Another form of hybrid vehicles is human power EV. These include such vehicles as the Sinclair C5, Twike, electric bicycles, and electric skateboards. Electrically powered bicycles use batteries, which have a limited capacity and thus a limited range, particularly when large amounts of power are utilized. This design limitation means that the use of the electric motor as an assist to pedal propulsion is more emphasized than is the case with an ICE. While costly, new types of lithium batteries along with electronic controls now offer users increased power and range while reducing overall weight. Newer electric motor bicycle designs are gaining increasing acceptance in countries with dense and aging populations and the ones concerned for environment [19–21,45–51].

HEVs combine the benefits of gasoline engines and electric motors and can be configured to obtain different objectives, such as improved fuel economy, increased power, or additional auxiliary power for electronic devices and power tools. They are primarily propelled by an ICE. However, they also convert energy normally wasted during coasting and braking into electricity, which is stored in a battery until needed by the electric motor. The electric is motor used to assist the engine when accelerating or hill climbing and in low-speed driving conditions where ICEs are less efficient [19–21]. Some HEVs also automatically shut off the engine when the vehicle comes to a stop and restart it when the accelerator is pressed. This prevents wasted energy from idling [19–21,43,48–51].

Electric- and gasoline-powered motorcycles and scooters of the same size and weight are roughly comparable in performance [19–21]. Electric machines, however, have better 0–60 acceleration, since they develop full torque immediately, and without a clutch, the torque is instantly available [19–21]. Electric motorcycles and scooters, however, do not provide distance range, since batteries cannot store the same amount of energy as a tank of gas [44,48–51]. Electric machines are most useful to daily commuters traveling a fixed distance [19–21]. Also, electric power trades off range against speed more dramatically than gasoline power. For instance, the range on a single charge for the Zev T10 LRC is reduced from 140 to 80 miles if the speed is increased from 55 to 70 mph [44–51,53].

10.3.3.2 Hybrid Family Vehicles

A hybrid family car contains an engine, batteries, and an electric motor–generator. The electric battery and motor provide the power to rapidly accelerate the car and provide power at low speeds. The battery is charged by recuperative braking (i.e., recovering the energy of forward motion when the car brakes) and by the ICE. The ICE operates at a constant speed and load under conditions to maximize the energy output per gallon of fuel. When the batteries are fully charged and the power demand is low, the engine is shut down. When the batteries are low or are rapidly drained, the engine is turned on to recharge the batteries and provide motive power. The efficiency of ICEs is a very strong function of engine speed and load. By operating the engine under efficient "base load" conditions and using the battery as an ES device to meet peak energy demands, the total fuel consumption per mile traveled is greatly reduced in a hybrid car [19–21,48–51].

The ICE became popular because it can economically deliver power rapidly over a wide range of conditions, including very high power levels for short periods of time. The hybrid engine eliminates this variable-power requirement. It allows for engines that are optimized for efficiency (not variable power levels) and can easily burn many types of fuel, including hydrogen. It also enables many other types of engines to become viable for transportation, including fuel cells. The design goal becomes an engine that can deliver a constant power continuously and efficiently, not wide variations in power output over very short time periods. Because the hybrid requires a much smaller engine, it is also the economically enabling technology that allows somewhat higher costs per power input if the engine efficiency is significantly higher. This change in requirements is the enabling technology for fuel cells and other engine technologies in vehicles [19–21,48–51].

When the term "hybrid vehicle" is used, it often refers to a HEV. These encompass such vehicles as the Saturn Vue, Toyota Prius, and Toyota Yaris, among others [30–38,48–51]. A petroleum-electric hybrid most commonly uses ICEs (generally gasoline or diesel engines, powered by a variety of fuels) and electric batteries to power the vehicle.

Some automobiles such as Ford Escape use multiple fuels in the same engine. These should be categorized as a part of dual-fuel-mode (or flexible-fuel-mode) automobiles (described in Section 10.2.1) and not hybrid in the true sense. In general, hybrid vehicles use two or more modes for propulsion, one of which need not be electric. Based on the nature of the propulsion, hybrid automobiles can also be divided into the following four categories:

1. Fluid power hybrid
2. Petro-air hybrid
3. Petro-hydraulic hybrid
4. Fuel cell-electric hybrid

Here, we briefly describe each of these categories.

Fluid power hybrid such as hydraulic and pneumatic hybrid vehicles use an engine to charge a pressure accumulator to drive the wheels via hydraulic or pneumatic (i.e., compressed air) drive units. In most cases, the engine is detached from the drivetrain merely only to change the energy accumulator. The transmission is seamless. A French company, MDI, designed a petro-air hybrid engine car. In this design, the engine uses a mixture of compressed air and gasoline injected into the cylinders [33]. A key aspect of the hybrid engine is the "active chamber," which heats air while supplying fuel thus doubling the energy output [19–21,30–38].

In petro-hydraulic hybrids, the energy recovery rate is high, and therefore the system is more efficient than battery charged hybrids, demonstrating a 60%–70% increase in energy economy [23,33]. Under tests undertaken by EPA, a hydraulic hybrid Ford Expedition gave high mileage per gallon [19–21,30–38]. Chrysler and EPA designed and developed an experimental petro-hydraulic hybrid powertrain, which is suitable for use in large passenger cars. In 2012, the new hydraulic powertrain was installed in an existing minivan [23,26,27,25]. UPS currently has several trucks in service with this technology [25,26].

A group of students in Minnesota [52] converted VW Beetle in petro-hydraulic hybrid and increased gas mileage from 32 to 75 mpg. EPA developed a test petro-hydraulic hybrid car, which achieved over 80 mpg on combined EPA city/highway driving cycles with acceleration of 0–60 mph in 8 s. While the petro-hydraulic system has faster and more efficient charge/discharge cycling and is cheaper than petro-electric hybrids, the accumulator size dictates total ES capacity and may require more space than a battery set.

A British company (Artemis Intelligent Power) made a breakthrough in the cost of expensive system components of petro-hydraulic car by introducing an electronically controlled hydraulic motor/pump, which is highly efficient at all speed ranges and loads. This made small applications of petro-hydraulic hybrids feasible [28,31,32]. The company converted a BMW car as a test bed to prove viability. The BMW 530i gave double the mpg in city driving compared to the standard car. Petro-hydraulic hybrids using well-sized accumulators involve the downsizing of the engine to average power usage and not peak power usage. Peak power is provided by the energy stored in the accumulator. A smaller more efficient constant speed engine reduces weight and provides space for a larger accumulator [19–21].

One petro-hydraulic design has claimed to return 130 mpg in tests by using a large hydraulic accumulator, which is also the structural chassis of the car [72]. In this design, the small hydraulic driving motors are incorporated within the wheel hubs. The hub motors eliminates the need for friction brakes, mechanical transmissions, drive shafts, and U joints, reducing costs and weight. Energy created by shock absorbers and kinetic braking energy that normally would be wasted assists in charging the accumulator. A small fossil-fuelled piston engine sized for average power use charges the accumulator. The accumulator is sized at running the car for 15 min when fully charged. Peugeot Citroën exhibited an experimental "Hybrid Air" engine [32,33], which used nitrogen gas compressed by energy harvested from braking or deceleration to power a hydraulic drive. This energy supplemented power from its conventional gasoline engine. Mileage was estimated to be about 80 mpg for city driving if installed in a Citroën C3 [19–21,30,32,33].

Finally, the fuel cell hybrid is generally an EV equipped with a fuel cell. The fuel cell and the electric battery are both power sources, making the vehicle a hybrid. Fuel cells use hydrogen as a fuel and power the electric battery when it is depleted. The Chevrolet Equinox FCEV, Ford Edge Hyseries Drive, and Honda FCX are some of the examples of a fuel cell/electric hybrid.

10.3.3.3   Hybrid Commercial and Military Vehicles

Generally extra costs of a hybrid system are compensated for by fuel savings and less emission of carbon dioxide [19–21,22–28]. Toyota, Ford, GM, and others introduced hybrid pickups and SUVs. Kenworth Truck Company recently introduced a hybrid-electric truck, called the Kenworth T270 Class 6, that is competitive for city usage [19–28]. FedEx and others are starting to invest in hybrid delivery type vehicles—particularly for city use where hybrid technology may pay off first [31–34].

Current manufacturers of diesel-electric hybrid buses (see Figure 10.3) include Alexander Dennis Limited, Azure Dynamics Corporation, and BAE Systems [19–28]. In the United Kingdom, Wrightbus has introduced a development of the London "Double-Decker," the Wright Pulsar Gemini HEV bus [19–28,53]. Since 1999, hybrid electric buses with gas turbine generators have been developed by several manufacturers in the United States and New Zealand, with the most successful design being the buses made by Designline of New Zealand. Mitsubishi Fuso developed a diesel engine hybrid bus using lithium batteries in 2002 for several Japanese cities. The Blue Ribbon City Hybrid bus was presented by Hino, a Toyota affiliate, in January 2005.

BAE Systems created a new product in its HybriDrive® family of heavy-duty hybrid-electric propulsion systems to address lowering emissions and increasing fuel and energy savings in the vocational truck market. This new system is scalable to meet a wide range of heavy-duty truck platforms, vocations, and duty cycles, allowing them to equip hybrid construction trucks, hybrid utility trucks, hybrid refuse trucks, and hybrid delivery trucks.

HybriDrive Parallel diesel-electric truck systems [53] use both simplified and proven components and controls. The system is based on a single electric machine integrated with the engine and the transmission and can be installed with minimal impact to the vehicle. Propulsion is enhanced through an optimized blending of power from a conventional power source and from the electrical power source. This hybrid-electric truck system's energy management and control capabilities ensure that all energy flow—such as propulsion and braking energy recuperation—occurs in the most efficient fashion, resulting in lower fuel consumption.

FIGURE 10.3   Hybrid New Flyer Metrobus.

The HybriDrive parallel-hybrid diesel truck system integrates with medium- to heavy-duty truck engines providing high power and high torque. New Flyer, Gillig, North American Bus Industries, and Nova Bus produced hybrid-electric buses using either BAE System's HybriDrive or Allison Transmission GM's electric drive system. The Whispering Wheel bus is another HEV [19–21,26–28,48–51].

Since 1985, the U.S. military has been testing serial hybrid Humvees [18,19] and has found them to deliver faster acceleration, a stealth mode with low thermal signature/near-silent operation, and greater fuel economy.

10.3.3.4 Hybrid Locomotives, Rapid Transit Buses, Cranes, Etc.

Hybrid powertrains use diesel-electric or turbo-electric to power railway locomotives, rapid transit buses (see Figure 10.4), etc. Typically, some form of heat engine (usually diesel) drives an electric generator or hydraulic pump, which powers one or more electric or hydraulic motors. There are advantages in distributing power through wires or pipes rather than mechanical elements especially when multiple drives—for example, driven wheels or propellers—are required. There is a power loss in the double conversion from typically diesel fuel to electricity to power an electric or hydraulic motor. With large vehicles, the advantages often outweigh the disadvantages especially as the conversion losses typically decrease with size. At present,

FIGURE 10.4 Bus Rapid Transit of Metz, a diesel-electric hybrid driving system by Van Hool. (From "Hybrid Vehicle," https://en.wikipedia.org/wiki/Hybrid_vehicle (last modified August, 5, 2015); "Hybrid Vehicle Drivetrain," https://en.wikipedia.org/wiki/Hybrid_vehicle_drivetrain (last modified July 22, 2015); Hybrid Powertrain technology—A glance at clean freight strategies.)

there is no or relatively little secondary ES capacity on most heavy vehicles, a design condition that is now changing [19–21,22–28,48–51].

Hybrid heavy vehicles are used all over the world. The high-capacity railcar built by the Canadian Company Bombardier is being used in France. This has dual mode (diesel and electric motors) and dual voltage capabilities (1,500 and 25,000 V) allowing it to be used on many different rail systems [8]. China converted G12 locomotive to hybrid locomotive by using a 200 kW diesel generator and batteries. Some electric trolleybuses can switch between an onboard diesel engine and overhead electrical power depending on conditions. In principle, this could be combined with a battery subsystem to create a true plug-in hybrid trolleybus.

A hybrid train engine with significant ES and energy regeneration capability was built by Japan as the KiHa E200. It utilized packs of lithium-ion batteries mounted on the roof to store recovered energy [19–21,22–28]. A more novel model of this concept is shown in Figure 10.5. In the United States, General Electric introduced a prototype railroad engine with their "Ecomagination" technology in 2007. They store energy in a large set of sodium nickel chloride (Na–NiCl$_2$) batteries to capture and store energy normally dissipated in dynamic braking or coasting downhill. They expect at least a 10% reduction in fuel use with this system [19–28,48–51].

Other models for diesel-electric locomotive include the Green Goat and Green Kid switching/yard engines built by Canada's Railpower Technologies. They utilize a large set of heavy-duty long-life (~10 year) rechargeable lead acid (PbA) batteries and 1000–2000 HP electric motors as the primary motive sources and a new clean burning diesel generator (~160 HP) for recharging the batteries that is used

FIGURE 10.5 East Japan Railway Company HB-E300 series.

only as needed. No power or fuel is wasted for idling, which is typically 60%–85% of the time for this type of locomotives. The diesel generator and battery package are normally built on an existing "retired yard" locomotive's frame for significant additional cost savings. The existing motors and running gear are all rebuilt and reused. Diesel fuel savings of 40%–60% and up to 80% pollution reductions are claimed over that of a "typical" older switching/yard engine. These locomotives have the same advantages as those that exist for hybrid cars for frequent starts and stops and idle periods [22–25].

In 2007 [19–28,48–51], Railpower Technologies working with TSI Terminal Systems built a hybrid diesel-electric power unit with battery storage for use in rubber tyred gantry (RTG) cranes. RTG cranes are typically used for loading and unloading shipping containers onto trains or trucks in ports and container storage yards. The energy used to lift the containers can be partially regained when they are lowered. Hybrid design reduced emission and diesel fuel consumption by 50%–70% [19–28,48–53].

10.3.3.5 Hybrid Ships, Boats, and Submarines

Submarines are one of the oldest widespread applications of hybrid technology, running on diesel engines while surfaced and switching to battery power when submerged. Both series-hybrid and parallel-hybrid drivetrains were used in the Second World War.

Ships with both mast-mounted sails and steam engines were an early form of hybrid vehicle. Newer hybrid ship-propulsion schemes include large towing kites manufactured by companies such as SkySails. Towing kites can fly at heights several times higher than the tallest ship masts, capturing stronger and steadier winds.

During the last decade, the Hornblower Hybrid [54] has redefined the profile of vessels on the San Francisco Bay. As the nation's first hybrid ferry, it uses energy from solar, wind, grid electric, and Tier 2 diesel generators to power the vessel. When the solar panels on top of the vessel absorb sunlight and energy is generated by the wind turbines, power is provided to charge 380 V DC batteries. Additional power is provided by the diesel generator for more efficient movement through the water. The vessel can operate on propulsion batteries alone for an hour providing a silent cruise around the bay.

Another boat, "the Greenline Hybrid," is powered by the sun, with diesel for higher speeds and longer ranges. With her great ease of handling, superb safety, and efficiency, the boat is an attractive hybrid boat. With her hybrid propulsion system, solar roof panels, unique, low-drag super displacement hull, she is a very efficient boat.

Greenline Hybrid [55] uses up to four times less fossil fuel than a powerboat in similar conditions, and it is completely emission-free in electric mode. The hybrid (diesel-electric) and solar-powered drive systems have proven themselves both reliable and cost-effective. The protected super displacement low-drag hull uses much less energy to move through the water. Not only does this innovation reduce fuel consumption, it also allows efficient electric propulsion using power supplied by the battery and solar energy.

10.3.3.6 Hybrid Aircraft

The Boeing fuel cell demonstrator [29,56,57] airplane has a proton exchange membrane fuel cell/lithium-ion battery hybrid system to power an electric motor, which is coupled to a conventional propeller. The fuel cell provides all power for the cruise phase of flight. During takeoff and climb, the two flight segments that require the most power, the power is drawn from light-weight lithium-ion batteries.

The demonstrator aircraft is a Dimona motor glider built by Diamond Aircraft Industries of Austria, which also carried out structural modifications to the aircraft. With a wingspan of 53 ft, the airplane is able to cruise at about 62 mph on power from the fuel cell [16,29]. Hybrid fan wings are created by two engines with the capability to autorotate and land like a helicopter [29,56,57].

A team of Boeing engineers dreamt up a hybrid plane that plugs in much like the Prius or Chevy Volt. The plane is called the Subsonic Ultra Green Aircraft Research (SUGAR) Volt, the concept in which Boeing aircraft will combine electric power with traditional fuel, vastly cutting down on in-flight emissions. Every time SUGAR Volt is at the gate, it will draw electricity from the airport's power grid, charging an array of batteries stored in the belly of the plane. Dual-turbine engines would be powered by traditional jet fuel for takeoff, but once the Volt reaches cruising altitude, the system will switch over to electrical power for the rest of the flight. By making the switch to electric power once the plane has reached cruising altitude, the entire flight from that point on could be made completely emission-free. In total, the SUGAR Volt will require only 30% of the fuel of a traditional aircraft [56].

Making the hybrid plane even more energy efficient are its wings, which stretch almost twice as long as today's typical aircraft. The extended wingspan will allow for greater lift, translating to shorter takeoffs and landings, and will fold up upon landing for easy parking at the airport gate. Boeing engineers are improving battery technology and the plane will not be completely ready until 2030 [29,56,57].

There are numerous other developments in building hybrid air vehicles all over the world. The long-endurance multi-intelligence vehicle (LEMV) from the U.S. Department of Defense is being redesigned as "Airlander"—in one of the historic airship sheds at Cardington, OH, with plans to further develop the vehicle and make a passenger flight in 2016 or, at the latest, 2017. An Australian-based company is working on a project to develop an air crane called the SkyLifter, a being capable of lifting up to 150 tons [57]. A Canadian start-up, Solar Ship, Inc., is developing solar-powered hybrid airship that can run on solar power alone. The idea is to create a viable platform that can travel anywhere in the world delivering cold medical supplies and other necessitates to locations in Africa and Northern Canada without needing any kind of fuel or infrastructure. The hope is that technology developments in solar cells and the large surface area provided by the hybrid airship are enough to make a practical solar-powered aircraft. Some key features of the "solarship" are that it can fly on aerodynamic lift alone without any lifting gas, and the solar cells along with the large volume of the envelope allow the hybrid airship to be reconfigured into a mobile shelter that can recharge batteries and other equipment [57].

New hybrid aircraft projects are for two different types of designs: dynastats and rotastats. The dynastat obtains additional lift by flying through the air. Configurations studied have included deltoid (triangular), lenticular (circular), or flattened hulls, or adding a fixed wing. Some of the important projects with this design are (1) hybrid air vehicles (HAV-3), (2) Lockheed Martin P-791, (3) Italian Nimbus EosXi, (4) SkyCat by ATG, (5) Russia's Thermoplan, and (6) Walrus HULA [56,57]. On the other hand, the rotastat obtains additional lift from powered rotors similar to a helicopter. Single-, twin-, and four-rotor designs have been studied. Some of the important projects with this design are (1) AeroLift CycloCrane, (2) Piasecki PA-97, and (3) SkyHook JHL-40 developed in the United States [56–57].

10.4 GRID-INTEGRATED VEHICLES

There are numerous types of hybrid or EVs, which are connected to smart grid for electricity exchange [43,58–63]. These grid-connected vehicles are becoming very popular. During the last decade, the plug-in EV (PEV) is becoming more and more common. It has the range needed in locations where there are wide gaps with no services. The batteries can be plugged into the house (mains) electricity for charging, as well as be charged while the engine is running.

Advanced hybrid vehicles may allow the battery to be recharged by connection to the electrical grid when the car is parked. For shorter trips, the hybrid car batteries provide the energy. For longer trips, after the battery is exhausted, the engine provides the energy. It has been estimated that if the battery can provide power for 20 miles, the fuel consumption in cars could be reduced to half compared to conventional vehicles. These advanced hybrids called PHEVs address the two major barriers that presently exist for using electricity in cars: (1) vehicle range and (2) battery recharging. Hybrid enhances vehicle range for mileage per fuel cycle. Battery recharging is not as rapid as filling of fuel tank. At the same time, recharging batteries overnight or while a person is at work is relatively easy. A PHEV allows slow battery recharging while providing the gasoline engine for propulsion if there is insufficient time to recharge batteries.

Today's hybrid cars are leading to the developments of PHEVs [48,59–64]. The first-generation PHEVs are now beginning road test. Consequently, the hybrid engine is potentially the enabling technology to couple transportation to a renewable energy and fuel system via the hybrid battery and H_2 via engines designed for H_2 fuels. It is forecasted that hydrogen and electricity will be the major noncarbon sources of energy for future vehicle industry.

10.4.1 PLUG-IN HYBRID ELECTRIC VEHICLE

The PHEV is usually a general fuel-electric (parallel or serial) hybrid with increased ES capacity, usually through a Li-ion battery, which allows the vehicle to drive on all-electric mode a distance that depends on the battery size and its mechanical layout (series or parallel). It may be connected to the main electricity supply at the end of the journey to avoid charging using the onboard ICE [43,48,58–72].

This concept is attractive to those seeking to minimize on-road emissions by avoiding—or at least minimizing—the use of ICE during daily driving. As with pure EVs, the total emissions saving, for example, in CO_2 terms, is dependent upon the energy source of the electricity generating company.

For some users, this type of vehicle may also be financially attractive so long as the electrical energy being used is cheaper than the petrol/diesel that they would have otherwise used. Some electricity suppliers also offer price benefits for off-peak night users, which may further increase the attractiveness of the plug-in option for commuters and urban motorists.

10.4.2 CONTINUOUSLY OUTBOARD RECHARGED ELECTRIC VEHICLE

Given suitable infrastructure, permissions, and vehicle design, battery EVs (BEVs) can be recharged while the user drives. The BEV establishes contact with an electrified rail, plate, or overhead wires on the highway via an attached conducting wheel or other similar mechanism. The BEV's batteries are recharged by this process on the highway and can then be used normally on other roads until the battery is discharged. Some of the battery-electric locomotives used for maintenance trains on the London Underground are capable of this mode of operation. Power is picked up from the electrified rails where possible, switching to battery power where the electricity supply is disconnected [43,48,58–72].

10.4.3 ELECTRIC VEHICLE GRID INTEGRATION PROJECT

The Electric Vehicle Grid Integration Project supports the development and implementation of electrified transportation systems, particularly those that integrate renewable-based vehicle charging systems [43,48,58–72].

PEVs—including all-EVs and PHEVs—provide a new opportunity to reduce oil consumption by drawing on power from the electric grid. To maximize the benefits of PEVs, the emerging PEV infrastructure must provide access to clean electricity generated from renewable sources, satisfy driver expectations, and ensure safety. Value creation from systems integration will be core to the success of PEVs.

National Renewable Energy Laboratory's (NREL) Electric Vehicle Grid Integration Team developed strategies and models to support the development of transportation electrification and the expansion of renewable generation through

1. Understanding vehicle use profiles, EV benefits, and battery life challenges
2. Integrating renewable resources (solar and wind) with vehicle charging
3. Developing and testing grid interoperability standards and exploring grid services technology opportunities

At NREL's "Vehicle Testing and Integration Facility (VTIF)," researchers collaborate with automakers, charging station manufacturers, utilities, and fleet operators to assess charging, communication, and control technology and modify PEVs so that it can create most effective PEV grid interactions and management.

NREL is working with the U.S. Department of Defense and the U.S. Army Corps of Engineers to develop specifications for a system that integrates photovoltaics, PEVs, and a renewable energy management unit with a microgrid system at Fort Carson, a large Army facility in Colorado. During this multiyear project, NREL will develop critical modeling tools to optimize the components needed to link vehicles to the microgrid. A microgrid that integrates renewable generation and vehicle ES with load management components offers energy security, cost savings, and reliability benefits. Through the coordination of generators and loads, the Fort Carson microgrid will make it possible to maintain electricity delivery to a portion of the facility that is critical to sustained operations [66–72].

The use of PHEVs represents a significant potential shift in the use of electricity and the operation of electric power systems. Electrification of the transportation sector could increase generation capacity as well as transmission and distribution requirements, especially if vehicles are charged during periods of high demand. NREL study evaluated several of these PHEV-charging impacts on utility system operations within the Xcel Energy Colorado service territory. NREL performed a series of simulations in which the expected electricity demand of a fleet of PHEVs was added to the projected utility loads under a variety of charging scenarios. These simulations provided insight into the potential grid impacts of PHEVs, with a focus on total system load, emissions, costs, and benefits.

Vehicle-to-grid (V2G) systems represent a means by which power capacity in parked vehicles can be used to generate electricity for the grid. This study [66] describes the first detailed and global analysis of the potential of V2G technologies over the long term (up to 2100) using a comprehensive energy-systems model. In this analysis, the potential for V2G systems to supply a number of electricity submarkets and concomitantly accelerate the diffusion of advanced vehicle technologies, including hybrid-electric and fuel cell drivetrains, is evaluated. The potential impact of V2G on the global energy system, particularly in terms of investment in conventional capacity, and the possible role of V2G-enabled vehicles in increasing the market penetration of renewable electricity generation technologies are also examined. Importantly, however, V2G technologies represent a paradigm shift on how the energy and mobility markets are related.

Even low levels of EV adoption will have a significant impact on utilities and the grid—a single EV plugged into a fast charger—can double a home's peak electricity demand. Consequently, it is crucial for utilities to manage EV charging. A smart grid must provide the visibility and control needed to protect components of the distribution network, such as transformers, from being overloaded by PEV charging and ensure electricity generating capacity is used most efficiently. With a smart grid, utilities can manage when and how EV charging occurs while adhering to customer preferences, collect EV-specific meter data, apply specific rates for EV charging, engage consumers with information on EV charging, and collect data for GHG abatement credits.

While PEVs offer the opportunity to shift transportation energy demands from petroleum to electricity, its broad adoption will require integration with other systems. While automotive experts work to reduce the cost of PEVs, fossil-fueled cars and trucks continue to burn hundreds of billions of gallons of petroleum each year—not only to get from point A to point B but also to keep passengers comfortable with air conditioning and heat.

At NREL's VTIF laboratory, engineers are developing strategies to address two separate but equally crucial areas of research: meeting the demands of electric vehicle grid integration (VGI) and minimizing fuel consumption related to vehicle climate control.

NREL researchers at VTIF instrument every class of on-road vehicle, conduct hardware and software validation for EV components and accessories, and develop analysis tools and technology that have the best chance for long-term sustainability.

They examine the interaction of building energy systems, utility grids, renewable energy sources, and PEVs, integrating energy management solutions, and maximizing potential GHG reduction, while smoothing the transition and reducing costs for EV owners [63–72].

NREL's collaboration with automakers, charging station manufacturers, utilities, and fleet operators to assess technologies using VTIF resources is designed to enable PEV communication with the smart grid and create opportunities for vehicles to play an active role in building and managing PEV grid integrations. Ultimately, this creates value for the vehicle owner and will help renewables be deployed faster and more economically, making the U.S. transportation sector more flexible and sustainable [63–72].

10.5 FUTURE OF MULTIFUEL, HYBRID, AND GRID-INTEGRATED VEHICLES

The reduction of carbon emission in the environment would be a major drive for the future vehicle industry. This means more efficiency in engine performance (i.e., more mileage per gallon of fuel) and less use of fossil fuel. In future biofuels, ethanol, biobutanol, and biodiesel will become more important and more used. The growth in natural gas will increase more use of CNG, NGL, and propane particularly for larger vehicles. In short, flex-fuel and multifuel cars will become more of a norm than exception. Future gas stations will provide the necessary infrastructure for the multiple fuels. Engine and fuel storage designs will be more and more adaptable to the use of multi fuels.

The penetration of hybrid vehicles in markets of two-wheel, four-wheel, commercial, and heavy vehicles, cranes, boats and ships, and even airplanes will significantly increase. As workings of hybrid car mature, the use of hybrid cars will also become a norm rather than an exception. Hybrid cars have a lot to offer with regard to the reduction in carbon emission and engine efficiency. Hybrid cars are also a gateway to the use of hydrogen and electricity, which will be the dominant sources of energy for all vehicles at the end of this century. The use of fuel cell in the car industry will also rapidly increase. Along with electricity, hydrogen will be the dominant fuel for car industry by the end of this century.

The next decade will bring a significant shift toward the electrification of transportation around the globe. Worldwide, more than 20 automakers have been planning to bring EVs to market since late 2010, and researchers expect sales of EVs to grow from 1% of the global market, or just under one million vehicles per year, to 6% by 2020. This sea change in transportation applies a new urgency to a wide variety of business, technical, and regulatory issues that must be addressed if the electrification of transportation is to succeed.

EVs and HEVs will fundamentally change how electric utilities do business. Although full electrification of the U.S. transportation system will take several decades, most utilities are well aware that even low levels of EV adoption can strain their existing infrastructure. Some are proactively preparing to address EV integration issues, while others will be reactively dealing with grid reliability problems as they arise.

Utilities are best positioned to manage the impact high-capacity EV charging will have on the grid. Either significant new infrastructure needs to be added to the distribution infrastructure (an EV can consume as much energy as an air-conditioned home—potentially doubling peak residential electricity demand) or these new EV loads need to be managed to avoid overlapping with existing residential usage patterns. Utilities taking an active role in planning and implementing an EV charging management solution will be well-positioned to benefit from the coming massive change in transportation industry.

Transportation accounts for more than 30% of the world's energy consumption and nearly 72% of global oil demand. Given the volatility of oil prices over the past decade, the political instability of oil producing nations, and the environmental damage caused by ICEs, governments increasingly are coming to view electric transport as essential to economic growth, energy independence, and GHG reduction.

The term vehicle grid integration or VGI encompasses the ways EVs can provide grid services. To that end, EVs must have capabilities to manage charging or support two-way interaction between vehicles and the grid. Managed charging refers to the technical capability to modulate the electric charging of the vehicle through delay, throttling to draw more or less electricity, or switching load on or off. Two-way interaction refers to the controlled absorption and discharge of power between the grid and a vehicle battery or a building and a vehicle battery. VGI is enabled through technology tools and products that provide reliable and dependable vehicle charging services to EV owners and potentially additional revenue opportunities, while reducing risks and creating cost savings opportunities for grid operators. Such tools might include technologies such as inverters, controls or chargers, or programs and products, such as time of use tariffs or bundled charging packages.

In a recent study of "Future VGI roadmap" [43,63–66,70–72], four use-case categories were considered:

1. Unidirectional power flow with one resource and unified actors
2. Unidirectional power flow with aggregated resources
3. Unidirectional power flow with fragmented actor objectives
4. Bidirectional power flow

The study indicated that the usage rate, amount, and pricing for these four cases must be managed independently. VGI roadmap indicated the need to coordinate VGI-related activities across the state or within the grid region. The study explored those aspects that affect EV's ability to offer grid services, rather than those efforts needed to support deployment of EVs. As the technical, economic, and policy issues associated with ES, the VGI Roadmap considered technologies beyond solely EVs, which might influence VGI implementation or value streams, such as ES or demand response (DR).

The VGI roadmap study concluded that

1. ES and DR have the potential to help implement VGI-related behaviors by assisting with net load reductions and charging appropriate management fees
2. Stationary ES and other DR may ultimately compete with EV-based VGI services
3. The development of a second-life battery market will likely influence the VGI-related behavior of customers

The future of grid-integrated hybrid or electrical vehicles will require more work on the smart grid. It is, however, clear that multifuel, hybrid, and grid-integrated vehicles are here to stay and they will be the most important parts of the future energy consumption by the vehicle industry. Both stationary and mobile energy consumptions will play equally important roles in different levels of energy and fuel systems integration.

REFERENCES

1. Dixon, P. Future of the automotive industry (auto trends), the National Auto Auction Association, 66th annual convention, September 2014, Boston, MA., http://www.globalchange.com/future-of-the-automotive-industry-auto-trends.htm (accessed August 10, 2015).
2. Pham, R., Review: Chevrolet Silverado 2500 HD with bi-fuel CNG option, a publication by Green Car Congress-Energy technologies, issues and policies for sustainable mobility, (14, July, 2014); http://www.typepad.com/services/trackback/6a00d8341c4fbe53ef01a3fd3153eb970b. (accessed December, 2014).
3. Young, K. Biofuels help environment, but they're hard to find. *The Vancouver Sun*, Vancouver, British Columbia, Canada (February 23, 2008). Retrieved September 16, 2008.
4. BAFF Bought Ethanol Cars, a report from BioAlcohol Fuel Foundation. Accessed April 2014. Archived from original on July 21, 2011.
5. Worldwide NGV statistics. *NGV Journal*. www.ngvjournal.com/worldwide-ngv-statistics/ (accessed on April 13, 2014).
6. Fargione, J., Hill, J., Tilman, D., Polasky, S., and Hawthorne, P. Land clearing and the biofuel carbon debt, *Science*, 319 (5867), 1235–1238. (February 29, 2008).
7. DENATRAN Frota per tipo/UF 2008 (file March 2008 in Portuguese), Departamento Nacional de Transito, Lisbon, Portugal. (accessed December 2008).
8. Pike Research. Pike Research predicts 68% jump in global CNG vehicle sales by 2016. AutoblogGreen (September 14, 2011). Retrieved September 26, 2011.
9. Honda Motor Company. Honda Announces First FCX Clarity Customers and World's First Fuel Cell Vehicle Dealership Network as Clarity Production Begins (June 16, 2008). Retrieved June 1, 2009.
10. Thomas, K. 'Flex-fuel' vehicles touted. *USA Today* (May 7, 2007). Retrieved September 15, 2008.
11. Gable, C. and Gable, S. Yellow E85 gas cap. About.com: Hybrid Cars & Alt Fuels. Archived from the original on October 5, 2008. Retrieved September 18, 2008.

12. National ethanol vehicle coalition. New E85 Stations. *NEVC FYI Newsletter*, 14(15) (September 8, 2008). Archived from the original on September 15, 2008. Retrieved September 15, 2008.

13. Li, Z., Gao, D., Chang, L., Liu, P., and Pistikopoulos, E. Coal derived methanol for hydrogen vehicles in china: Energy, environment, and economic analysis for distributed reforming. *Chem. Eng. Res. Des.*, 88, 73–80 (2010).

14. Flex fuel vehicles: Advantages and disadvantages. A Report from Green Cars (February 15, 2012). Volvo multi-fuel car run on 5 different fuels. Editorial of streetdirectory.com (2012).

15. Robinson, T. Multi fuel vehicles & American fuel vehicles. A Report on Pickens Plans (February 2008).

16. "Alternative fuel vehicle," *Wikipedia*, https://en.wikipedia.org/wiki/Alternative_fuel_vehicle (last modified July 8, 2015).

17. Boretti, A. Novel engine concepts for multi-fuel military vehicles. SAE technical paper 2012-01-1514 (2012). doi:10.4271/2012-01-1514.

18. Reign, G. Volvos multi-fuel car run on 5 different fuels. Editorials, Automobiles, Mileage Guide of streetdirectory.com (2012).

19. "Hybrid Vehicle," *Wikipedia*, https://en.wikipedia.org/wiki/Hybrid_vehicle (last modified August, 5, 2015)

20. "Hybrid Vehicle Drivetrain," *Wikipedia*, https://en.wikipedia.org/wiki/Hybrid_vehicle_drivetrain (last modified July 22, 2015)

21. Hybrid Powertrain technology—A glance at clean freight strategies. Smart way transport partnership, a publication by EPA 420-F09-035, Washington, DC (2009). www.epa.gov/smart way.

22. The World's First Hybrid Train Officially Enters Commercial Service in France, CleanMPG forums, http://news.yahoo.com/s/afp/20071009/lf_afp/francecanadatran sportrailenvironment_071009144322/);http://www.cleanmpg.com/photos/data/501/Bombardier_AGC_Hybrid_Train_-_Bibi.jpgAnne Froger - Bombardier (accessed October 9, 2007 and retrieved January 13, 2012).

23. Japan to launch first hybrid trains. *The Sydney Morning Herald*. AP digital (July 29, 2007). Retrieved April 30, 2013.

24. Rail Power To Supply Y TSI Terminal Systems Inc. With hybrid power plants for rubber tyred gantry cranes (PDF) (Press release) (October 10, 2006). Archived from the original on February 28, 2008.

25. Railpower to supply TSI Terminal Systems Inc. With hybrid power plants for rubber tyred gantry cranes (Press release). RailPower Technologies Corporation (October 10, 2006).

26. Kenworth Unveils T270 Class 6 Hybrid Truck Targeted at Municipal, Utility Applications (Press release). Kenworth Truck Company (March 21, 2007). Archived from the original on March 1, 2007.

27. Komarow, S. Military hybrid vehicles could boost safety, mobility. *USA Today* (February 13, 2006).

28. XM1124 Hybrid electric HMMWV. GlobalSecurity.Org. Retrieved November 17, 2008.

29. Manned airplane powered by fuel cell makes flight tests (METALS/POLYMERS/CERAMICS). *Adv. Mater. Process.*, 9(1), 165.6 (June 2007). Expanded Academic ASAP. Gale. Gale Document Number: A166034681.

30. Toyota Europe News. Worldwide Prius sales top 3-million mark; Prius family sales at 3.4 million. *Green Car Congress* (July 3, 2013). Retrieved July 3, 2013.

31. Proefrock, P. (March 25, 2010). Hybrid Hydraulic Drive Vehicle Promises 170 MPG. Inhabitat. Retrieved April 22, 2013.

32. Hanlon, M. Chrysler announces development of hydraulic hybrid technology for cars. Gizmag.com (January 26, 2011). Retrieved April 22, 2013.

33. Lewis, T. Peugeot's hybrid air: The car of the future that runs on air. *The Observer The Guardian* (March 23, 2013). Retrieved March 25, 2013.
34. David, J. Compressing gas for a cheaper, simpler hybrid. *The New York Times* (March 1, 2013). Retrieved March 2, 2013.
35. Cobb, J. Americans Buy Their 3,000,000th Hybrid. HybridCars.com (November 4, 2013). Retrieved November 17, 2013.
36. TMC Press Release. Cumulative Sales of TMC Hybrids Top 2 Million Units in Japan, Toyota (November 8, 2012). Retrieved November 8, 2012.
37. Hyundai Unveils Elantra LPI HEV at Seoul Motor Show. *Hyundai Global News* (April 2, 2009). Retrieved March 23, 2010.
38. Worldwide Prius Cumulative Sales Top 2M Mark; Toyota Reportedly Plans Two New Prius Variants for the US by End of 2012. *Green Car Congress* (October 7, 2010). Retrieved October 7, 2010.
39. TMC to Sell Approaching Vehicle Audible System for 'Prius'. Toyota Motor Company News Release (August 24, 2010). Retrieved August 25, 2010.
40. Dignan, L. Hybrid, electric vehicles to become louder for pedestrian safety. SmartPlanet.com (December 16, 2010). Retrieved December 17, 2010.
41. Hybrid Cars. Hybridcars.com (2006). Retrieved December 9, 2009 from Hybrid Battery Toxicity.
42. "Moped," *Wikipedia*, https://en.wikipedia.org/wiki/Moped (last modified July 30, 2015).
43. Nichols, W. Ford tips hybrids to overshadow electric cars. *Business Green* (June 25, 2012). Retrieved October 16, 2012 (By June 2012 Ford had sold 200,000 full hybrids in the U.S. since 2004).
44. Du, Z., Cheong, K., Li, P., and Chase, T. Fuel economy comparisons of series, parallel and HMT hydraulic hybrid architectures. In Paper presented at *American Control Conference (ACC)*, Washington, DC (June 17–19, 2013).
45. "Hybrid Bicycle," *Wikipedia*, https://en.wikipedia.org/wiki/Hybrid_bicycle (last modified July 9, 2015).
46. "Vespa," *Wikipedia*, https://en.wikipedia.org/wiki/Vespa (last modified July 29, 2015).
47. "Scooter," *Wikipedia*, https://en.wikipedia.org/wiki/Scooter_(motorcycle) (last modified July 24, 2015).
48. "Plug in Hybrid," *Wikipedia*, https://en.wikipedia.org/wiki/Plug-in_hybrid (last modified August 5, 2015).
49. "Types of Motorcycles," *Wikipedia*, https://en.wikipedia.org/wiki/Types_of_motorcycles (last modified July 29, 2015).
50. "Electric Motor Cycles and Scooters," *Wikipedia*, https://en.wikipedia.org/wiki/Electric_motorcycles_and_scooters (last modified August 5, 2015).
51. "Hybrid Electric Vehicle," *Wikipedia*, https://en.wikipedia.org/wiki/Hybrid_electric_vehicle (last modified July 26, 2015).
52. Post, T. Minnesota students create super hybrid cars. *MPR* Minnesota Public Radio (2009), http://www.mprnews.org/story/2009/07/22/hybrid-prius. (Accessed August 10, 2015.)
53. BAE System. HybriDrive propulsion system, 2010, http://www.baesystems.com/product/BAES_019588/hybridrive-propulsion-systems (Accessed August 10, 2015.)
54. Hornblower hybrid, a report from Cruise Green (2013).
55. Greenline Hybrid, an announcement from Greenline Hybrid Co., Fort Lauderdale, Florida, http://Greenlinehybrid.com/ Greenline Hybrid (2013).
56. Owano, N. SUGAR volt: Boeing puts vision to work hybrid electric aircraft. Energy and Green Tech Report (December 30, 2012).
57. "Hybrid Airship," *Wikipedia*, https://en.wikipedia.org/wiki/Hybrid_airship (last modified July 2, 2015).
58. King, D. Neighborhood Electric Vehicle Sales To Climb. Edmunds.com, AutoObserver (June 20, 2011). Retrieved February 5, 2012.

59. Roberts, G. U.S.: Washington Nissan Leaf owner clocks up 100,000 miles. just-auto (December 17, 2013). Retrieved January 15, 2014.
60. Cobb, J. Nissan Sells 99,999th Leaf. HybridCars.com (January 19, 2014). Retrieved January 20, 2014.
61. Crippen, A. Warren Buffett's Electric Car Hits the Chinese Market, But Rollout Delayed For U.S. & Europe. *CNBC* (December 15, 2008). Retrieved December 2008.
62. Balfour, F. China's First Plug-In Hybrid Car Rolls Out. *Business Week* (December 15, 2008). Retrieved December 2008.
63. How the smart grid enables utilities to integrate electric vehicles. *Whitepaper*, Silver Spring Networks, Silver Spring, MD (2013), www.silverspringnet.com.
64. "Vehicle-to-Grid," *Wikipedia*, https://en.wikipedia.org/wiki/Vehicle-to-grid (last modified August 1, 2015).
65. California vehicle grid integration (VGI) roadmap: Enabling vehicle based grid services. The California ISO Prepared by the governor's interagency working group on zero mission vehicles under California executive order B–16–2012. (February 2014).
66. Merkel, T. Electric vehicle grid integration. A Report by Vehicles and Fuels Research Division, NREL, Golden, CO (2014).
67. Gearhart, C., Gonder, J., and Markel, T. Connectivity and convergence: Transportation for the 21st century. *IEEE Electrification Magazine* (2014).
68. Markel, T. *PEV Grid Integration Research: Vehicles, Buildings, and Renewables Working Together* (2014).
69. Breitenbach, A. *Vehicle Testing and Integration Facility* (2012).
70. Simpson, M. and Markel, T. *Plug-in Electric Vehicle Fast Charge Station Operational Analysis with Integrated Renewables* (2012).
71. Simpson, M. *Mitigation of Vehicle Fast Charge Grid Impacts with Renewables and Energy Storage* (2012).
72. Markel, T., Kuss, M., and Simpson, M. *Value of Plug-in Vehicle Grid Support Operation* (2010).
73. What is WINSmartEV, A Report by UCLA Smart Grid Energy Research Center, UCLA, Los Angeles, CA (2010–2014), also obtained from info@smartgrid.ucla.edu.
74. The Regulatory Assistance Project, The International Council on Clean Transportation, and M.J. Bradley & Associates. Electric vehicle grid integration in the US, Europe and china-challenges and choices for electricity and transportation policy. Published Jointly by The Regulatory Assistance Project; The International Council on Clean Transportation and M.J. Bradley & Associates (July 2013).

Index

T - #0325 - 071024 - C13 - 234/156/21 - PB - 9780367377342 - Gloss Lamination